Contents

Preface xi

Mapping to City and Guilds 2330 Certificate in Electrotechnical Technology xii

1 Basic information and calculations 1
Units 1
Indices 4
Simple algebra 6
Formulae or equations 7
Manipulation or transposition of formulae 8
The theorem of Pythagoras 12
Basic trigonometry 13
Areas and volumes 15

2 Electricity 16
Molecules and atoms 16
Potential difference 19
Electron flow and conventional current flow 19
Conductors and insulators 19
Electrical quantities 19
Ohm's law 27
Electricity and the human body 28
Types and sources of supply 29
Voltage bands 29
Measuring current and voltage 30
Components of a circuit 31
Self-assessment questions 31

3 Resistance, current and voltage, power and energy 33
Resistance 33
Voltage drop 35
Power 45
Electrical energy 51
Tariffs 52
Measuring power and energy 52

	Water heating	53
	Efficiency	54
	Self-assessment questions	55
4	**Electromagnetism**	58
	Magnetism	58
	Electromagnetism	59
	Application of magnetic effects	63
	Drawing the waveform of an alternating quantity	68
	Addition of waveforms	69
	Root-mean-square (r.m.s.) value	70
	Average value	71
	Three-phase a.c. generator	71
	Inductance	72
	Induced e.m.f. due to change in flux	73
	Self-inductance	74
	Mutual inductance	74
	Time constant	76
	Graphical derivation of current growth curve	76
	Energy stored in a magnetic field	78
	Inductance in a.c. circuits	78
	Resistance and inductance in series (R–L circuits)	81
	Impedance	81
	Resistance and inductance in parallel	83
	Power in a.c. circuits	84
	Transformers	87
	Self-assessment questions	92
5	**Capacitors and capacitance**	94
	Capacitors	94
	Capacitance	95
	Dimensions of capacitors	95
	Capacitors in series	96
	Capacitors in parallel	97
	Capacitors in d.c. circuits	98
	Capacitance in a.c. circuits	99
	Capacitive reactance	99
	Resistance and capacitance in series	100
	Resistance and capacitance in parallel	101
	Working voltage	101
	Applications of capacitors	101
	Self-assessment questions	101
6	**Resistance, inductance and capacitance in installation work**	103
	Power factor improvement	108
	Self-assessment questions	117

Electrical Installation Work

To my wife

Electrical Installation Work

Fifth Edition

BRIAN SCADDAN

AMSTERDAM • BOSTON • HEIDELBERG • LONDON
NEW YORK • OXFORD • PARIS • SAN DIEGO
SAN FRANCISCO • SINGAPORE • SYDNEY • TOKYO

Newnes is an imprint of Elsevier

Newnes
An imprint of Elsevier
Linacre House, Jordan Hill, Oxford OX2 8DP
30 Corporate Drive, Burlington, MA 01803

First published 1992
Reprinted 1993
Second edition 1996
Reprinted 1996, 1997
Third edition 1998
Reprinted 1999 (twice), 2000, 2001
Fourth edition 2002
Reprinted 2003, 2004
Fifth edition 2005

British Library Cataloguing in Publication Data
A catalogue record for this book is available from the British Library

Library of Congress Cataloguing in Publication Data
A catalogue record for this book is available from the Library of Congress

ISBN 0 7506 6619 6

For information on all Newnes publications visit our website at http://books.elsevier.com

Typeset by Integra Software Services Pvt. Ltd, Pondicherry, India
www.integra-india.com
Printed and bound in Great Britian

7 Three-phase circuits 118
Star and delta connections 118
The neutral conductor 118
Current and voltage distribution 119
Measurement of power in three-phase systems 123
Self-assessment questions 124

8 Motors and generators 126
Direct-current motors 126
Alternating-current motors 136
Starters 145
Installing a motor 149
Fault location and repairs to a.c. machines 152
Power factor of a.c. motors 156
Motor ratings 158
Self-assessment questions 161

9 Cells and batteries 162
General background 162
The primary cell 162
The secondary cell 163
Cell and battery circuits 168
Self-assessment questions 171

10 Illumination and ELV lighting 172
Light sources 172
Calculation of lighting requirements 182
Self-assessment questions 192

11 Electricity, the environment and the community 194
Environmental effects of the generation of electricity 194
New developments 198
The purpose and function of the National Grid 198
Generation, transmission and distribution systems 199
The aesthetic effects of the siting of generation and transmission plant 200

12 Health and safety 202
The Health and Safety at Work Act 1974 203
Electricity at Work Regulations 1989 205
Personal Protective Equipment Regulations (PPE) 207
Construction (Design and Management) Regulations (CDM) 207
Control of Substances Hazardous to Health Regulations (COSHH) 207
The Building Regulations 208
General safety 208

	The mechanics of lifting and handling	210
	Work, load and effort	213
	Access equipment	215
	The joining of materials	219
	Fire safety	220
	Electrical safety	221
	First aid	223
13	**The electrical contracting industry**	225
	Sample specification	229
	Cost of materials and systems	234
14	**Installation materials and tools**	239
	Cables	239
	Jointing and terminations	243
	Plastics	248
	Conduit	249
	Trunking	255
	Traywork	260
	Fixing and tools	260
	Comparison of systems	265
	Self-assessment questions	266
15	**Installation circuits and systems**	268
	Lighting circuits	268
	Lighting layouts	270
	Power circuits	271
	Space heating systems	274
	Thermostats	276
	Installation systems	278
	Industrial installations	278
	Multi-storey commercial or domestic installations	282
	Off-peak supplies	284
	Alarm and emergency systems	286
	Call systems	290
	Emergency lighting systems	291
	Central heating systems	292
	Extra-low-voltage lighting	294
	Choice of system	295
16	**Earthing and bonding**	304
	Earth: what it is, and why and how we connect to it	304
	Earth electrode resistance	307
	Earthing systems	309
	Earth fault loop impedence	312
	Residual current devices	315

Supplementary bonding 316
Points to note 318
Self-assessment questions 319

17 Protection 320
Protection 320
Control 335
Points to note (IEE Reglations) 339

18 Circuit and design 341
Design procedure 341
Design current 341
Nominal setting of protection 342
Correction factors 342
Current-carrying capacity 345
Choice of cable size 346
Voltage drop 346
Shock risk 347
Thermal constraints 347
Example of circuit design 351
Design problem 357

19 Testing 358
Measurement of electrical quantities 358
Measurement of current 360
Measurement of voltage 362
Instruments in general 363
Selection of test instruments 365
Approved test lamps and indicators 366
Accidental RCD operation 367
Calibration, zeroing and care of instruments 367
Initial inspection 368
Testing continuity of protective conductors 370
Testing continuity of ring final circuit conductors 372
Testing insulation resistance 375
Special tests 377
Testing polarity 378
Testing earth fault loop impedance 379
External loop impedance Z_e 381
Prospective fault current 381
Testing earth electrode resistance 382
Functional testing 384
Periodic inspection 386
Certification 388
Inspection and testing 390
Fault finding 394

20 Basic electronics technology 395
Electronics components 395
Semi-conductors 397
Rectification 400
Electronics diagrams 405
Electronics assembly 406
Self-assessment questions 407

Answers to self-assessment questions 409

Index 415

Preface

This book is intended for the trainee electrician who is working towards NVQs, gaining competences in various aspects of installation work.

It covers both installation theory and practice in compliance with the 16th edition of the *IEE Wiring Regulations*, and also deals with the electrical contracting industry, the environmental effects of electricity and basic electronics.

Much of the material in this book is based on my earlier series, *Modern Electrical Installation for Craft Students*, but it has been rearranged and augmented to cater better for student-centred learning programmes. Self assessment questions and answers are provided at the ends of chapters.

Since January 1995, the UK distribution *declared* voltages at consumer supply terminals have changed from 415 V/240 V ± 6% to 400 V/230 V + 10% – 6%. As there has been no physical change to the system, it is likely that measurement of voltages will reveal little or no difference to those before, nor will they do so for some considerable time to come. Hence I have used both the old and the new values in many of the examples in this book.

Also, BS 7671 2001 now refers to PVC as thermosetting (PVC). I have, however, left the original wording as all in the industry will recognize this more easily.

Brian Scaddan

Mapping to City and Guilds 2330 Certificate in Electrotechnical Technology

Level	Unit	Unit title	Out-come	Outcome title	Book chapters
2	1	Working effectively and safely in an electrotechnical environment	1	Identify the legal responsibilities of employers and employees and the importance of health and safety in the working environment	Ch. 12: Health and safety
			2	Identify the occupational specialisms	Ch. 11: Electricity, the environment and the community Ch. 13: The electrical contracting industry
			3	Identify sources of technical information	Ch. 1: Basic information and calculations Ch. 2: Electricity Ch. 3: Resistance, current and voltage, power and energy Ch. 4: Electromagnetism Ch. 5: Capacitors and capacitance Ch. 6: Resistance, inductance and capacitance in installation work Ch. 7: Three-phase circuits Ch. 14: The mechanics of lifting and handling
	2	Principles of electrotechnology	1	Describe the application of basic units	Ch. 8: Motors and generators
			2	Describe basic scientific concepts	Ch. 8: Motors and generators
			3	Describe basic electrical circuitry	Ch. 8: Motors and generators
			4	Identify tools, plant, equipment and materials	Ch. 14: Installation materials and tools Ch. 15: Installation circuits and systems Ch. 6: Access equipment
	3	Application of health and safety and electrical principles	1	Safe systems of working	Ch. 1: Basic information and calculations Ch. 2: Electricity Ch. 3: Resistance, current and voltage, power and energy Ch. 4: Electromagnetism Ch. 5: Capacitors and capacitance Ch. 6: Resistance, inductance and capacitance in installation work Ch. 7: Three-phase circuits Ch. 10: Illumination and ELV lighting Ch. 14: Installation materials and tools Ch. 15: Installation circuits and systems Ch. 6: Access equipment Ch. 14: The mechanics of lifting and handling

2	Use technical information	Ch. 1: Basic information and calculations Ch. 2: Electricity Ch. 3: Resistance, current and voltage, power and energy Ch. 4: Electromagnetism Ch. 5: Capacitors and capacitance Ch. 6: Resistance, inductance and capacitance in installation work Ch. 7: Three-phase circuits Ch. 10: Illumination and ELV lighting Ch. 14: Installation materials and tools Ch. 15: Installation circuits and systems Ch. 6: Access equipment Ch. 14: The mechanics of lifting and handling
3	Electrical machines and a.c. theory	Ch. 1: Basic information and calculations Ch. 2: Electricity Ch. 3: Resistance, current and voltage, power and energy Ch. 4: Electromagnetism Ch. 5: Capacitors and capacitance Ch. 6: Resistance, inductance and capacitance in installation work Ch. 7: Three-phase circuits Ch. 10: Illumination and ELV lighting Ch. 14: Installation materials and tools Ch. 15: Installation circuits and systems Ch. 6: Access equipment Ch. 14: The mechanics of lifting and handling
4	Polyphase systems	Ch. 11: Electricity, the environment and the community Ch. 13: The electrical contracting industry Ch. 18: Circuits and design
5	Overcurrent, short circuit and earth fault protection	Ch. 11: Electricity, the environment and the community Ch. 13: The electrical contracting industry Ch. 18: Circuits and design

Continued

Mapping to City and Guilds 2330 Certificate in Electrotechnical Technology—*(Continued)*

Level	Unit	Unit title	Out-come	Outcome title	Book chapters
	4	Installation (Buildings and structures)	1	Regulations and related information for electrical installations	Ch. 1: Basic information and calculations Ch. 2: Electricity Ch. 3: Resistance, current and voltage, power and energy Ch. 4: Electromagnetism Ch. 5: Capacitors and capacitance Ch. 6: Resistance, inductance and capacitance in installation work Ch. 7: Three-phase circuits Ch. 11: Electricity, the environment and the community Ch. 12: Health and safety Ch. 13: The electrical contracting industry Ch. 14: Installation materials and tools Ch. 15: Installation circuits and systems Ch. 6: Access equipment Ch. 14: The mechanics of lifting and handling
			2	Purpose and application of specifications and data	None listed by Brian at proposal
			3	Types of electrical installations	Ch. 1: Basic information and calculations Ch. 2: Electricity Ch. 3: Resistance, current and voltage, power and energy Ch. 4: Electromagnetism Ch. 5: Capacitors and capacitance Ch. 6: Resistance, inductance and capacitance in installation work Ch. 7: Three-phase circuits Ch. 11: Electricity, the environment and the community Ch. 12: Health and safety Ch. 13: The electrical contracting industry Ch. 14: Installation materials and tools Ch. 15: Installation circuits and systems Ch. 6: Access equipment Ch. 14: The mechanics of lifting and handling
			4	Undertake electrical installation	None listed by Brian at proposal

3	1	Application of health and safety and electrical principles (Stage 3)	1	Comply with statutory regulations and organisational requirements
3			2	Apply safe working practices and follow accident and emergency procedures
3			3	Effective working practices
3			4	Understand the functions of electrical components
3			5	Understand electrical supply systems, protection and earthing
3			6	Understand the functions of electrical machines and motors
3	2	Installation (Buildings and structures) – Inspection, testing and commissioning	1	Use safe, effective and efficient working practices to complete electrical installations
3			2	Select appropriate working methods and use tools, equipment and instruments for inspection testing and commissioning
3	3	Installation (Buildings and structures) – Fault diagnosis and rectification	1	Use safe, effective and efficient working practices to undertake fault diagnosis
3			2	Carry out commissioning to restore systems, components and equipment to working order

1 Basic information and calculations

Units

A unit is what we use to indicate the measurement of a quantity. For example, a unit of *length* could be an *inch* or a *metre* or a *mile*, etc.

In order to ensure that we all have a common standard, an international system of units exists known as the SI system. There are six basic SI units from which all other units are derived.

Basic units

Quantity	Symbol	Unit	Symbol
Length	l	Metre	m
Mass	m	Kilogram	kg
Time	s	Second	s
Current	I	Ampere	A
Temperature	t	Degree kelvin	K
Luminous intensity	I	Candela	cd

Conversion of units

Temperature

Kelvin (K) = 0°C + 273.15

Celsius (°C) = K – 273.15

$$\text{Celsius } (^\circ\text{C}) = \frac{5}{9}(^\circ\text{F} - 32)$$

$$\text{Fahrenheit } (^\circ\text{F}) = \left(\frac{9^\circ\text{C}}{5}\right) + 32$$

Boiling point of water at sea level = 100°C or 212°F
Freezing point of water at sea level = 0°C or 32°F
Normal body temperature = 36.8°C or 98.4°F

Length

millimetre (mm);
centimetre (cm);
metre (m);
kilometre (km).

To obtain	multiply	by
mm	cm	10^1
	m	10^3
	km	10^6
cm	mm	10^{-1}
	m	10^2
	km	10^5
m	mm	10^{-3}
	cm	10^{-2}
	km	10^3
km	mm	10^{-6}
	cm	10^{-5}
	m	10^{-3}

Area

square millimetre (mm^2);
square centimetre (cm^2);
square metre (m^2);
square kilometre (km^2);
also, $1\ km^2 = 100$ hectares (ha).

To obtain	multiply	by
mm^2	cm^2	10^2
	m^2	10^6
	km^2	10^{12}
cm^2	mm^2	10^{-2}
	m^2	10^4
	km^2	10^{10}
m^2	mm^2	10^{-6}
	cm^2	10^{-4}
	km^2	10^6
km^2	mm^2	10^{-12}
	cm^2	10^{-10}
	m^2	10^{-6}

Volume

cubic millimetre (mm^3);
cubic centimetre (cm^3);
cubic metre (m^3);

To obtain	multiply	by
mm^3	cm^3	10^3
	m^3	10^9
cm^3	mm^3	10^{-3}
	m^3	10^6
m^3	mm^3	10^{-9}
	cm^3	10^{-6}

Capacity

millilitre (ml);
centilitre (cl);
litre (l);
also, 1 litre of water has a mass of 1 kg.

To obtain	multiply	by
ml	cl	10^1
	l	10^3
cl	ml	10^{-1}
	l	10^2
l	ml	10^{-3}
	cl	10^{-2}

Mass

milligram (mg);
gram (g);
kilogram (kg);
tonne (t).

To obtain	multiply	by
mg	g	10^3
	kg	10^6
	t	10^9
g	mg	10^{-3}
	kg	10^3
	t	10^6
kg	mg	10^{-6}
	g	10^{-3}
	t	10^3
t	mg	10^{-9}
	g	10^{-6}
	kg	10^{-3}

Multiples and submultiples of units

Name	Symbol	Multiplier	Example
tera	T	10^{12} (1 000 000 000 000)	terawatt (TW)
giga	G	10^{9} (1 000 000 000)	gigahertz (GHz)
mega	M*	10^{6} (1 000 000)	megawatt (MW)
kilo	k*	10^{3} (1000)	kilovolt (kV)
hecto	h	10^{2} (100)	hectogram (hg)
deka	da	10^{1} (10)	dekahertz (daHz)
deci	d	10^{-1} (1/10 th)	decivolt (dV)
centi	c	10^{-2} (1/100 th)	centimetre (cm)
milli	m*	10^{-3} (1/1000 th)	milliampere (mA)
micro	μ*	10^{-6} (1/1 000 000 th)	microvolt (mV)
nano	n	10^{-9} (1/1 000 000 000 th)	nanowatt (nW)
pico	p*	10^{-12} (1/1 000 000 000 000 th)	picofarad (pF)

*Multiples most used in this book.

Indices

It is very important to understand what *Indices* are and how they are used. Without such knowledge, calculations and manipulation of formulae are difficult and frustrating.

So, what are *Indices*? Well, they are perhaps most easily explained by example. If we multiply two identical numbers, say 2 and 2, the answer is clearly 4, and this process is usually expressed thus:

$2 \times 2 = 4$

However, another way of expressing the same condition is

$2^2 = 4$

The upper 2 simply means that the lower 2 is multiplied by itself. The upper 2 is known as the indice. Sometimes this situation is referred to as 'Two *raised to the power of* two'. So, 2^3 means 'Two multiplied by itself *three* times'.

i.e. $2 \times 2 \times 2 = 8$

Do not be misled by thinking that 2^3 is 2×3.

$2^4 = 2 \times 2 \times 2 \times 2 = 16$ (*not* $2 \times 4 = 8$)

$24^2 = 24 \times 24 = 576$ (*not* $24 \times 2 = 48$)

Here are some other examples:

$3^3 = 3 \times 3 \times 3 = 27$

$9^2 = 9 \times 9 = 81$

$4^3 = 4 \times 4 \times 4 = 64$

$10^5 = 10 \times 10 \times 10 \times 10 \times 10 = 100\,000$

A number by itself, say 3, has an invisible indice, 1, but it is not shown. Now, consider this: $2^2 \times 2^2$ may be rewritten as $2 \times 2 \times 2 \times 2$, or as 2^4 which means that the indices 2 and 2 or the invisible indices 1 have been added together. So the rule is, when multiplying, *add* the indices.

Examples

$$4 \times 4^2 = 4^1 \times 4^2 = 4^3 = 4 \times 4 \times 4 = 64$$

$$3^2 \times 3^3 = 3^5 = 3 \times 3 \times 3 \times 3 \times 3 = 243$$

$$10 \times 10^3 = 10^4 = 10 \times 10 \times 10 \times 10 = 10\,000$$

Let us now advance to the following situation:

$$10^4 \times \frac{1}{10^2} \text{ is the same as } \frac{10^4}{10^2} = \frac{10 \times 10 \times 10 \times 10}{10 \times 10}$$

Cancelling out the tens

$$\frac{\cancel{10} \times \cancel{10} \times 10 \times 10}{\cancel{10} \times \cancel{10}}$$

we get

$$10 \times 10 = 10^2$$

which means that the indices have been *subtracted*, i.e. 4 – 2. So the rule is, when dividing, *subtract* the indices.

How about this though: 4 – 2 is either 4 subtract 2 or 4 add –2, and remember, the addition of indices goes with multiplication, so from this we should see that 10^4 divided by 10^2 is the same as 10^4 multiplied by 10^{-2}.

So,

$$\frac{1}{10^2} \text{ is the same as } 10^{-2}$$

Examples

$$\frac{1}{3^4} = 3^{-4} \quad \frac{1}{2^6} = 2^{-6}$$

and conversely,

$$\frac{1}{10^{-2}} = 10^2$$

Hence we can see that indices may be moved above or below the line providing the sign is changed.

Examples

1 $$\frac{10^6 \times 10^7 \times 10^{-3}}{10^4 \times 10^2} = \frac{10^{13} \times 10^{-3}}{10^6}$$

$$= \frac{10^{10}}{10^6} = 10^{10} \times 10^{-6} = 10^4 = 10\,000$$

2 $$\frac{10^4 \times 10^{-6}}{10} = 10^4 \times 10^{-6} \times 10^{-1}$$

$$= 10^4 \times 10^{-7} = 10^{-3} = \frac{1}{10^3}$$

$$= \frac{1}{1000} = 0.001$$

Self-assessment questions

1 Write, in numbers, 'eight raised to the power of four'.
2 Addition of indices cannot be used to solve $3^2 \times 2^3$. Why?
3 What is $\frac{10}{10}$ equal to?
4 Replace $\frac{10}{10}$ using the addition of indices. Write down the answer using indices.
5 What is the answer to $3^1 \times 3^{-1}$, as a single number and using indices?
6 What is 8^0 equal to?
7 Solve the following:

(a) $$\frac{3^2 \times 3^{-1} \times 3^3}{3^6 \times 3^{-2}}$$

(b) $10^{-6} \times 10^3 \times 10^4 \times 10^0$

(c) $$\frac{5^5 \times 5^{-7}}{5^{-4}}$$

Simple algebra

Algebra is a means of solving mathematical problems using letters or symbols to represent unknown quantities. The same laws apply to algebraic symbols as to real numbers.

Hence: if *one ten* times *one ten* = 10^2, then *one X* times *one X* = X^2

i.e. $X \times X = X^2$

In algebra the multiplication sign is usually left out. So, for example $A \times B$ is shown as AB and $2 \times Y$ is shown as $2Y$. This avoids the confusion of the multiplication sign being mistaken for an X. Sometimes a dot (.) is used to replace the multiplication sign. Hence $3.X$ means 3 times X, and $2F.P$ means 2 times F times P.

The laws of indices also apply to algebraic symbols.

e.g. $X.X = X^2 \quad Y^2.Y^2 = Y^4 \quad \frac{1}{X} = X^{-1} \quad \frac{1}{Y^3} = Y^{-3}$ etc.

Addition and subtraction are approached in the same way:

e.g. $X + X = 2X$

$3X - X = 2X$

$10P - 5P = 5P$

$4M + 6M + 2F = 10M + 2F$

Also with multiplication and division:

e.g. $X.X = X^2$

$3M.2M = 6M^2$

$$\frac{14P}{7} = 2P$$

$$\frac{10Y}{2Y} = 5$$

Formulae or equations

A formula or equation is an algebraic means of showing how a law or rule is applied. For example, we all know that the money we get in our wage packets is our gross pay less deductions. If we represent each of these quantities by a letter say W for wages, G for gross pay, and D for deductions, we can show our pay situation as

$W = G - D$

Similarly, we know that if we travel a distance of 60 km at a speed of 30 km per hour, it will take us 2 hours. We have simply divided distance (D) by speed (S) to get time (T), which gives us the formula

$$T = \frac{D}{S}$$

Manipulation or transposition of formulae

The equals sign (=) in a formula or equation is similar to the pivot point on a pair of scales (Fig. 1.1).

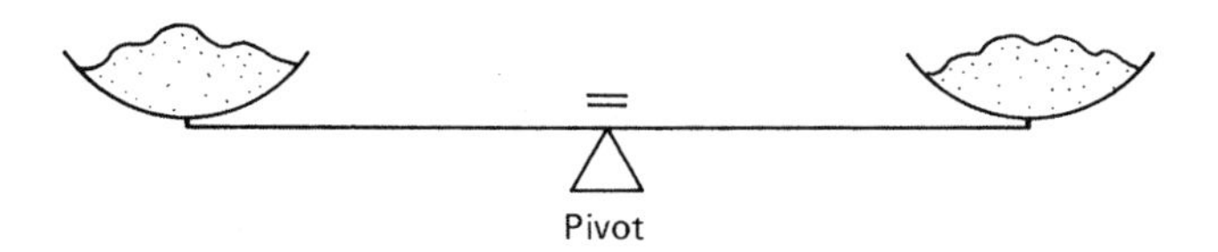

Fig. 1.1

If an item is added to one side of the scales, they become unbalanced, so an identical weight needs to be added to the other side to return the scales to a balanced condition. The same applies to a formula or equation, in that whatever is done on one side of the equals sign must be done to the other side.

Consider the formula $X + Y = Z$.

If we were to multiply the left-hand side (LHS) by, say 2, we would get $2X + 2Y$, but in order to ensure that the formula remains correct, we must also multiply the right-hand side (RHS) by 2, hence $2X + 2Y = 2Z$.

Formulae may be rearranged (transposed) such that any symbol can be shown in terms of the other symbols. For example, we know that our pay formula is $W = G - D$ but if we know our wages and our gross pay how do we find the deductions? Clearly we need to transpose the formula to show D in terms of W and G. Before we do this, however, let us consider the types of formula that exist.

There are three types:

(a) Pure addition/subtraction
(b) Pure multiplication/division
(c) Combination of (a) and (b).

Other points to note are:

1 A symbol on its own with no sign is taken as being positive, i.e. K is $+K$.
2 Symbols or groups of symbols will be on either the top or the bottom of each side of an equation, for example

$$\frac{A}{B} = \frac{M}{P}$$

A and M are on the top, B and P are on the bottom. In the case of, say,

$$X = \frac{R}{S}$$

X and R are on the top line and S is on the bottom.

(Imagine X to be divided by 1, i.e. $\frac{X}{1}$.)

3 Formulae are usually expressed with a single symbol on the LHS, e.g. $Y = P - Q$, but it is still correct to show it as $P - Q = Y$.
4 Symbols enclosed in brackets are treated as one symbol. For example, $(A + C + D)$ may, if necessary, be transposed as if it were a single symbol.

Let us now look at the simple rules of transposition.

(a) Pure addition/subtraction

Move the symbol required to the LHS of the equation and move all others to the RHS. *Any move needs a change in sign.*

Example

If $A - B = Y - X$, what does X equal?

Move the $-X$ to the LHS and change its sign. Hence,

$$X + A - B = Y$$

Then move the A and the $-B$ to the RHS and change signs. Hence,

$$X = Y - A + B$$

Example

If $M + P = R - S$, what does R equal?

We have

$$- R + M + P = - S$$
$$\therefore - R = - S - M - P$$

However, we need R, not $- R$, so simply change its sign, but remember to do the same to the RHS of the equation. Hence,

$$R = S + M + P$$

So we can now transpose our wages formula $W = G - D$ to find D:

$$W = G - D$$
$$D + W = G$$
$$D = G - W$$

(b) Pure multiplication/division

Move the symbol required *across* the equals sign so that it is on the *top* of the equation and move all other symbols away from it, *across* the equals sign but in the opposite position, i.e. from top to bottom or vice versa. Signs are not changed with this type of transposition.

Example

If

$$\frac{A}{B} = \frac{C}{D}$$

what does D equal?

Move the D from bottom RHS to top LHS. Thus,

$$\frac{A.D}{B} = \frac{C}{1}$$

Now move A and B across to the RHS in opposite positions. Thus,

$$\frac{D}{1} = \frac{C.B}{A} \quad \text{or simply} \quad D = \frac{C.B}{A}$$

Example

If

$$\frac{X.Y.Z}{T} = \frac{M.P}{R}$$

what does P equal?

As P is already on the top line, leave it where it is and simply move the M and R. Hence,

$$\frac{X.Y.Z.R}{T.M} = P$$

which is the same as

$$P = \frac{X.Y.Z.R}{T.M}$$

(c) Combination transposition

This is best explained by examples.

Example

If

$$\frac{A(P + R)}{X.Y} = \frac{D}{S}$$

what does S equal?

We have

$$\frac{S.A(P + R)}{X.Y} = D$$

Hence,

$$S = \frac{D.X.Y}{A(P + R)}$$

Example

If

$$\frac{A(P + R)}{X.Y} = \frac{D}{S}$$

what does R equal?

Treat $(P + R)$ as a single symbol and leave it on the top line, as R is part of that symbol. Hence,

$$(P + R) = \frac{D.X.Y}{A.S}$$

Remove the brackets and treat the RHS as a single symbol. Hence,

$$P + R = \left(\frac{D.X.Y}{A.S}\right)$$

$$R = \left(\frac{D.X.Y}{A.S}\right) - P$$

Self-assessment questions

1 Write down the answers to the following:

(a) $X + 3X =$
(b) $9F - 4F =$
(c) $10Y + 3X - 2Y + X =$
(d) $M.2M =$
(e) $P.3P.2P =$
(f) $\dfrac{12D}{D} =$
(g) $\dfrac{30A}{15} =$
(h) $\dfrac{X^3}{X^2} =$

2 Transpose the following to show X in terms of the other symbols:

(a) $X + Y = P + Q$
(b) $F - D = A - X$
(c) $L - X - Q = P + W$
(d) $2X = 4$
(e) $XM = PD$
(f) $\dfrac{A}{X} = W$
(g) $\dfrac{B}{K} = \dfrac{H}{2X}$

(h) $A.B.C. = \dfrac{MY}{X}$

(i) $X(A + B) = W$

(j) $\dfrac{M + N}{2X} = \dfrac{P}{R}$

The theorem of Pythagoras

Pythagoras showed that if a square is constructed on each side of a right-angled triangle (Fig. 1.2), then the area of the large square equals the sum of the areas of the other two squares.

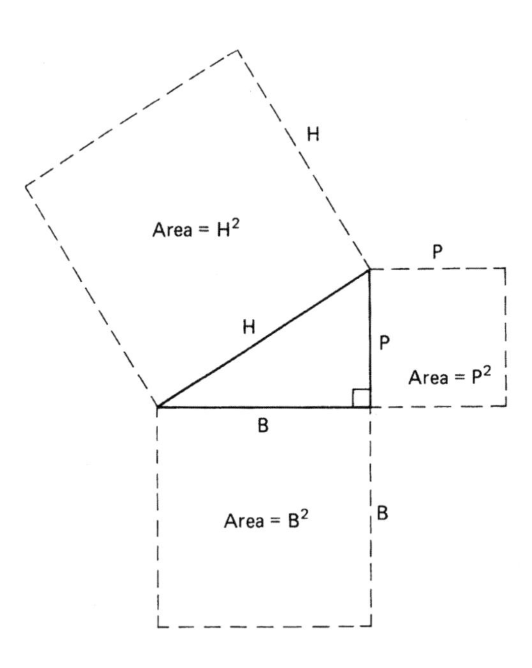

Fig. 1.2

Hence: 'The square on the hypotenuse of a right-angled triangle is equal to the sum of the squares on the other two sides.' That is,

$$H^2 = B^2 + P^2$$

Or, taking the square root of *both* sides of the equation

$$H = \sqrt{B^2 + P^2}$$

Or, transposing

$$B = \sqrt{H^2 - P^2}$$

$$\text{or } P = \sqrt{H^2 - B^2}$$

Examples

1 From Fig. 1.3 calculate the value of H:

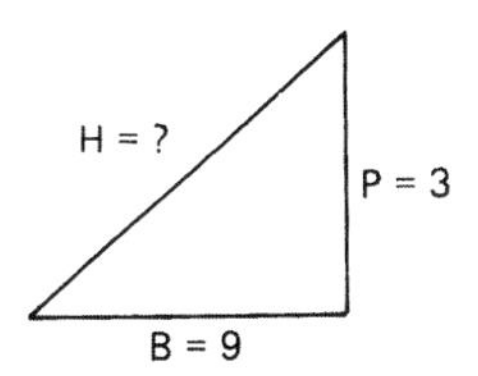

Fig. 1.3

$$H = \sqrt{B^2 + P^2}$$
$$= \sqrt{3^2 + 9^2}$$
$$= \sqrt{9 + 81}$$
$$= \sqrt{90}$$
$$= 9.487$$

2 From Fig. 1.4 calculate the value of B:

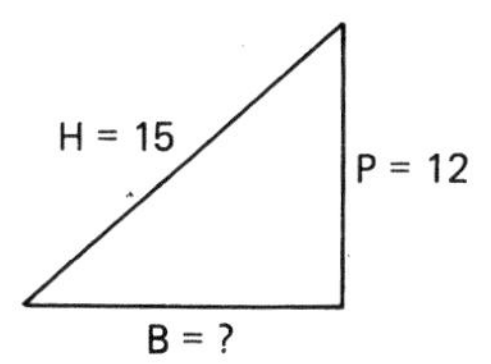

Fig. 1.4

$$B = \sqrt{H^2 - P^2}$$
$$= \sqrt{15^2 - 12^2}$$
$$= \sqrt{225 - 144}$$
$$= \sqrt{81}$$
$$= 9$$

Basic trigonometry

This subject deals with the relationship between the sides and angles of triangles. In this section we will deal with only the very basic concepts.

Consider the right-angled triangle shown in Fig. 1.5. *Note*: Unknown angles are usually represented by Greek letters, such as alpha (α), beta (β), phi (ϕ), theta (θ), etc.

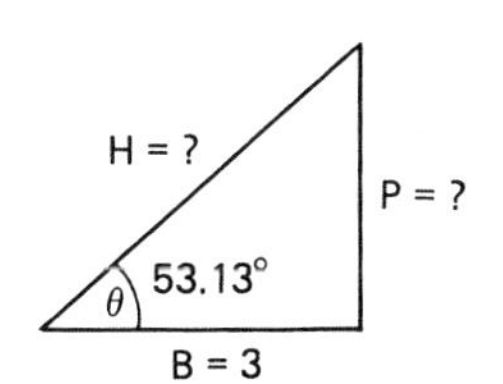

Fig. 1.5

There are three relationships between the sides H (hypotenuse), P (perpendicular), and B (base), and the base angle θ. These relationships are known as the *sine*, the *cosine* and the *tangent* of the angle θ, and are usually abbreviated to sin, cos and tan.

The *sine* of the base angle θ,

$$\sin\theta = \frac{P}{H}$$

The *cosine* of the base angle θ,

$$\cos\theta = \frac{B}{H}$$

The *tangent* of the base angle, θ,

$$\tan\theta = \frac{P}{B}$$

The values of sin, cos and tan for all angles between 0° and 360° are available either in tables or, more commonly now, by the use of a calculator.

How then do we use trigonometry for the purposes of calculation? Examples are probably the best means of explanation.

Examples

1 From the values shown in Fig. 1.5, calculate P and H:

$$\cos \theta = \frac{B}{H}$$

Transposing,

$$H = \frac{B}{\cos \theta} = \frac{3}{\cos 53.13^\circ}$$

From tables or calculator, $\cos 53.13^\circ = 0.6$

$$\therefore H = \frac{3}{0.6} = 5$$

Now we can use sin or tan to find P:

$$\tan \theta = \frac{P}{B}$$

Transposing,

$$P = B.\tan \theta$$

$$\tan \theta = \tan 53.13^\circ = 1.333$$

$$\therefore P = 3 \times 1.333 = 4\ (3.999)$$

2 From the values shown in Fig. 1.6, calculate α and P:

$$\cos \alpha = \frac{B}{H} = \frac{6}{12} = 0.5$$

We now have to find the angle whose cosine is 0.5. This is usually written $\cos^{-1} 0.5$. The –1 is not an *indice*; it simply means 'the angle whose sin, cos or tan is . . .'

So the angle $\alpha = \cos^{-1} 0.5$.

We now look up the tables for 0.5 or use the INV cos or ARC cos, etc., function on a calculator. Hence,

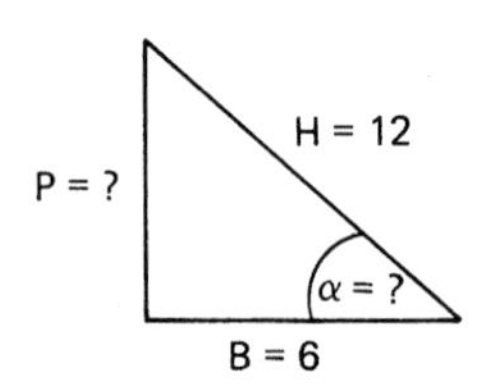

Fig. 1.6

$$\alpha = 60^\circ$$

$$\sin \alpha = \frac{P}{H}$$

Transposing,

$$P = H.\sin \alpha$$

$$= 12.\sin 60^\circ$$

$$= 12 \times 0.866$$

$$= 10.4$$

Self-assessment questions

1 What kind of triangle enables the use of Pythagoras' theorem?
2 Write down the formula for Pythagoras' theorem.
3 Calculate the hypotenuse of a right-angled triangle if the base is 11 and the perpendicular is 16.
4 Calculate the base of a right-angled triangle if the hypotenuse is 10 and the perpendicular is 2.
5 Calculate the perpendicular of a right-angled triangle if the hypotenuse is 20 and the base is 8.
6 What is the relationship between the sides and angles of a triangle called?
7 For a right-angled triangle, write down a formula for:
(a) The sine of an angle.
(b) The cosine of an angle.
(c) The tangent of an angle.
8 A right-angled triangle of base angle 25° has a perpendicular of 4. What is the hypotenuse and the base?
9 A right-angled triangle of hypotenuse 16 has a base of 10. What is the base angle and the perpendicular?
10 A right-angled triangle of base 6 has a perpendicular of 14. What is the base angle and the hypotenuse?

Areas and volumes

Areas and volumes are shown in Fig. 1.7.

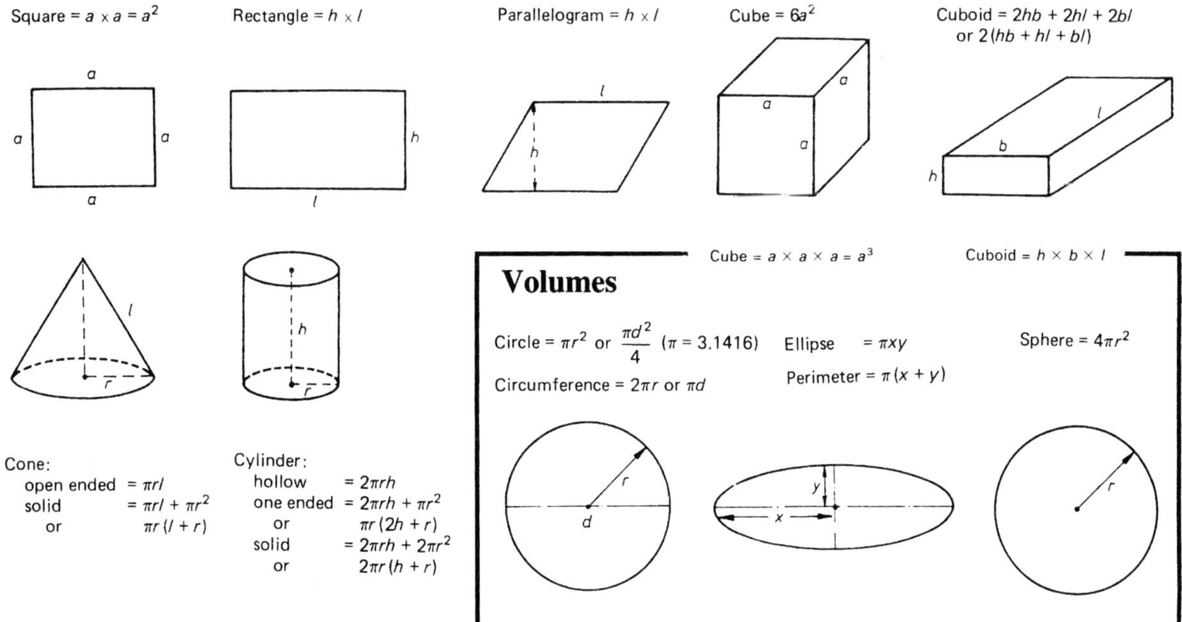

Fig. 1.7

2 Electricity

What is electricity? Where does it come from? How fast does it travel? In order to answer such questions, it is necessary to understand the nature of substances.

Molecules and atoms

Every substance known to man is composed of molecules which in turn are made up of atoms. Substances whose molecules are formed by atoms of the same type are known as *elements*, of which there are known to be, at present, more than 100 (Table 2.1).

Substances whose molecules are made up of atoms of different types are known as compounds. Hence, water, which is a compound, comprises two hydrogen atoms (H) and one oxygen atom (O), i.e. H_2O. Similarly, sulphuric acid has two hydrogen, one sulphur and four oxygen atoms: hence, H_2SO_4.

Molecules are always in a state of rapid motion, but when they are densely packed together this movement is restricted and the substance formed by these molecules is stable, i.e. a *Solid*. When the molecules of a substance are less tightly bound there is much free movement, and such a substance is known as a *liquid*. When the molecule movement is almost unrestricted the substance can expand and contract in any direction and, of course, is known as a *gas*.

The atoms which form a molecule are themselves made up of particles known as protons, neutrons, and electrons. Protons are said to have a positive (+ve) charge, electrons a negative (–ve) charge, and neutrons no charge. Since neutrons play no part in electricity at this level of study, they will be ignored from now on.

So what is the relationship between protons and electrons; how do they form an atom? The simplest explanation is to liken an atom to our Solar System, where we have a central star, the Sun, around which are the orbiting planets. In the tiny atom, the protons form the central nucleus and the electrons are the orbiting particles. The simplest atom is that of hydrogen which has one proton and one electron (Fig. 2.1).

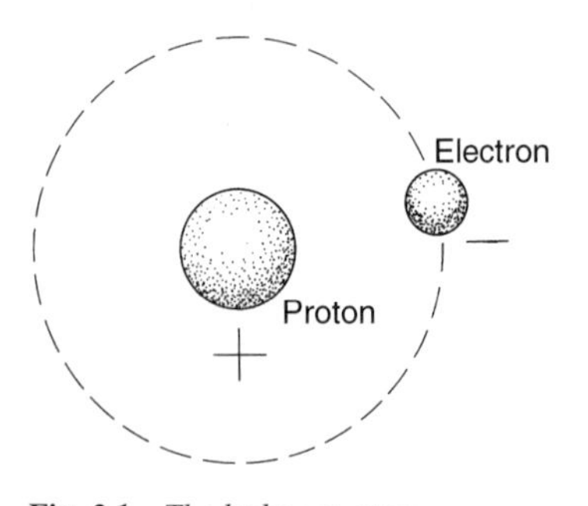

Fig. 2.1 *The hydrogen atom*

The atomic number (Table 2.1) gives an indication of the number of electrons surrounding the nucleus for each of the known elements. Hence, copper has an atomic number of 29, indicating that it has 29 orbiting electrons.

Electrons are arranged in layers or clouds at varying distances from the nucleus (like the rings around Saturn); those nearest the nucleus are more

Table 2.1 Elements

Atomic number	Name	Symbol	Atomic number	Name	Symbol
1	Hydrogen	H	52	Tellurium	Te
2	Helium	He	53	Iodine	I
3	Lithium	Li	54	Xenon	Xe
4	Beryllium	Be	55	Caesium	Cs
5	Boron	B	56	Barium	Ba
6	Carbon	C	57	Lanthanum	La
7	Nitrogen	N	58	Cerium	Ce
8	Oxygen	O	59	Praseodymium	Pr
9	Fluorine	F	60	Neodymium	Na
10	Neon	Ne	61	Promethium	Pm
11	Sodium	Na	62	Samarium	Sm
12	Magnesium	Mg	63	Europium	Eu
13	Aluminium	Al	64	Gadolinium	Gd
14	Silicon	Si	65	Terbium	Tb
15	Phosphorus	P	66	Dysprosium	Dy
16	Sulphur	S	67	Holmium	Ho
17	Chlorine	Cl	68	Erbium	Er
18	Argon	A	69	Thulium	Tm
19	Potassium	K	70	Ytterbium	Yb
20	Calcium	Ca	71	Lutecium	Lu
21	Scandium	Sc	72	Hafnium	Hf
22	Titanium	Ti	73	Tantalum	Ta
23	Vanadium	V	74	Tungsten	W
24	Chromium	Cr	75	Rhenium	Re
25	Manganese	Mn	76	Osmium	Os
26	Iron	Fe	77	Iridium	Ir
27	Cobalt	Co	78	Platinum	Pt
28	Nickel	Ni	79	Gold	Au
29	Copper	Cu	80	Mercury	Hg
30	Zinc	Zn	81	Thallium	Tl
31	Gallium	Ga	82	Lead	Pb
32	Germanium	Ge	83	Bismuth	Bi
33	Arsenic	As	84	Polonium	Po
34	Selenium	Se	85	Astatine	At
35	Bromine	Br	86	Radon	Rn
36	Krypton	Kr	87	Francium	Fr
37	Rubidium	Rb	88	Radium	Ra
38	Strontium	Sr	89	Actinium	Ac
39	Yttrium	Y	90	Thorium	Th
40	Zirconium	Zr	91	Protoactinium	Pa
41	Niobium	Nb	92	Uranium	U
42	Molybdenum	Mo	93	Neptunium	Np
43	Technetium	Tc	94	Plutonium	Pu
44	Ruthenium	Ru	95	Americium	Am
45	Rhodium	Rh	96	Curiam	Cm
46	Palladium	Pd	97	Berkelium	Bk
47	Silver	Ag	98	Californium	Cf
48	Cadmium	Cd	99	Einsteinium	Es
49	Indium	In	100	Fermium	Fm
50	Tin	Sn	101	Mendelevium	Md
51	Antimony	Sb	102	Nobelium	No

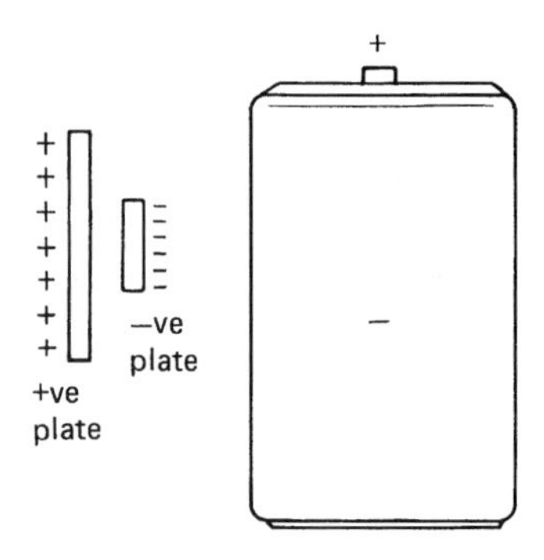

Fig. 2.2

strongly held in place than those farthest away. These distant electrons are easily dislodged from their orbits and hence are free to join those of another atom whose own distant electrons in turn may leave to join other atoms, and so on. These wandering or *random* electrons that move about the molecular structure of the material are what makes up electricity.

So, then, how do electrons form electricity? If we take two dissimilar metal plates and place them in a chemical solution (known as an electrolyte) a reaction takes place in which electrons from one plate travel across the electrolyte and collect on the other plate. So one plate has an excess of electrons which makes it more –ve than +ve, and the other an excess of protons which makes it more +ve than –ve. What we are describing here, of course, is a simple cell or battery (Fig. 2.2).

Now then, consider a length of wire in which, as we have already seen, there are electrons in random movement (Fig. 2.3).

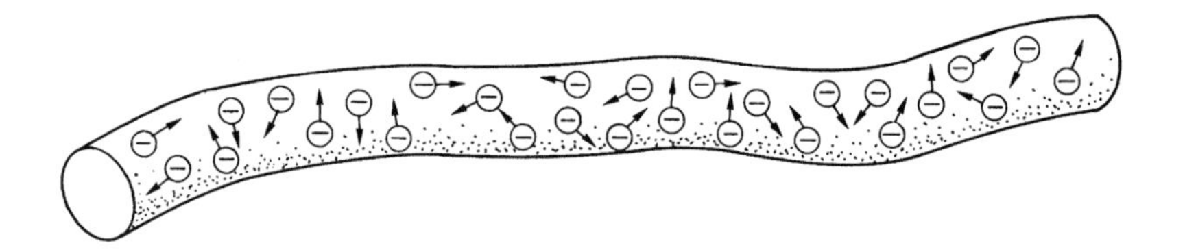

Fig. 2.3

If we now join the ends of the wire to the plates of a cell the excess electrons on the –ve plate will tend to leave and return to the +ve plate, encouraging the random electrons in the wire to *drift* in the same direction (Fig. 2.4). This drift is what we know as electricity. The process will continue until the chemical action of the cell is exhausted and there is no longer a difference, +ve or –ve, between the plates.

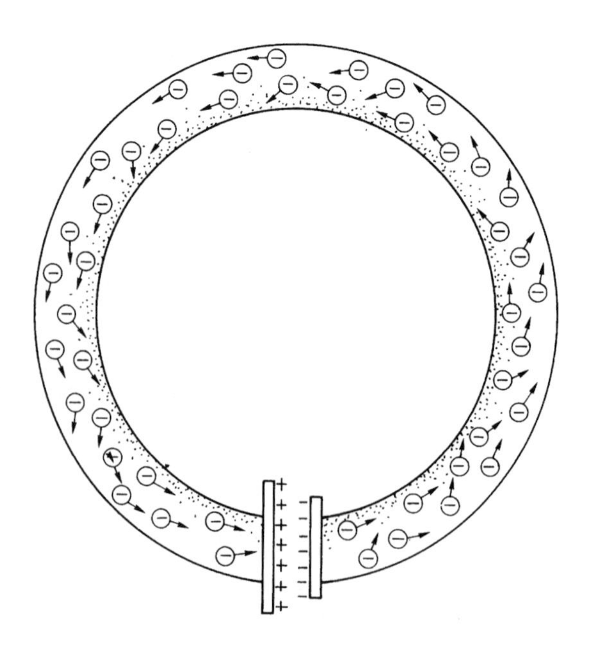

Fig. 2.4

Potential difference

Anything that is in a state whereby it may give rise to the release of energy is said to have *potential*. For example, a ball held above the ground has potential in that if it were let go, it would fall and hit the ground. So, a cell or battery with its +ve and –ve plates has potential to cause electron drift. As there is a difference in the number of electrons on each of the plates, this potential is called the *potential difference* (p.d.).

Electron flow and conventional current flow

As we have seen, if we apply a p.d. across the ends of a length of wire, electrons will drift from –ve to +ve. In the early pioneering days, it was incorrectly thought that electricity was the movement of +ve protons and, therefore, any flow was from +ve to –ve. However, as the number of proton charges is the same as the number of electron charges, the convention of electric current flow from +ve to –ve has been maintained.

Conductors and insulators

Having shown that electricity is the general drift of random electrons, it follows that materials with large numbers of such electrons give rise to a greater drift than those with few random electrons. The two different types are known as conductors and insulators. Materials such as P.V.C., rubber, mica, etc., have few random electrons and therefore make good *insulators*, whereas metals such as aluminum, copper, silver, etc., with large numbers make good *conductors*.

Electrical quantities

The units in which we measure electrical quantities have been assigned the names of famous scientific pioneers, brief details of whom are as follows. (Others will be detailed as the book progresses.)

André Marie Ampère (1775–1836)

French physicist who showed that a mechanical force exists between two conductors carrying a current.

Charles Augustin de Coulomb (1746–1806)

French military engineer and physicist famous for his work on electric charge.

Georg Simon Ohm (1789–1854)

German physicist who demonstrated the relationship between current, voltage, and resistance.

Allessandro Volta (1745–1827)

Italian scientist who developed the electric cell, called the 'voltaic pile', which comprised a series of copper and zinc discs separated by a brine-soaked cloth.

Electric current: symbol, I; unit, ampere (A)

This is the flow or drift of random electrons in a conductor.

Electric charge or quantity: symbol, Q; unit, coulomb (C)

This is the quantity of electricity that passes a point in a circuit in a certain time. One coulomb is said to have passed when one ampere flows for one second:

$$Q = I \times T$$

Electromotive force (e.m.f.): symbol, E; unit, volt (V)

This is the total potential force available from a source to drive electric current around a circuit.

Potential difference (p.d.): symbol, V; unit, volt (V)

Often referred to as 'voltage' or 'voltage across', this is the actual force available to drive current around a circuit.

The difference between e.m.f. and p.d. may be illustrated by the *pay* analogy used in Chapter 1. Our gross wage (e.m.f.) is the total available to use. Our net wage (p.d.) is what we actually have to spend after deductions.

Resistance: symbol, R; unit, ohm (Ω)

This is the opposition to the flow of current in a circuit.

When electrons flow around a circuit, they do not do so unimpeded. There are many collisions and deflections as they make their way through the complex molecular structure of the conductor, and the extent to which they are impended will depend on the material from which the conductor is made and its dimensions.

Resistivity: symbol, ρ; unit, μΩ mm

If we take a sample of material in the form of a cube of side 1 mm and measure the resistance between opposite faces (Fig. 2.5), the resulting value is called the *resistivity* of that material.

This means that we can now determine the resistance of a sample of material of any dimension. Let us suppose that we have a 1 mm cube of material of resistivity, say, 1 ohm (Fig. 2.6a). If we double the length of that

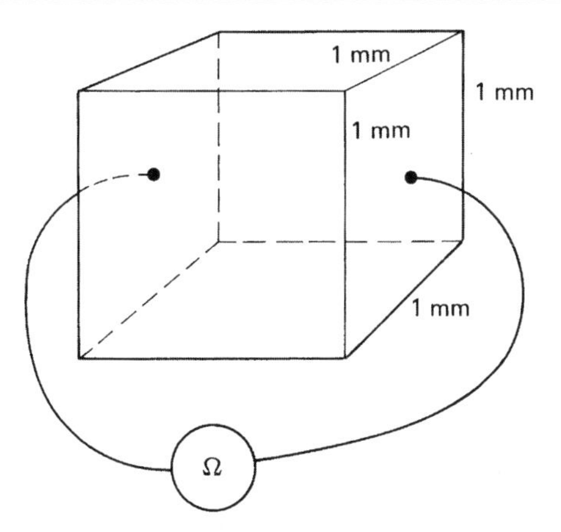

Fig. 2.5

sample, leaving the face area the same (Fig. 2.6b), the resistance now measured would be 2 ohms, i.e. the resistance has doubled. If, however, we leave the length the same but double the face area (Fig. 2.6c), the measured value would now be 0.5 ohms, i.e. the resistance has halved.

Hence we can now state that whatever happens to the length of a conductor also happens to its resistance, i.e. *resistance is proportional to length*, and whatever happens to the cross-sectional area has the opposite effect on the resistance, i.e. *resistance is inversely proportional to area.*
So,

$$\text{resistance } R = \frac{\text{resistivity } \rho \times \text{length } l}{\text{area } a}$$

$$R = \frac{\rho \times l}{a}$$

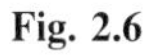
Fig. 2.6

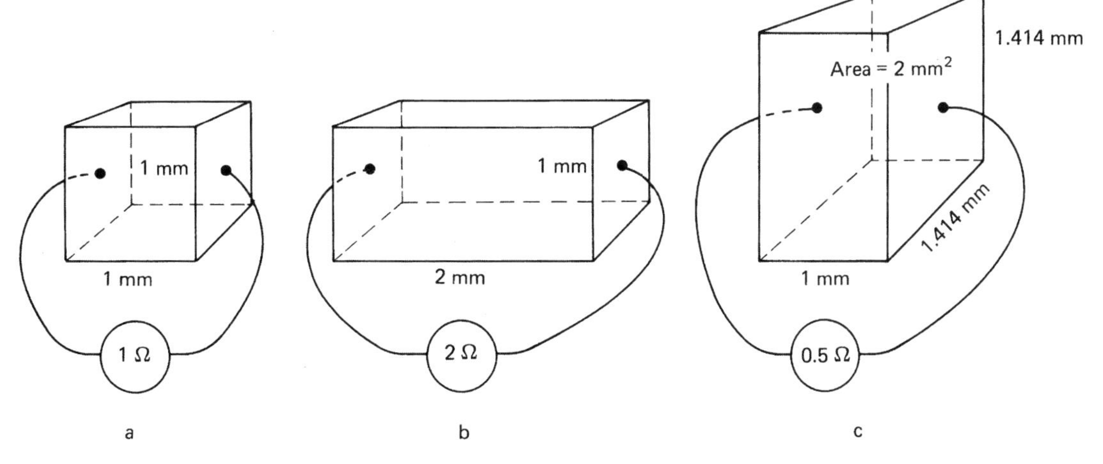

Table 2.2 Examples of resistivity

Material	Resistivity, ρ (μΩ mm at 20 °C)
Copper, International Standard	17.2
Copper, annealed	16.9 to 17.4
Copper, hard drawn	17.4 to 18.1
Aluminium, hard drawn	28
Silver, annealed	15.8
Silver, hard drawn	17.5
Platinum	117
Tungsten	56
Eureka (constantan)	480
German silver (platinoid)	344
Manganin	480

In practice, the resistance across a 1 mm cube of a material is extremely small, in the order of millionths of an ohm (μΩ) as Table 2.2 shows.

Example

Calculate the resistance of a 50 m length of copper conductor of cross-sectional area (c.s.a.) 2.5 mm^2, if the resistivity of the copper used is 17.6 μΩ mm.

Note: All measurements should be of the same type, i.e. resistivity, microohm millimetres; length, *millimetres*; c.s.a. square *millimetres*. Hence the 10^3 to convert metres to millimetres.

$$R = \frac{\rho \times l}{a}$$

$$\therefore R = \frac{17.6 \times 10^{-6} \times 50 \times 10^3}{2.5}$$

$$= \frac{17.6 \times 10^{-3} \times 50}{2.5}$$

$$= 17.6 \times 10^{-3} \times 20$$

$$= 352 \times 10^{-3}$$

$$= 0.352$$

Example

Calculate the resistivity of aluminium if a 100 m length of conductor of c.s.a. 4 mm^2 has a measured resistance of 0.7 Ω.

$$R = \frac{\rho \times l}{a}$$

$$\therefore \rho = \frac{R \times a}{l}$$

$$= \frac{0.7 \times 4}{100 \times 10^3}$$

$$= 7 \times 10^{-1} \times 4 \times 10^{-5}$$

$$= 28 \times 10^{-6}$$

$$= 28\,\mu\Omega\,\text{mm}$$

Reference to Table 2.2 will show that values of resistivity are based on a conductor temperature of 20°C which clearly suggests that other temperatures would give different values. This is quite correct. If a conductor is heated, the molecules vibrate more vigorously making the passage of random electrons more difficult, i.e. the conductor resistance increases. On the other hand, a reduction in temperature has the opposite effect, and hence a decrease in conductor resistance occurs. The amount by which the resistance of a conductor changes with a change in temperature is known as the temperature coefficient of resistance.

Temperature coefficient: symbol α; unit, ohms per ohm per °C (Ω/Ω/°C)

If we were to take a sample of conductor that has a resistance of 1 ohm at a temperature of 0°C, and then increase its temperature by 1°C, the resulting increase in resistance is its temperature coefficient. An increase of 2°C would result in twice the increase, and so on. Therefore the new value of a 1 ohm resistance which has had its temperature raised from 0°C to t°C is given by $(1 + \alpha t)$. For a 2 ohm resistance the new value would be $2 \times (1 + \alpha t)$, and for a 3 ohm resistance, $3 \times (1 + \alpha t)$ etc. Hence we can now write the formula:

$$R_f = R_0\,(1 + \alpha t)$$

where: R_f is the final resistance; R_0 is the resistance at 0°C; α is the temperature coefficient; and t°C is the change in temperature.

For a change in temperature between any two values, the formula is:

$$R_2 = \frac{R_1\,(1 + \alpha t_2)}{(1 + \alpha t_1)}$$

where: R_1 is the initial resistance; R_2 is the final resistance; t_1 is the initial temperature; and t_2 is the final temperature.

The value of temperature coefficient for most of the common conducting materials is broadly similar, ranging from 0.0039 to 0.0045 Ω/Ω/°C, that of copper being taken as 0.004 Ω/Ω/°C.

Example

A sample of copper has a resistance of 10 Ω at a temperature of 0°C. What will be its resistance at 50°C?

$$R_f = R_0 (1 + \alpha\, t)$$

$R_f = ?$

$R_0 = 10$

$t = 50$

$\alpha = 0.004\ \Omega/\Omega/°C$

$$\therefore R_f = 10(1 + 0.004 \times 50)$$

$$= 10(1 + 0.2)$$

$$= 10 \times 1.2$$

$$= 12 \text{ ohms}$$

Example

A length of tungsten filament wire has a resistance of 200 Ω at 20°C. What will be its resistance at 600°C? ($\alpha = 0.0045\ \Omega/\Omega/°C$).

$$R_2 = \frac{R_1 (1 + \alpha t_2)}{(1 + \alpha t_1)}$$

$R_2 = ?$

$R_1 = 200$

$t_1 = 20°C$

$t_2 = 600°C$

$\alpha = 0.0045\ \Omega/\Omega/°C$

$$\therefore R_2 = \frac{200(1 + 0.0045 \times 600)}{(1 + 0.0045 \times 20)}$$

$$= \frac{200(1 + 2.7)}{(1 + 0.09)}$$

$$= \frac{200 \times 3.7}{1.09}$$

$$= 679 \text{ ohms}$$

There are certain conducting materials such as carbon and electrolytes whose resistances display an *inverse* relationship with temperature, i.e. their resistances *decrease* with a rise in temperature, and vice versa. These conductors are said to have *negative* temperature coefficients. Carbon, for example, is used for the brushes in some types of motor, where friction causes the brushes to become very hot. In this way current flow to the motor is not impeded.

We have already learned that random electrons moving in the same direction (electric current) through the molecular structure of a conductor experience many collisions and deflections. The energy given off when this happens is in the form of heat; hence the more electrons the more heat and thus the greater the resistance. So current flow can, itself, cause a change in conductor resistance.

The water analogy

Consider a tap and a length of hose. With the tap just turned on, only a trickle of water will issue from the hose (Fig. 2.7a). Turn the tap further and more water will flow (Fig. 2.7b). Hence pressure and flow are *proportional*. Leave

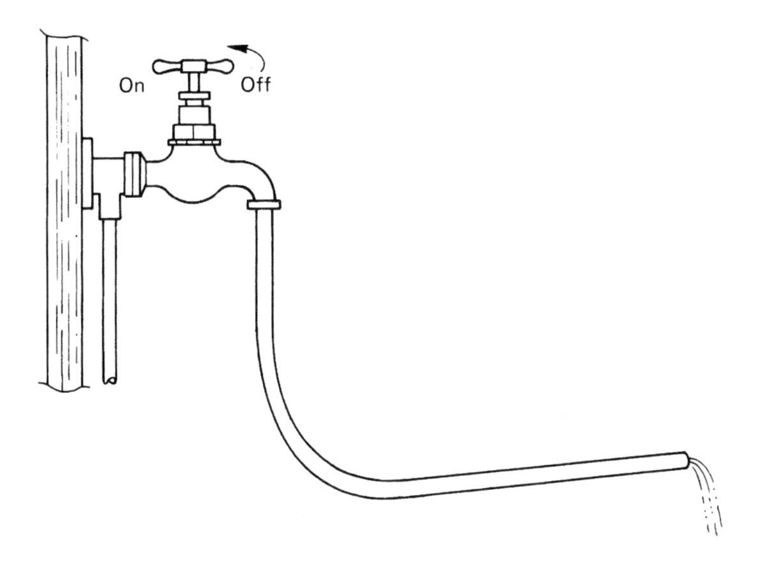

Fig. 2.7a

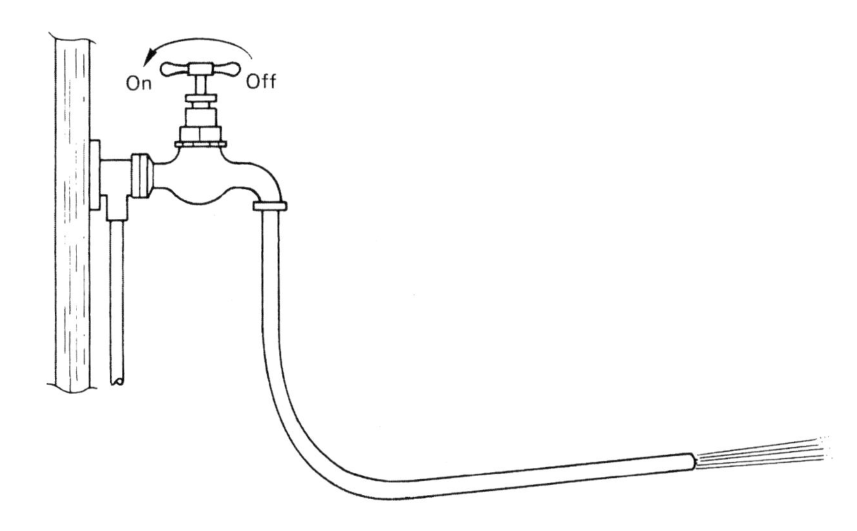

Fig. 2.7b

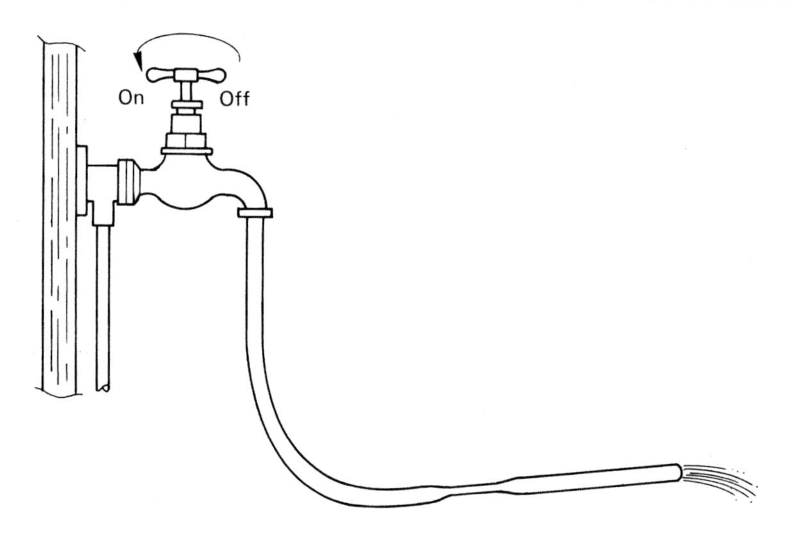

Fig. 2.7c

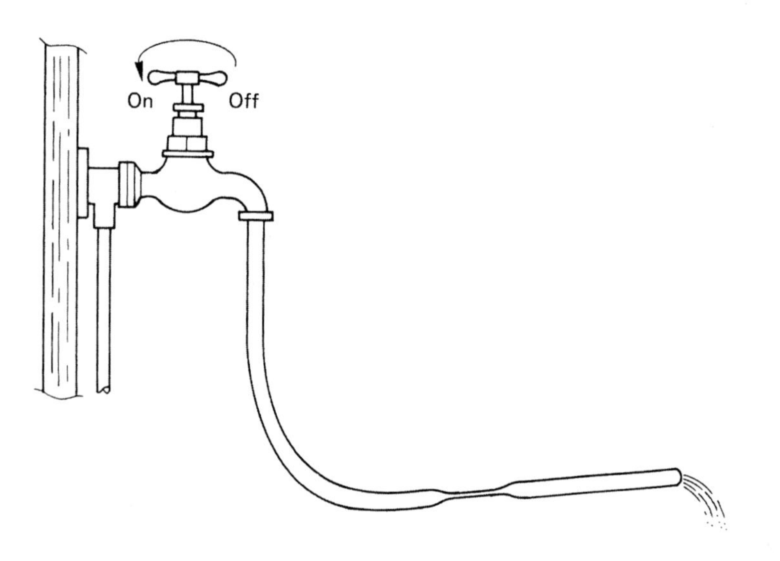

Fig. 2.7d

the tap in this position and squeeze the pipe: less water will flow (Fig. 2.7c). Increase the opposition by squeezing more and even less water will flow (Fig. 2.7d). Hence opposition and flow are *inversely proportional*.

Now, for an electric circuit, replace the tap with some source of electricity supply, change the hose to a conductor, and the constriction in the hose into added resistance. The flow of water becomes the current. We will now have the same effect, in that a small voltage will only give rise to a small current (Fig. 2.8a), an increase in voltage produces a greater current (Fig. 2.8b), and a constant voltage but with an increase in resistance results in reduced current flow (Fig. 2.8c and d).

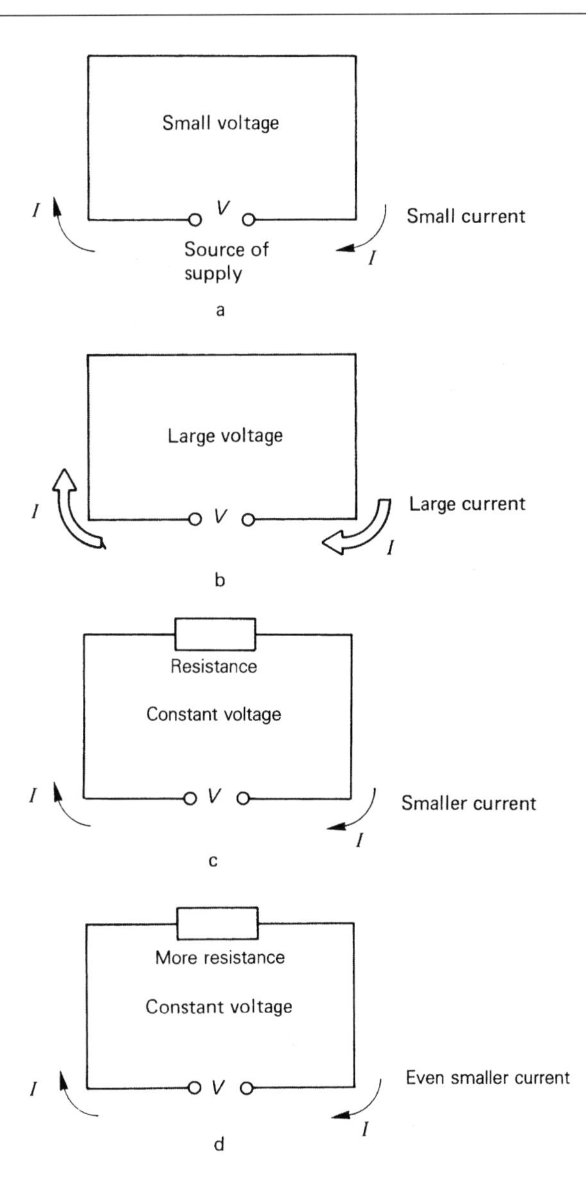

Fig. 2.8

Ohm's law

Georg Simon Ohm demonstrated the relationships we have just seen, and stated them in his famous law which is: 'The current in a circuit is proportional to the circuit voltage and inversely proportional to the circuit resistance, at constant temperature.'

So, we can show Ohm's law by means of the formula:

$$I = \frac{V}{R}$$

Also, transposing,

$$R = \frac{V}{I} \text{ and } V = I \times R$$

Example

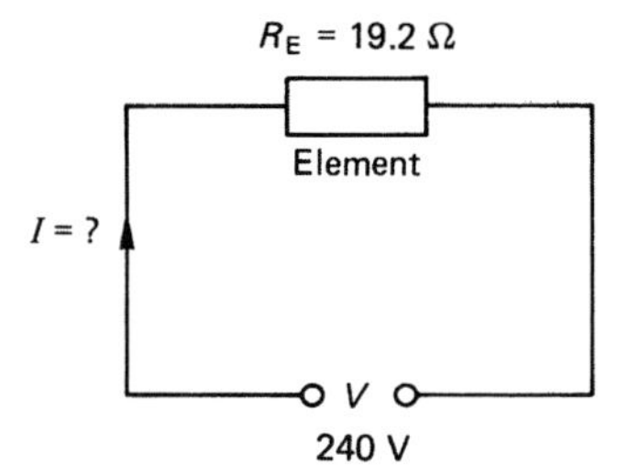

Fig. 2.9

A 240 V electric heating element has a measured resistance of 19.2 ohms. Calculate the current that will flow. (*Note*: Whenever possible, draw a diagram, no matter how simple; this will help to ensure that correct values are assigned to the various circuit quantities.)

$$I = \frac{V}{R}$$

$$= \frac{240}{19.2}$$

$$= 12.5 \text{ A}$$

Example

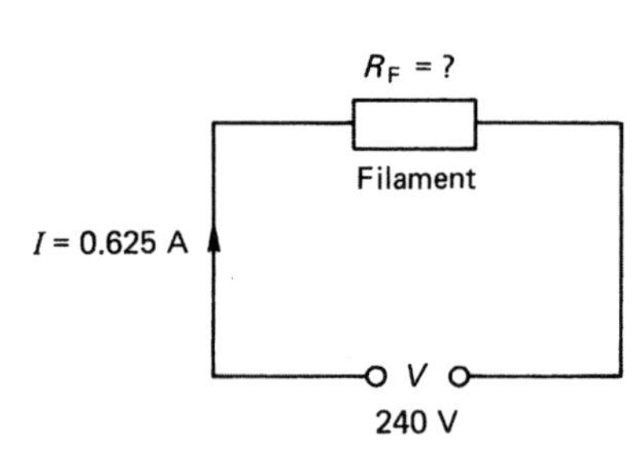

Fig. 2.10

What is the resistance of an electric lamp filament, if it draws a current of 0.625 A from a 240 V supply?

$$R = \frac{V}{I}$$

$$= \frac{240}{0.625}$$

$$= 384 \text{ ohms}$$

Example

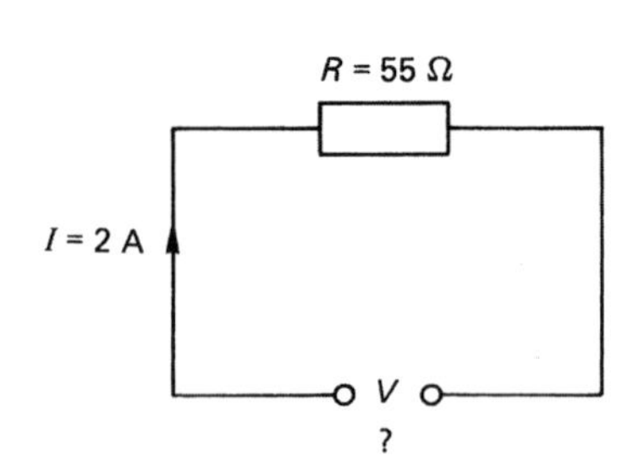

Fig. 2.11

What voltage would be required to cause a current of 2 A to flow through a resistance of 55 ohms?

$$V = I \times R$$

$$= 2 \times 55$$

$$= 110 \text{ V}$$

Electricity and the human body

Water is a conductor of electricity and since the human body is made up of a high proportion of water, it follows that it also is a conductor. However, unlike other materials we have dealt with so far, there is no exact value for body resistivity, and therefore body resistance can vary, not only between individuals but between values for each person. Depending on whether the body is dry, moist, or wet, the value measured between hands or between hands and feet can be anywhere between 1000 Ω and 10 000 Ω. As we have just seen from Ohm's law, the current flowing through a body will depend on the voltage and the body resistance. Different levels of current will have different effects, the worst occurring when the heart goes out of rhythm and will not return to normal. This condition is known as ventricular fibrillation, and will often result in death. Table 2.3 is a guide to the various levels of shock current and their effects on the body.

Table 2.3

Current (mA)	Effect
1–2	Perception level, no harmful effects
5	Throw-off level, painful sensation
10–15	Muscular contraction begins, cannot let go
20–30	Imapired breathing
50 and above	Ventricular fibrillation and death

It will be seen that 50 mA (0.05 A) is considered to be the minimum *lethal* level of shock current, so, if a person's body resistance was as low as 1000 ohms, the voltage required to cause this current to flow would be

$$V = I \times R$$
$$= 50 \times 10^{-3} \times 1000$$
$$= 50\,\text{V}$$

Note this voltage, it is important.

Types and sources of supply

There are only two types of electricity supply, *direct current* (d.c.) and *alternating current* (a.c.). D.c. is obtained from cells and batteries, d.c. generators, or electronically derived from a.c. (rectification). A.c. is obtained from a.c. generators.

The methods of producing a.c. and d.c. supplies are discussed in later chapters.

Voltage bands

Extra low *Not* exceeding 50 V a.c. or 120 V d.c. between conductors or conductors and earth (Fig. 2.12a).

Low Exceeding extra low, but not exceeding 1000 V a.c. or 1500 V d.c. between conductors; or 600 V a.c. or 900 d.c. between conductors and earth (Fig. 2.12b).

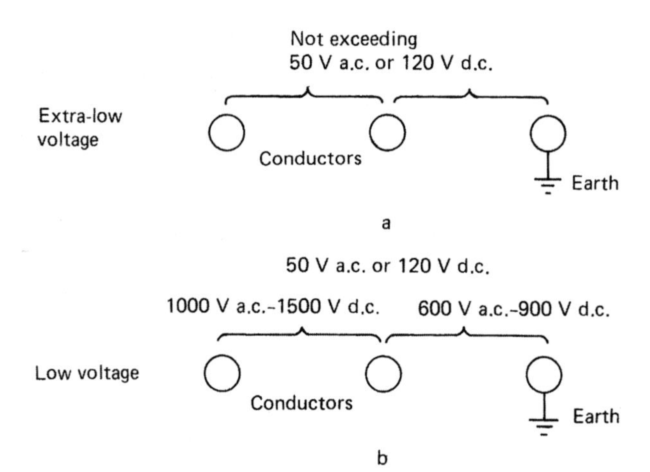

Fig. 2.12

Measuring current and voltage

As current flows *through* a conductor, it seems logical to expect that any instrument used to measure current would need to have that current flowing *through* it. This is known as a *series* connection (Fig. 2.13).

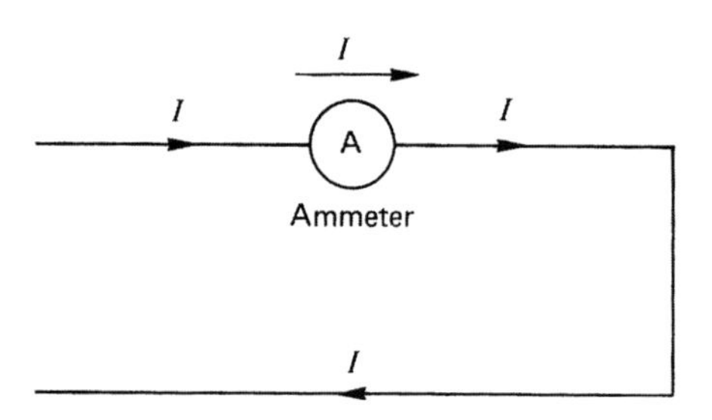

Fig. 2.13

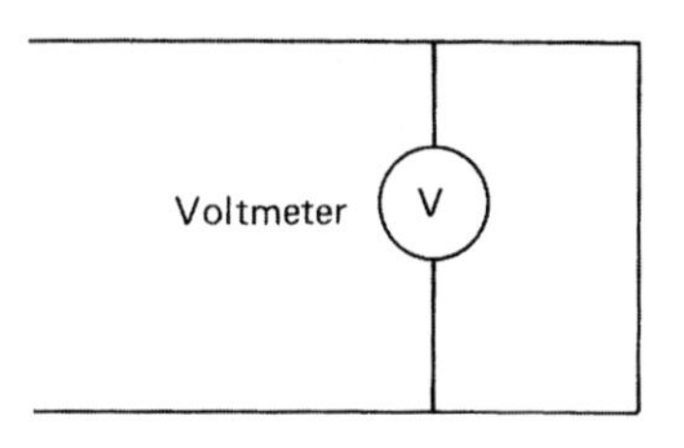

Fig. 2.14

Voltage, on the other hand, is a measure of potential difference *between* or *across* two points, and hence a voltage measuring instrument would need to be connected *between* or *across* two points in a circuit. This is known as a *parallel* connection (Fig. 2.14).

So, the arrangement of instruments to measure the current and voltage associated with a circuit supplying, say, a lamp would be as shown in Fig. 2.15.

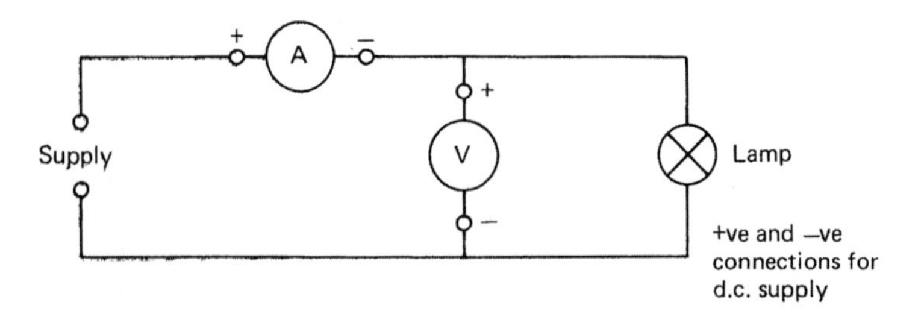

Fig. 2.15

In order to allow all the current needed by the lamp to flow, the ammeter should not impede that flow, and hence should have a very *low* resistance. To ensure that very little of the current needed is diverted away from the lamp via the voltmeter, it should have a very *high* resistance.

Components of a circuit

An electric circuit (Fig. 2.16) comprises the following main components:

1 A source of supply, a.c. or d.c.
2 A fuse or circuit breaker which will cut off supply if too much current flows. This is called circuit *protection*.
3 Conductors through which the current will flow. Two or more conductors embedded in a protective sheathing is called a cable.
4 A switch with which to turn the supply on or off. This is called circuit *control*.
5 A device which needs current to make it work. This is known as the *load*.

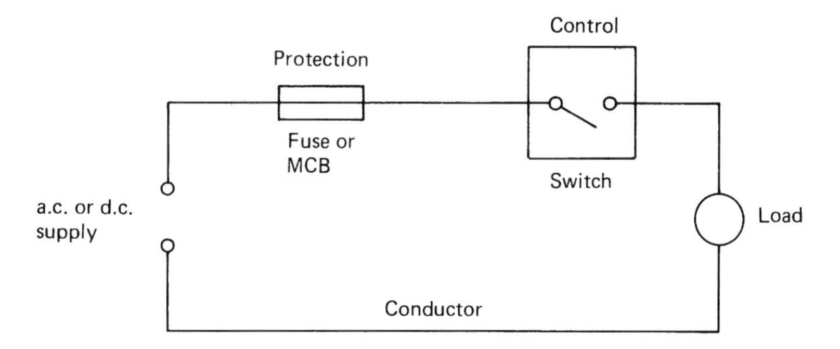

Fig. 2.16

Self-assessment questions

1 Protons and electrons have equal and opposite charges. What are they and which is which?
2 What does an atomic number give an indication of?
3 What are random electrons?
4 (a) What is the difference between a conductor and an insulator?
(b) Give three examples of each.
5 In what units are the following quantities measured?
(a) Current
(b) Potential difference
(c) Resistance
(d) Charge
(e) Resistivity.
6 What length of $4.0\,mm^2$ copper conductor, having a resistivity of $17\,\mu\Omega\,mm$ would have a measured resistance of $0.68\,\Omega$?
7 A 100 m length of aluminium wire ($\rho = 28\,\mu\Omega\,mm$) has a resistance of 0.7 ohms. What is its cross-sectional-area?

8 A length of copper wire has a resistance of 0.5 ohms at 0°C. What would be its resistance at 50°C if the temperature coefficient is 0.004 Ω/Ω/°C?

9 A coil of aluminium wire has a resistance of 54 ohms at 20°C. What would be its resistance at 70°C (α = 0.004 Ω/Ω/°C)?

10 What would happen to the current in a circuit if:
(a) The voltage is constant and the resistance is doubled?
(b) The voltage is doubled and the resistance is constant?
(c) The voltage and the resistance were both trebled?

11 Ignoring any effects of temperature, what would happen to circuit resistance if the current was increased?

12 State Ohm's law.

13 Solve the following circuit problems:
(a) I = 10 A, V = 240 V, R = ?
(b) R = 38.4 Ω, V = 240 V, I = ?
(c) R = 60 Ω, I = 0.4 A, V = ?

14 At 110 V, what body resistance would allow the accepted lethal level of shock current to flow?

15 What do the letters a.c. and d.c. stand for?

16 What are the limits of a.c. extra low and low voltage?

17 Draw a labelled diagram of the connections of instruments used to measure the current and voltage in an a.c. circuit supplying a lamp. Include on the diagram circuit protection and control.

18 In such a circuit as described in question 17 above, the instrument readings were 0.5 A and 12 A. What is the resistance of the lamp?

3 Resistance, current and voltage, power and energy

Resistance

We should know by now what resistance is and how it affects current flow. However, a circuit may contain many resistances connected in various ways, and it is these connections we are to consider now.

Resistance in series

Remember how we connected an ammeter in *series* in a circuit so that the current could flow through it? Two or more resistances or resistors connected in the same way are said to be connected in *series*. It is like squeezing our hosepipe in several places (Fig. 3.1).

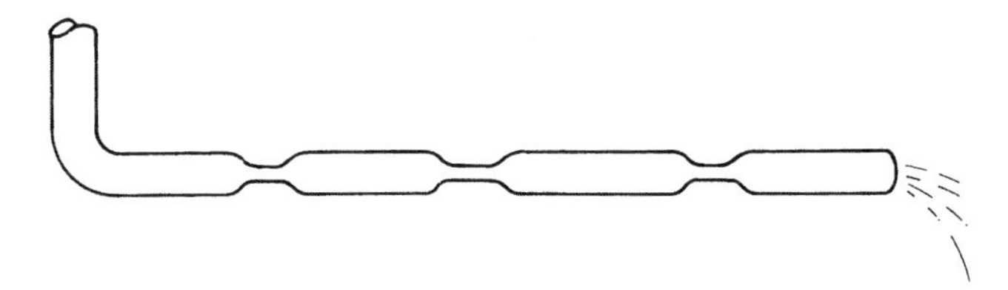

Fig. 3.1

The more depressions we make in the pipe, the less water will flow – similarly with resistance (Fig. 3.2).

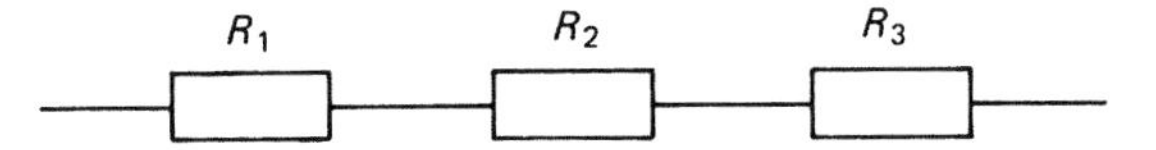

Fig. 3.2

Less current will flow with, for example, R_3 connected than with only R_1 and R_2. Hence we can see that the more resistances that are connected in series, the smaller the current flow. So, the total resistance R_T of a number of resistances in *series* is

$$R_T = R_1 + R_2 + R_3 \text{ etc.}$$

As we have seen in Chapter 1, a conductor will have a resistance to the flow of current. Hence if we take a length of conductor and add another length to it (*series connection*) the resistance will increase. So resistance is proportional to conductor length. This is important to remember, because if we supply a load with too long a length of cable, the current flow may be reduced to such an extent that the load may not work properly.

A load has a fixed resistance, so a reduced current due to cable resistance means that there must be less voltage across the load. This voltage loss is called *voltage drop* and is a very important topic.

Consider the following circuit (Fig. 3.3):

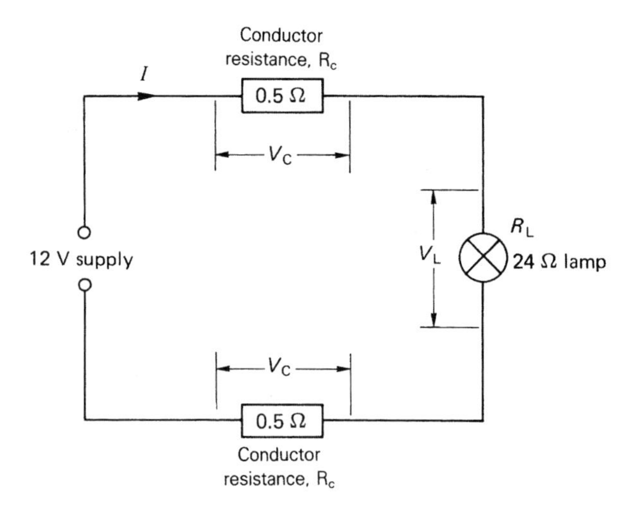

Fig. 3.3

The conductor resistances and the lamp resistance are all in series

$\therefore$ Cable resistance $= 0.5 + 0.5 = 1\,\Omega$

$$\therefore \text{Total series resistance} = R_C + R_L$$
$$= 1 + 24$$
$$= 25\,\Omega$$

$$\therefore \text{Circuit current } I = \frac{V}{R}$$
$$= \frac{12}{25}$$
$$= 0.48\text{ A}$$

This current flows *through all the series* components, and where we have current flow we must have potential difference. So, each of these components will have a p.d. across them. Hence, the p.d. across each conductor is

$$V_C = I \times R_C \text{ (Ohm's law)}$$
$$= 0.48 \times 0.5$$
$$= 0.24\text{ V}$$

$\therefore$ Total p.d. across cable $= 2 \times 0.24$

$= 0.48\,\text{V}$

The p.d. across the lamp is

$$V_L = I \times R_L$$
$$= 0.48 \times 24$$
$$= 11.52\,\text{V}$$

Now, if we add these p.d.s together, we get

$0.48 + 11.52 = 12\,\text{V}$ (the supply voltage)

Hence we have lost 0.48 V due to the cable resistance, and the lamp has to work on 11.52 V not 12 V. The 0.48 V lost is the *voltage drop* due to the cable.

If we were to make the cable, say, six times longer, its resistance would become $6 \times 1 = 6\,\Omega$.

$\therefore$ Circuit resistance $= 6 + 24$

$= 30\,\Omega$

$$\therefore \text{Total current } I = \frac{V}{R}$$
$$= \frac{12}{30}$$

$= 0.4\,\text{A}$ (less than before)

$\therefore$ Cable voltage drop $= I \times R_C$

$= 0.4 \times 6$

$= 2.4\,\text{V}$

The p.d. across lamp is

$$V_L = I \times R_L$$
$$= 0.4 \times 24$$
$$= 9.6\,\text{V}$$

Check: $2.4\,\text{V} + 9.6\,\text{V} = 12\,\text{V}$.

So, we can now state that, for a series circuit, the voltages across all the components add up to the supply voltage. That is,

$V_T = V_1 + V_2 + V_3$ etc.

Voltage drop

In order to ensure that loads are not deprived of too much of their operating voltage due to cable resistance, the IEE Wiring Regulations recommend that the voltage drop in a circuit should not exceed 4% of the voltage at the origin of the circuit. Hence, for a 230 V single-phase supply, the voltage drop should not exceed 4% of 230:

$$\frac{4 \times 230}{100} = 9.2\,\text{V}$$

Example

A circuit (Fig. 3.4) comprises two heating elements each of 28.8 ohms resistance. These are connected in series across a 230 V supply. If the supply cable has a resistance of 2.4 ohms calculate:

(a) the total circuit current
(b) the total circuit resistance
(c) the cable voltage drop
(d) the voltage available across each heater.

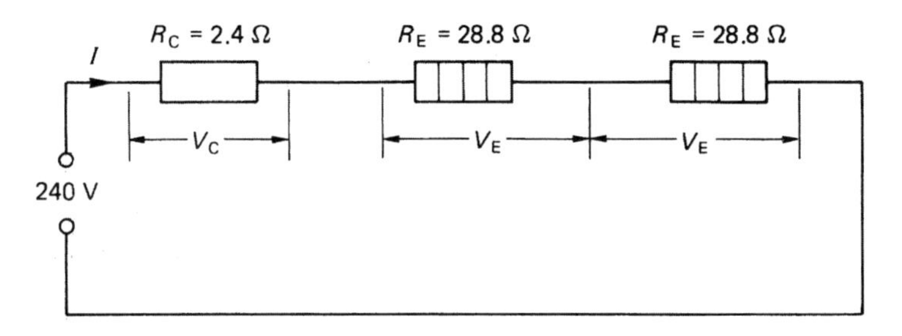

Fig. 3.4

(a) Total resistance:

$$R = R_C + R_E + R_E$$
$$= 2.4 + 28.8 + 28.8$$
$$= 60\,\text{ohms}$$

(b) Total current:

$$I = \frac{V}{R}$$
$$= \frac{230}{60}$$
$$= 3.833\,\text{A}$$

(c) Cable voltage drop:

$$V_C = I \times R_C$$
$$= 3.833 \times 2.4$$
$$= 9.2\,\text{V}$$

(d) Voltage across each element:

$$V_E = I \times R_E$$
$$= 3.833 \times 28.8$$
$$= 110.4\,\text{V}$$

Check:

$$V = V_C + V_E + V_E$$
$$= 9.2 + 110.4 + 110.4$$
$$= 230\,\text{V}$$

Resistance in parallel

Remember the connection of a voltmeter *across* the ends of a load? This was said to be connected in *parallel*.

Similarly, if we connect one or more conductors across the ends of another conductor (Fig. 3.5), these are said to be wired in parallel.

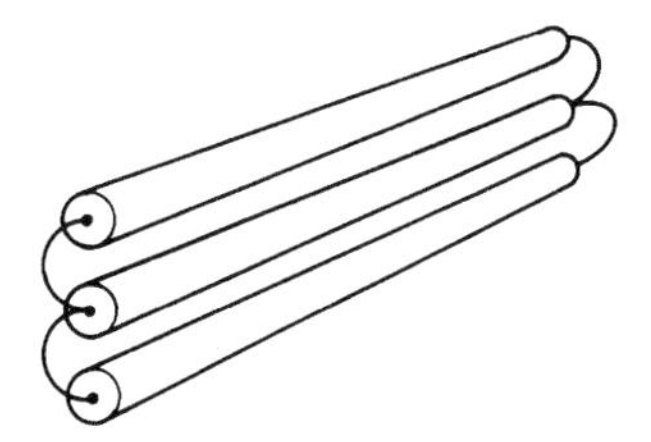

Fig. 3.5

If each of these conductors has the same c.s.a., then the effect of three wired in parallel is that of having one conductor three times as large as any one. As we have seen in Chapter 2, an increase in c.s.a. results in a corresponding decrease in resistance, so parallel connections reduce resistance.

Note: Wiring two cables in parallel, or, more simply, changing one for a larger size, will reduce resistance and hence lessen voltage drop.

In the IEE Regulations, there are tables giving the values of voltage drop for various types and sizes of conductor. These values are given in millivolts (mV) for every ampere (A) that flows along a length of 1 metre (m), i.e. mV/A/m. So, we should be able to check on our original comments that resistance, and hence voltage drop, reduces with an increase of c.s.a. For example, a 10.0 mm^2 conductor should have ten times less of a voltage drop than a 1.0 mm^2 conductor. Reference to table 4D1B, column 3 (IEE Regs) confirms this, the millivolt drop for 1.0 mm^2 being 44 mV and that for 10.0 mm^2 being 4.4 mV.

Addition of resistances in parallel

Consider, say, three resistances connected in parallel across a supply Fig. 3.6.

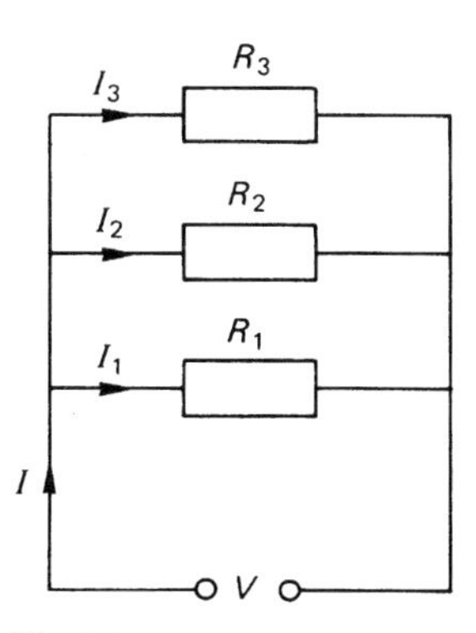

Fig. 3.6

It will be seen from the figure that, unlike a series circuit, the same supply voltage is being applied across each resistance. In this case it is the current flowing through each resistance that is different. (Unless, of course, all the resistances are the same value.) The total current, therefore, is the sum of each of the individual currents:

$$I = I_1 + I_2 + I_3$$

If we now use Ohm's law to convert I to V/R we get:

$$\frac{V}{R} = \frac{V}{R_1} + \frac{V}{R_2} + \frac{V}{R_3}$$

Now, dividing right through by V, we get:

$$\frac{V}{V.R} = \frac{V}{V.R_1} = \frac{V}{V.R_2} + \frac{V}{V.R_3}$$

The V cancels out, leaving:

$$\frac{1}{R} = \frac{1}{R_1} + \frac{1}{R_2} + \frac{1}{R_3} \text{ etc.}$$

Example

Referring to Fig. 3.7, calculate the total resistance of the circuit.

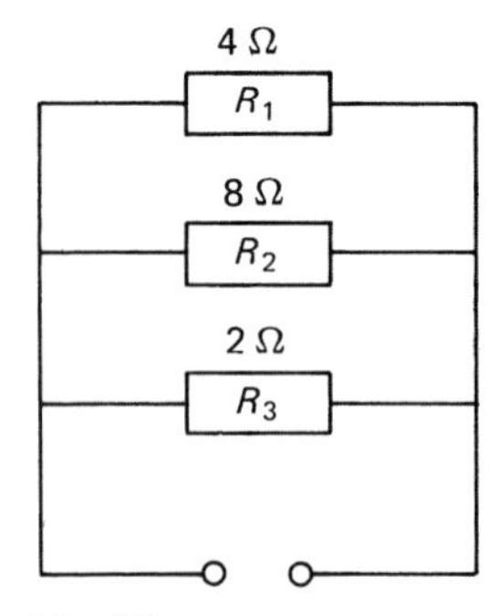

Fig. 3.7

$$\frac{1}{R} = \frac{1}{R_1} + \frac{1}{R_2} + \frac{1}{R_3}$$

$$\frac{1}{R} = \frac{1}{4} + \frac{1}{8} + \frac{1}{2}$$

$$\frac{1}{R} = 0.25 + 0.125 + 0.5$$

$$\frac{1}{R} = 0.875$$

Transposing,

$$R = \frac{1}{0.875}$$

$$= 1.143 \text{ ohms.}$$

Note: The total resistance in parallel is less than the smallest resistance in the circuit.

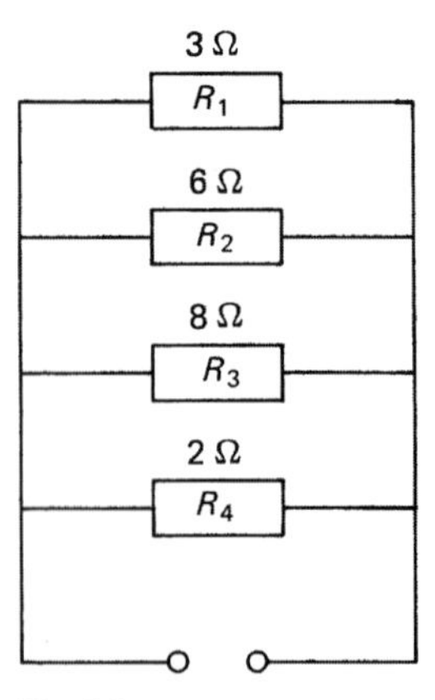

Fig. 3.8

There are some short-cuts to the calculation of resistance in parallel. These are as follows:

1 For a number of resistances identical in value, the total resistance is the value of any one, divided by the number of resistances. That is, twelve 24 ohm resistances in parallel will have a total resistance of 24/12 = 2 ohms.
2 For any two resistances in parallel, the total may be found by dividing their *product* by their *sum*. Hence a 6 ohm and a 3 ohm resistance in parallel would have a total of

$$\frac{\text{Product}}{\text{Sum}} = \frac{6 \times 3}{6 + 3} = \frac{18}{9} = 2 \text{ ohms}$$

This method can be used for more than two resistances, by simply doing two at a time.

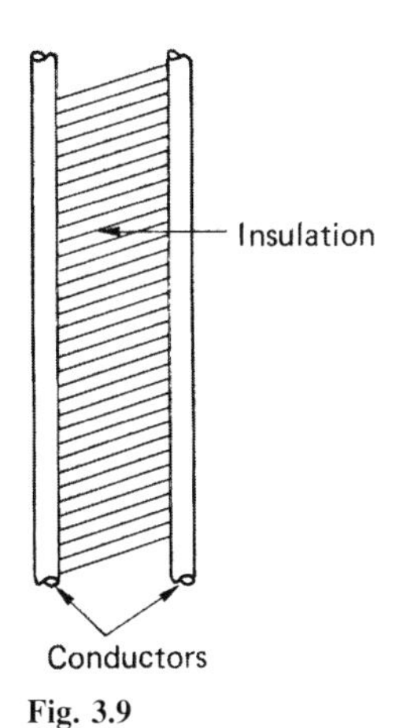

Fig. 3.9

Example

Determine the total resistance of the circuit shown in Fig. 3.8.

Use product/sum for R_1 and R_2, then for R_3 and R_4 and then again for the two totals.

Hence,

$$\text{total for } R_1 \text{ and } R_2 = \frac{3 \times 6}{3 + 6} = 2\,\text{ohms}$$

$$\text{total for } R_3 \text{ and } R_4 = \frac{8 \times 2}{8 + 2} = 1.6\,\text{ohms}$$

$$\therefore \text{Overall total} = \frac{2 \times 1.6}{2 + 1.6} = 0.89\,\text{ohms}$$

Insulation resistance

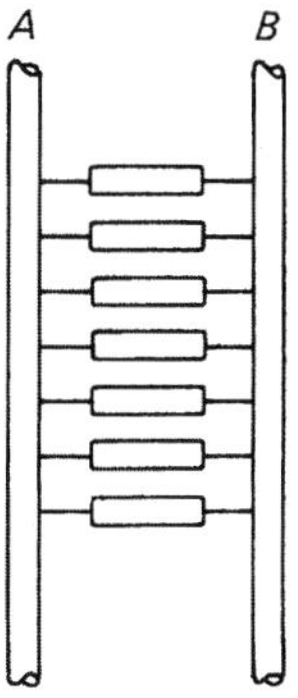

As we should already know now, insulation is a very poor conductor and this presents a very high resistance to the flow of current. Consider then two short lengths of conductor, A and B, separated by insulation (Fig. 3.9).

Current is inhibited from flowing from A to B due to the insulation which could be said to comprise an infinite number of very high resistances in parallel (Fig. 3.10).

A B

Fig. 3.10

If we now extend the length of these conductors and insulation, the effect is that of adding extra parallel resistance to that insulation, and as we have just seen, the greater the number of parallel resistances, the smaller the total resistance. So an increase in cable length results not only in an increase of conductor resistance, but in a *decrease* in insulation resistance.

The insulation resistance of an installation must, of course, be very high, usually in the order of many millions of ohms, i.e. megohms (MΩ). The IEE Regulations require that the minimum acceptable value is 0.5 MΩ for circuits up to 500 V.

Current distribution

Example

From Fig. 3.11 determine (a) the total resistance, (b) the total current and (c) the current in each resistance.

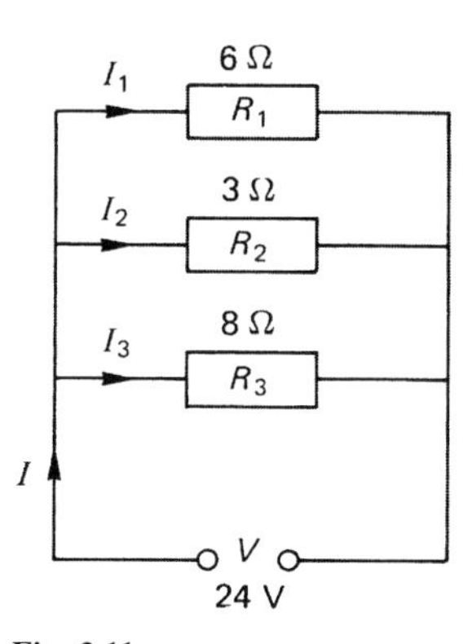

Fig. 3.11

(a) Using the product/sum method:

$$\text{Total of } R_1 \text{ and } R_2 = \frac{6 \times 3}{6 + 3} = 2 \text{ ohms}$$

Using this value with R_3:

$$\text{Total } R = \frac{2 \times 8}{2 + 8}$$

$$= 1.6 \text{ ohms}$$

(b) Total current:

$$I = \frac{V}{R} = \frac{24}{1.6} = 15 \text{ A}$$

(c) Current in R_1:

$$I_1 = \frac{V}{R_1} = \frac{24}{6} = 4 \text{ A}$$

Current in R_2:

$$I_2 = \frac{V}{R_2} = \frac{24}{3} = 8 \text{ A}$$

Current in R_3:

$$I_3 = \frac{V}{R_3} = \frac{24}{8} = 3 \text{ A}$$

Check: Total current I should be:

$$I = I_1 + I_2 + I_3$$
$$= 4 + 8 + 3 = 15 \text{ A}$$

Combined series – parallel connections

Current-using pieces of equipment in installation circuits are connected across the supply and are therefore wired in parallel. The cable supplying such equipment, however, is connected in series with the parallel arrangement (Fig. 3.12).

To calculate the total resistance of the circuit, the parallel network must be worked out first and this total added to the series cable resistance.

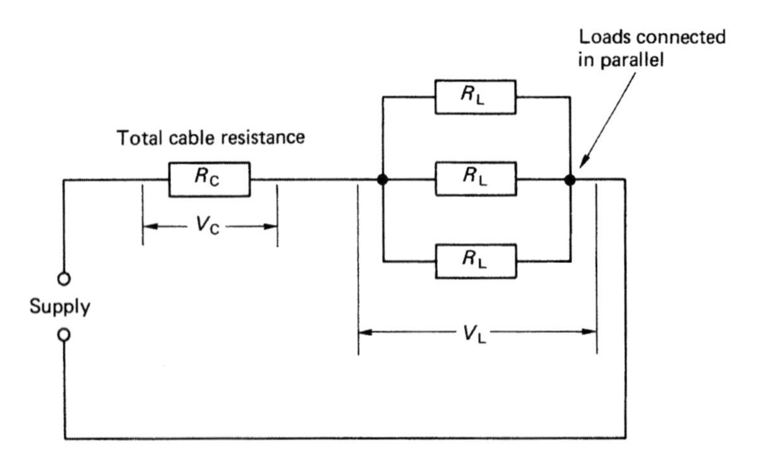

Fig. 3.12

Example

Three lamps A, B and C, having resistances of 1440 Ω, 960 Ω and 576 Ω, are connected to a 240 V supply by a cable of resistance 2 Ω (Fig. 3.13). Calculate (a) the total circuit resistance, (b) the total current, (c) the cable voltage drop, (d) the voltage across the lamps and (e) the current drawn by each lamp.

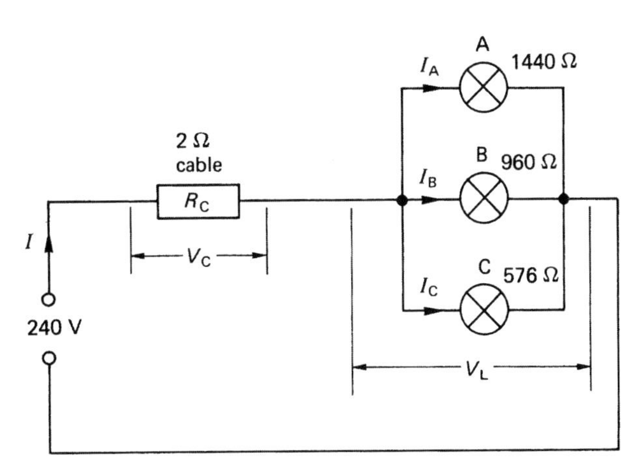

Fig. 3.13

(a) Total resistance of A, B and C in parallel:

$$\text{A and B product/sum} = \frac{1440 \times 960}{1440 + 960}$$
$$= 576 \text{ ohms}$$

This is the same value as lamp C.

$$\therefore \text{ the total with C} = \frac{576}{2} = 288 \text{ ohms}$$

So the resistance of the lamp circuit is 288 ohms.

$$\text{Total resistance of circuit } R = R_C + R_L$$
$$= 2 + 288$$
$$= 290 \text{ ohms}$$

(b) Total current:

$$I = \frac{V}{R} = \frac{240}{290} = 0.8276 \text{ A}$$

(c) Cable voltage drop:

$$V_L = I \times R_C = 0.8276 \times 2 = 1.655 \text{ V}$$

(d) Voltage across lamps:

$$V_C = V - V_C$$
$$= 240 - 1.655 = 238.35 \text{ V}$$

(*Check*: V_L also equals $I \times R_L = 0.8276 \times 288 = 238.35$ V.)

(e) Current through A:

$$I_A = \frac{V_L}{R_A} = \frac{238.34}{1440} = 0.1655\,\text{A}$$

Current through B:

$$I_B = \frac{V_L}{R_B} = \frac{238.34}{960} = 0.2483\,\text{A}$$

Current through C:

$$I_C = \frac{V_L}{R_C} = \frac{238.34}{576} = 0.4138\,\text{A}$$

(*Check*: $I_A + I_B + I_C$ should equal total current I. Hence, 0.1655 + 0.2483 + 0.4138 = 0.8276 A, correct.)

The following examples show other resistance combinations not usually encountered in ordinary installation work, but which, nevertheless, need to be understood.

Example

Calculate the total resistance of the circuit shown in Fig. 3.14.

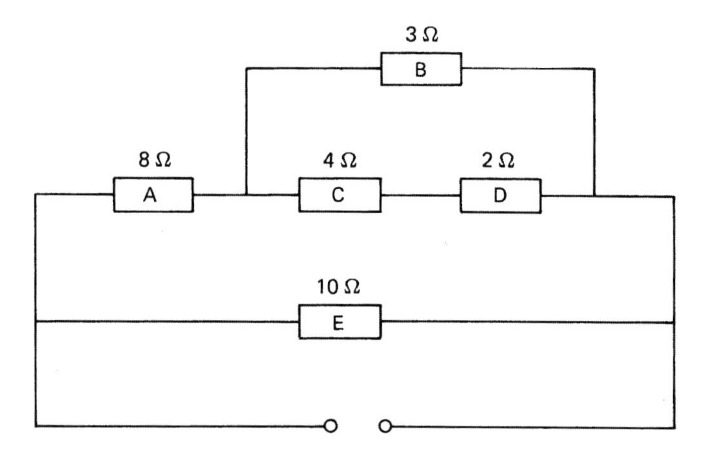

Fig. 3.14

It is important to recognize which resistances are either in series or in parallel with other resistances. That is,

1 Two or more resistances can only be in series if they are connected end to end with *no* joins from other resistances between them.
2 Two or more resistances can only be in parallel if the ends of each are *directly* connected to the ends of any other.

So, in Fig. 3.14 there is no single resistance in parallel with any other, and only C and D are directly in series:

∴ total of C and D in series = 4 + 2 = 6 ohms

So these two could be replaced by a single resistance of 6 ohms (Fig. 3.15).
Now we should see that B and CD are in parallel:

$$\therefore \text{ total of B and CD } = \frac{\text{product}}{\text{sum}} = \frac{3 \times 6}{3 + 6} = 2 \text{ ohms}$$

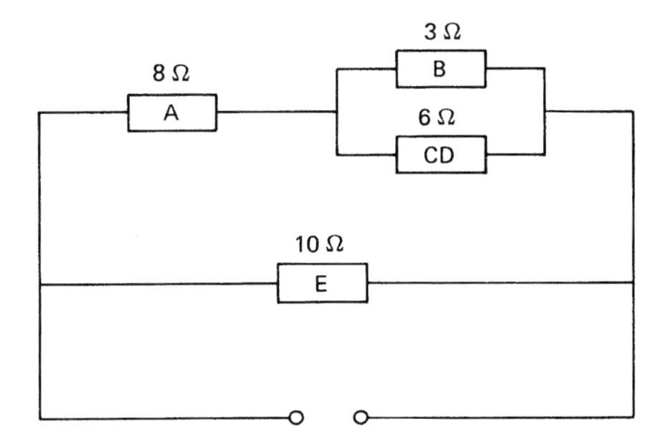

Fig. 3.15

So B and CD could be replaced by a 2 ohm resistance (Fig. 3.16).
Now, A and BCD are in series:

$$\therefore \text{ total of A and BCD } = 8 + 2 = 10 \text{ ohms}$$

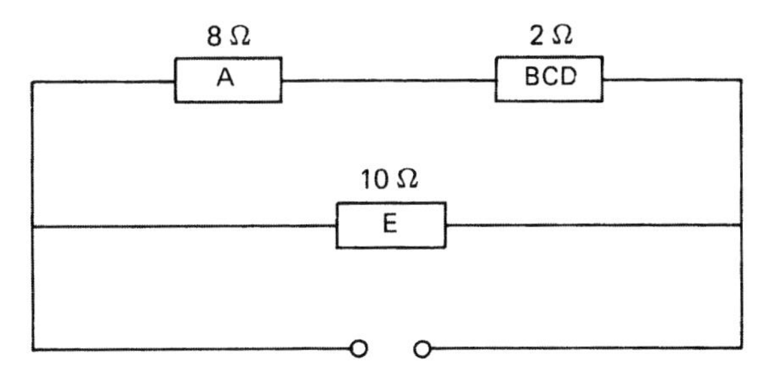

Fig. 3.16

Once again these can be replaced by a single 10 ohm resistance (Fig. 3.17).

$$\therefore \text{ total of E and ABCD in parallel is clearly } \frac{10}{2} = 5 \text{ ohms}$$

So, total resistance of original circuit is 5 ohms.

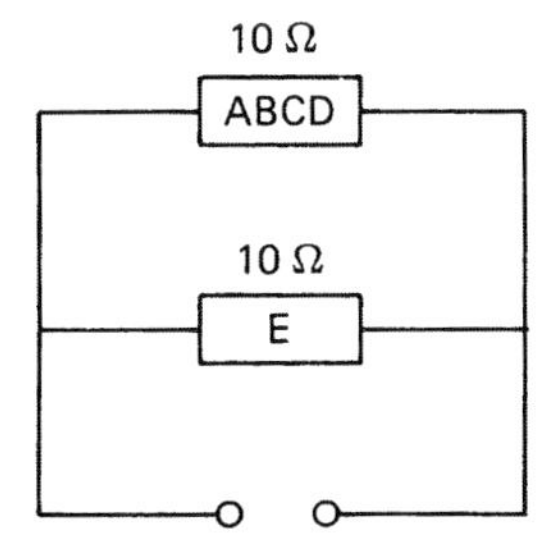

Fig. 3.17

Example

Calculate the total resistance in Fig. 3.18, the voltage at point X, and the current through resistance F.

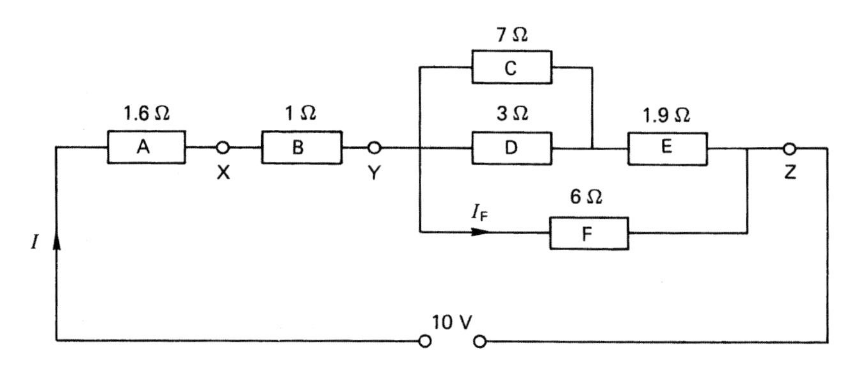

Fig. 3.18

Total of A and B in series = 1.6 + 1 = 2.6 ohms

$$\text{Total of C and D in parallel} = \frac{7 \times 3}{7 + 3} = 2.1 \text{ ohms}$$

So circuit becomes as in Fig. 3.19.

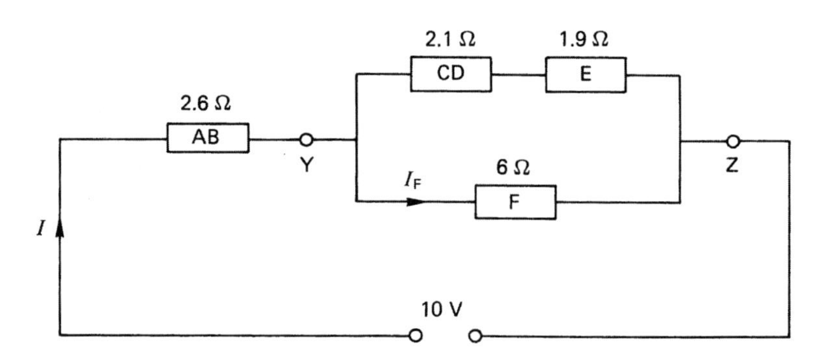

Fig. 3.19

Total of CD and E in series = 2.1 + 1.9 = 4 ohms

So, circuit is now as Fig. 3.20.

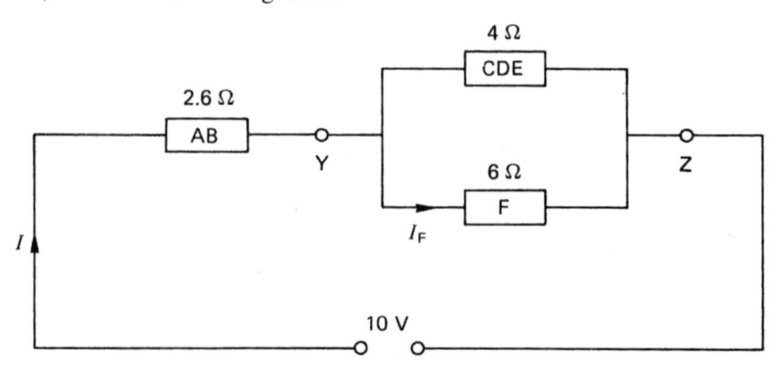

Fig. 3.20

Total of CDE and F in parallel $= \dfrac{4 \times 6}{4 + 6} = 2.4\,\text{ohms}$

So, circuit becomes as in Fig. 3.21.

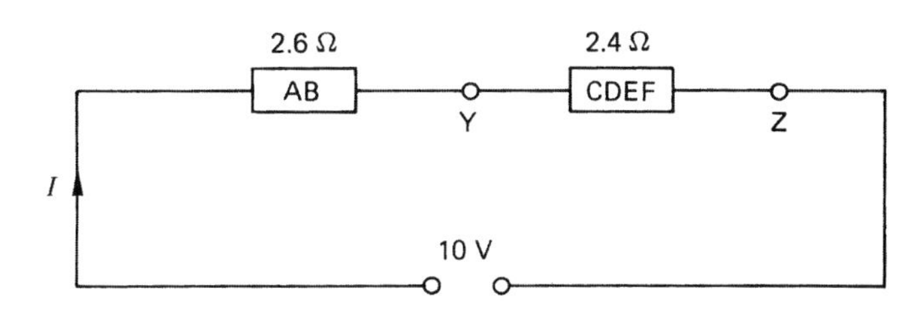

Fig. 3.21

Total of AB and CDEF in series $= 2.6 + 2.4 = 5\,\text{ohms}$

As shown in Fig. 3.22

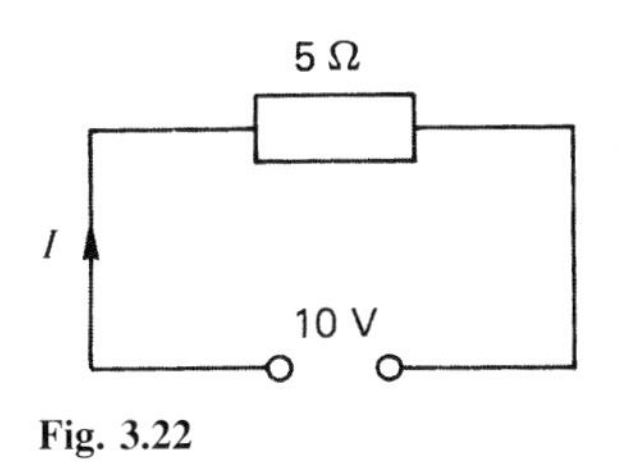

Fig. 3.22

Total circuit resistance $= 5\,\text{ohms}$

Total current $I = \dfrac{V}{R} = \dfrac{10}{5} = 2\,\text{A}$

The voltage V_A dropped across A $= I \times R_A$

$= 2 \times 1.6$

$= 3.2\,\text{V}$

Since the supply voltage is 10 V, then the voltage at X is

$10 - 3.2 = 6.8\,\text{V}$

To find the current through F, we need to know the p.d. across it, i.e. across YZ. From Fig. 3.21 the total resistance between Y and Z is CDEF which is 2.4 ohms and the current flowing through it is the total current of 2 A.

$\therefore$ the p.d. across YZ $= I \times \text{CDEF}$

$= 2 \times 2.4$

$= 4.8\,\text{V}$

$\therefore I_F = \dfrac{V_F}{R_F} = \dfrac{4.8}{6} = 0.8\,\text{A}$

POWER: symbol, *P*; unit, watt (W)

Someone with plenty of energy has the potential to convert that energy into work. The rate at which that energy is converted is called *power*. Hence power is the rate of energy conversion, i.e.

$$P = \frac{\text{energy}}{\text{time}}$$

Two engineering pioneers, Watt and Joule, give their names to the units of power and energy.

James Watt (1736–1819)

British engineer who invented the improved steam engine and introduced horsepower (hp) as a means of measuring power.

James Prescott Joule (1818–89)

British scientist and engineer best known for his mechanical equivalent of heat and his work on the heating effect of an electric current.

Energy is measured in joules and of course the SI unit of time is the second:

$$\therefore \text{power in watts} = \frac{\text{joules}}{\text{seconds}}$$

Transposing, we get

$$\text{Joules} = \text{watts} \times \text{seconds}$$

$$\therefore 1 \text{ joule} = 1 \text{ watt second}$$

Joule showed by experiment that heat energy was produced when a current flowed through a resistance for a certain time, and from these experiments it was shown that

$$\text{joules} = I^2 \times R \times \text{seconds}$$

But we have already seen that

$$\text{joules} = \text{watts} \times \text{seconds}$$

$\therefore$ watts must equal $I^2 \times R$

Hence electrical power

$$P = I^2 \times R$$

By using Ohm's law, we can develop two other equations for P. Hence,

$$I = \frac{V}{R}$$

Therefore, by replacing I with V/R, we get

$$P = \frac{V^2}{R^2} \times R$$

$$\therefore P = \frac{V^2}{R}$$

Similarly,

$$R = \frac{V}{I}$$

$$\text{so, } P = I^2 \times \frac{V}{I}$$

$$\therefore P = I \times V$$

Hence power is

$$P = I^2R \text{ or } P = \frac{V^2}{R} \text{ or } P = IV$$

Nearly all of us are familiar with power. It is all around us in our homes, e.g. a 60 *watt* lamp or a 3 *kilowatt* fire; in fact all appliances should have ratings of power and voltage marked on them.

The rated values are extremely important. A 60 W 240 V lamp will dissipate 60 W only if connected to a 240 V supply. The resistance of the lamp when working is of course unaffected by the voltage and we can calculate this resistance from the lamp's rated values. For a 60 W 240 V lamp:

$P = 60\,\text{W}$; $V = 240\,\text{V}$; R_L = lamp resistance

$$P = \frac{V^2}{R_L}$$

$$\therefore R_L = \frac{V^2}{P}$$

$$= \frac{240^2}{60}$$

$$= \frac{57\,600}{60}$$

$$R_L = 960\,\Omega$$

If we therefore supply the lamp from a 200 V source the power dissipated by the lamp would be

$$P = \frac{V^2}{R}$$

$$= \frac{200^2}{960}$$

$$= \frac{40\,000}{960}$$

$$P = 41.67\,\text{W}$$

Example

Two 100 W 240 V lamps are wired in series across a 240 V supply. Calculate the power dissipated by each lamp and the total power dissipated.

Both lamps have the same resistance:

$$P = \frac{V^2}{R_L}$$

$$R_L = \frac{V^2}{P}$$

$$= \frac{57\,600}{100}$$

$$R_L = 576\,\Omega$$

It is clear from Fig. 3.23 that as both resistances are the same, then the voltage across each is the same:

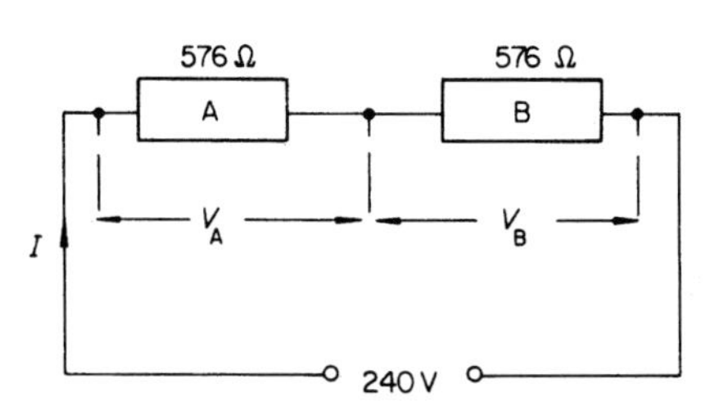

Fig. 3.23

$$V_A = V_B = \frac{240}{2}$$

$$= 120\,\text{V}$$

$$P_A = \frac{{V_A}^2}{R_A}$$

$$= \frac{14\,400}{576}$$

$$P_A = 25\,\text{W}$$

$$\therefore P_B = 25\,\text{W}$$

$$\text{Total resistance } R_{total} = R_A + R_H$$

$$= 576 + 576$$

$$= 1152\,\Omega$$

$$\therefore \text{Total power } P_{total} = \frac{V^2}{R}$$

$$= \frac{57\,600}{1152}$$

$$P = 50\,\text{W}$$

which is $P_A + P_B$

Thus the individual powers dissipated in a series circuit may be added to find the total power dissipated.

Also notice that when two identical lamps are connected in series the power dissipated by each is a quarter of its original value, i.e. 100 W down to 25 W.

Example

Two lamps A and B are connected in series across a 240 V supply. Lamp A is rated at 40 W 240 V and lamp B at 60 W 240 V. Calculate the power dissipated by each lamp and the total power dissipated (Fig. 3.24).

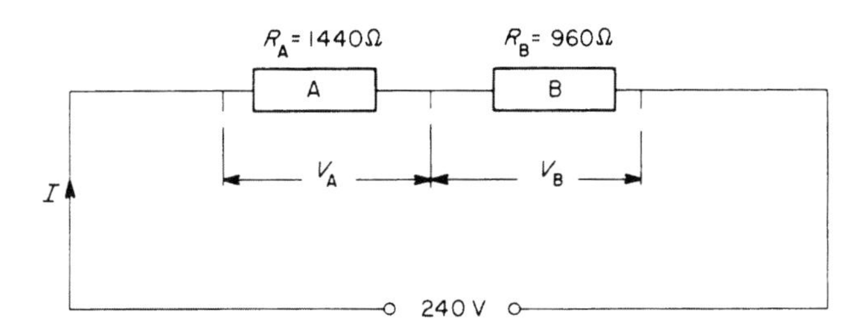

Fig. 3.24

Lamp A

$$R_A = \frac{V^2}{P_A} = \frac{57\,600}{40}$$

$$R_A = 1440\,\Omega$$

Lamp B

$$R_B = \frac{V^2}{P_B} = \frac{57\,600}{60}$$

$$R_B = 960\,\Omega$$

In this case we need not calculate V_A and V_B. As I is common to A and B we could calculate power from I^2R. To find I:

$$R = R_A + R_B$$

$$= 1440 + 960$$

$$R = 2400\,\Omega$$

and

$$I = \frac{V}{R}$$

$$= \frac{240}{2400}$$

$$I = 0.1\text{ A}$$

$$P_A = I^2R_A$$

$$= 0.01 \times 1440$$

$$P_A = 14.4\text{ W}$$

$$P_B = I^2 R_B$$
$$= 0.01 \times 960$$
$$P_B = 9.6\,W$$
$$P_{total} = P_A + P_B$$
$$= 14.4 + 9.6$$
$$P_{total} = 24\,W$$

Check: P_{total} is also equal to I^2R:

$$P_{total} = 0.01 \times 2400$$
$$= 24\,W$$

It can be seen from this example that a considerable amount of power can be lost by series connections. It will also be obvious that incorrect selection of cables can cause a loss of power to the equipment they are supplying.

However, these power losses can be put to some useful purpose, in an electric grill for example, which has three heat settings (Fig. 3.25).

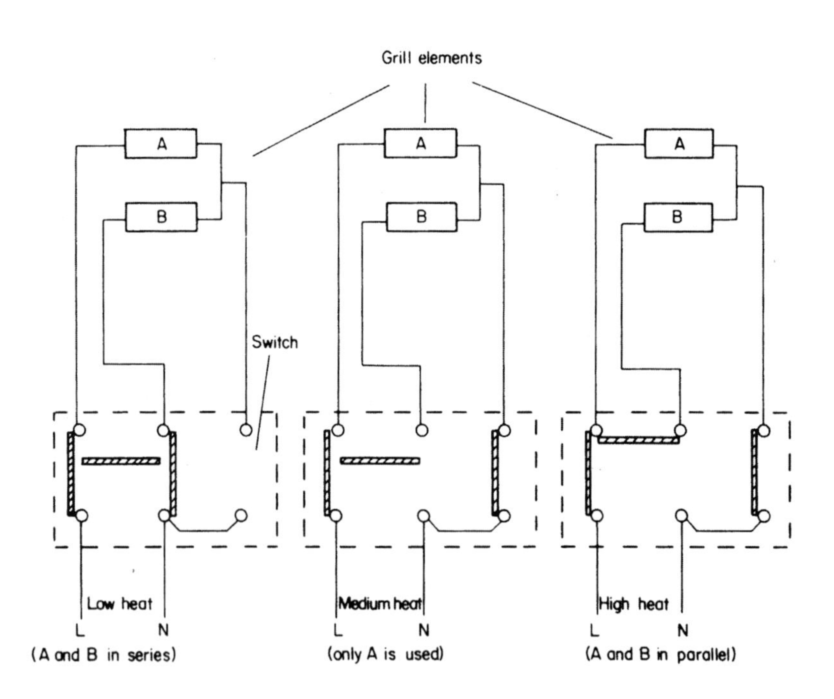

Fig. 3.25 *Three-heat switch*

If each element were rated at 1 kW 240 V, then on high heat A and B would both have their rated voltage and would both dissipate their rated power, i.e. 1 kW each, giving a 2 kW total. On medium heat only one element is used, giving 1 kW total. On low heat, as A and B are in series each will dissipate 250 W (a quarter of 1 kW), giving a total of 500 W.

Electrical energy: symbol, *W*, unit kWh

Energy is the ability to do work and is measured in joules (J). As we have already seen,

$$1 \text{ joule} = 1 \text{ watt second}$$
$$= \text{power} \times \text{time}$$

Generally, in considering domestic loads we measure the power in kilowatts and time in hours:

$$\therefore \text{Energy} = \text{kW} \times \text{hours}$$
$$= \text{kWh (kilowatt hours)}$$

To convert watt seconds to kWh:

$$\text{kWh} = \frac{\text{watt seconds}}{1000 \times 3600}$$

Also

$$\text{kWh} = \frac{\text{joules}}{3\,600\,000}$$

The kilowatt hour is often referred to as a *unit* of electricity, i.e.

6 kWh = 6 units

which could be

6 kW for 1 hour

or 3 kW for 2 hours

or 2 kW for 3 hours

or 1 kW for 6 hours, etc.

As previously mentioned, domestic and industrial appliances are connected in parallel to the supply and therefore the energy they each consume may simply be added together to determine the total energy used.

Example

A domestic consumer has the following daily loads connected to the supply:

Five 60 W lights for 4 hours

Two 3 kW electric fires for 2 hours

One 3 kW water heater for 3 hours

One 2 kW kettle for 1/2 hour

Calculate the energy consumed in 1 week.

Lights: $E = 5 \times 60 \times 4 = 1200 = 1.2\,\text{kWh}$

Fires: $E = 2 \times 3 \times 2 = 12.0\,\text{kWh}$

Water heater: $E = 1 \times 3 \times 3 = 9.0\,\text{kWh}$

Kettle: $E = 1 \times 2 \times 0.5 = 1.0\,\text{kWh}$

Total for 1 day $= 23.2\,\text{kWh}$

$\therefore$ Energy expended in 1 week $= 7 \times 23.2$

$= 162.4\,\text{kWh}$

Tariffs

Tariffs are simply charges made by the electricity boards for the use of electricity. There are several types of tariff available depending on the kind of installation, i.e. domestic, commercial, etc. However, the basis of them all is a charge per unit (kWh) consumed.

Example

If the cost per unit of electrical energy in the previous example is 6 p per unit, calculate the cost to the consumer over a 13 week period (1 quarter).

Total energy consumed per week = 162.4 units (kWh)

$\therefore$ Cost at 6 p per unit $= 6 \times 162.4$

$= 974.4\,\text{p}$

$= £9.74$

$\therefore$ Cost for 13 weeks $= 13 \times £9.74$

$= £126.62$

Measuring power and energy

A *wattmeter* is connected as shown in Fig. 3.26. It is basically a combination of an ammeter and a voltmeter, and it measures the product of current and voltage:

P (watts) $= I \times V$

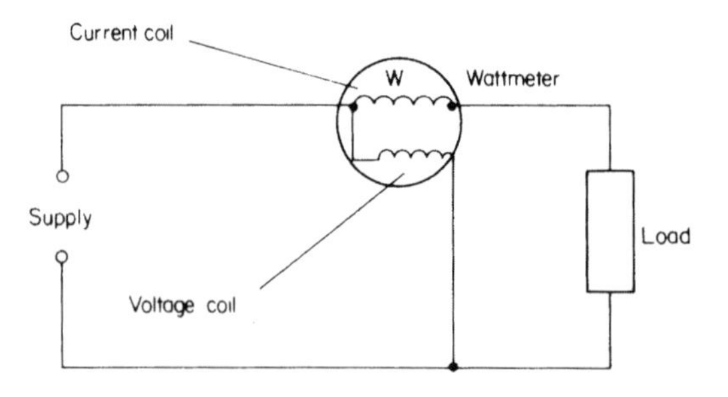

Fig. 3.26

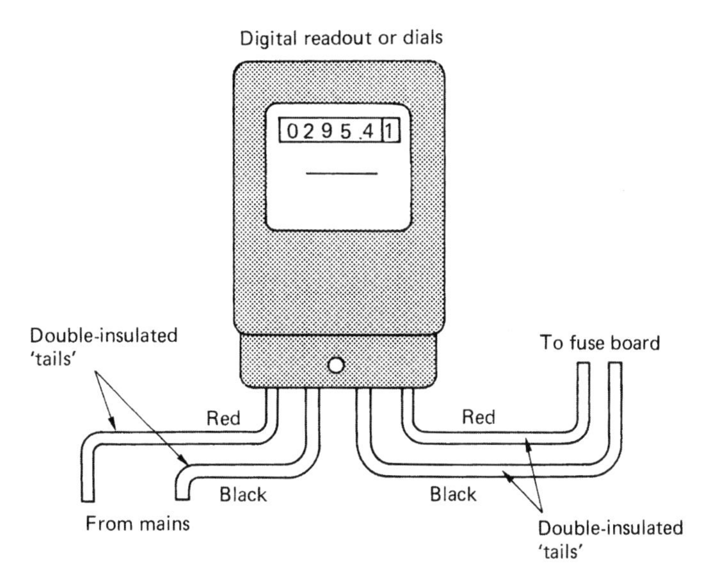

Fig. 3.27 *Energy meter*

An *energy meter* is similar to a wattmeter and its connections are the same. However, it is designed to show the number of kilowatt hours of energy used. It is familiar to most of us as our electricity meter (Fig. 3.27).

Water heating

We have just discussed electricity energy and showed that

$$\text{kWh} = \frac{\text{joules}}{3\,600\,000}$$

Joule in his experiments showed that 4.2 J of electrical energy = 1 calorie of heat energy. Hence it required 4.2 J of electrical energy to raise the temperature of 1 g of water though 1°C

or 4.2 J/gm/°C

or 4200 J/kg/°C

This value is called the *specific heat* of water (SH), i.e. if 2 kg of water was raised through 2°C then the amount of electrical energy required to do this would be

$2 \times 2 \times 4200 = 16\,800\,\text{J}$

Hence

joules = mass (kg) × change in temp. (°C) × SH of water (4200)

But

$$\text{kWh} = \frac{\text{joules}}{3\,600\,000}$$

$$\therefore \text{Heat output in kWh} = \frac{\text{mass} \times \text{change in temp.} \times \text{SH}}{3\,600\,000}$$

If a system is 100% efficient then whatever we put into that system we get the same out (i.e. there are no losses).

Example

How long will it take a 2 kW 240 V kettle to raise the temperature of 2 litres of water from 8°C to boiling point? Assume 100% efficiency. (SH of water = 4200 J/kg/°C and 1 litre of water has a mass of 1 kg.)

$$\text{kWh}_{\text{output}} = \frac{\text{mass} \times \text{change in temp.} \times \text{SH}}{3\,600\,000}$$

$$= \frac{2 \times 92 \times 4200}{3\,600\,000}$$

$$= 0.215\,\text{kWh}$$

As the system is 100% efficient, then

$$\text{kWh}_{\text{input}} = \text{kWh}_{\text{output}}$$

and, as 2 kW is the input power dissipated by the element, then

$$2 \times \text{hours} = 0.215$$

$$\therefore \text{hours} = \frac{0.215}{2}$$

$$= 0.1075 \text{ hours}$$

$$= 6.45 \text{ minutes}$$

Efficiency

The efficiency of a system is the ratio of the output to the input:

$$\therefore \text{Percentage efficiency} = \frac{\text{output} \times 100}{\text{input}}$$

Example

Calculate the efficiency of a water heater if the output in kilowatt hours is 24 kWh and the input energy is 30 kWh.

$$\text{Efficiency (\%)} = \frac{\text{output}}{\text{input}} \times 100$$

$$= \frac{24}{30} \times 100$$

$$= 80\%$$

Example

Calculate for how long a 3 kW immersion heater must be energized to heat 137 litres of water from 10°C to 70°C. The efficiency of the system is 80%. (SH of water = 4200 J/kg/°C and 1 litre has a mass of 1 kg.)

$$\text{kWh}_{\text{output}} = \frac{\text{mass} \times \text{change in temp.} \times \text{SH}}{3\,600\,000}$$

$$= \frac{137 \times 60 \times 4200}{3\,600\,000}$$

$$= 9.59\,\text{kWh}$$

But

$$\text{Efficiency} = \frac{\text{output}}{\text{input}} \times 100$$

$$\therefore 80 = \frac{9.59 \times 100}{\text{input kWh}}$$

$$\therefore \text{kWh}_{\text{input}} = \frac{9.59 \times 100}{80}$$

$$= 12\,\text{kWh}$$

As the input power is 3 kWh, then

$$\text{kWh}_{\text{input}} = \text{kW}_{\text{input}} \times \text{hours}$$

$$\therefore 12 = 3 \times \text{hours}$$

$$\therefore \text{Hours} = \frac{12}{3}$$

$$\therefore \text{Time} = 4\ \text{hours}$$

If the tank were lagged to prevent as many losses as possible, the efficiency would improve, reducing the time taken and hence ensuring greater economy.

Self-assessment questions

1 A 240 V electric iron has a resistance of 96 Ω and is connected to a socket outlet by a twin cable, each conductor of which has a resistance of 0.1 Ω. If the total resistance of the cable from the fuse board to the socket is 0.8 Ω, calculate the total resistance of the whole circuit.

2 If the total resistance of the three lamps in the diagram is 4128 Ω, calculate the resistance of lamp A.

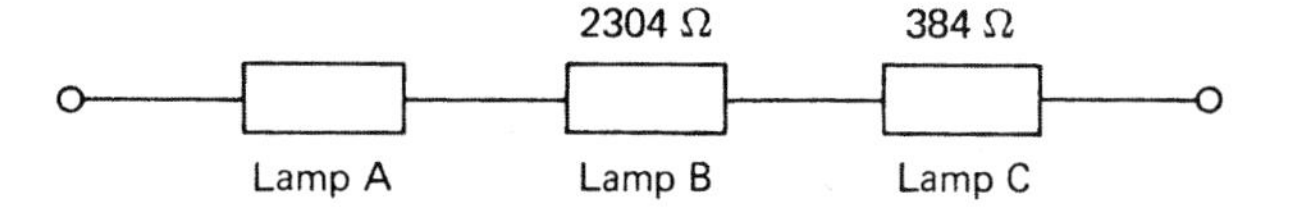

3 An electric kettle, an iron and a food mixer have element resistances of 28.8 Ω, 96 Ω and 576 Ω respectively. If they are all connected in parallel, calculate the total resistance.

4 A 2.5 mm^2 twin copper cable 20 m long supplies a heating appliance having three elements each of resistance 57.6 Ω. If the elements are arranged such that two are in series and the third is in parallel with these two, calculate: (a) the resistance of the cable; and (b) the total resistance of the whole circuit. (ρ for copper is 17 μΩ mm.)

5 (a) What is the maximum permissible voltage drop on an installation, as recommended by the IEE Regulations?

(b) A length of cable supplying a cooker at 230 V has a resistance of 0.24 Ω. If the cooker has a total resistance of 5.76 Ω, calculate the voltage drop along the cable. Is this value permissible?

6 If an electric fire of resistance 28.8 Ω, an immersion heater of resistance 19.2 Ω, a small electric kettle of resistance 57.6 Ω and a toaster of 115.2 Ω are connected to a 230 V domestic power circuit, calculate the current taken by each appliance and the total current drawn from the supply.

7 What would be the resistance of and the current drawn by the following when connected to a 240 V supply: (a) a 3 kW 240 V immersion heater; (b) a 600 W 240 V food mixer; (c) a 1 kW 240 V electric fire; and (d) a 40 W 240 V filament lamp?

8 Three lamps have the following rated values: 60 W 240 V; 100 W 200 V; and 40 W 100 V. Calculate the power dissipated by each if they are connected in series across a 240 V supply.

9 From the diagram, calculate the total resistance, the current in each branch, the voltage at X and Y, and the p.d. across XY.

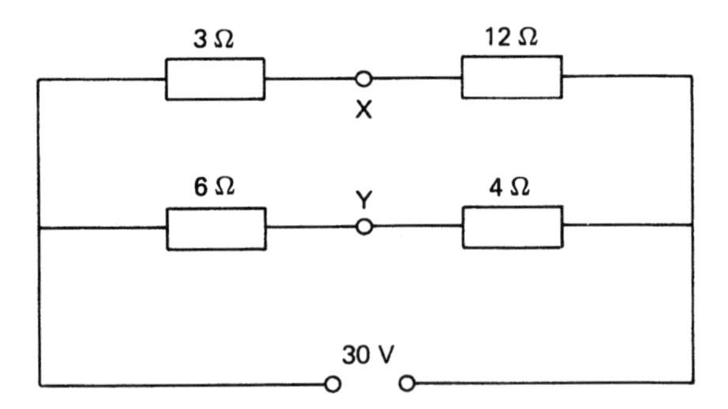

10 A domestic consumer has the following loads connected each day (supply voltage 230 V).

Five 100 W lamps for 4 hours
One 3 kW immersion heater for 2 hours
One 10 kW cooker for $1\frac{1}{2}$ hours
Two 3 kW electric fires for 3 hours
Sundry appliances taking 12 A for $\frac{3}{4}$ hour

If electricity costs 7p per unit, calculate the cost for one quarter (13 weeks).

11 It is required to raise the temperature of 1.5 litres of water in a kettle from 5°C to 100°C in 3.5 minutes. Assuming 100% efficiency, calculate the nearest size of element required to do this. (SH of water = 4200 J/kg/°C.)

12 An immersion tank contains 110 litres of water at 15°C. An immersion element supplied from a 240 V source takes 12.5 A when energized and heats the water to 78°C. If the system is 90% efficient, calculate for how long the element is energized. (SH of water = 4200 J/kg/°C.)

13 A small 25 litre boiler is completely filled with water at 12°C. The heating element has a resistance of 28.8 Ω and is connected to a 240 V supply. If the boiler raises the temperature of the water to boiling point in 97 minutes, calculate the efficiency of the system.

4 Electromagnetism

Before we deal with this major subject, it would be sensible to discuss basic magnetism briefly.

Magnetism

We are all familiar with simple magnets and have probably all seen the lines of force traced on paper with iron filings. The quantity of lines of force that come out from a magnet is called the *flux* and is measured in *webers* (Wb).

Wilhelm Edward Weber (1804–91)

German scientist famous for his work in the measurement of electrical quantities.

Flux density: symbol, *B*; unit, tesla (T)

Just as population densities are measured in people per km^2, flux density is measured in flux per m^2 or Wb/m^2. This unit, however, is known as the *tesla* (T).

$$\therefore B(\mathrm{T}) = \frac{\Phi\ (\mathrm{Wb})}{a\ (\mathrm{m}^2)}$$

Nickola Tesla (1856–1943)

Yugoslavian electrical engineer renowned for his work on a.c. generation and distribution.

Example

A motor field pole has an area of $60\,\mathrm{cm}^2$. If the pole carries a flux of 0.3 Wb, calculate the flux density.

$B = ?;\ \Phi = 0.3\,\mathrm{Wb};\ a = 60\,\mathrm{cm}^2 = 0.006\,\mathrm{m}^2$

$$B = \frac{\Phi}{a}$$

$$B = \frac{0.3}{0.006}$$

$$= 50\,\mathrm{T}$$

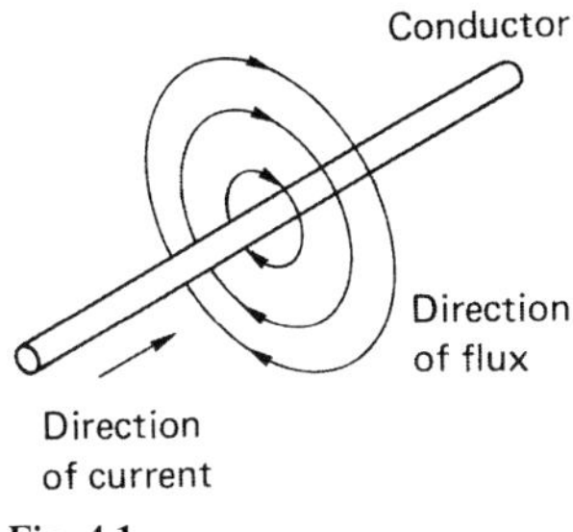

Fig. 4.1

Electromagnetism

Field around a conductor carrying a current

When a conductor carries a current, a magnetic field is produced around that conductor. This field is in the form of concentric circles along the whole length of the conductor. The direction of the field depends on the direction of the current – clockwise for a current flowing away from the observer and anticlockwise for a current flowing towards the observer. In order to show these directions, certain signs are used (Fig. 4.2).

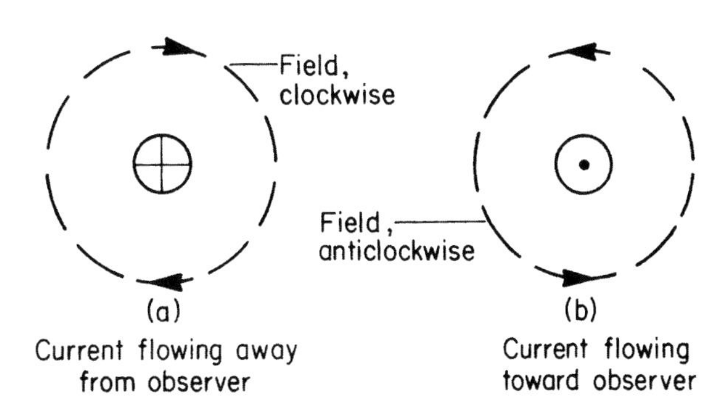

Fig. 4.2

The screw rule

In order quickly to determine the direction of the magnetic field around a current-carrying conductor, the screw rule may be applied (Fig. 4.3).

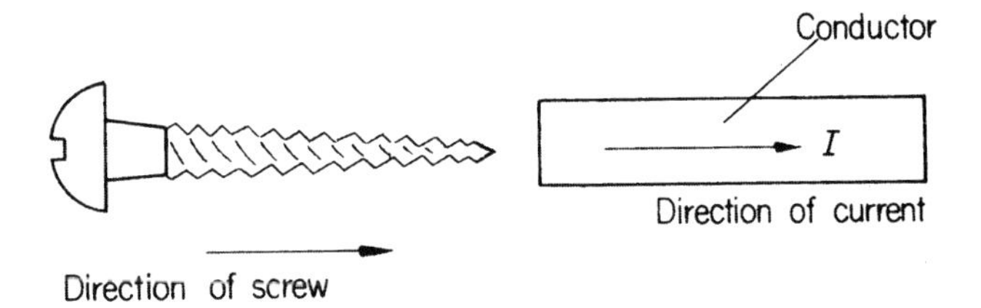

Fig. 4.3

Imagine a screw being twisted into or out of the end of a conductor in the same direction as the current. The direction of rotation of the screw will indicate the direction of the magnetic field.

Force between current-carrying conductors

If we place two current-carrying conductors side by side there will exist a force between them due to the flux. The direction of this force will depend on the directions of the current flow (Fig. 4.4).

In Fig. 4.4a there is more flux between the conductors than on either side of them, and they will be forced apart.

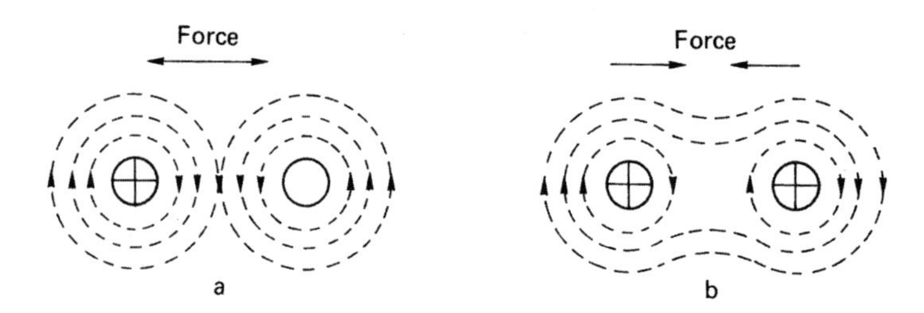

Fig. 4.4

In Fig. 4.4b, the flux between the conductors is in opposite directions and tends to cancel out leaving more flux on the outside of the conductors than in between them, so they will be forced together.

The direction of movement can be found using Fleming's left-hand rule.

Fleming's left-hand rule

If the thumb, first and second fingers of the left hand are placed at right angles to one another (Fig. 4.5), they indicate:

First finger Field
seCond finger Current
thuMb Motion

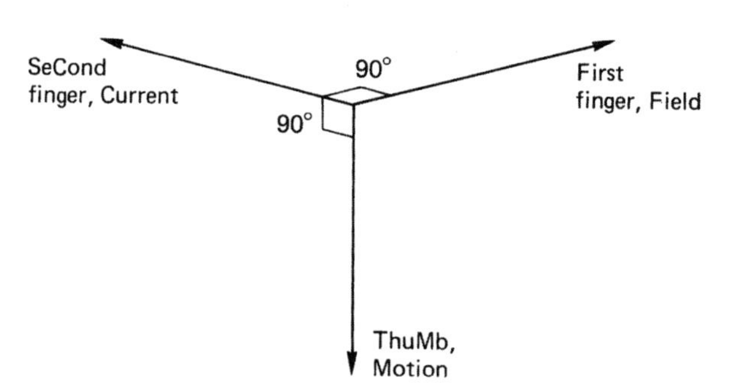

Fig. 4.5

Force on a conductor carrying a current in a magnetic field

If we arrange for a current-carrying conductor to be placed at right angles to a magnetic field, a force will be exerted on that conductor (Fig. 4.6). This force is measured in newtons.

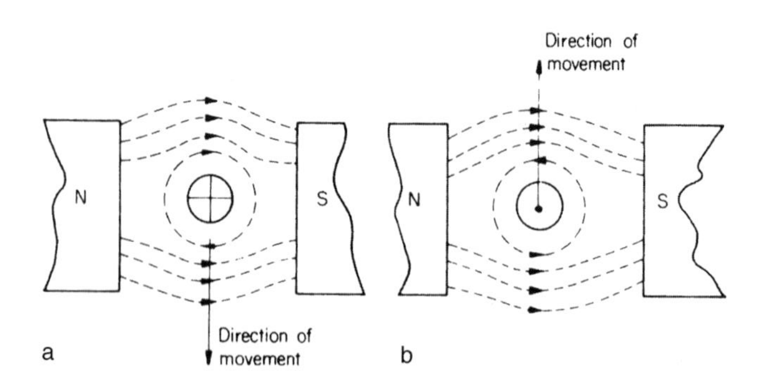

Fig. 4.6

Sir Isaac Newton (1642–1727)

English scientist of considerable fame, known especially for his work on force, mass, motion and momentum.

In Fig. 4.6a the flux above the conductor is greater than the flux below, and the conductor is forced downwards. In Fig. 4.6b the current and hence the field around the conductor is opposite to that in Fig. 4.6a and the conductor is forced upwards.

The magnitude of this force is dependent on three things:

1 the current flowing in the conductor (I);
2 the density of the magnetic field (B); and
3 the length of the conductor in the magnetic field (l).

$$\therefore F \text{ (newtons)} = B \text{ (teslas)} \times l \text{ (metres)} \times I \text{ (amperes)}$$

Example

Calculate the force exerted on a conductor 40 cm long carrying a current of 100 A at right angles to a magnetic field of flux density 0.25 T.

$F = ?;\ B = 0.25\,\text{T};\ l = 40\,\text{cm} = 0.4\,\text{m};\ I = 100\,\text{A}$

$$F = B \times l \times I$$

$$= 0.25 \times 0.4 \times 100$$

$$F = 10\,\text{N}$$

Example

A circular magnetic field has a diameter of 20 cm and a flux of 149.6 mWb. Calculate the force exerted in a conductor 21 cm long lying at right angles to this field if the current flowing is 15 A.

$$B = \frac{\Phi}{a}$$

But

$$a = \frac{\pi\, d^2}{4}$$

$\pi = 3.1416;\ d = 20\,\text{cm} = 0.2\,\text{m};\ B = ?;\ \Phi = 149.6\,\text{mWb} = 149.6 \times 10^{-3}\text{Wb}$

$$\therefore a = \frac{3.146 \times 0.2 \times 0.2}{4}$$

$$= 0.031\ 416\,\text{m}^2$$

$$B = \frac{\Phi}{a}$$

$$= \frac{149.6 \times 10^{-3}}{0.031\ 416} = 4.762\,\text{T}$$

$F = ?$; $B = 4.762\,\text{T}$; $l = 21\,\text{cm} = 0.21\,\text{m}$; $I = 15\,\text{A}$

$$F = B \times l \times I$$

$$= 4.762 \times 0.21 \times 15$$

$$F = 15\,\text{N}$$

The e.m.f. induced in a moving conductor

We have seen that passing a current through a conductor in a magnetic field produced a movement of that conductor. If we were to reverse the process and physically move the conductor through a magnetic field, such that it cut across the flux, then a current would flow in that conductor (Fig. 4.7).

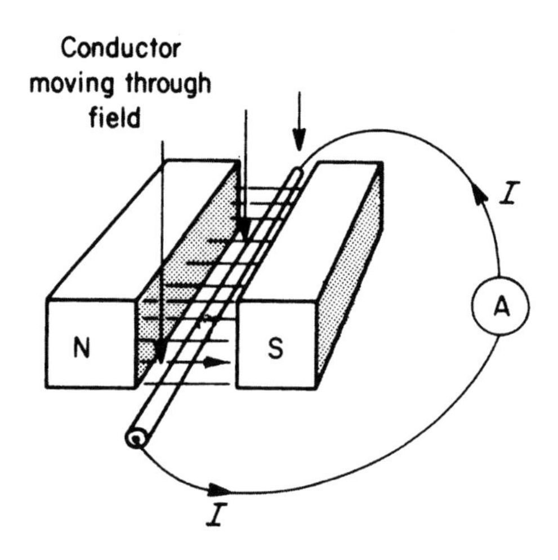

Fig. 4.7

As we have already seen in Chapter 2, a pressure is required for a current to flow. Therefore, if a current flows (Fig. 4.7) then an e.m.f. must be producing it. This e.m.f. is called an *induced e.m.f.* and its direction is the same as that of the current flow. This direction can be determined by using Fleming's right-hand rule.

Fleming's right-hand rule

If the thumb, first and second fingers of the right hand are arranged at right angles to one another (Fig. 4.8) they indicate:

First finger, Field (north to south)
seCond finger, Current (and e.m.f.)
thuMb, Motion

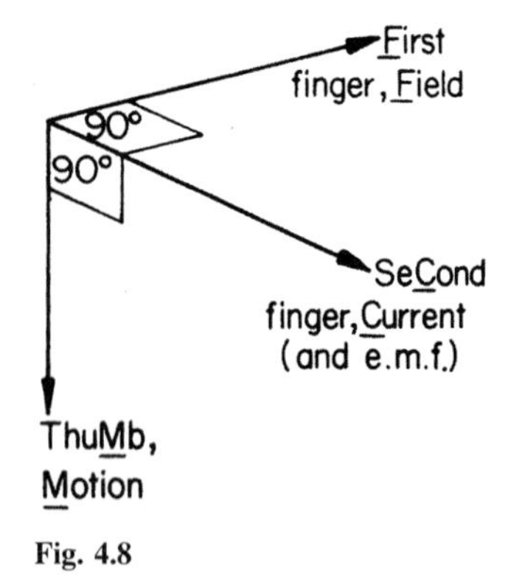

Fig. 4.8

The magnitude of the induced e.m.f. depends upon:

1 the flux density of the field (B);
2 the length of the conductor (l); and
3 the velocity at which the conductor cuts across the flux (v).

$$\therefore E \text{ (volts)} = B \text{ (teslas)} \times l \text{ (metres)} \times v \text{ (metres/second)}$$

Example

A conductor 15 cm long is moved at 20 m/s perpendicularly through a magnetic field of flux density 2 T. Calculate the induced e.m.f.

E = ?; B = 2 T; l = 15 cm = 0.15 m; v = 20 m/s

$$E = B \times l \times v$$

$$= 2 \times 0.15 \times 20$$

$$E = 6\,\text{V}$$

Application of magnetic effects

There are several major areas in which use is made of the magnetic effects of electric current. These are measuring instruments and motors (discussed in later chapters), solenoids, electromagnets, inductors and transformers.

Solenoid

If we wind a conductor on to a hollow cardboard cylinder or former and pass a current through it, the whole assembly will act like a magnet, having a north and south pole, and an iron rod will be drawn inside the solenoid when it is energized. This effect may be used in various ways.

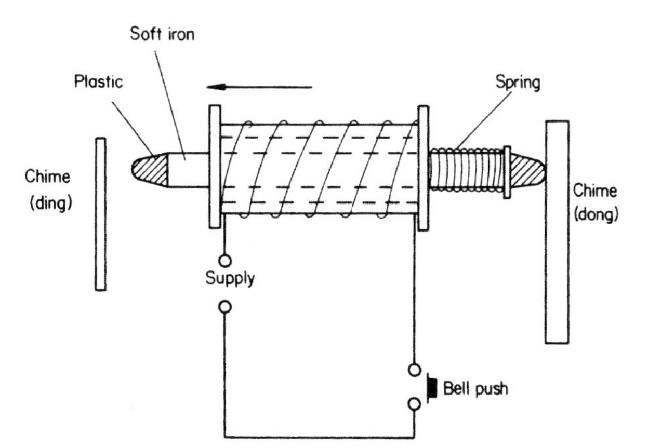

Fig. 4.9 *'Ding-dong' chimes for d.c. or a.c. supplies*

One example is a chime bell (Fig. 4.9). When the bell push is depressed, the solenoid is energized, and the soft iron rod, with plastic end inserts, is attracted by the magnetic field, in the direction shown, and the chime (ding) will sound. When the bell push is released the spring will return the rod with enough force to sound the chime (dong) again.

Electromagnet

If the coil of the solenoid is now wound on an iron core, when energized, the coil will cause the core to become a magnet. This effect is used in many different ways, for example in bells, relays, contractors, telephones, or circuit-tripping mechanisms.

Fig. 4.10 shows how a simple trembler bell works. When the bell push is operated, the electromagnetic is energized and the iron armature is attracted to it, the striker hitting the bell. This action, however, breaks the circuit at A and the electromagnet de-energizes. The spring returns the striker and armature to their original position, completing the circuit, and the electromagnet is energized again.

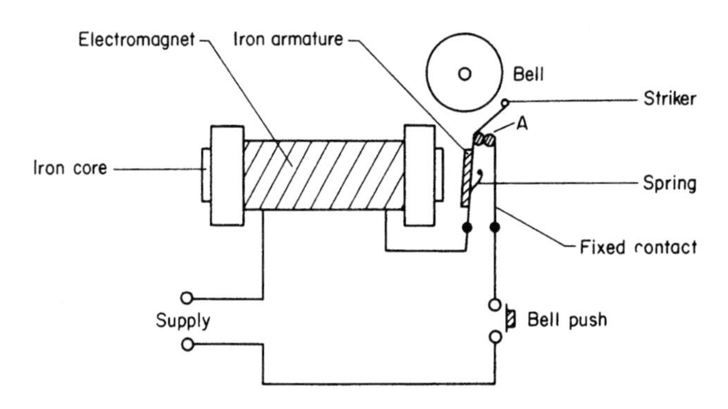

Fig. 4.10

Another device making use of the same principle is the electromagnetic trip. Its latching mechanism is similar to that of the thermal trip, but the bimetal is replaced by an iron armature (Fig. 4.11). The current supplying the load flows through the coil and through the closed contacts. The coil is designed to allow a certain value of current to flow, and beyond this amount the core of the electromagnet is magnetized sufficiently to attract the armature, causing the circuit to be interrupted.

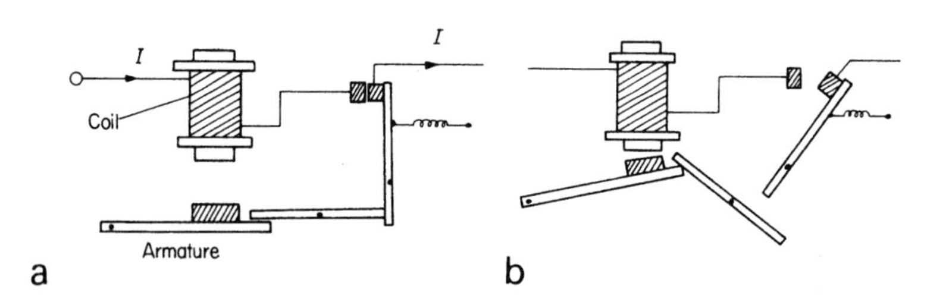

Fig. 4.11

Direct-current generator

We have just seen how an e.m.f. is induced in a conductor when it is moved through a magnetic field.

This effect is the principle which enables a simple generator to work (Fig. 4.12). A single loop conductor arranged as shown in this figure has its ends connected to a simple commutator, which comprises two copper segments insulated from each other. The commutator and loop are fixed to a central

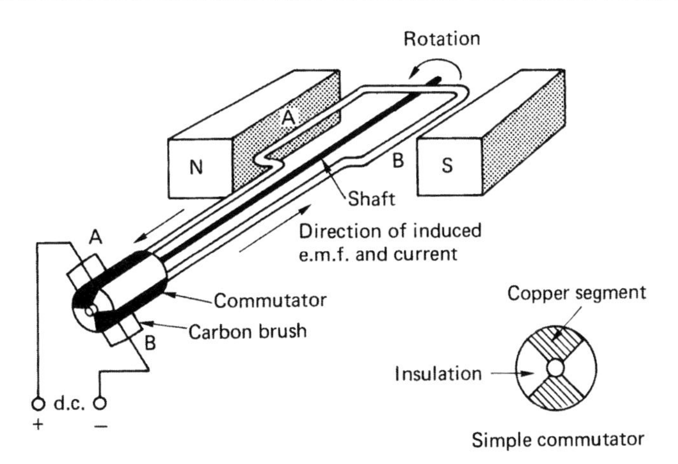

Fig. 4.12 *Simple single-loop d.c. generator*

shaft which enables the whole assembly to be freely rotated. Two fixed carbon brushes bear on the surface of the commutator, enabling an external load to be connected to the generator.

When the loop is rotated in the direction shown, side A will travel downwards and side B upwards. From Fleming's right-hand rule (p. 62), the current and hence the induced e.m.f. will be in the direction shown, making brush A +ve and B –ve. After 180° of revolution, side A will be travelling upwards and side B downwards, but the induced e.m.f. stays in the same direction (right-hand rule) and so the polarity at the brushes remains unchanged. We have therefore generated a d.c. source of supply.

In the larger type of generator the magnetic field is provided by electromagnets rather than permanent magnets. The single loop is replaced by many such loops held in slots in an iron core. This arrangement is called an *armature*. The commutator fixed to the armature shaft has of course many insulated segments to which the ends of all the loops are connected.

The armature is laminated, i.e. made up of many thin sheets insulated from one another. This is done to reduce eddy currents.

Eddy currents

As we have seen, cutting magnetic flux with a conductor induces an e.m.f. in it, and if the conductor is made part of a complete circuit, a current will flow.

The iron of an armature as well as its conductors cut the flux and small currents are induced in the armature core, which, when circulating together can cause it to heat up. This effect is overcome to a large extent by laminating the core. This confines the currents to each lamination and prevents a large circulating current from building up.

Alternating-current generator

The principle of a.c. generation is the same as that for d.c. The ends of the loop in this case, however, are terminated in slip rings, not a commutator (Fig. 4.13).

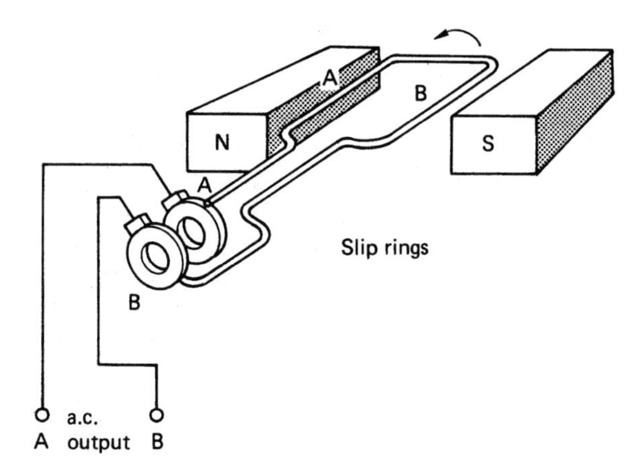

Fig. 4.13 *Simple single-loop a.c. generator*

As the loop rotates, side A will have an e.m.f. induced first in one direction and then in the other. Therefore, as side A is permanently connected to a slip ring A, this ring will be alternatively +ve and –ve. The same process applies to ring B. The generated supply output is therefore alternating.

In practice, large generators are arranged such that the magnetic field rotates, its flux cutting across the armature conductors (Fig. 4.14). In a generator of this type, the armature is made stationary and called the *stator*, while the rotating magnetic field, in the form of an electromagnet, is called the *rotor*. This arrangement is preferred to that in Fig. 4.14, because of the excessive cost of providing slip rings and brushes capable of handling the large output currents.

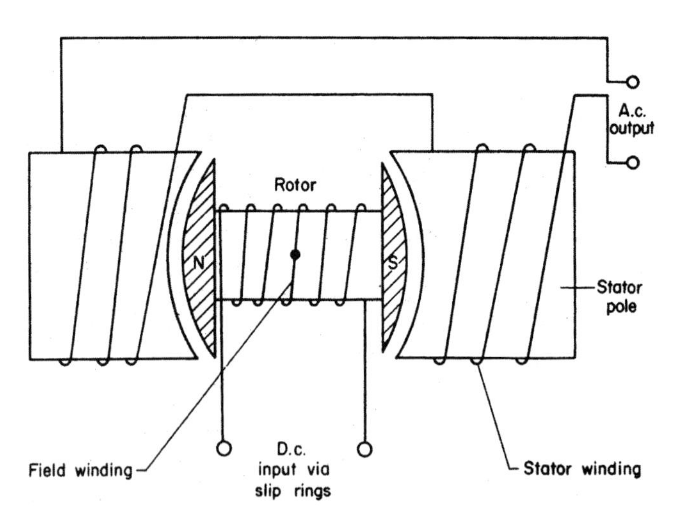

Fig. 4.14 *Single-phase a.c. generator*

The single-phase a.c. waveform

Fig. 4.15 shows a cross-section through a single-loop generator. It will be seen that in the vertical position AB, the loop sides are cutting no flux and hence no e.m.f. is induced. However, as the loop rotates, more and more flux is cut

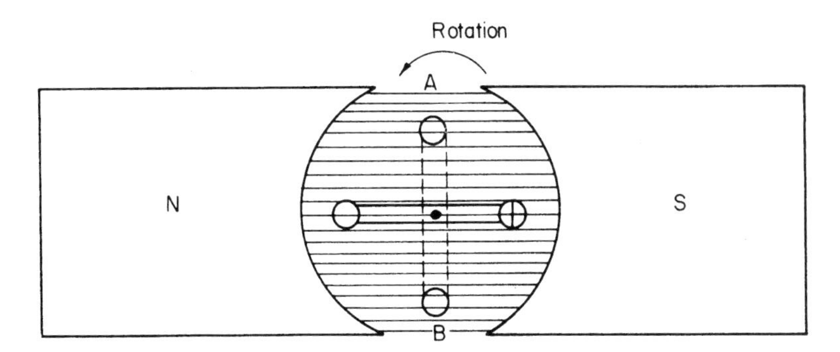

Fig. 4.15

and hence more and more e.m.f. is being induced, up to a maximum in the horizontal position. Further rotation causes the e.m.f. to fall to zero again. This rise and fall of the e.m.f. can be traced graphically.

Note: Since the current will flow in the same direction as the induced e.m.f., it too will rise and fall in time with the e.m.f. The current and the e.m.f. are said to be in *phase* with one another.

Fig. 4.16a shows a single conductor rotating in a magnetic field; after each 30° of revolution the conductor is at positions 1, 2, 3, 4, etc. The horizontal axis of Fig. 4.16a is the circular path taken by the conductor simply opened out to form a straight line, each 30° linear space corresponding to each 30° angle of movement. The vertical axis represents the magnitude of the induced e.m.f. As the induced e.m.f. depends on the amount of flux being cut, which itself depends on the position of the conductor, then the magnitude of the e.m.f. can be represented by the conductor position.

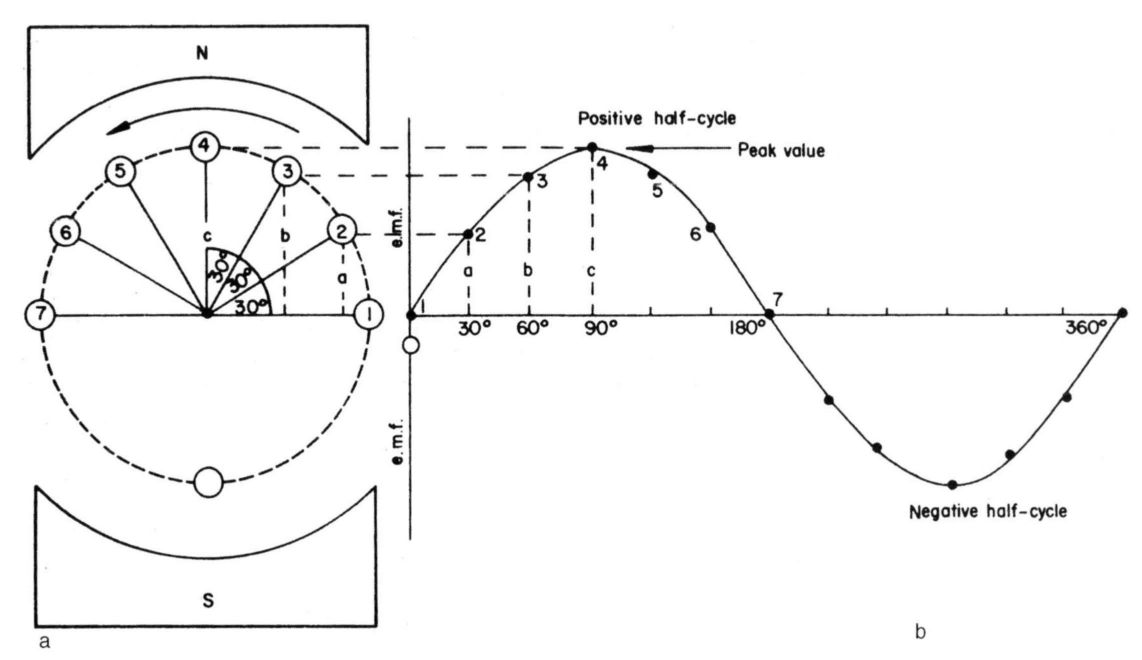

Fig. 4.16 *Single conductor rotating in a magnetic field: one complete cycle*

Hence each 30° position of conductor rotation (Fig. 4.16a) can be represented by an e.m.f. at each 30° space of linear movement (Fig. 4.16b).

The resulting graph indicates the e.m.f. induced in one complete revolution of the conductor. This waveform is called a *sine wave*, and any quantity that has a wave of that nature is called a *sinusoidal quantity.*

Fig. 4.16b shows the variation of the e.m.f. during one revolution of the conductor and is termed *one cycle.*

Frequency: symbol, f; unit hertz (Hz) (= cycles per second)

The number of complete cycles which occur in one second is called the *frequency* (Fig. 4.17). In the British Isles, the frequency of the supply is 50 cycles per second or 50 Hz.

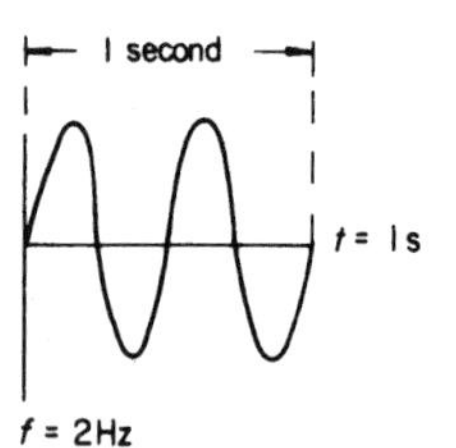

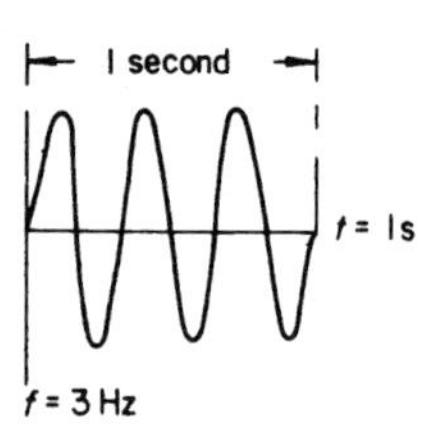

Fig. 4.17

Heinrich Rudolf Hertz (1857–94)

German physicist who demonstrated the transmission of electromagnetic waves.

Drawing the waveform of an alternating quantity

The following example shows how the waveform of an alternating quantity can be drawn.

Example

Draw the waveform of a sinusoidal e.m.f. having a maximum or peak value of 90 V, and from it determine the value of the e.m.f. after one-third of a cycle.

First a suitable scale must be chosen and then a circle with a radius representing 90 V is drawn (Fig. 4.18). (This radius representing 90 V is called a *phasor*, phasors simply being straight lines drawn to scale to represent electrical quantities.) We will choose a scale in which 1 cm = 30 V so that 90 V will be represented by 3 cm.

Now the circle is divided up into 30° segments. If the scale permits, it can be divided into 15° segments: the more points on the waveform the easier it is to draw, and the more accurate the result.

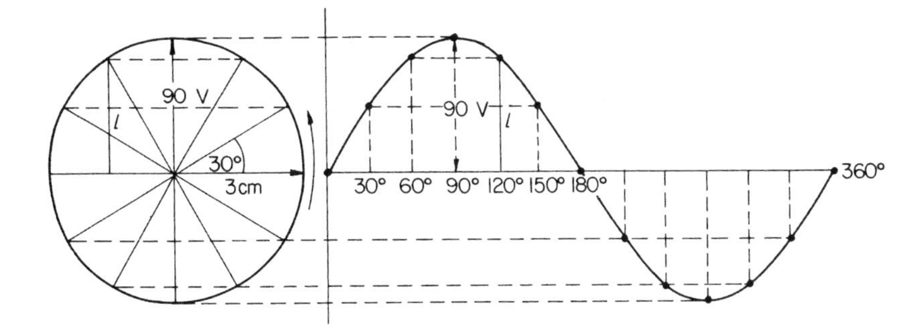

Fig. 4.18

After one-third of a cycle the conductor has moved 1/3 of 360°, i.e. 120°, and the vertical measurement l, to scale, is the value of induced e.m.f. at that point. By measurement,

$l = 2.6\,\text{cm}$

$= 2.6 \times 30$

$= 78\,\text{V}$

Addition of waveforms

Sometimes two or more voltages or currents are acting simultaneously in a.c. circuits, and they may not act together. Under these circumstances they are referred to as being *out of phase*.

Whether the voltages or currents are in or out of phase, their combined effect or *resultant* can be shown.

Example

Two sinusoidal voltages A and B of peak values 90 V and 60 V respectively act together in a circuit. If voltage B lags behind voltage A by 70°, draw the two waveforms on the same axis and show the resultant voltage. What is its peak value?

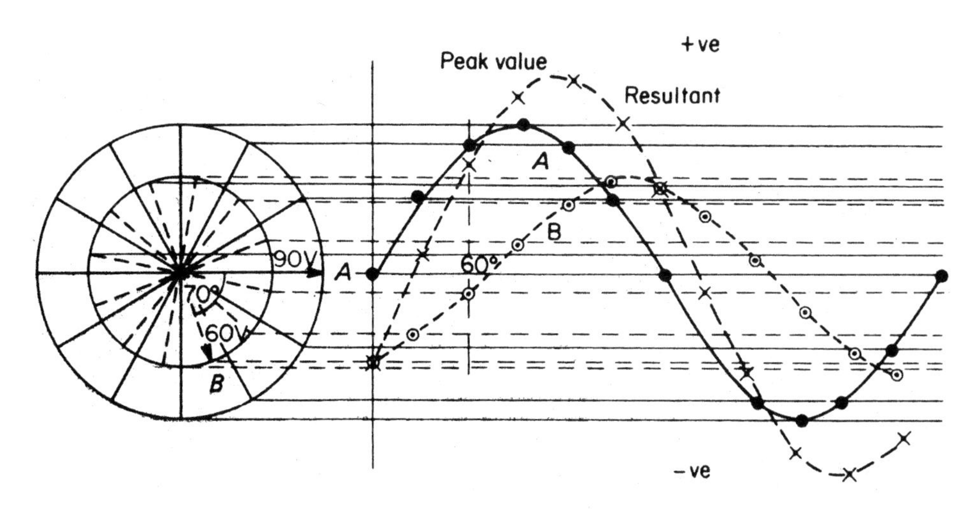

Fig. 4.19

The construction is carried out as follows (Fig. 4.19).

1 Choose a suitable scale.
2 Draw two circles, one inside the other, of radius corresponding to 90 V and 60 V.
3 Starting at A, divide the larger circle into 30° segments.
4 Starting at B, divide the smaller circle into 30° segments.
5 Draw each waveform.

The resultant waveform is drawn by adding the value of waves A and B at each 30° interval, e.g. at 60° B has a value of –4 mm and A of + 25 mm.

$\therefore A + B$ at 60° $= -4 + 25 = 21$ mm

This is the value of the resultant at 60°. By measurement, the peak value of the resultant is 4 cm.

$\therefore$ Peak voltage of resultant $= 4 \times 30 = 120$ V

Root-mean-square (r.m.s.) value

As we saw in Chapter 3, a current passing through a resistance has a heating effect, the magnitude of which we measure in watts, and we have also seen that

$$P = I^2 \times R$$

and

$$P = \frac{V^2}{R}$$

so that the power (or heat) dissipated in a resistance is proportional to either the square of the current or the voltage. Take, say, an a.c. current of peak value I amperes, draw its waveform over half a cycle and then square the value of the current at each 30° or 15° spacing (instantaneous values), then draw a waveform using these squared values. The resulting wave will represent the power dissipated. The average or mean heating effect will therefore be the sum of all the instantaneous values divided by the number of instantaneous values,

$$\therefore P = I^2 = \frac{i_1^2 + i_2^2 + i_3^2 + i_4^2 \ldots}{\text{number of values}}$$

$$\therefore I = \sqrt{\frac{i_1^2 + i_2^2 + i_3^2 + i_4^2 \ldots}{\text{number of values}}}$$

(in other words, the *root* of the *mean* of the *squares*). This can be shown mathematically to be 0.7071 of the peak value,

$$\therefore I_{\text{r.m.s.}} = I_{\max} \times 0.7071$$

where $I_{\text{r.m.s.}}$ is the root-mean-square current and $I_{\max}$ is the peak value of the current.

If we connect a resistance to a d.c. supply and draw a d.c. current equal to the value of the a.c. r.m.s. value, the heating effect will be the same in both cases. We can therefore define the r.m.s. value of alternating current or voltage as 'that value of alternating current or voltage which will give the same heating effect as the same value of direct current or voltage', i.e.

10 A (r.m.s.) = 10 A (d.c.)

Note: Unless otherwise stated, all values of voltage and current quoted on a.c. equipment are given as r.m.s. values.

It is interesting to note that the peak value of our 240 V domestic supply is

$$V_{\text{r.m.s.}} = V_{\max} \times 0.7071$$

$$\therefore V_{\max} = \frac{V_{\text{r.m.s.}}}{0.7071} = \frac{240}{0.7071} = 339.4\text{ V}$$

It is also interesting that the r.m.s. value of an alternating quantity is achieved when the conductor has rotated through an angle of 45° (Fig. 4.20). It will be seen that

$$v = V \times \sin 45° = V \times 0.7071$$

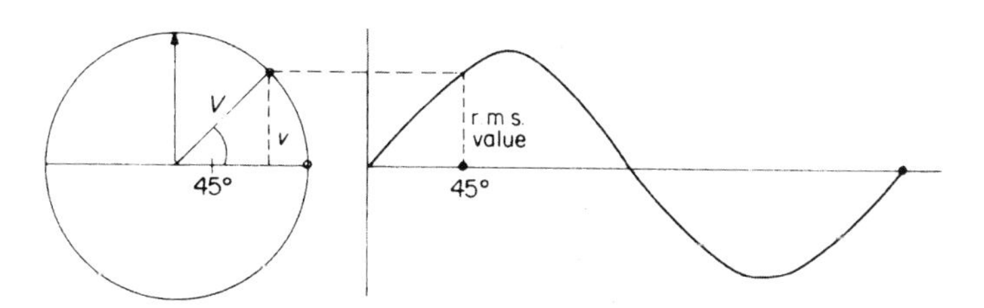

Fig. 4.20

Average value

The average value of all the instantaneous values that make up a sine wave is given by

$$I_{\text{average}} = I_{\max} \times 0.637$$

Three-phase a.c. generator

The principle of generating a three-phase supply is the same as that for a single phase. In this case, however, the stator poles are arranged 120° apart Fig. 4.21. With the rotor in the position shown in the diagram, the induced e.m.f. is at a maximum in the red phase, is increasing in the yellow phase, and

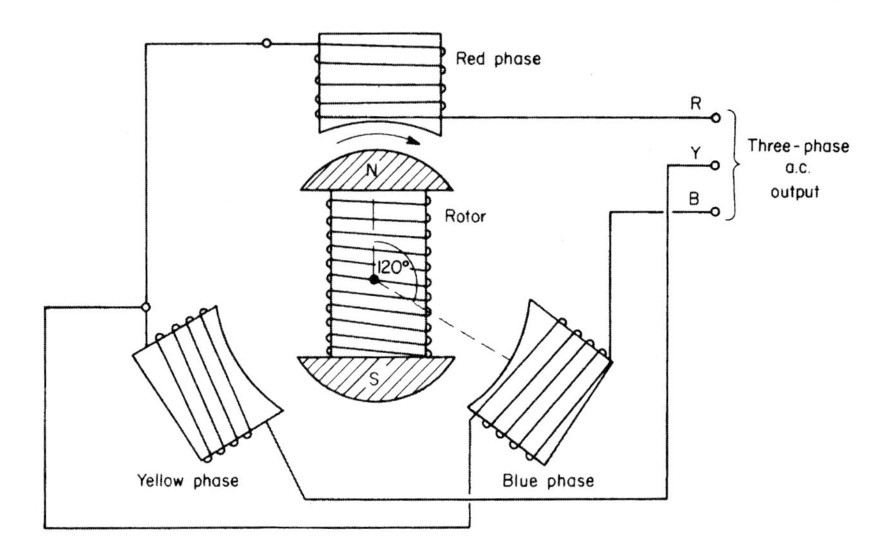

Fig. 4.21 *Simple three-phase star-connected a.c. generator*

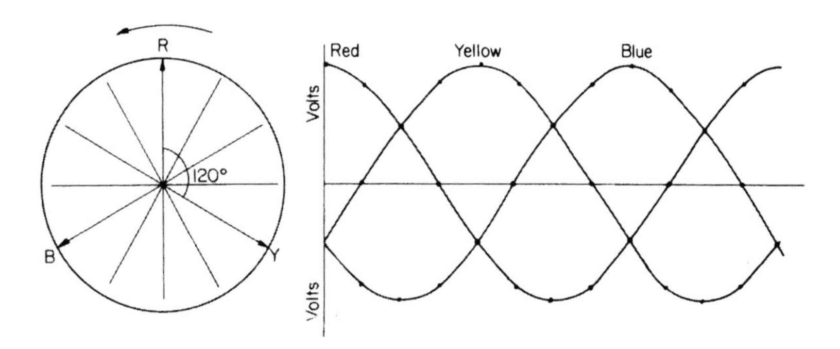

Fig. 4.22

is decreasing in the blue phase. There are therefore three waveforms, each 120° apart (Fig. 4.22). If we were to find the resultant of, say, the red and yellow phases first and then add this to the blue phase, we would find that the total resultant was zero.

Hence, in a balanced (all phases equal) three-phase system the resultant voltage and current is zero. As we will see later, this fact is important in distribution systems and in the design of three-phase motors.

Inductance: symbol, L; unit, henry (H)

Let us consider the effect of forming a coil from a length of wire, and connecting it to a d.c. source of supply. Fig. 4.23 shows the distribution of the magnetic lines of force, or flux, produced by such a circuit. We know that if we wind the same coil on to an iron core, the lines of force tend to be confined to that core and the flux is much greater (Fig. 4.24), and that when a conductor is cut by magnetic lines of force, a current, and hence an e.m.f., is produced in that conductor. Consider then, what happens when the switch S is first closed (Fig. 4.24).

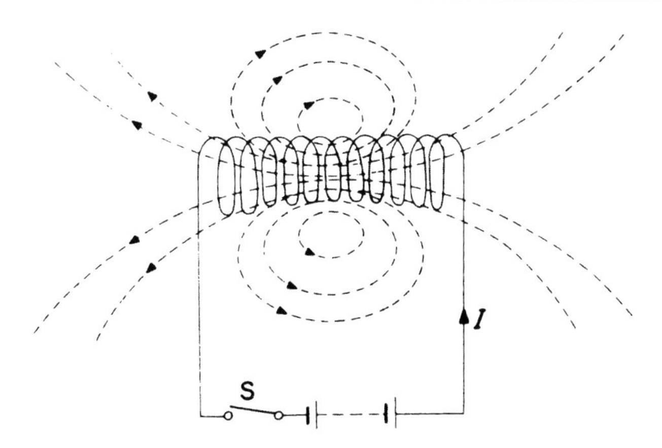

Fig. 4.23

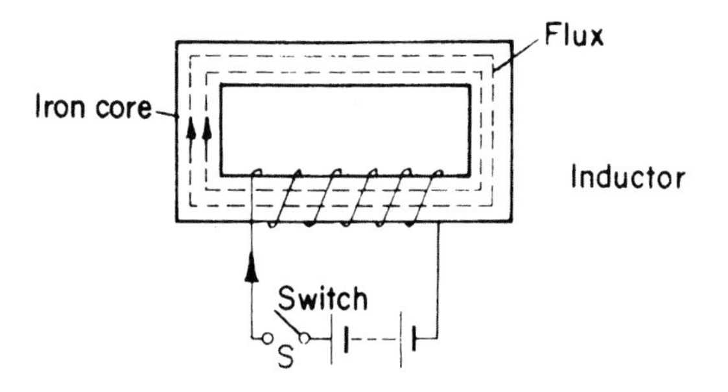

Fig. 4.24

As the current increases from zero to a maximum, the flux in the core also increases, and this growing magnetic field cuts the conductors of the coil, inducing an e.m.f. in them. This e.m.f., called the *back e.m.f.*, operates in the reverse direction to the supply voltage and opposes the change in the circuit current that is producing it. The effect of this opposition is to slow down the rate of change of current in the circuit.

When the switch S is opened, the current falls to zero and the magnetic field collapses. Again, lines of force cut the conductors of the coil inducing an e.m.f. in them. In this case, the e.m.f. appears across the switch contacts in the form of an arc.

Induced e.m.f. due to change in flux

The average value of the induced e.m.f. in a circuit such as the one shown in Fig. 4.24 is dependent on the rate of change of flux, and the number of turns of the coil. Hence the average induced e.m.f.

$$E = -\frac{\Phi_2 - \Phi_2}{t} \times N \text{ volts}$$

The minus sign indicates that the e.m.f. is a back e.m.f., and is opposing the rate of change of current.

Example

The magnetic flux linking the 1800 turns of an electromagnet changes from 0.6 mWb to 0.5 mWb in 50 ms. Calculate the average value of the induced e.m.f. (E = induced e.m.f.)

$\Phi_2 = 0.6\,\text{mWb};\ \Phi_1 = 0.5\,\text{mWb};\ t = 50\,\text{ms};\ N = 1800$

$$E = -\frac{(\Phi_2 - \Phi_1)}{t} \times N$$

$$= -\frac{(0.6 - 0.5)}{50 \times 10^{-3}} \times 10^{-3} \times 1800$$

$$= -\frac{0.1 \times 1800}{50} = \frac{18}{5}$$

$$= -3.6\,\text{V}$$

Self-inductance

Self-inductance is the property of a coil in which a change of current, and hence a change of flux, produces an e.m.f. in that coil.

The average induced e.m.f. in such a circuit is given by:

$$E = \frac{-L(I_2 - I_1)}{t}\ \text{volts}$$

The inductance L can be calculated from:

$$L = \frac{N\Phi}{I}$$

where N = number of turns; Φ = flux in webers; and I = current.

The unit of inductance

The unit of inductance is the *henry* (symbol H), and is defined as follows: 'A circuit is said to possess an inductance of 1 henry when an e.m.f. of 1 volt is induced in that circuit by a current changing at the rate of 1 ampere per second.'

Joseph Henry (1797–1878)

American electrophysicist whose work with magnetism led him to discover self-inductance.

Mutual inductance: symbol, *M*; unit, henry (H)

Let us consider the effect of winding two coils on the same iron core (Fig. 4.25).

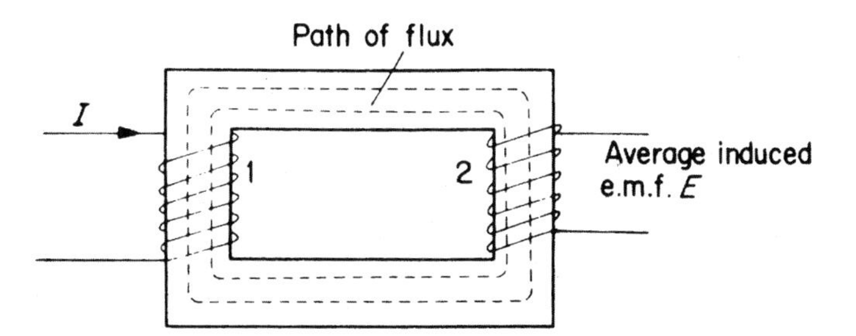

Fig. 4.25

A change of current in coil 1 produces a change of flux which will link with coil 2, thus inducing an e.m.f. in that coil. These two coils are said to possess the property of *mutual inductance*, which is defined as: 'A mutual inductance of 1 henry exists between two coils when a uniformly varying current of 1 ampere/second in one coil produces an e.m.f. of 1 volt in the other coil.'

If a change of current $(l_2 - l_1)$, in the first coil, induces an average e.m.f. E in the second coil, then:

$$E = -\frac{M(I_2 - I_1)}{t} \text{ volts}$$

But as E can also be expressed as

$$E = -\frac{(\Phi_2 - \Phi_1)N}{t} \text{ volts}$$

Then:

$$\frac{M(I_2 - I_1)}{t} = \frac{(\Phi_2 - \Phi_1)}{t} \times N$$

$$\therefore M = \frac{(\Phi_2 - \Phi_1)}{(I_2 - I_1)} \times N \text{ henrys}$$

Example

Two coils A and B have a mutual inductance of 0.5 H. If the current in coil A is varied from 6 A to 2 A, calculate the change in flux if coil B is wound with 500 turns.

$M = 0.5\,\text{H}$; $I_2 = 6\,\text{A}$; $I_1 = 2\,\text{A}$; $N = 500$; $(\Phi_2 - \Phi_1) = ?$

$$M = \frac{(\Phi_2 - \Phi_1)}{(I_2 - I_1)} \times N$$

$$(\Phi_2 - \Phi_1) = \frac{M \times (I_2 - I_1)}{N} = \frac{0.5 \times (6 - 2)}{500}$$

$$= \frac{0.5 \times 4}{500} = \frac{2}{500} = 4\,\text{mWb}$$

Time constant: symbol, T

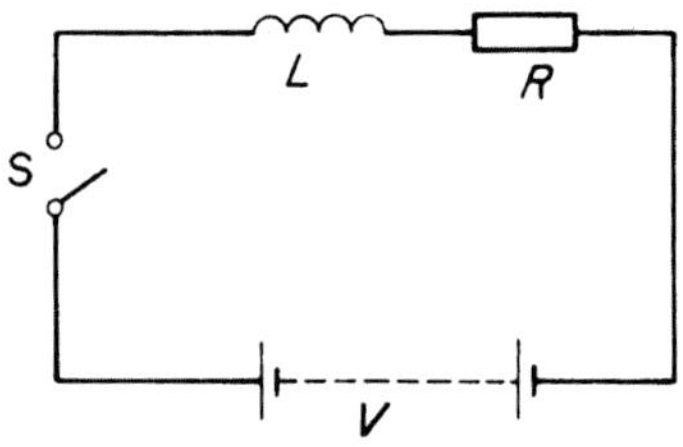

Fig. 4.26 *Typical inductive circuit*

When considering inductive circuits it is useful to represent the inductance and resistance of a coil as separate entities, on a circuit diagram. A typical inductive circuit is shown in Fig. 4.26.

When switch S is closed the current increases from zero to a steady maximum, given by:

$$I_{max} = \frac{V}{R} \text{ amperes}$$

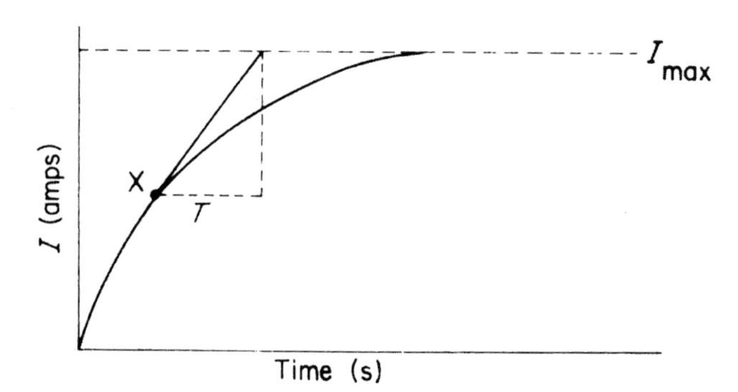

Fig. 4.27 *Growth of current in an inductive circuit*

Fig. 4.27 shows the growth of current in an inductive circuit. At any instant, say X, on the growth curve, if the rate of growth of current at that instant is such that if it continued to increase at that rate, it would reach its maximum value in L/R seconds, then this period of time is called the time constant and is given by:

$$T = \frac{L}{R} \text{ seconds}$$

Graphical derivation of current growth curve

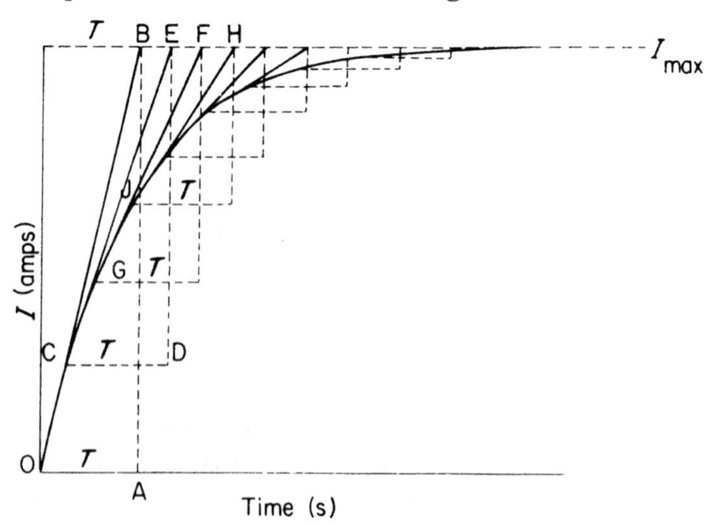

Fig. 4.28 *Graphical representation of current growth curve*

Construction (Fig. 4.28)

1 Select suitable scales for the two axes.
2 Draw the dotted line corresponding to the value of the maximum current (from $I = V/R$).
3 Along the time axis, mark off OA corresponding to the time constant T (from $T = L/R$).
4 Draw the perpendicular AB.
5 Join OB.
6 Select a point C close to O along OB.
7 Draw CD = T horizontally.
8 Draw the perpendicular DE.
9 Join CE.
10 Repeat procedures 6 to 9 for the line CE, and continue in the same manner as shown in the figure.
11 Join all the points O, C, G, J, etc., to form the growth curve.

The more points that are taken, the more accurate the final curve will be.

Example

A coil having a resistance of 25 Ω and an inductance of 2.5 H is connected across a 50 V d.c. supply. Derive the curve of the current growth graphically (Fig. 4.29).

$$I_{max} = \frac{V}{R} = \frac{50}{25} = 2\,\text{A; time constant, } T = \frac{L}{R} = \frac{2.5}{25} = 0.1\,\text{s}$$

Scales: 10 cm = 1 A and 10 cm = 0.2 s

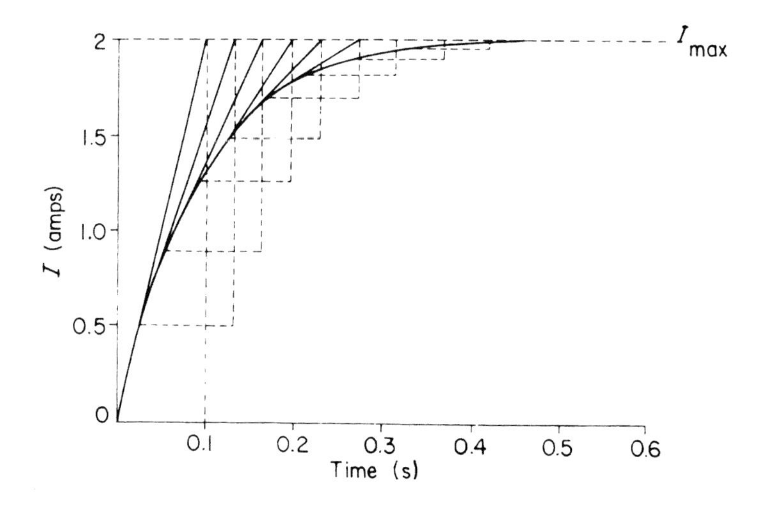

Fig. 4.29

Derivation of curve of current decay

The curve of current decay is constructed in the same manner as the growth curve, but in reverse as is shown in Fig. 4.30.

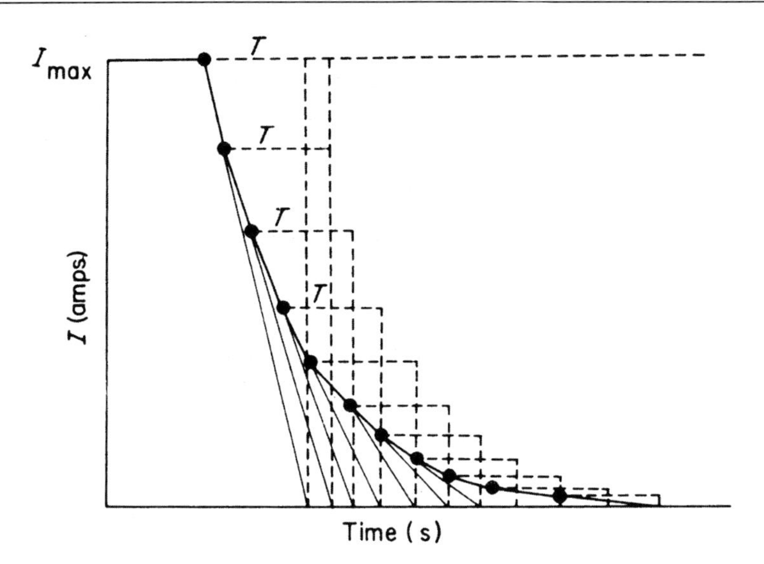

Fig. 4.30

Energy stored in a magnetic field

As we have already seen, opening an inductive circuit produces an arc across the switch contacts. This arc is the dissipation of the magnetic energy which was stored in the coil; the value of this energy can be calculated from

$W = \frac{1}{2} \times L \times I^2$ joules

Example

When carrying a current of 1.2 A, each field coil of a generator has an inductance of 2.5 H. Calculate the value of the energy stored in each coil.

W = energy stored; $I = 1.2$ A; $L = 2.5$ H

$W = \frac{1}{2} \times L \times I^2$

$$= \frac{2.5 \times 1.2 \times 1.2}{2}$$

$= 1.8$ J

Inductance in a.c. circuits

Inductive reactance: symbol, – X_L; unit, ohm (Ω)

Let us now consider the effect of supplying an iron-cored coil of negligible resistance with an alternating current and voltage.

In this instance the current, and therefore the magnetic field, is building up and collapsing (in the case of a 50 Hz supply) 50 times every second and hence a continual alternating back e.m.f. is produced. As we have seen at the beginning of the chapter, the back e.m.f. opposes the change in circuit current which is producing that e.m.f. Therefore, under a.c. conditions the e.m.f. produces a *continual* opposition to the current (much in the same way as resistance does in a resistive circuit). This opposition is called the *inductive reactance* (symbol X_L and is measured in ohms. X_L is given by

$$X_L = 2\pi fL\ \Omega$$

where f = frequency in hertz and L = inductance in henrys.

Example

Calculate the inductive reactance of a coil of inductance 0.5 H when connected to a 50 Hz supply,

$X_L = ?$; $f = 50\,\text{Hz}$; $L = 0.5\,\text{H}$

$$\begin{aligned} X_L &= 2\pi fL \\ &= 2\pi \times 50 \times 0.5 \\ &= 2\pi \times 25 \\ &= 50\pi \\ &= 157.1\,\Omega \end{aligned}$$

When an a.c. supply is given to a *pure* inductance the principles of Ohm's law may be applied, i.e. $V = I.X_L$.

Example

Calculate the current taken by a coil of inductance 0.8 H when connected to a 100 V, 50 Hz supply.

$X_L = ?$; $V = 100\,\text{V}$; $f = 50\,\text{Hz}$; $L = 0.8\,\text{H}$ and $I = ?$

In order to find the current, the formula $V = I \times X_L$ must be used; therefore the value of X_L must be calculated first.

$$\begin{aligned} X_L &= 2\pi fL \\ &= 2\pi \times 50 \times 0.8 \\ &= 80\pi \\ &= 251.36\,\Omega \end{aligned}$$

$$V = I \times X_L$$

$$\begin{aligned} \therefore I &= \frac{V}{X_L} \\ &= \frac{100}{251.36} \\ &= 0.398\,\text{A} \end{aligned}$$

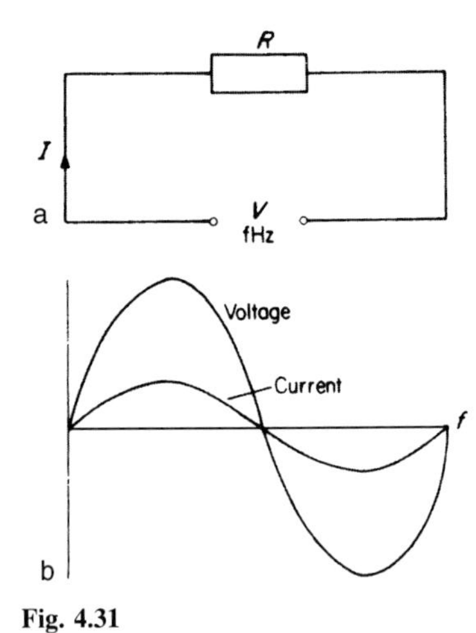

Fig. 4.31

Representation of current by a phasor diagram

In a purely resistive circuit (Fig. 4.31a) only the magnitude of the current is opposed by the resistance, and as the current and voltage alternate at the same time they are said to be *in phase*. Fig. 4.31 b shows the waveforms of current and voltage in a resistive circuit.

In a purely inductive circuit (Fig. 4.32a) the rate of change of current is opposed by the reactance of the coil, and the effect of this opposition is to make the current lag behind the applied voltage or be *out of phase* by 90°. The waveforms of current and voltage in a purely inductive circuit are shown in Fig. 4.32b.

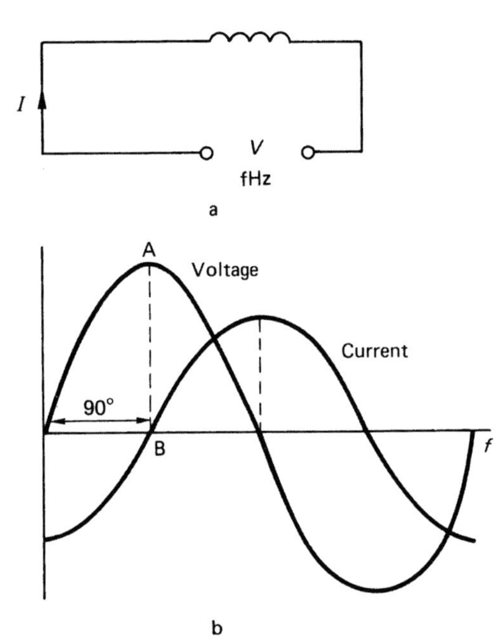

Fig. 4.32

The current lags the voltage by 90°, as V has reached its maximum at point A when current is at zero, point B.

We can represent this effect by means of *phasors* (scaled lines representing electrical quantities). Fig. 4.33 shows the phasor representation of current and voltage in a purely resistive circuit. Fig. 4.34 shows the phasor representation of current and voltage is a purely inductive circuit.

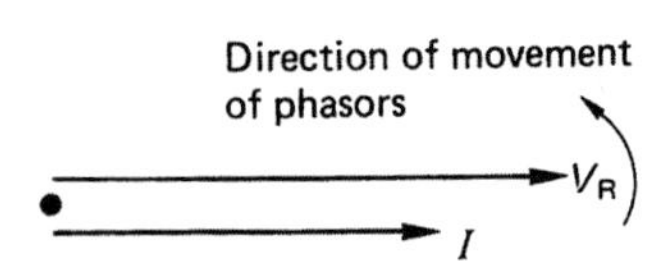

Fig. 4.33 *Voltage and current in phase*

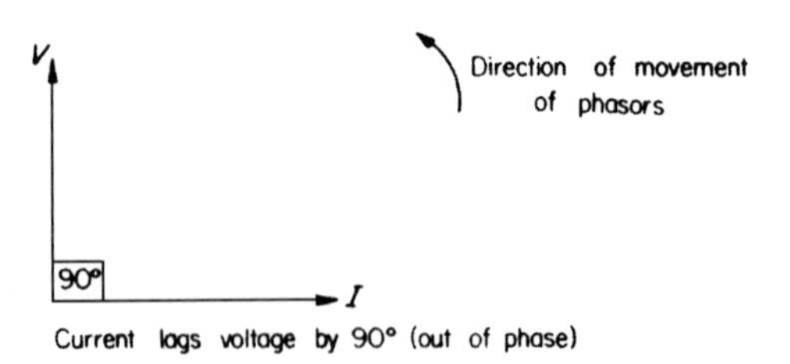

Fig. 4.34

Resistance and inductance in series (R–L circuits)

Consider a coil which has inductance and resistance as shown in Fig. 4.35.

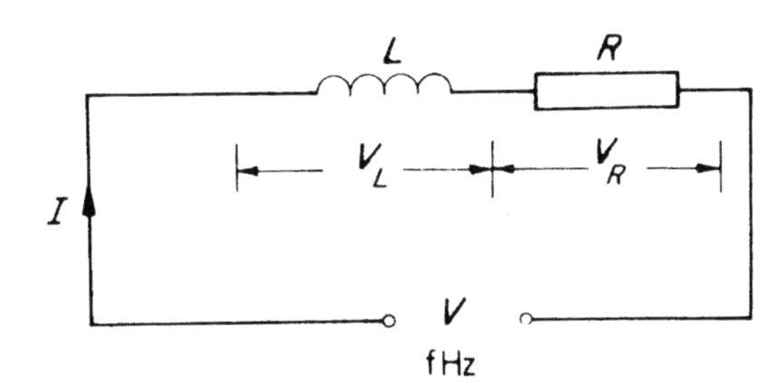

Fig. 4.35

It is clear that the applied voltage, V, comprises the voltage across L, V_L, and the voltage across R, V_R, the current remaining common. However, unlike purely resistive circuits, we cannot merely add V_L to V_R to obtain V. The reason for this is that in the inductive part of the circuit the common current I is out of phase with V_L and in the resistive part I is in phase with V_R. V_L and V_R can only, therefore, be added graphically (or by phasors) as in Fig. 4.36.

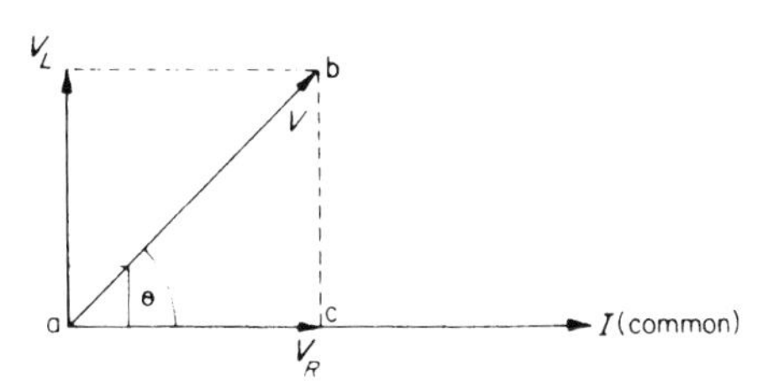

Fig. 4.36

By construction we can see that V_R is in phase with I and V_L is 90° out of phase with I, and that the resultant is the applied voltage V. θ is the number of degrees that I lags behind the applied voltage V.

From Fig. 4.36 it can be seen that the triangle abc is right angled, having ab = V, bc = V_L and ac = V_R. Hence by Pythagoras' theorem:

$$\text{ab} = \sqrt{\text{ac}^2 + \text{bc}^2}$$

$$\therefore V = \sqrt{V_R^2 + V_L^2}$$

Note: This formula need not be remembered; it is simply a useful check after a value has been obtained with the aid of a phasor diagram.

Impedance: symbol, Z; unit, ohm (Ω)

It is clear that there are two separate oppositions to the flow of current in an R–L circuit, one due to resistance and the other due to reactance.

The combination of these oppositions is called the *impedance* of the circuit: its symbol is Z and it is measured in ohms.

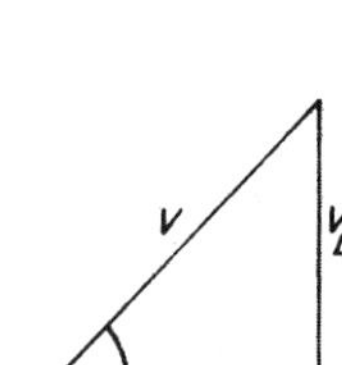

Fig. 4.37

Impedance may be defined as the *total* opposition offered by the components in that circuit. Ohm's law may once again be applied:

$$Z = \frac{V}{I}$$

where V is the applied voltage of the whole circuit.

Impedance triangle

From Fig. 4.37, it can be seen that a triangle can represent all the voltages in the circuit.

If we now divide these voltages by the current I which is common to all components in the circuit the triangle can be shown as in Fig. 4.38.

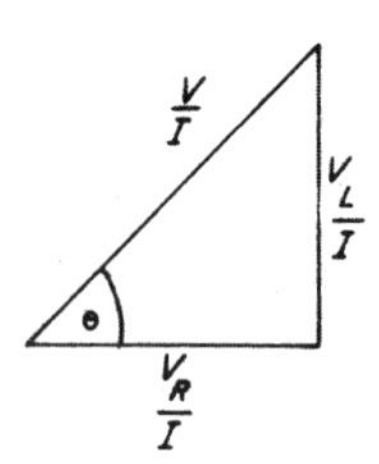

Fig. 4.38

$$\frac{V}{I} = Z \quad \frac{V_L}{I} = X_L \quad \frac{V_R}{I} = R$$

Therefore the triangle can be shown as in Fig. 4.39. This triangle is called the *impedance triangle*. Applying Pythagoras' theorem:

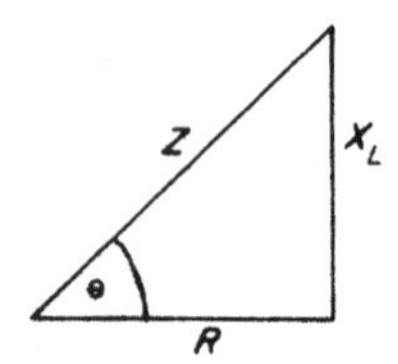

Fig. 4.39

$$Z = \sqrt{R^2 + X_L^2}$$

Example

A choke coil has a resistance of $6\,\Omega$ and an inductance of 25.5 mH. If the current flowing in the coil is 10 A when connected to a 50 Hz supply (Fig. 4.40), find the supply voltage V.

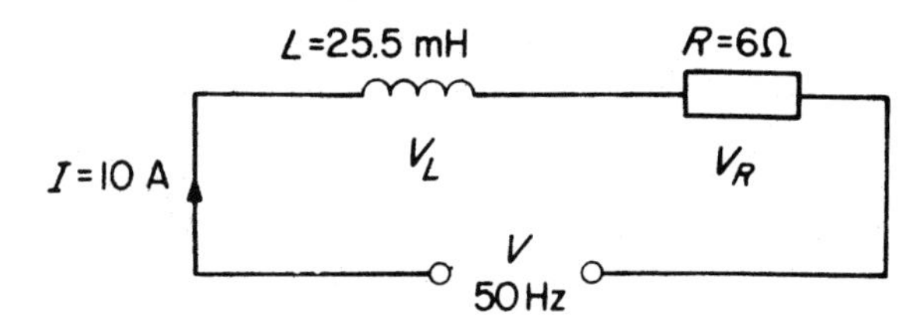

Fig. 4.40

In order to solve the problem by the use of phasors it is necessary to know the values of V_L and V_R. Hence,

$$V_R = I \times R = 10 \times 6 = 60\,\text{V}$$

$$V_L = I \times X_L$$

$$X_L = 2\pi fL$$

$$= 2\pi \times 50 \times 25.5 \times 10^{-3}$$

$$= 8\,\Omega$$

$$\therefore V_L = I \times X_L$$

$$= 10 \times 8$$

$$= 80\,\text{V}$$

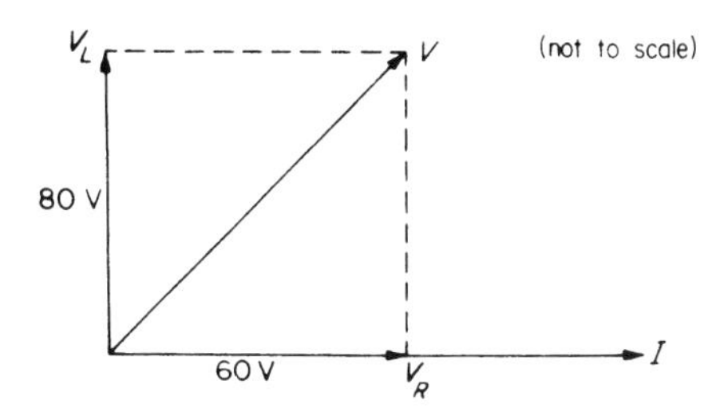

Fig. 4.41

By phasors we can draw Fig. 4.41 (scale 1 cm = 10 V).

By measurement V = 100 V. Check by Pythagoras' theorem:

$$Z = \sqrt{R^2 + X_L^2}$$

$$= \sqrt{6^2 + 8^2}$$

$$= \sqrt{100}$$

$$= 10\,\Omega$$

$$V = I \times Z$$

$$\therefore V = 10 \times 10 = 100\,\text{V}$$

Resistance and inductance in parallel

Again, unlike resistive circuits, currents in parallel branches of an R–L circuit cannot simply be added to find the total current. It will be seen from Fig. 4.42a that the common quantity in the circuit is the voltage. This is used as the reference phasor (as the current was in the series circuit).

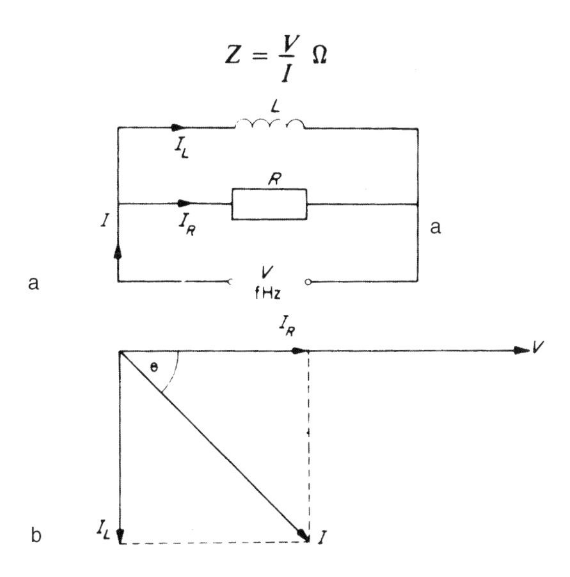

Fig. 4.42

The current in the resistive branch I_R is in phase with the applied voltage and the current in the inductive branch I_L lags the applied voltage by 90°. The resultant of these two currents is the supply current I (Fig. 4.42b). The impedance Z is given by:

$$Z = \frac{V}{I}\ \Omega$$

Example

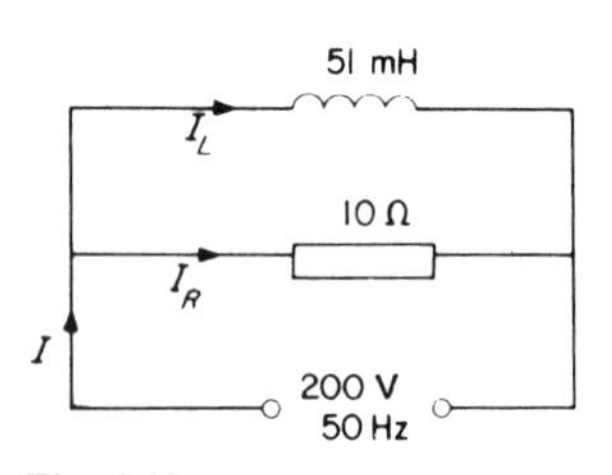

Fig. 4.43

A non-resistive inductor of 51 mH is connected to a non-inductive resistor of 10 Ω across a 200 V, 50 Hz supply. What is the value of the supply current and impedance of the circuit (Fig. 4.43)?

In order to find I, the values of I_R and I_L must be found:

$$I_L = \frac{V}{X_L} \text{ and } I_R = \frac{V}{R}$$

$$X_L = 2\pi fL$$

$$= 2\pi \times 50 \times 51 \times 10^{-3}$$

$$= 16\ \Omega$$

$$\therefore I_L = \frac{V}{X_L} = \frac{200}{16}$$

$$= 12.5\ \text{A}$$

$$I_R = \frac{V}{R} = \frac{200}{10}$$

$$= 20\ \text{A}$$

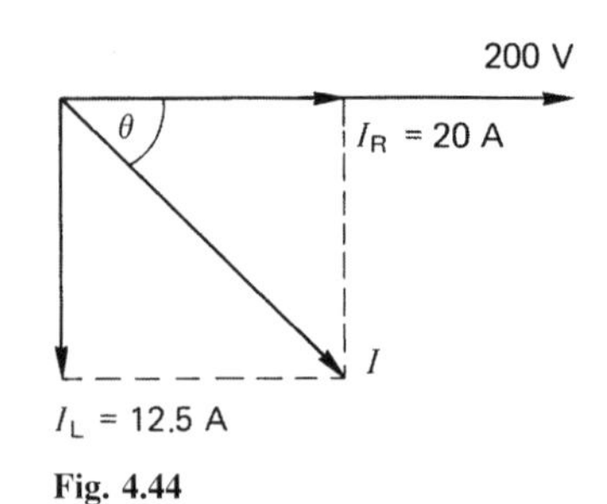

Fig. 4.44

By phasor diagram (Fig. 4.44) $I = 23.7$ A:

$$Z = \frac{V}{I}$$

$$= \frac{200}{23.7}$$

$$= 8.45\ \Omega$$

Power in a.c. circuits

All the power (watts) in a circuit is dissipated in the circuit resistance (Chapter 3). The purely inductive part of the circuit consumes *no* power; it only provides a magnetic field.

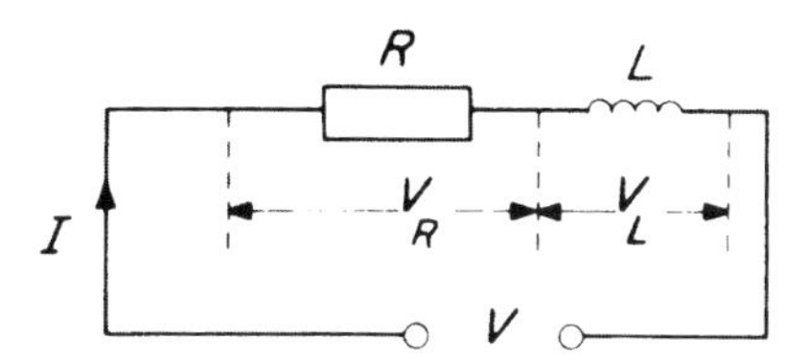

Fig. 4.45

Consider the circuit shown in Fig. 4.45. We have already seen the phasor diagram of the voltage in the circuit and developed a voltage triangle from it. This time we *multiply* the voltage by the common current (Fig. 4.46).

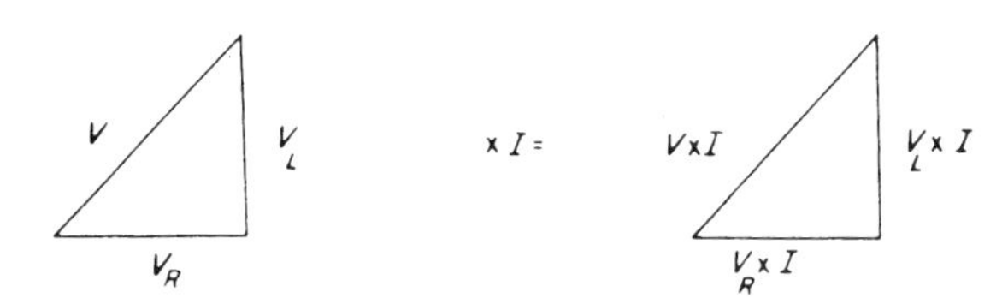

Fig. 4.46

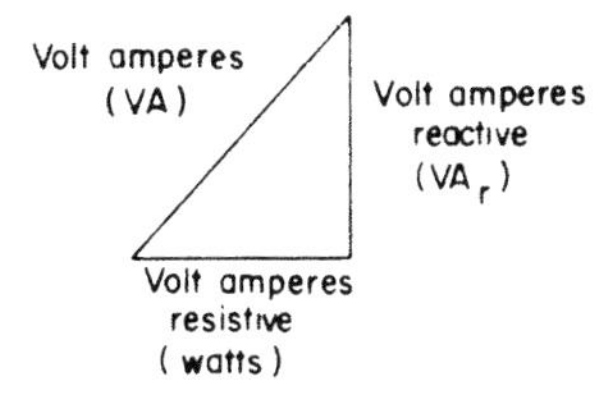

Fig. 4.47

But voltage × current = power. Therefore the triangle becomes as shown in Fig. 4.47.

This is called the *power triangle* but as the inductive part of the circuit consumes no power, it is called the *wattless component of power (VAr)*. The resistive part is called the *wattful component* or *true power* and the combination of the two is known as the *apparent power* (VA). The power triangle is usually shown in terms of kVA, kW and kVA (Fig. 4.48).

The relationship between the true power and the apparent power is very important.

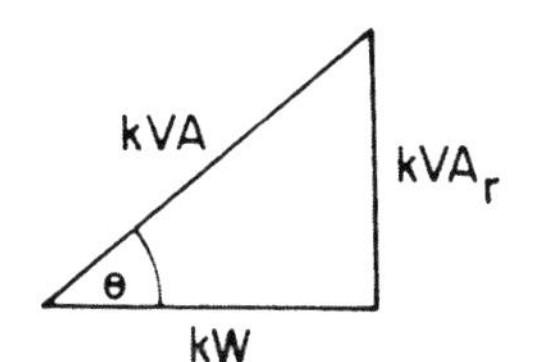

Fig. 4.48

Power factor

The ratio of the kW (true power) to the kVA (apparent power) is called the power factor (PF):

$$\text{PF} = \frac{\text{kW}}{\text{kVA}}$$

By trigonometry

$$\frac{\text{kW}}{\text{kVA}} = \cos\theta$$

$$\therefore \text{PF} = \cos\theta$$

As the voltage, impedance, and power triangles have the same angles, cos θ is either

$$\frac{V_R}{V} \text{ or } \frac{R}{Z} \text{ or } \frac{\text{kW}}{\text{kVA}}$$

all of which equal the power factor.

As the original triangle was formed from the phasor diagram (Fig. 4.36), θ is the angle between the current and the supply voltage, and therefore PF may be defined as: 'the cosine of the angle of phase difference between the current and the applied voltage'.

Power factors in inductive circuits are termed *lagging* as the current lags the voltage.

When the true power equals the apparent power, the PF of 1 is usually referred to as *unity*. Under these circumstances there would be no wattless power (kVA*r*) and the current taken by the circuit would be at a minimum. This is clearly an ideal situation.

The beer analogy

This is a useful way to explain the power factor. Fig. 4.49 shows a pint beer glass with the main body of beer and the head. Although the glass is full, part of it is useless (remember this is only an analogy) and the true amount of beer is less than a pint. A ratio of true to apparent beer would indicate how much head there was. So, if this ratio (pint factor) were 1 or unity there would be no head, and a PF of 0.5 would mean half beer and half head. Clearly, it is better to have a PF close to unity.

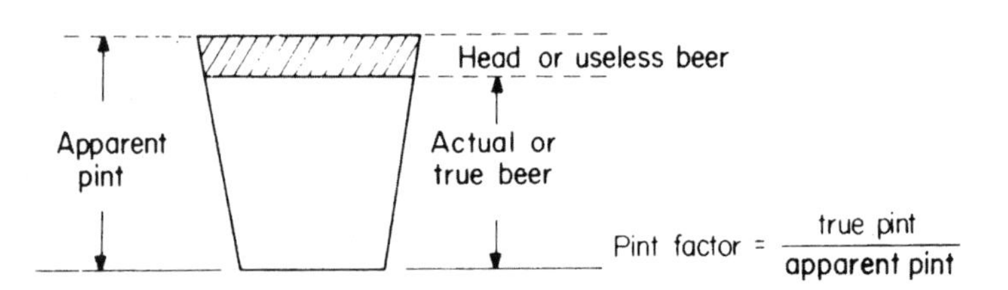

Fig. 4.49 *Beer analogy*

All large plant (motors, transformers, etc.) are rated in kVA, unlike most domestic appliances which are rated in kW. The reason for this is best explained by an example.

Example

If a heating appliance has a power rating of 1 kW at 240 V it will take a current of

$$I = \frac{P}{V}$$

$$= \frac{1000}{240}$$

$$= 4.17\,\text{A}$$

But if a motor has a power rating of 1 kW at 240 V and the motor windings cause of PF of 0.6, then as

$$\text{PF} = \frac{\text{kW}}{\text{kVA}}$$

$$\text{kVA} = \frac{\text{kW}}{\text{PF}}$$

$$= \frac{1}{0.6}$$

$$= 1.667\,\text{kVA}$$

$$= 1667\,\text{VA}$$

and since

$$\text{current} = \frac{\text{volt amperes}}{\text{volts}}$$

$$I = \frac{1667}{240}$$

$$= 7\,\text{A}$$

Had the cable supplying the motor been rated on the kW value it would clearly have been undersized.

Applications of inductance

Inductors, or chokes as they are more popularly called, are used in many areas of modern technology. In electrical installation work the main application is in fluorescent lighting, where the choke is open-circuited across the ends of the tube to cause it to strike. This effect is discussed further in Chapter 10. Motor windings are also inductances.

Transformers

As we have seen earlier, two coils that are wound on the same iron core have the property of mutual inductance, because a change in flux, and hence in e.m.f., in one coil produces, via the iron core, a corresponding change in the other coil.

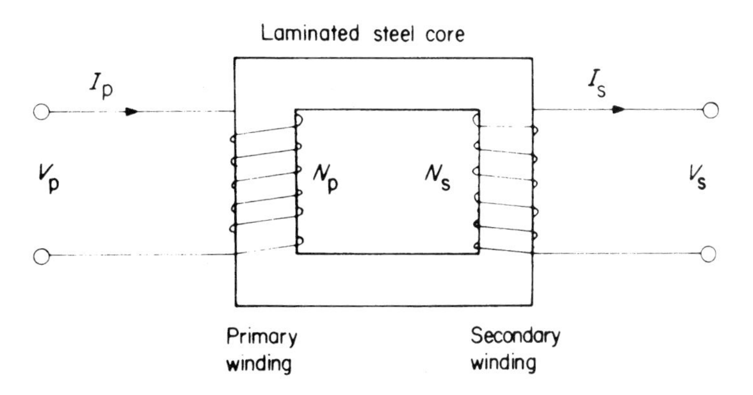

Fig. 4.50 *Simple transformer*

If we take the same arrangement and apply an alternating voltage to one coil, it will induce an alternating e.m.f. in the other coil; this is called the *transformer effect*. The coil or winding to which the supply is connected is called the *primary* and the winding from which the induced voltage is taken is called the *secondary* (Fig. 4.50).

The relationship between the voltage, current and number of turns for each winding is as follows:

$$\frac{V_p}{V_s} = \frac{N_p}{N_s} = \frac{I_s}{I_p}$$

where: V_p = primary voltage; I_p = primary current; N_p = primary turns; V_s = secondary voltage; I_s = secondary current, and N_s = secondary turns.

Transformers which have a greater secondary voltage are called *step-up* transformers, while those with a smaller secondary voltage are called *step-down* transformers.

Example

A single-phase step-down transformer has 796 turns on the primary and 365 turns on the secondary winding. If the primary voltage is 240 V calculate the secondary voltage. Also calculate the secondary current if the primary current is 10 A.

$$\frac{V_p}{V_s} = \frac{N_p}{N_s}$$

$$\therefore V_s = \frac{V_p \times N_s}{N_p}$$

$$= \frac{240 \times 365}{796}$$

$$= 110\,\text{V}$$

$$\text{also} \quad \frac{V_p}{V_s} = \frac{I_s}{I_p}$$

$$\therefore I_s = \frac{I_p \times V_p}{V_s}$$

$$= \frac{10 \times 240}{110}$$

$$= 21.82\,\text{A}$$

Note the larger secondary current. The secondary winding would need to have a larger conductor size than the primary winding to carry this current. If the transformer were of the step-up type the secondary current would be smaller.

Types of transformer

Double wound

This type is constructed as shown in Fig. 4.50. Two electrically separate coils are wound on to a common silicon steel core.

The core is laminated to lessen the effects of eddy currents and silicon steel is preferred, as there are few losses due to hysteresis. These losses are dealt with later in this section.

The double-wound transformer is the commonest form of transformer, and has a wide range of applications.

Auto-transformer

In this type of transformer a single coil is wound on to a steel core, the primary and secondary windings being part of one winding (Fig. 4.51).

The main use of this type of transformer is in the grid system. When 400 000 V (400 kV) has to be transformed (stepped down) to 132 kV, huge transformers are required. Auto-transformers are used mainly because, as there is only one winding, a great saving in copper and hence expense is achieved.

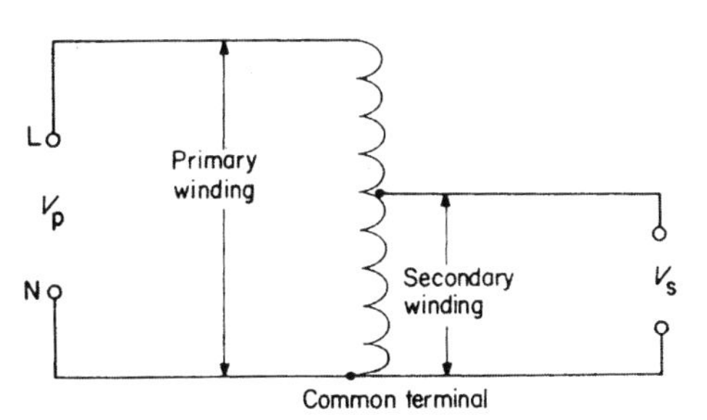

Fig. 4.51 *Auto-transformer*

The main disadvantage in using auto-transformers for applications such as bells or train sets, etc., is that the primary and secondary windings are *not* electrically separate and a short circuit on the upper part of the winding (see Fig. 4.51) would result in the whole of the primary voltage appearing across the secondary terminals.

The same ratio applies between the voltages, currents and number of turns.

The current transformer

The action of this transformer is the same as those previously discussed. It is a step-up (voltage) transformer and is used extensively for taking measurements. The most common form is the bar primary type (Fig. 4.52).

It is clearly impracticable to construct an instrument to measure currents as high as, say, 200–300 A, so a current transformer is used to step down the secondary current to a value which can be measured on a standard instrument.

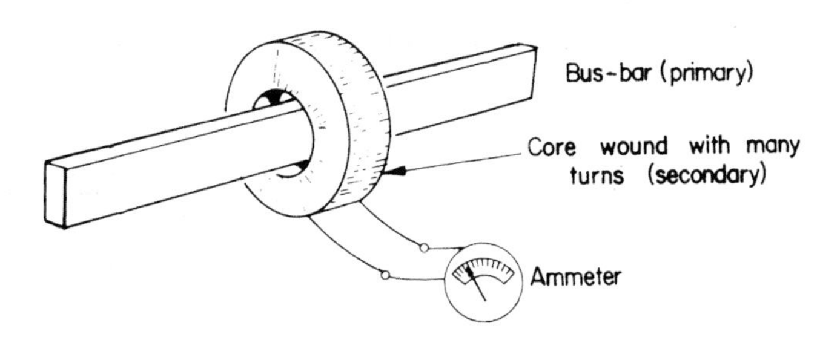

Fig. 4.52 *Bar primary transformer*

Example

A current of 300 A flowing in a bus-bar needs to be measured. The ammeter available has a maximum rating of 0.5 A. How many turns on the secondary of a current transformer would be required to measure the primary current, where $N_p = 1$ (single bus-bar), $I_p = 300\,A$ and $I_s = 0.5\,A$ (instrument rating).

$$\frac{I_s}{I_p} = \frac{N_p}{N_s}$$

$$N_s = \frac{N_p \times I_p}{I_s}$$

$$N = \frac{1 \times 300}{0.5}$$

$$= 600 \text{ turns}$$

Transformer losses

Ideally, the power input to a transformer ($I_p \times V_p$) should equal the power output ($I_s \times V_s$). However, there are power losses which reduce the efficiency. These losses are copper, eddy current and hysteresis losses.

Copper ($I^2 R$) loss

Current flowing in the copper windings causes a heating loss.

Eddy current loss

This loss is caused by alternating currents which are induced magnetically in the core. They are reduced by laminating.

Hysteresis loss

This is an energy loss due to the changing magnetism in the core.

If we take a sample of unmagnetized iron (Fig. 4.53), wind a coil on it and pass a current through the winding, the core will become magnetized. The density of the flux will depend on the current and the number of turns. The product of the current and the turns is called the *magnetizing force* (H).

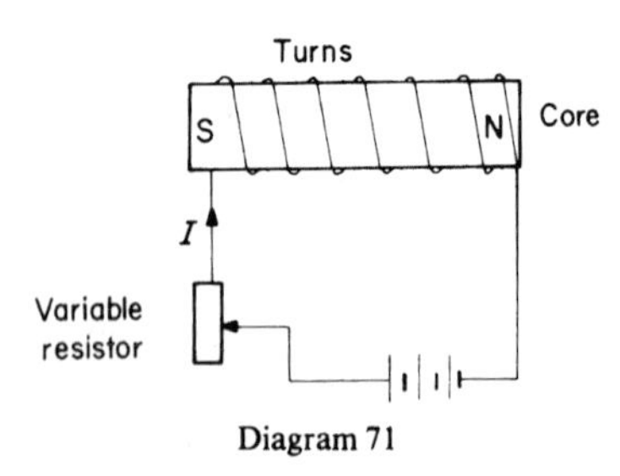

Fig. 4.53

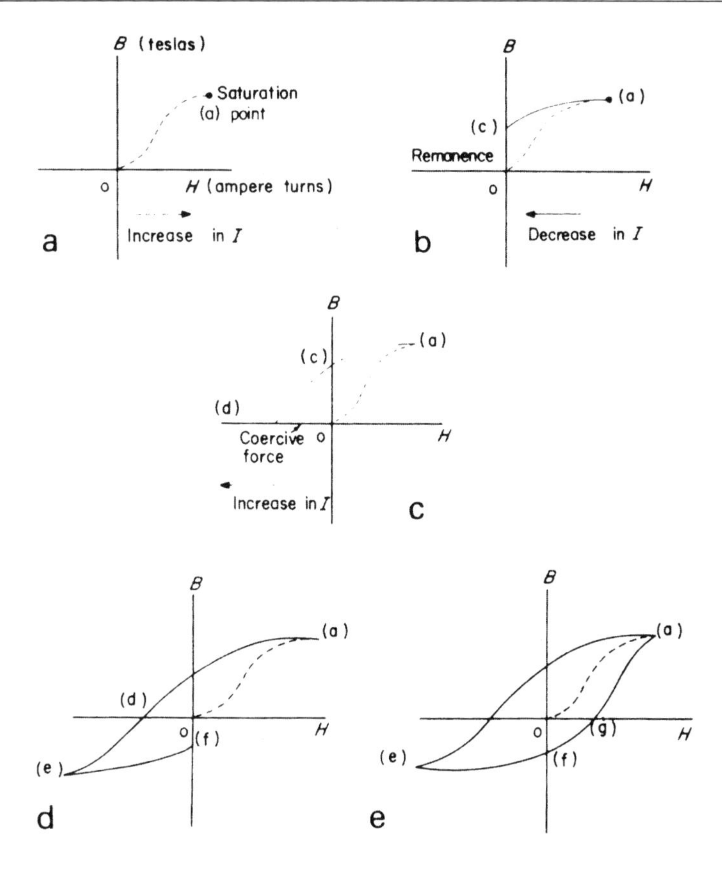

Fig. 4.54

Clearly, once the coil has been wound it is only the current that is variable, and if it is increased the core becomes magnetized. A graph of this effect is shown in Fig. 4.54a–e. The current in the coil is increased from zero to saturation point. Beyond this, an increase in current does not increase the magnetism (Fig. 4.54a). If the current is now decreased to zero (see Fig. 4.54b), its path is along (a,c) not (a,o), leaving the core slightly magnetized. This remaining magnetism (o,c) is called the residual magnetism or *remanence*.

If the polarity of the supply is now reversed, and the current increased again, the current follows (c,d) (see Fig. 4.54c). Clearly some force has been used to reduce the remanence to zero (o,d). This force is called the *coercive force*. This would not have occurred if the current had followed the original route (o,a). Hence, energy has been used to overcome the remanence. This is an *energy loss*.

If the current is further increased in the same direction, saturation point will be reached again (e) and the core will have reversed polarity. Decreasing the current to zero will result in a remanence (o,f) which has to be overcome (Fig. 4.54d).

Another reversal of polarity and an increase in current will result in a coercive force (o,g) being used (energy loss). A further increase in current will bring the curve back to (a) (Fig. 4.54). This complete curve is called the *hysteresis loop* (Fig. 4.55).

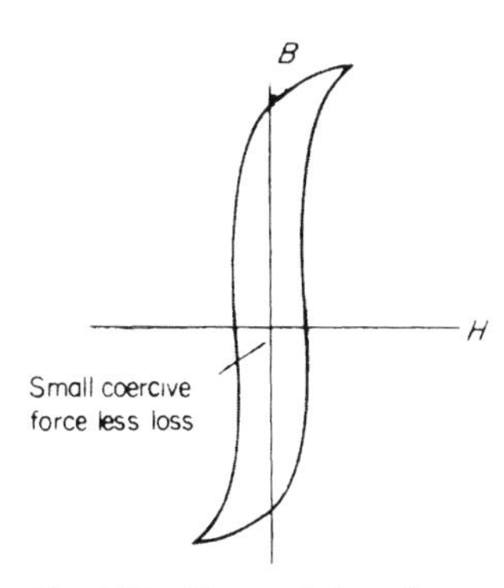

Fig. 4.55 *Hysteresis loop for transformer core*

If the supply to the coil were alternating current, the polarity would be changing constantly and there would be a continual energy loss. It can be minimized to some extent by using a silicon steel core, the remanence of which is easily overcome.

Self-assessment questions

1 A magnetic flux density of a circular field is 27 T. If the flux is 108 mWb, calculate the area of the field.
2 A magnetic field has a flux of 54 mWb and a cross-sectional area of 9 cm^2. Calculate the field's flux density.
3 A conductor 11 cm long is lying at right angles to a magnetic field of flux density 10 T; if the conductor carries a current of 10.91 A, calculate the force on the conductor.
4 A conductor 8 cm long connected to a 50 V d.c. supply is situated at right angles to a magnetic field of flux 30 mWb and cross-sectional area 15 cm^2. If the force exerted on the conductor is 16 N, calculate the resistance of the conductor.
5 With the aid of diagrams, explain why a force is exerted on a conductor carrying a current in a magnetic field.
6 Give three examples of how the magnetic effect of an electric current may be used, using explanatory diagrams.
7 With the aid of sketches, explain the difference between simple d.c. and a.c. generators.
8 Draw a diagram showing the arrangement of a typical single-phase a.c. generator and explain its action.
9 An alternating current has a peak value of 50 A. Draw to scale the sine wave of this current over half a cycle and from it determine the value of the current after 70°. What is the r.m.s. value of the current?
10 Explain the meaning of the term 'root-mean-square value.'
11 What is the meaning of the term 'frequency'? Illustrate your answer with sketches.
12 What is meant by three-phase generation?
13 If a coil has an e.m.f. of 6 V induced in it by a flux changing from zero to 36 mWb in 0.18 s, calculate the number of turns on the coil.
14 (a) What is self-inductance? (b) A relay coil of 300 turns produces a flux of 5 mWb when carrying a current of 1.5 A. Calculate the inductance of the coil.
15 An iron-cored coil having an inductance of 0.1 H and a resistance of 1.25 Ω is connected to a 25 V d.c. supply. Calculate the circuit time constant and the maximum current. Draw to scale the curve of the current decay when the supply is switched off. What will be the value of current after 0.15 s?
16 (a) What is the effect of opening an inductive circuit? How can this effect be used? (b) The energy stored in a coil is 2 J; if the inductance of the coil is 160 mH, calculate the coil current.
17 An inductor of negligible resistance has an inductance of 100 mH and an inductive reactance of 31.42 Ω when connected to an a.c. supply. Calculate the supply frequency.

18 A coil of inductance 1528 mH and negligible resistance takes a current of 0.5 A when connected to a 50 Hz supply. Calculate the value of the supply voltage.

19 An inductor has a resistance of 16 Ω and an inductive reactance of 12 Ω. If the current flowing in the circuit is 12 A, find, by means of a phasor diagram, the value of the supply voltage.

20 A pure inductance which has a resistance of 12 Ω is wired in parallel with a resistance of 8 Ω across a 240 V supply. Calculate the current in each component and determine the value of the supply current by means of a phasor diagram.

21 (a) Define the term *power factor.* (b) A circuit consists of a resistance and an inductance in series. The voltage across the resistance is 192 V, and the power factor is 0.8. Determine the value of the supply voltage and the voltage across the inductance.

22 An ammeter, a voltmeter and a wattmeter are to be connected into the circuit supply a single-phase motor. Draw a diagram showing how these instruments would be connected. If the readings obtained were 240 V, 1.25 A and 180 W, calculate the power factor of the motor.

23 (a) Explain the action of a transformer. (b) A double-wound transformer has a primary voltage of 240 V and a secondary voltage of 110 V. If there are 720 primary turns, calculate the number of turns on the secondary.

24 With the aid of a diagram explain what an *auto-transformer* is.

25 List the losses which occur in a transformer. What steps can be taken to overcome them?

5 Capacitors and capacitance

Capacitors

A capacitor consists of two metal plates separated by an insulator, called a *dielectric*; the whole assembly is able to store electricity. This store is in the form of an excess of electrons on one plate and a deficiency on the other. In this state the capacitor is said to be charged. The charge is achieved by applying a voltage across the plates.

The use of water is once again excellent for an analogy. Consider Fig. 5.1, which shows a water-filled system where pressure on the plunger P causes the flexible diaphragm D to distort. In this way, energy is stored in the diaphragm, since when the plunger pressure is removed, the diaphragm will cause the plunger to return to normal.

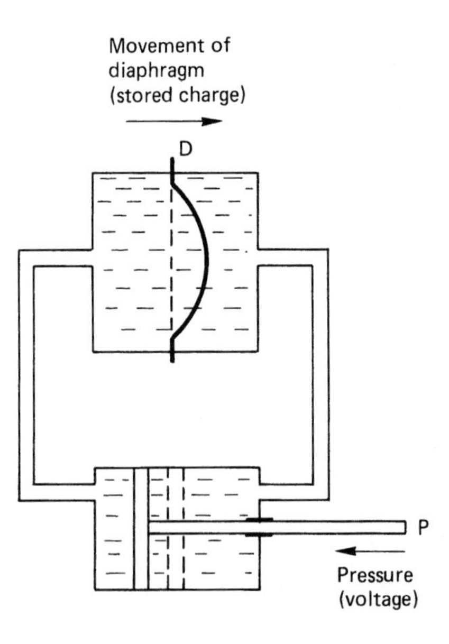

Fig. 5.1

The type of capacitor commonly used in installation work is the electrolytic capacitor. This consists of plates of metal foil placed on either side of a waxed paper dielectric like a sandwich (Fig. 5.2). It is manufactured in a long strip, rolled up and sealed into a metal container.

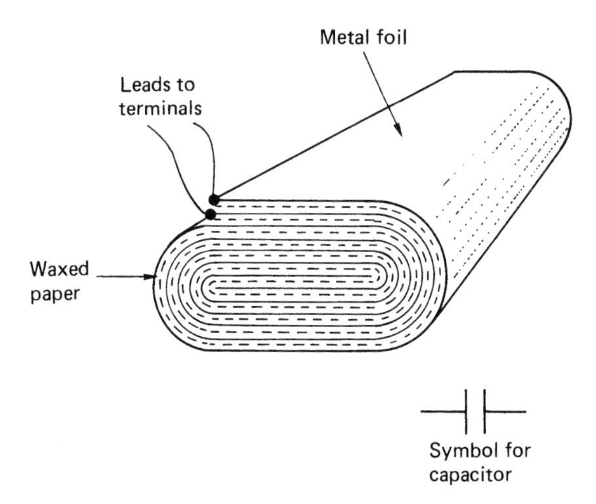

Fig. 5.2

Capacitance: symbol, C; unit, farad (F)

The unit of capacitance is the *farad* and may be defined as: 'the capacitance of a capacitor which requires a potential difference of 1 volt to maintain a charge of 1 coulomb on that capacitor.'

Hence,

charge = capacitance × voltage

$Q(\text{C}) = C(\text{F}) \times V(\text{V})$

Michael Faraday (1791–1867)

British physicist and chemist known as the 'Father of electricity'. He experimented in many different areas of physics but is probably best known for his discovery of electromagnetic induction and hence the transformer.

Example

Calculate the charge on a 50 μF capacitor when it is connected across a 200 V d.c. supply:

$$\begin{aligned} Q &= C \times V \\ &= 50 \times 10^{-6} \times 200 \\ &= 0.01\,\text{C} \end{aligned}$$

Dimensions of capacitors

If we take a simple parallel plate capacitor with an air dielectric, measure its capacitance, and then move the plates further apart, we find that the capacitance is smaller when measured a second time. We can therefore state that an increase in dielectric thickness (d) causes a decrease in capacitance. *Capacitance is inversely proportional to dielectric thickness*

$$C \propto \frac{1}{d}$$

If, however, we were to keep the dielectric thickness constant and to vary the area of the plates (a), we would find that a change in plate area would cause a corresponding change in capacitance. The larger the plate area the larger the capacitance etc. *Capacitance is directly proportional to plate area*:

$$C \propto a$$

Combining these two effects we can see that

$$C \propto \frac{a}{d}$$

Capacitors in series

Consider the effect of connecting three similar capacitors in (Fig. 5.3).

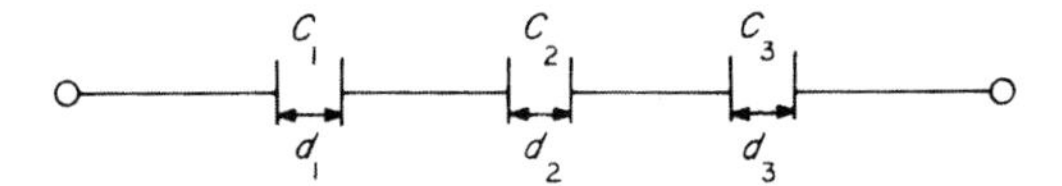

Fig. 5.3

We know that

$$C \propto \frac{1}{d}$$

$$\therefore d \propto \frac{1}{C}$$

So:

$$d_1 \propto \frac{1}{C_1} \quad d_2 \propto \frac{1}{C_2} \quad d_3 \propto \frac{1}{C_3}$$

If we were to combine all the dielectrics, we would have one capacitor of dielectric thickness d_t and capacitance C:

$$d_t \propto \frac{1}{C}$$

But

$$d_t = d_1 + d_2 + d_3$$

$$= \frac{1}{C_1} + \frac{1}{C_2} + \frac{1}{C_3}$$

But

$$d_t = \frac{1}{C}$$

$$\therefore \frac{1}{C} = \frac{1}{C_1} + \frac{1}{C_2} + \frac{1}{C_3} \ldots \text{etc.}$$

Just as the current is common to all parts of a series resistive circuit, so charge is common in a series capacitive circuit. Therefore:

$Q = CV \quad Q = C_1V_1 \quad Q = C_2V_2 \quad Q = C_3V_3 \ldots$ etc.

Example

Three capacitors of 60 μF, 40 μF and 24 μF are connected in series across a 500 V d.c. supply. Calculate the total capacitance and the charge on each capacitor.

$$\frac{1}{C} = \frac{1}{C_1} + \frac{1}{C_2} + \frac{1}{C_3}$$

$$= \frac{1}{60} + \frac{1}{40} + \frac{1}{24}$$

$$\frac{1}{C} = \frac{1}{12}$$

$$\therefore\ C = 12\,\mu\text{F}$$

Q is common to each capacitor:

$$\therefore Q = C\,V$$

$$= 12 \times 10^{-6} \times 500$$

$$= 6 \times 10^{-3}$$

$$= 6\,\text{mC}$$

Capacitors in parallel

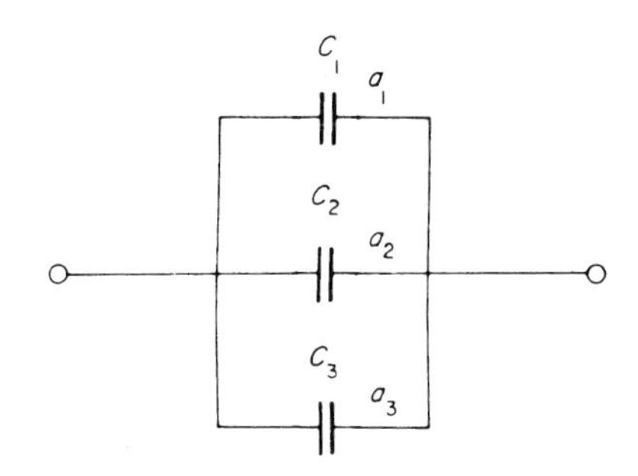

Fig. 5.4

Let us arrange three similar capacitors in parallel (see Fig. 5.4). We know that $C \propto a$. Therefore $C_1 \propto a_1$, $C_2 \propto a_2$ and $C_3 \propto a_3$.

As the plates connected to either side of the supply are common, we could replace the arrangement with one capacitor C of plate area a_T.

$$\therefore a_T = a_1 + a_2 + a_3$$

$$\therefore a_T = C_1 + C_2 + C_3$$

But $C \propto a_T$

$$\therefore C = C_1 + C_2 + C_3$$

In this case it is the voltage that is common and the charge Q behaves like the current in a parallel resistive circuit. So:

$Q = CV \quad Q_1 = C_1V \quad Q_2 = C_2V \quad Q_3 = C_3V$

Example

Three capacitors of 60 μF, 40 μF and 24 μF are connected in parallel across a 500 V supply. Calculate the total capacitance, the total charge and the charge on each capacitor.

$$C = C_1 + C_2 + C_3$$
$$= 60 + 40 + 24$$
$$= 124\,\mu\text{F}$$

$$\text{Total charge } Q = C \times V$$
$$= 124 \times 10^{-6} \times 500$$
$$= 62\,\text{mC}$$
$$Q_1 = C_1\,V$$
$$= 60 \times 10^{-6} \times 500$$
$$= 30\,\text{mC}$$
$$Q_2 = C_2\,V$$
$$= 40 \times 10^{-6} \times 500$$
$$= 20\,\text{mC}$$
$$Q_3 = C_3\,V$$
$$= 24 \times 10^{-6} \times 500$$
$$= 12\,\text{mC}$$

Energy stored in a capacitor

The amount of energy stored in a capacitor is expressed in joules and is given by

$$W = \tfrac{1}{2}\,CV^2$$

Capacitors in d.c. circuits

A capacitor connected across a d.c. supply is shown in Fig. 5.5a. The curves of the current and voltage in the circuit are shown in Fig. 5.5b.

As the capacitor begins to charge, its voltage increases until it is equal to the supply voltage. At the same time the charging current decreases. When the

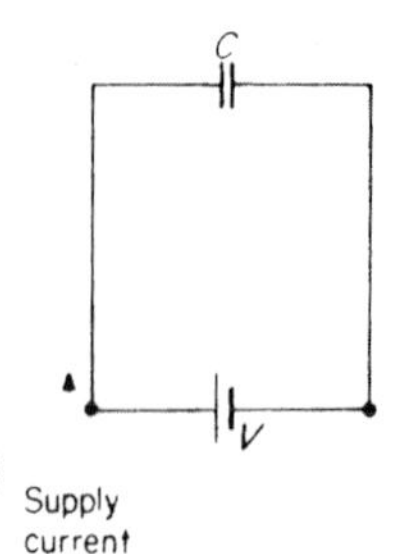

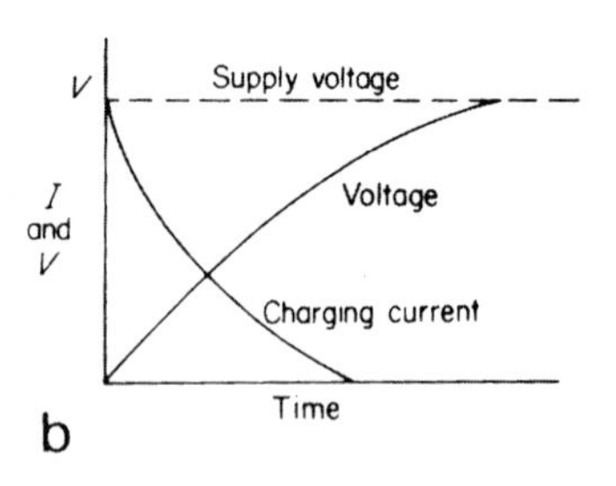

Fig. 5.5 *Charging: (a) capacitor connected across d.c. supply (b) curves of discharge voltage and current;*

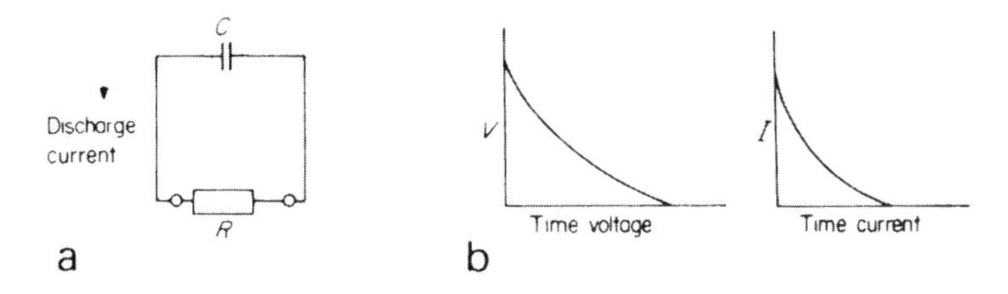

Fig. 5.6 *Discharging: (a) charged capacitor connected across a resistor; (b) curves of discharge voltage and current*

supply voltage and the capacitor voltage are equal the current in the circuit will be zero.

Fig. 5.6a shows the charged capacitor connected across a resistor. (Fig. 5.6b) shows the curves of the discharge voltage and current.

Curves of current and voltage change

These curves are plotted in the same manner as those in inductive circuits (Chapter 4).

Maximum charging or discharging current:

$$I = \frac{V}{R}$$

Time constant:

$$T = CR$$

Capacitance in a.c. circuits

In an a.c. circuit a capacitance has the effect of opposing the voltage, thus causing the circuit current to *lead*. In a purely capacitive circuit the current leads the voltage by 90°. The waveforms and phasors of such a circuit are shown in Fig. 5.7.

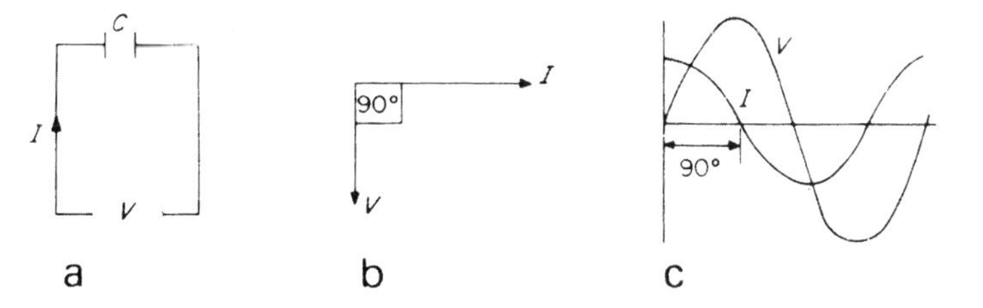

Fig. 5.7 *(a) Circuit diagram; (b) phasor diagram; (c) waveforms*

Capacitive reactance: symbol, X_C; unit, ohm (Ω)

The opposition offered by a capacitor in an a.c. circuit is called the *capacitive reactance* and is given by:

$$X_c = \frac{1}{2\pi fC}$$

where: X_c = capacitive reactance (Ω); f = frequency of supply (Hz); and C = capacitance (F).

As in the case of inductive reactance, Ohm's law may be applied, i.e.

$$V = I \times X_c$$

Example

A purely capacitive circuit of 31.8 μF is connected to a 240 V, 50 Hz supply. Calculate the capacitive reactance and the circuit current.

$$X_c = \frac{1}{2\pi fC}$$

$$= \frac{1}{2\pi \times 50 \times 31.8 \times 10^{-6}}$$

$$= \frac{10^6}{100\pi \times 31.8}$$

$$= 100\,\Omega$$

$$V = I \times X_c$$

$$\therefore I = \frac{V}{X_c}$$

$$= \frac{240}{100}$$

$$= 2.4\,\text{A}$$

Resistance and capacitance in series

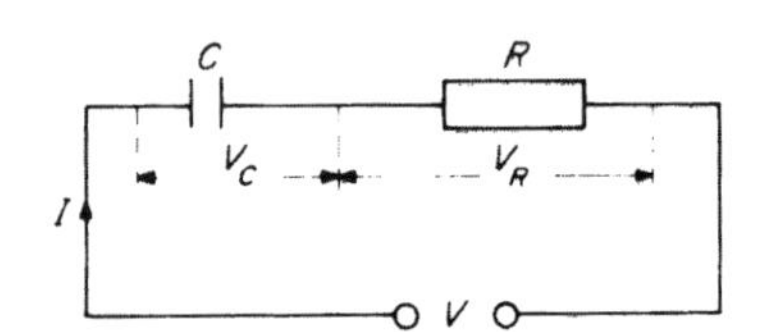

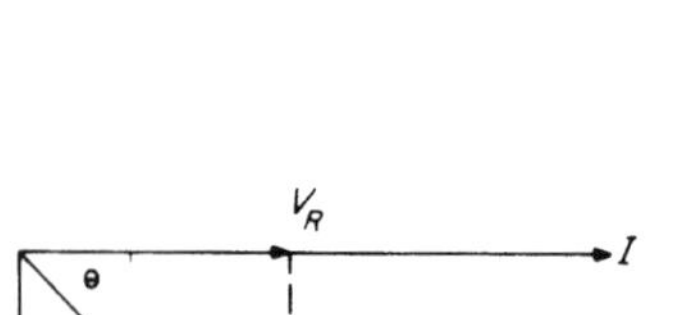

Fig. 5.8 *Capacitor in series with resistor*

Fig. 5.8 shows a capacitor in series with a resistor:

As the current leads the voltage across the capacitor and is in phase with the voltage across the resistor, the phasor diagram may be drawn as shown in Fig. 5.9.

Example

A capacitor of 159 μF is connected in series with a non-inductive resistor of 15 Ω across a 50 Hz supply. If the current drawn is 10 A calculate X_c, the voltage across each component and find by means of a phasor diagram the value of the supply voltage.

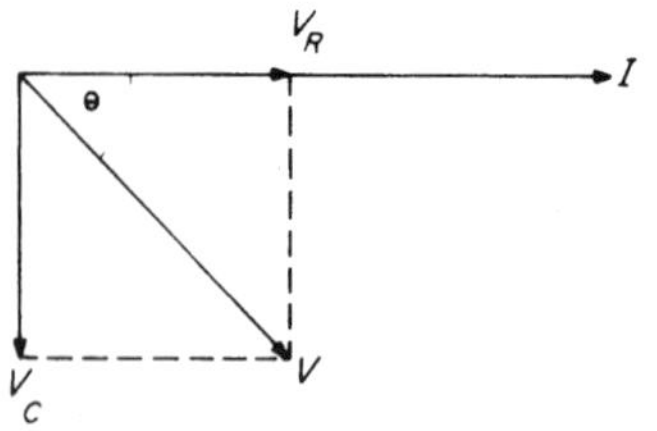

Fig. 5.9

$$X_c = \frac{1}{2\pi fC}$$

$$= \frac{1}{2\pi \times 50 \times 159 \times 10^{-6}} = \frac{10^6}{100\pi \times 159}$$

$$X_c = 20\,\Omega$$

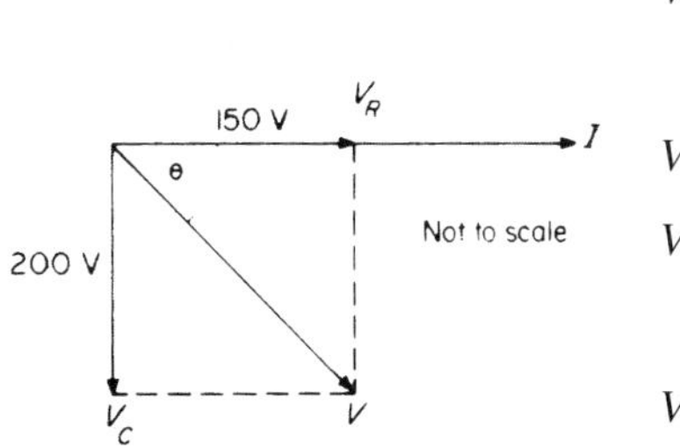

Fig. 5.10

$$V_c = I \times X_c$$
$$= 10 \times 20$$
$$V_c = 200\,\text{V}$$
$$V_R = I \times R$$
$$= 10 \times 5$$
$$V_R = 150\,\text{V}$$

By measurement (Fig. 5.10) V will be found to be 250 V.

Resistance and capacitance in parallel

In this case it is the voltage that is common and the currents that are added by phasors (see Figs 5.11 and 5.12).

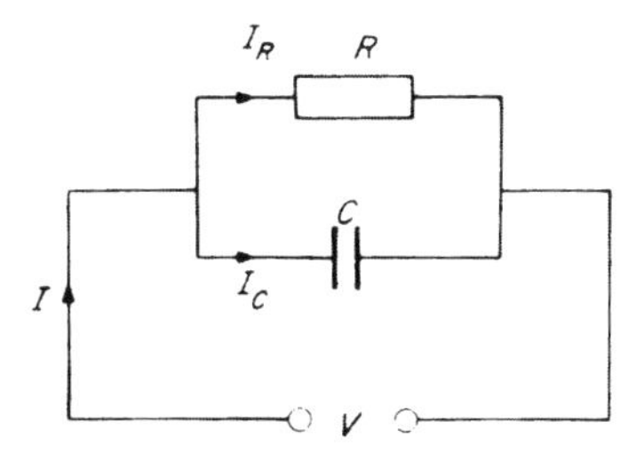

Fig. 5.11

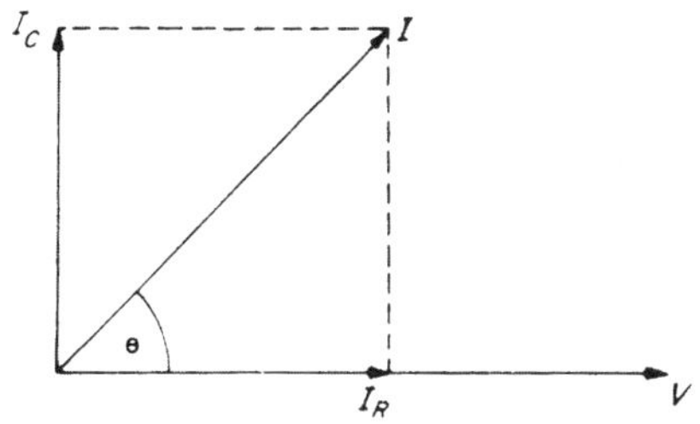

Fig. 5.12

Working voltage

Every capacitor has the value of its working voltage marked on it. Beyond this value the dielectric would break down and the capacitor would be useless.

If we refer back to Fig. 5.1, should the pressure on the plunger be too great, the diaphragm would puncture and the system rendered useless.

Applications of capacitors

Capacitors are used extensively in electrical engineering. In the field of installation work, they are mainly used for motor starting, power factor correction and radio interference suppression and to minimize the stroboscopic effects in fluorescent lighting circuits. Their use in power factor correction and fluorescent lighting is dealt with in Chapter 10.

Self-assessment questions

1 A capacitor has a value of 73 μF and is connected across a 100 V supply. Calculate the charge on the capacitor.

2 Three capacitors of 20 μF, 80 μF and 16 μF are connected in series across a 240 V supply. Calculate the charge.

3 A variable capacitor has a capacitance of 100 μF when the distance between the plates is 1 mm. What will the capacitance be if the plates are adjusted to be 10 mm apart?

4 Four capacitors of 10 μF, 20 μF, 15 μF and 5 μF are connected in parallel across a 240 V supply. Calculate the total capacitance, the total charge, and the charge across the 20 μF capacitor.

5 A capacitor has a value of 150 μF and a plate area of 60 cm^2. What would be the plate area of a similar type of capacitor of 200 μF and the same dielectric thickness?

6 A parallel plate capacitor has a value of 636 μF. At what value of supply frequency would the reactance be 5 Ω?

7 A resistor of 6 Ω is connected in series with a capacitor of 398 μF, the current drawn being 24 A. Calculate the voltage across each component and find the supply voltage by means of a phasor diagram (f = 50 Hz).

8 Draw the waveforms and phasor diagram for an a.c. current and voltage in a purely capacitive circuit.

9 A 127.3 μF capacitor is connected in parallel with a 50 Ω resistor across a 240 V supply. Calculate the current taken by each component. Determine the value of the supply current (f = 50 Hz).

10 Explain with the aid of a sketch the construction of an electrolytic capacitor. What is meant by the term *working voltage*?

6 Resistance, inductance and capacitance in installation work

In Chapters 4 and 5 we have discussed separately the effects of inductance and capacitance in an a.c. circuit. Here we consider how these effects may be applied, and in some cases, combined in a.c. circuits.

Let us first refresh our memories regarding the phasor diagrams for R, L and C.

Pure resistance (Fig. 6.1)

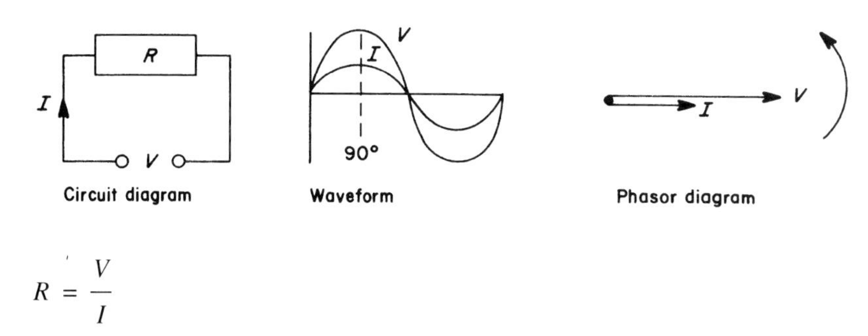

Fig. 6.1

$$R = \frac{V}{I}$$

Pure inductance (Fig. 6.2)

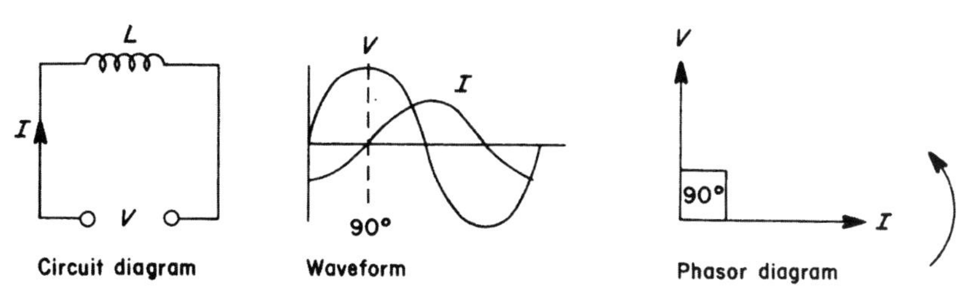

Fig. 6.2

$$X_L = \frac{V}{I}$$

and

$$X_L = 2\pi fL$$

Pure capacitance (Fig. 6.3)

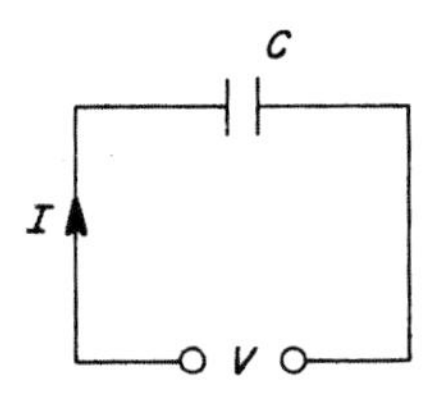

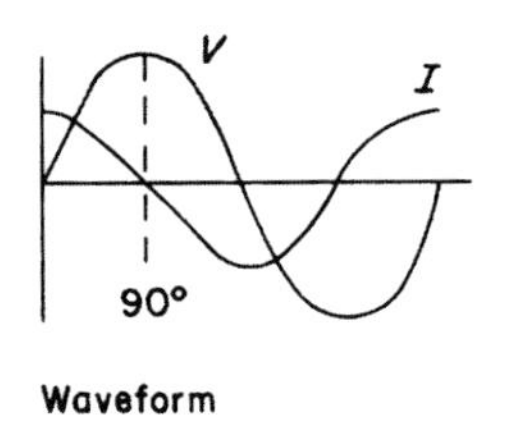

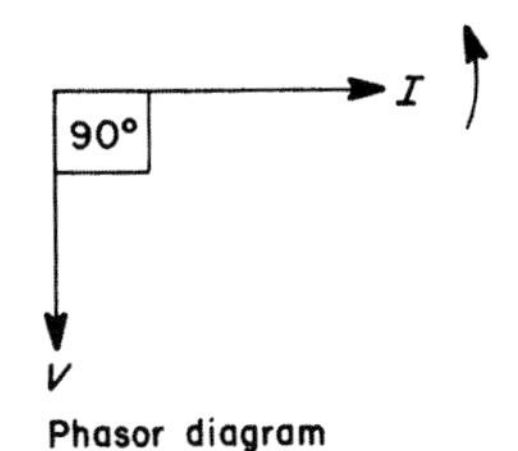

Fig. 6.3

$$X_C = \frac{V}{I}$$

and

$$X_C = \frac{1}{2\pi fC}$$

R and L in series (Fig. 6.4)

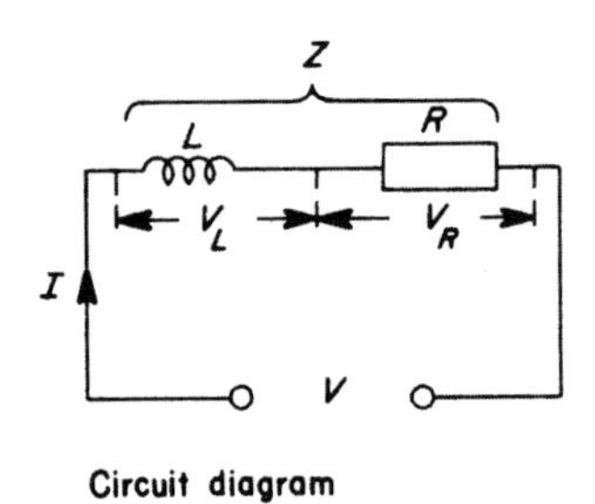

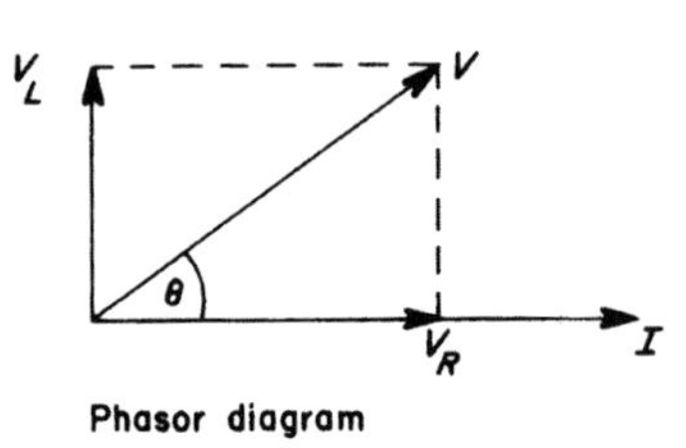

Fig. 6.4

$$Z = \frac{V}{I}$$

From the phasor diagram of voltages (Fig. 6.4), an impedance triangle may be formed (Fig. 6.5).

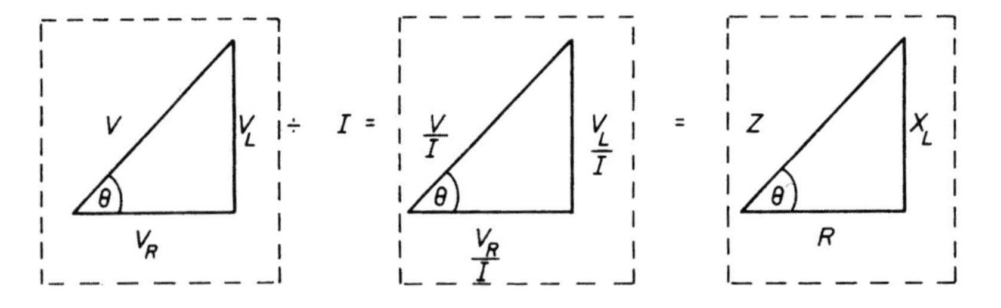

Fig. 6.5

By Pythagoras' theorem:

$$Z = \sqrt{R^2 + X_L^2}$$

Also

$$\cos\theta = \frac{R}{Z} = \text{power factor (PF)}$$

R and C in series (Fig. 6.6)

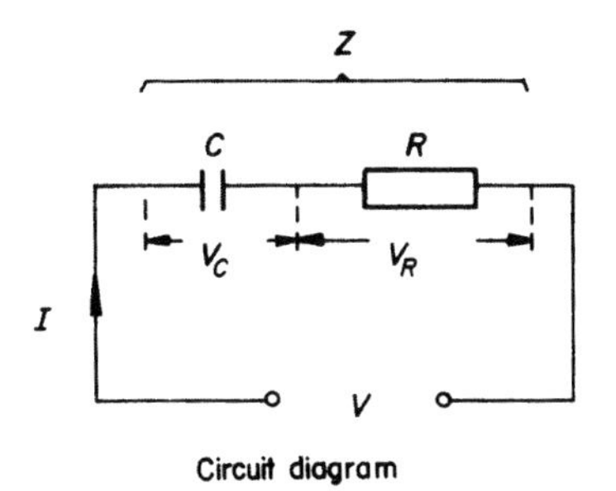

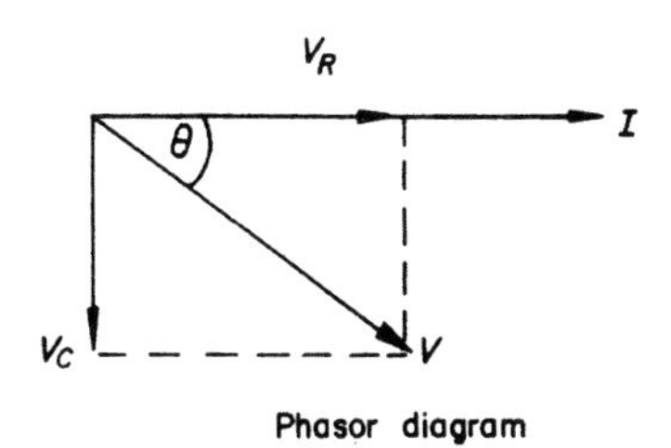

Fig. 6.6

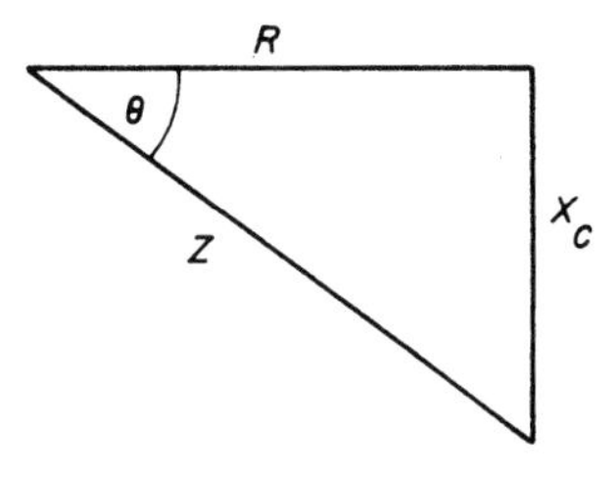

Fig. 6.7

It is clear that a similar impedance triangle may be formed, as shown in Fig. 6.7.

$$\therefore Z = \sqrt{R^2 + X_C^2}$$

and

$$\cos\theta = \frac{R}{Z} = \text{PF}$$

Now we can begin to combine these separate phasor diagrams.

R, L and C in series (Fig. 6.8)

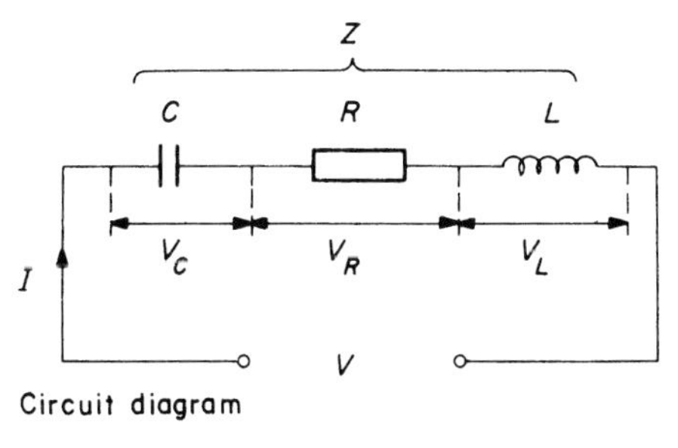

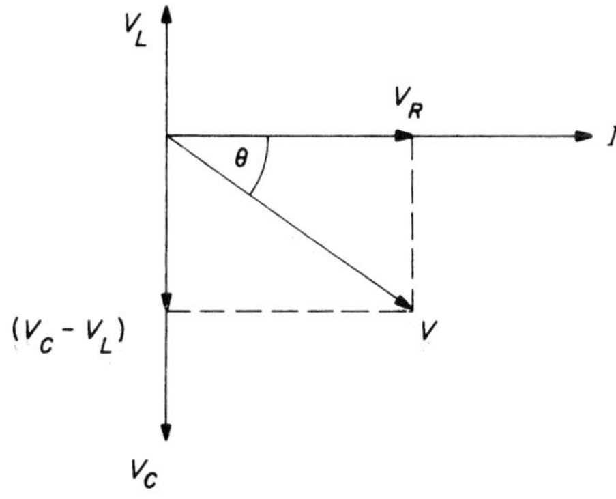

Fig. 6.8

The impedance triangle will be shown in Fig. 6.9.

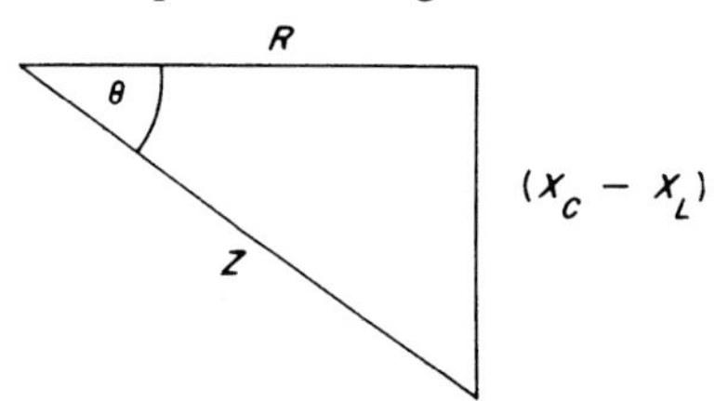

Fig. 6.9

$$\therefore Z = \sqrt{R^2 + (X_C - X_L)^2}$$

or, if V_L is greater than V_C

$$Z = \sqrt{R^2 + (X_L - X_C)^2}$$

and

$$\cos\theta = \frac{R}{Z} = \text{PF}$$

The following problems will be solved using different methods.

Example

From the circuit shown in Fig. 6.10, determine the value of the supply voltage and the power factor.

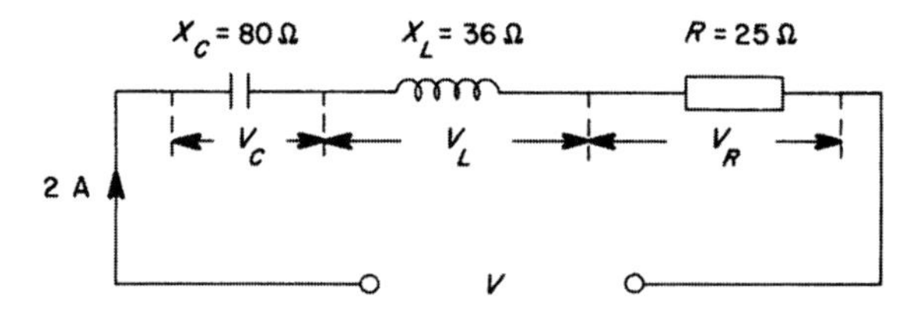

Fig. 6.10

Method 1, by phasors

$$V_C = I \times X_C = 2 \times 80 = 160\,\text{V}$$
$$V_L = I \times X_L = 2 \times 36 = 72\,\text{V}$$
$$V_R = I \times R = 2 \times 25 = 50\,\text{V}$$

From 6.11:

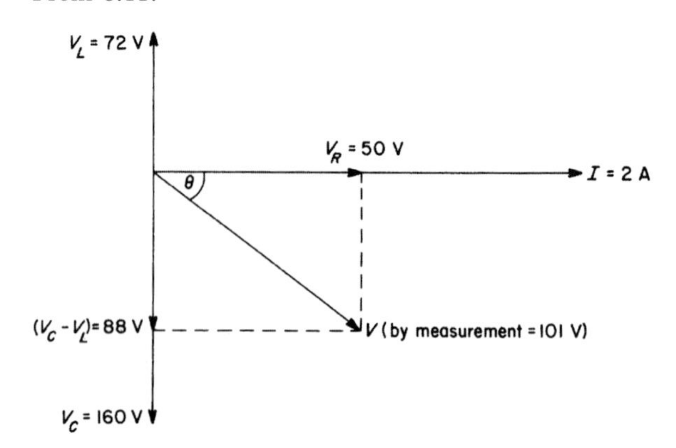

Fig. 6.11

$V = 101\,\text{V}$

By measurement

$\theta = 60.3°$

From cosine tables

$\cos\theta = 0.495$

$\text{PF} = 0.495$ leading

Method 2, using the theorem of Pythagoras (impedance)

$$Z = \frac{V}{I}$$

But

$$\begin{aligned} Z &= \sqrt{R^2 + (X_C - X_L)^2} \\ &= \sqrt{25^2 + (80 - 36)^2} \\ &= \sqrt{25^2 + 44^2} \\ &= \sqrt{2561} \\ &= 50.6\,\Omega \end{aligned}$$

$$\therefore 50.6 = \frac{V}{2}$$

$$\begin{aligned} \therefore V &= 2 \times 50.6 \\ &= 101.2\,\text{V} \end{aligned}$$

$$\text{PF} = \cos\theta = \frac{R}{Z} = \frac{25}{50.6}$$

$$= 0.49 \text{ leading}$$

Method 3, using the theorem of Pythagoras (voltage)

From Fig. 6.12:

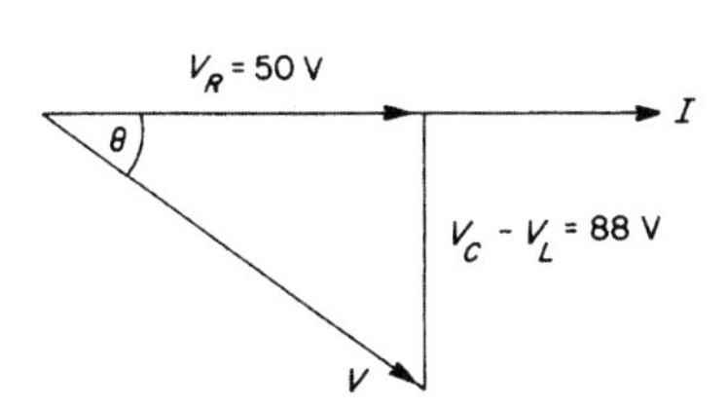

Fig. 6.12

$$\begin{aligned} V &= \sqrt{V_R^2 + (V_C - V_L)^2} \\ &= \sqrt{50^2 + 88^2} \\ &= \sqrt{10\,244} \end{aligned}$$

$$V = 101.2$$

$$\text{PF} = \cos\theta = \frac{\text{base}}{\text{hypotenuse}}$$

$$= \frac{50}{101.2}$$

$$= 0.49 \text{ leading}$$

Method 4, using trigonometry

From Fig. 6.13:

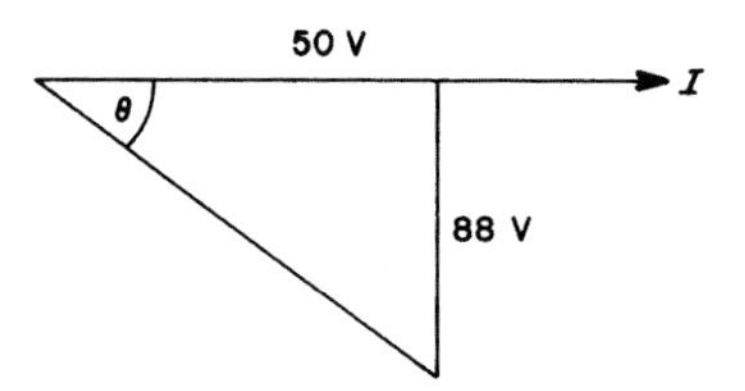

Fig. 6.13

$$\tan \theta = \frac{\text{perpendicular}}{\text{base}}$$

$$= \frac{88}{50}$$

$$= 1.76$$

From tangent tables,

$$\theta = 60.4°$$

$$\cos \theta = \frac{\text{base}}{\text{hypotenuse}} = \frac{50}{V}$$

$$\therefore V = \frac{50}{\cos \theta}$$

From cosine tables,

$$\cos 60.4° = 0.49 \text{ (PF)}$$

$$\therefore V = \frac{50}{0.49}$$

$$= 101.2\,\text{V}$$

$$\text{PF} = \cos \theta = 0.49 \text{ leading}$$

Power factor improvement

The magnetic effect of an inductor has many uses. It is, however, in equipment such as motors and fluorescent lighting that its effect on the power factor is substantial enough to cause concern, and make it necessary to improve the power factor.

Consider the diagram of a resistive and inductive circuit (Fig. 6.14).

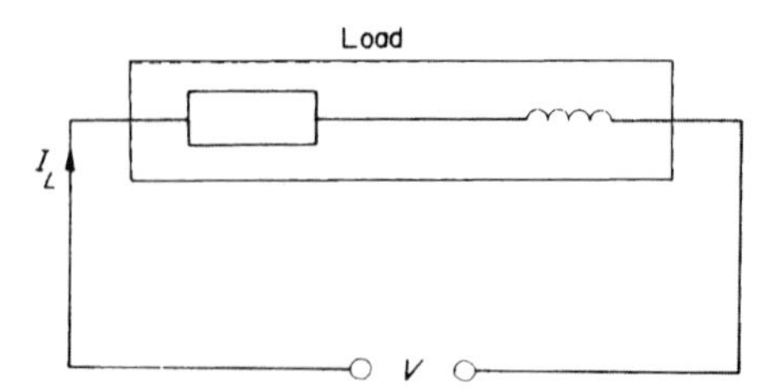

Fig. 6.14

The phasor diagram for this circuit is shown in Fig. 6.15.

We can show the supply current and voltage as in Fig. 6.16.

If we redraw this phasor diagram so that the voltage is drawn horizontally it becomes as shown in Fig. 6.17.

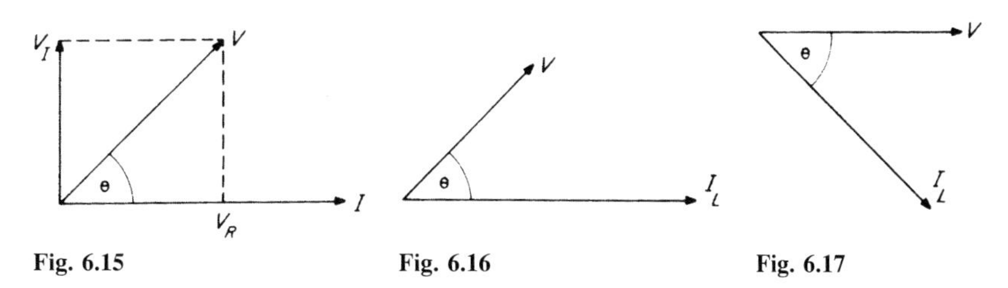

Fig. 6.15 **Fig. 6.16** **Fig. 6.17**

If we now connect a variable capacitor across the supply terminals of the original load we have the result shown in Fig. 6.18.

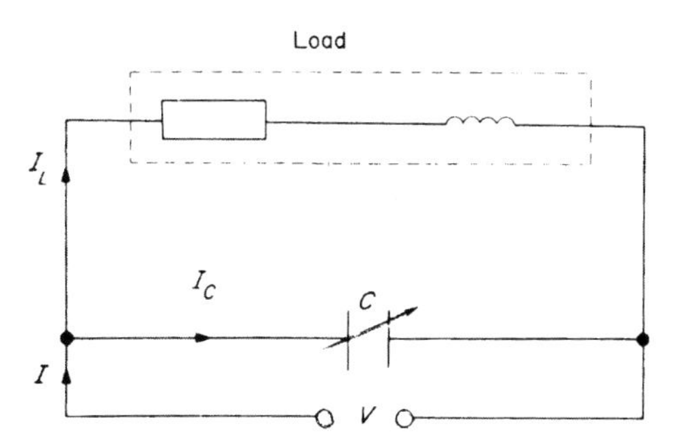

Fig. 6.18

It is a parallel circuit and the voltage is common to both branches. We can therefore draw a current phasor diagram (Fig. 6.19). As V is common, we can combine both diagrams (Fig. 6.20).

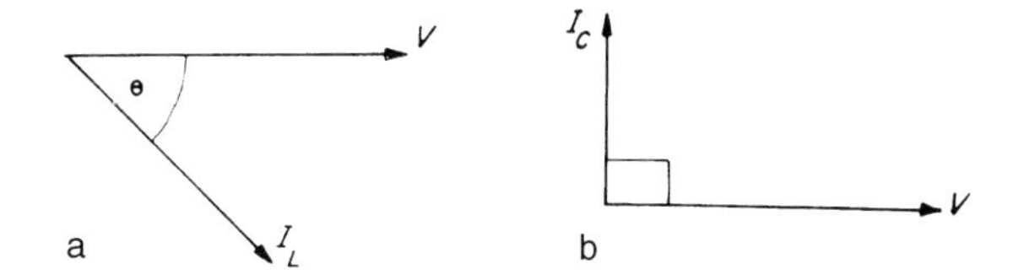

Fig. 6.19 *(a) Phasor diagram for load; (b) phasor diagram for capacitor*

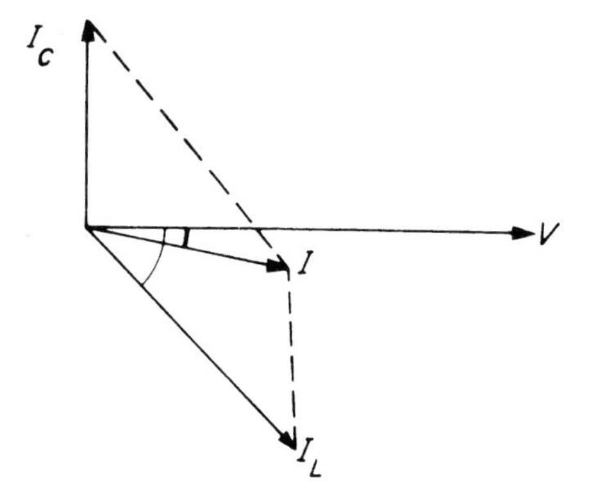

Fig. 6.20

The phasor resultant of I_L and I_C (Fig. 6.20) is clearly I, which is of course the supply current. Note that it is smaller than the load current I_L and that the angle between I and V is smaller than that between I_L and V. The closer an angle is to zero the nearer its cosine is to unity. Therefore, the addition of the capacitor has improved the power factor of the system.

The actual current taken by the load does not change; it is the total supply current that decreases. This means that smaller supply cables may be used. With industrial loads, the supply authority's transformer and switchgear as well as their cable may be reduced in size.

In order to encourage power factor (PF) improvement the supply authorities make a higher charge to consumers who do not correct or improve their PF to a suitable level (usually about 0.95 lagging). It is not usual to improve the PF much beyond this point as the cost of providing extra capacitance required to gain a small decrease in current is uneconomic.

Capacitors are the most popular method of improving the PF although synchronous motors are used occasionally (this is discussed in greater detail in Chapter 8).

PF improvement capacitors may be fitted to individual plant or in banks connected to the supply intake terminals. The first method is more popular as the banked type needs automatic variation as plant is switched on and off.

Example

A 240 V single-phase motor takes a current of 8 A and has a power factor of 0.7 lagging. A capacitor is connected in parallel with the motor and takes a current of 3.1 A.

Draw a scaled phasor diagram of the currents in the circuit and find the value of the supply current, and the new PF.

From Fig. 6.21:

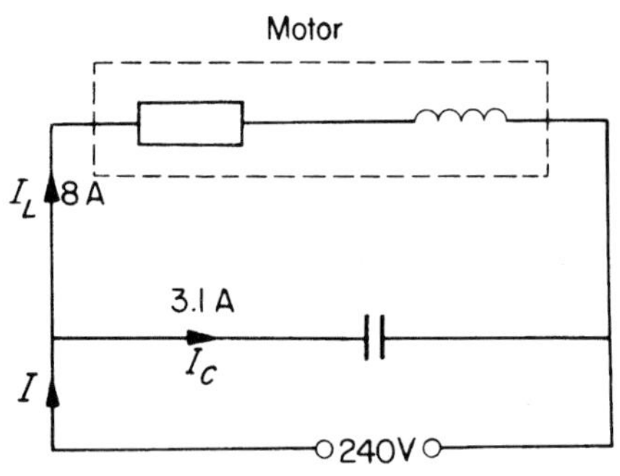

Fig. 6.21

$$I_1 = 8\text{ A}$$

$$\text{PF} = 0.7$$

$$\text{as PF} = \cos\theta$$

$$\text{Then } \cos\theta = 0.7$$

$$\theta = 45.5°$$

Therefore the phasor diagram for the motor is as shown in Fig. 6.22.

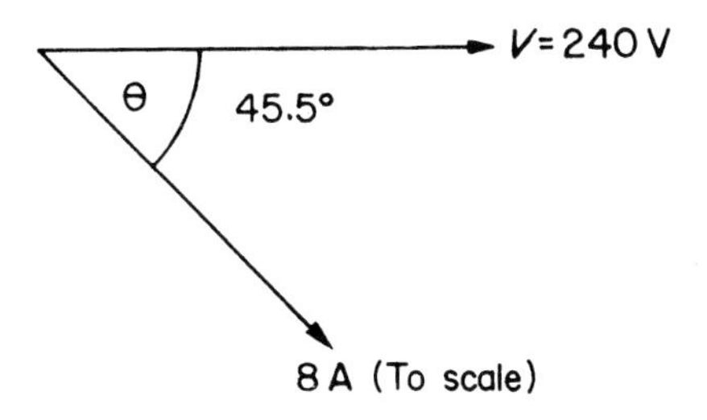

Fig. 6.22

From the question,

$I_c = 3.1$ A

I_c leads V by 90°

Therefore the phasor diagram for the capacitor is Fig. 6.23a. Combining both phasor diagrams, we have Fig. 6.23b.

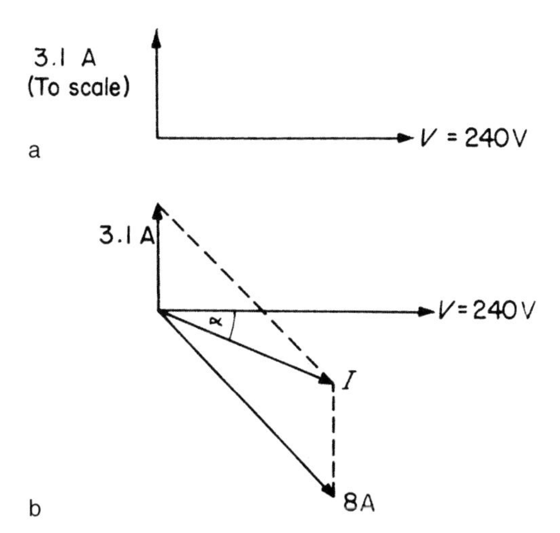

Fig. 6.23

By measurement:

$I = 6.2$ A

$= 24.5°$

$\therefore$ PF $= \cos \alpha$

PF $= \cos 24.5°$

$= 0.91$ lagging

Example

A 240 V, 50 Hz single-phase motor takes a current of 8 A at a power factor of 0.65 lagging. Determine the value of capacitor required to improve the PF to 0.92 lagging. What is the value of the new supply current?

Method 1, by phasors (Fig. 6.24)

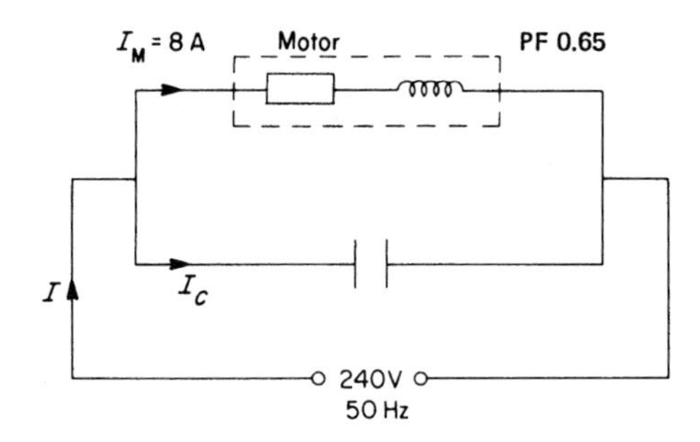

Fig. 6.24

Old PF = cos θ = 0.65

From cosine tables,

θ = 49.5°

New PF = cos α = 0.92

From cosine tables,

α = 23°

The phasor diagram is now drawn to scale (Fig. 6.25). By measurement,

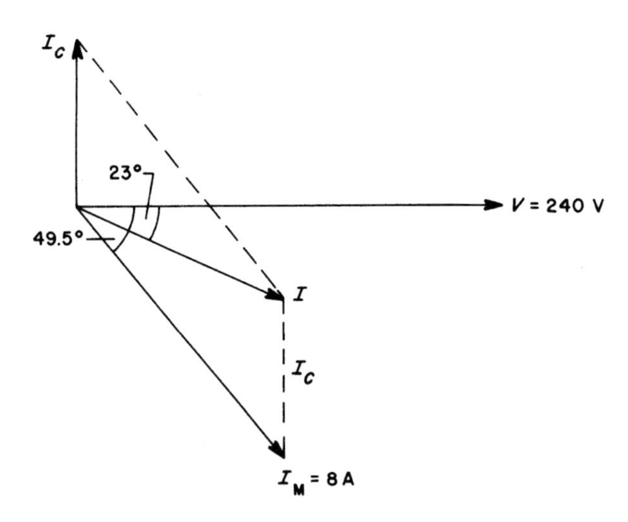

Fig. 6.25

$I_C = 3.9\,\text{A}$

and

$I = 5.6\,\text{A}$

To find the capacitance required,

$$I_C = \frac{V}{X_C}$$

$$\therefore X_C = \frac{V}{I_C}$$

$$= \frac{240}{3.9}$$

$$= 61.54\,\Omega$$

But

$$X_C = \frac{1}{2\pi fC}$$

$$\therefore C = \frac{1}{2\pi fX_C}$$

$$= \frac{1}{2\pi \times 50 \times 61.54}$$

$$= 51.7\,\mu\text{F}$$

Method 2, by trigonometry

In this method, I_C is found by calculating lengths PR and PQ in Fig. 6.26 and subtracting (PR – PQ = QR = I_C).

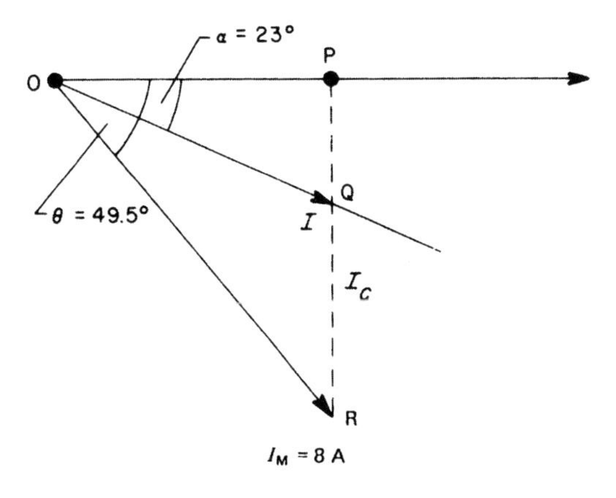

Fig. 6.26

Triangle OPR:

$$\cos\theta = \frac{\text{base}}{\text{hypotenuse}} = \frac{\text{OP}}{\text{OR}}$$

$$\therefore \text{OP} = \text{OR} \times \cos\theta$$

$$= 8 \times 0.65$$

$$\text{OP} = 5.2\,\text{A}$$

This is called the active or horizontal component of I_M.

$$\sin\theta = \frac{\text{perpendicular}}{\text{hypotenuse}} = \frac{\text{PR}}{\text{OR}}$$

$$\therefore \text{PR} = \text{OR} \times \sin\theta$$

$$= 8 \times 0.76$$

$$\text{PR} = 6.08 \text{ A}$$

This is called the reactive or vertical component of I_M.

Triangle OPQ:

$$\tan\alpha = \frac{\text{perpendicular}}{\text{base}} = \frac{\text{PQ}}{\text{OP}}$$

$$\therefore \text{PQ} = \text{OP} \times \tan\alpha$$

$$= 5.2 \times 0.424$$

$$\text{PQ} = 2.2 \text{ A}$$

$$\therefore \text{QR} = I_C = (\text{PR} - \text{PQ})$$

$$= 6.08 - 2.2$$

$$I_C = 3.88 \text{ A}$$

Calculation of C is as for method 1. To find current I:

$$\cos\alpha = \frac{\text{base}}{\text{hypotenuse}} = \frac{\text{OP}}{\text{OQ}}$$

$$\therefore \text{OQ} = I = \frac{\text{OP}}{\cos\alpha}$$

$$= \frac{5.2}{0.92}$$

$$I = 5.65 \text{ A}$$

Further examples of PF correction appear in Chapters 8 and 10.

The use of the previous methods of drawing and calculation may also be applied to power.

Power in a.c. circuits

The phasor diagram of voltages in an a.c. series circuit can be used to give a power triangle (Fig. 6.27).

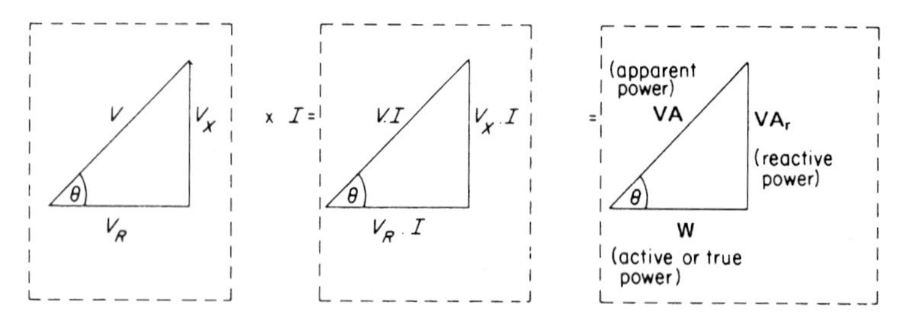

Fig. 6.27

$$\cos\theta = \frac{W}{VA} = PF$$

Power may be added by phasor diagram or calculated by trigonometry.

Example

The following loads are connected to a factory supply: 5 kVA at 0.75 PF lagging; 8 kW at a PF of unity; 6.8 kVA at 0.6 PF lagging. Determine the total load taken from the supply and the overall PF.

Method 1, by phasor diagram (Fig. 6.28)

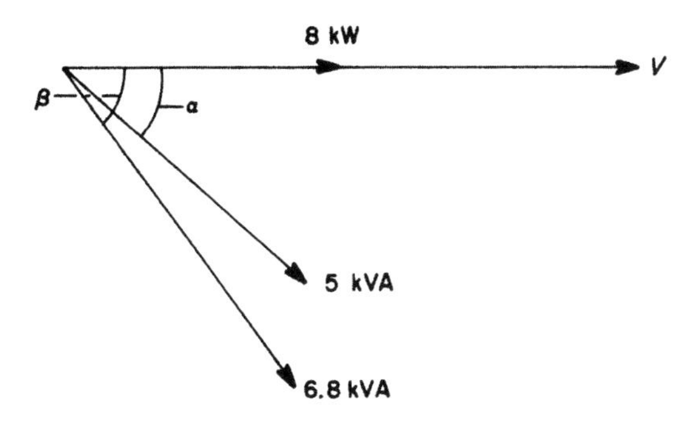

Fig. 6.28

$\cos\alpha = 0.75$

$\therefore \alpha = 41.4°$

$\cos\beta = 0.6$

$\therefore \beta = 53.1°$

From Fig. 6.29:

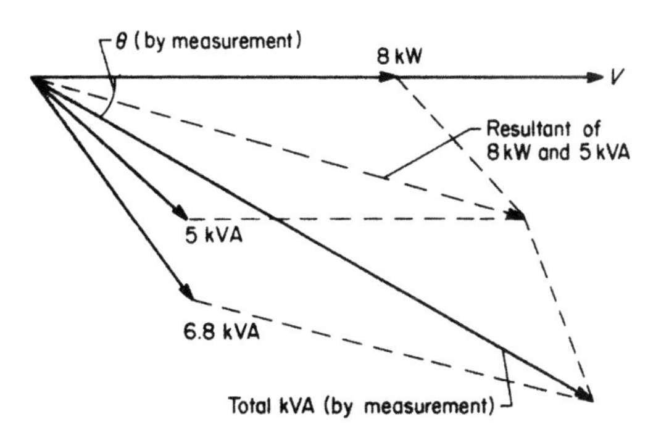

Fig. 6.29

Total kVA = 18.1 kVA

$\theta = 29°$

$\therefore \cos\theta = PF = 0.874$

Method 2, by trigonometry (Fig. 6.30)

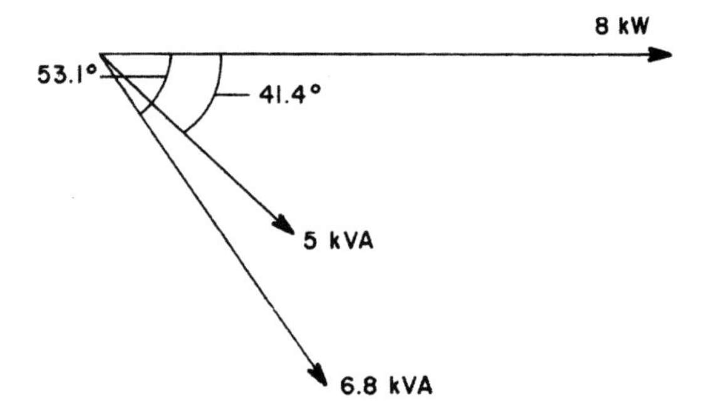

Fig. 6.30

Active component of 8 kW = 8 kW

Active component of 5 kVA = 5 × cos 41.4°
= 5 × 0.75
= 3.75 kW

Active component of 6.8 kVA = 6.8 × cos 53.1°
= 6.8 × 0.6
= 4.08 kW

Total of active components = 8 + 3.75 + 4.08
= 15.83 kW

Reactive component of 8 kW = 0

Reactive component of 5 kVA = 5 × sin 41.4°
= 5 × 0.66
= 3.3 kVA_r

Reactive component of 6.8 kVA = 6.8 × sin 53.1°
= 6.8 × 0.8
= 5.44 kVA_r

As both 5 kVA and 6.8 kVA have lagging power factors, their reactive components are added (Fig. 6.31).

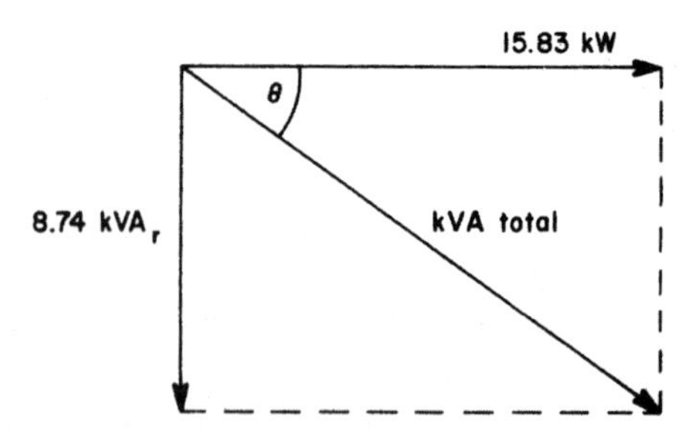

Fig. 6.31

$$\therefore \text{Total reactive component} = 0 + 3.3 + 5.44$$

$$= 8.74\,\text{kVA}_r$$

$$\tan\theta = \frac{8.74}{15.83} = 0.55$$

$$\therefore \theta = 28.9°$$

$$\therefore \text{PF} = \cos\theta = 0.875$$

Also

$$\cos\theta = \frac{15.83}{\text{kVA}}$$

$$\therefore \text{kVA} = \frac{15.83}{\cos\theta}$$

$$= \frac{15.83}{0.875}$$

$$= 18\,\text{kVA at } 0.875\,\text{PF lagging}$$

Self-assessment questions

1 A 240 V single-phase motor takes a current of 10 A and has a working power factor of 0.5 lagging. Draw a scaled phasor diagram and from it determine the value of capacitor current required to improve the PF to 0.9 lagging. Calculate the value of the capacitor.

2 Determine the value of the voltage and the power factor in the circuit shown in the diagram.

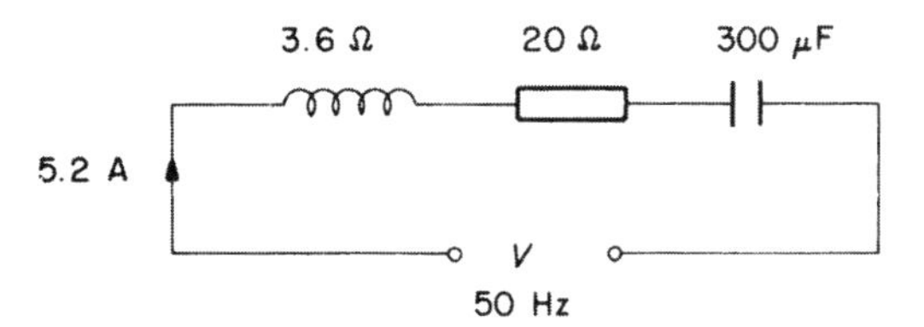

3 A 240 V, 50 Hz single-phase motor takes 6 A at 0.56 PF lagging. Determine the value of capacitor required to improve the PF to unity.

4 A 240 V, 50 Hz fluorescent lamp unit takes a current of 0.6 A at a PF of 0.45 lagging. Calculate the capacitance required to correct the PF to 0.92 lagging.

5 Two 240 V, 50 Hz single-phase motors A and B are connected in parallel. Motor A takes a current of 8.6A at 0.75 PF lagging and the total current taken from the supply is 16 A at 0.6 lagging. Calculate the current and PF of motor B.

6 A consumer has the following loads connected to his or her supply: 3 kVA at 0.8 lagging; 4 kW at a PF of unity; and 5 kVA at 0.5 lagging. Calculate the total load in kVA and the overall PF.

7 Three-phase circuits

As we have seen in Chapter 4, a three-phase supply comprises three wave forms each separated by 120° and the resultant waveform is zero. Let us now consider how we can utilize this supply and how we connect to it.

Star and delta connections

Fig. 7.1a and b show the two main ways of connecting three-phase equipment.

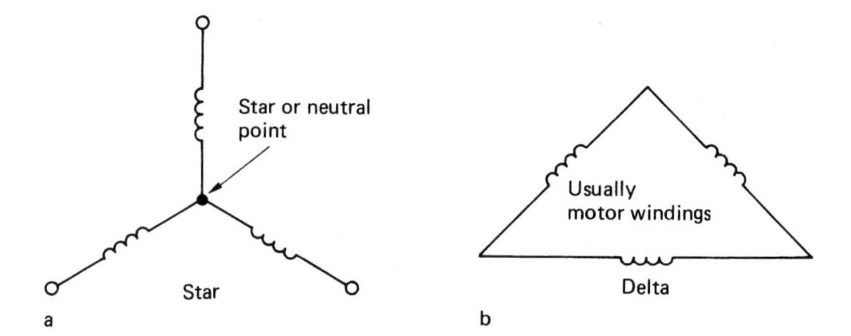

Fig. 7.1 *The two main ways of connecting three-phase equipment: (a) star and (b) delta*

The neutral conductor

Fig. 7.2 shows the simple system of a star-connected load fed from a star-connected supply. The addition of the conductor between the star points converts the system into what is known as a 'three-phase four-wire system'. We can see that the currents supplied by the generator flow along the lines, through the load, and return via the neutral conductor.

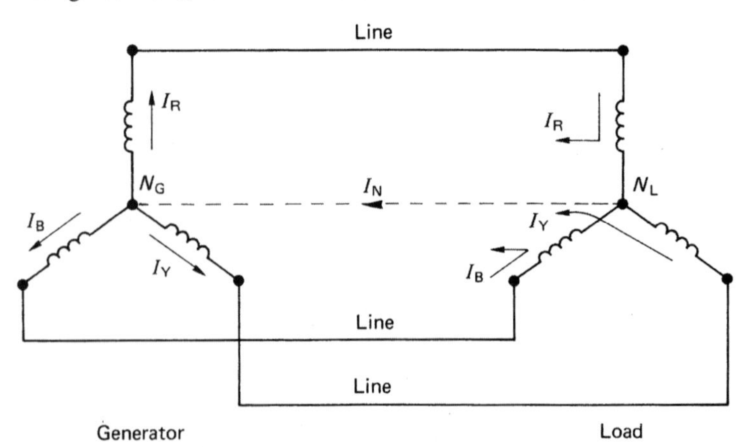

Fig. 7.2

However, we have already seen that all currents in a *balanced* three-phase system add up to zero:

$\therefore I_R = I_Y + I_B = I_N = 0$

Hence the current flowing in the neutral is zero. Also, since no current flows between the star points, they must both be at the same potential, which is also zero. The star point of a transformer is earthed, as earth is also at zero volts.

One reason for the connection of the neutral conductor is to provide a path for currents if the system became unbalanced. Another is that it enables single-phase loads to be connected to a three-phase system. The windings of most three-phase motors are connected in delta as the phase windings are perfectly balanced and no neutral is needed.

Current and voltage distribution

Balanced three-phase a.c. systems (Fig. 7.3)

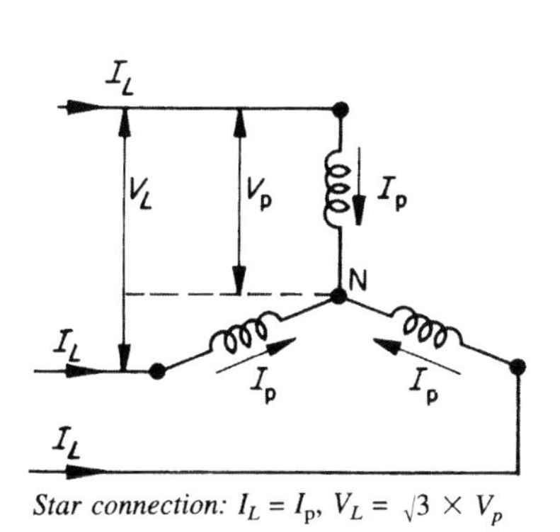

Star connection: $I_L = I_p$, $V_L = \sqrt{3} \times V_p$

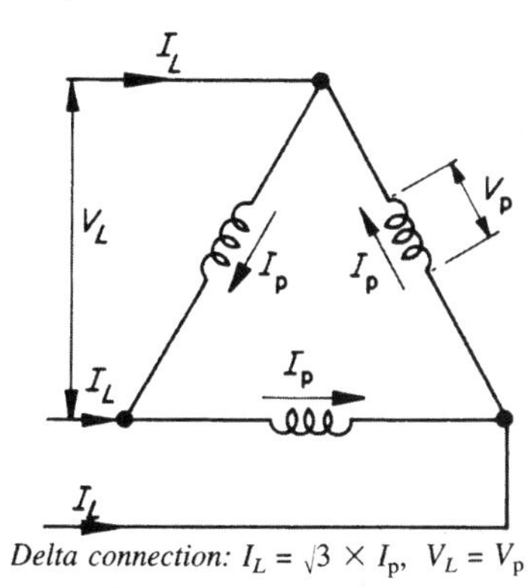

Delta connection: $I_L = \sqrt{3} \times I_p$, $V_L = V_p$

Fig. 7.3

Use of phasors

Currents in three-phase systems may be added by the use of phasors. For balanced systems, the resultant current will, of course be zero (Fig. 7.4).

The resultant of I_R and I_Y is I_{RY} which is clearly equal and opposite to I_B and hence the overall resultant will be zero.

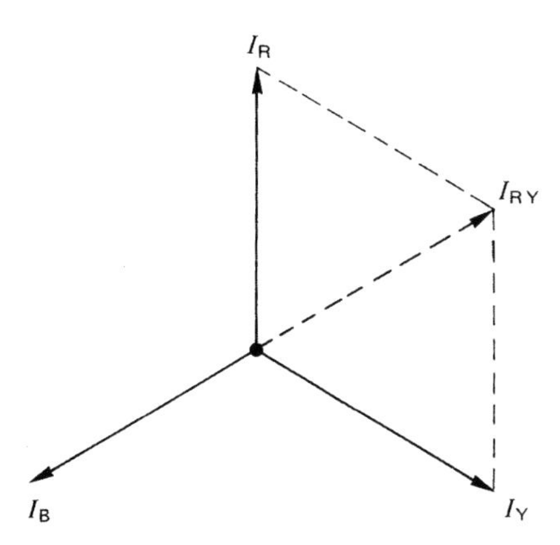

Fig. 7.4

Unbalanced three-phase a.c. systems (Fig. 7.5)

In an unbalanced system the resultant current will be the neutral current (Fig. 7.5).

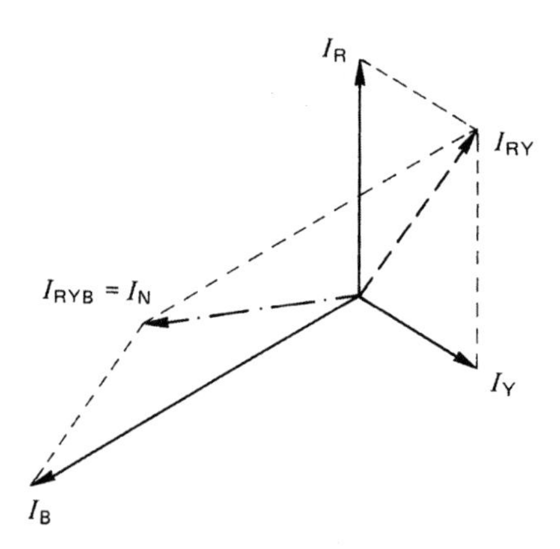

Fig. 7.5

The resultant of I_R and I_Y is I_{RY}, and the resultant of this (I_{RY}) and I_B is I_{RYB} which is of course the neutral current I_N.

Example

The currents measured in a factory three-phase supply were as follows: red phase, 70 A; yellow phase, 50 A; Blue phase, 40 A. Determine, using a phasor diagram, the magnitude of the neutral current.

By measurement (Fig. 7.6):

$I_N = 28$ A

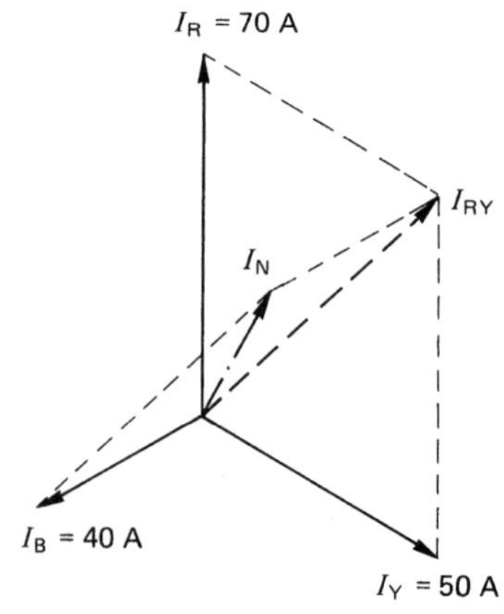

Fig. 7.6

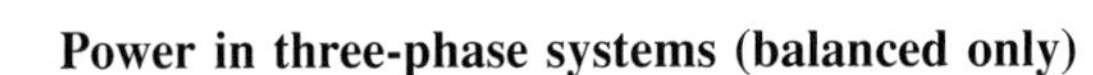

Power in three-phase systems (balanced only)

Power in a star-connected system

Since we are considering balanced systems, the total power is three times the power in one phase, and as

$$\text{PF} = \frac{\text{W}}{\text{VA}}$$

then

$$\text{W} = \text{VA} \times \text{PF}$$

$$\therefore \text{W} = V_P \times I_P \times \text{PF}$$

But

$$\text{V}_P = \frac{V_L}{\sqrt{3}} \text{ and } I_p = I_L$$

$$\therefore W = \frac{V_L}{\sqrt{3}} \times I_L \times \text{PF}$$

$$\therefore \text{ Total power } P = 3 \times \frac{V_L}{\sqrt{3}} \times I_L \times \text{PF}$$

$$= \sqrt{3}\ V_L \times I_L \times \text{PF}$$

Delta-connected system

$$\text{Again W} = \text{VA} \times \text{PF}$$

$$\therefore \text{W} = V_p \times I_p \times \text{PF}$$

But

$$V_p = V_L \text{ and } I_p = \frac{I_L}{\sqrt{3}}$$

$$\therefore \text{W} = V_L \times \frac{I_L}{\sqrt{3}} \times \text{PF}$$

$$\therefore \text{ Total power } P = 3 \times V_L \times \frac{I_L}{\sqrt{3}} \times \text{PF}$$

$$P = \sqrt{3}\ V_L \times I_L \times \text{PF}$$

which is the same as for a star connection.

Hence for either star or delta connections the total power in watts is given by

$$P \text{ (watts)} = \sqrt{3}\ V_L \times I_L \times \text{PF}$$

Since

$$\text{PF} = \frac{\text{W}}{\text{VA}}$$

then

$$\text{VA} = \frac{\text{W}}{\text{PF}}$$

$$\therefore \text{ Three-phase VA} = \sqrt{3}\ V_L \times I_L$$

$$\therefore \text{ Line current} = \frac{\text{VA}}{\sqrt{3}\ V_L}$$

Example

A 15 kW, 415 V, balanced three-phase delta-connected load has a power factor of 0.8 lagging. Calculate the line and phase currents.

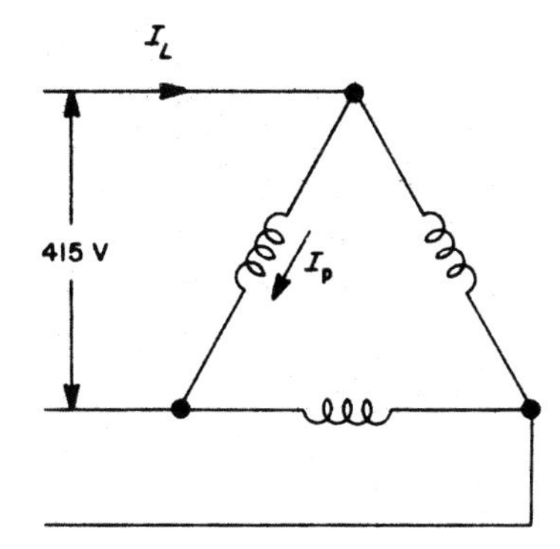

Fig. 7.7

From Fig. 7.7:

$$P = \sqrt{3}\ V_L \times I_L \times \text{PF}$$

$$\therefore I_L = \frac{P}{\sqrt{3}\ V_L \times \text{PF}}$$

$$= \frac{15\,000}{\sqrt{3} \times 415 \times 0.8}$$

$$I_L = 26 \text{ A}$$

For delta connections

$$I_L = \sqrt{3}\, I_p$$

$$\therefore I_p = \frac{I_L}{\sqrt{3}}$$

$$= \frac{26}{\sqrt{3}}$$

$$I_p = 15\,\text{A}$$

Example

Three identical loads each having a resistance of 10 Ω and an inductive reactance of 20 Ω are connected first in star and then in delta across a 415 V, 50 Hz three-phase supply. Calculate the line and phase currents in each case.

$$Z \text{ for each load} = \sqrt{R^2 + X_L^2}$$

$$= \sqrt{10^2 + 20^2}$$

$$= \sqrt{500}$$

$$= 22.36\,\Omega$$

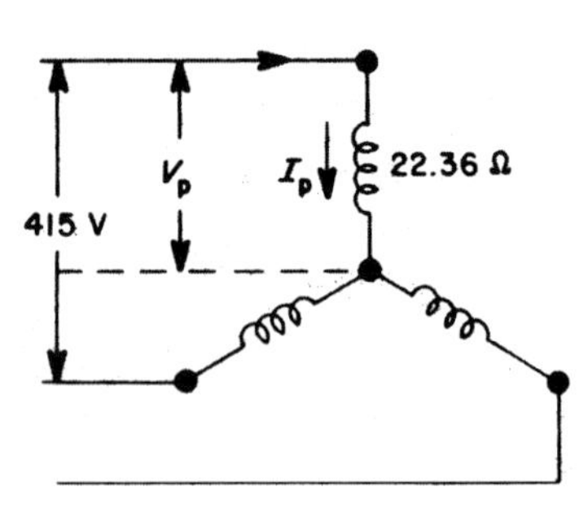

Fig. 7.8

For star connection (Fig. 7.8),

$$V_p = \frac{V_L}{\sqrt{3}} = 240\,\text{V}$$

$$I_L = I_p = \frac{V_L}{Z}$$

$$= \frac{240}{22.36}$$

$$I_L = I_P = \mathit{10.73\,A}$$

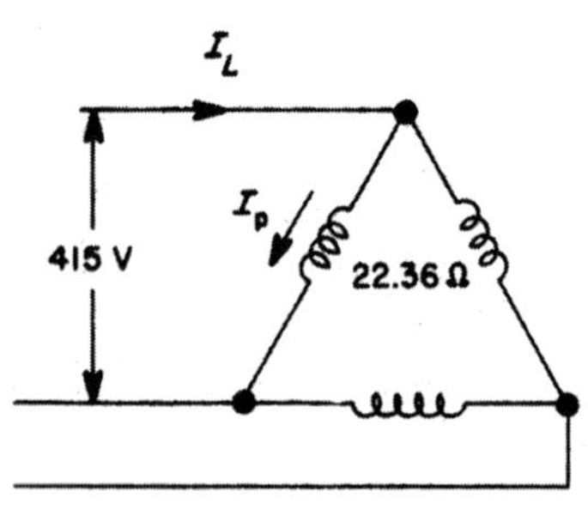

Fig. 7.9

For delta connection (Fig. 7.9),

$$V_L = V_p = 415\,\text{V}$$

$$\therefore I_p = \frac{V_p}{22.36}$$

$$= \frac{415}{22.36}$$

$$I_p = 18.56\,\text{V}$$

and

$$I_L = \sqrt{3}\, I_p$$

$$= \sqrt{3} \times 18.56$$

$$I_L = 32.15\,\text{A}$$

Measurement of power in three-phase systems

Single-phase circuit

Fig. 7.10 shows the instruments required to determine the powers in watts and volt amperes, and the power factor of a single-phase circuit.

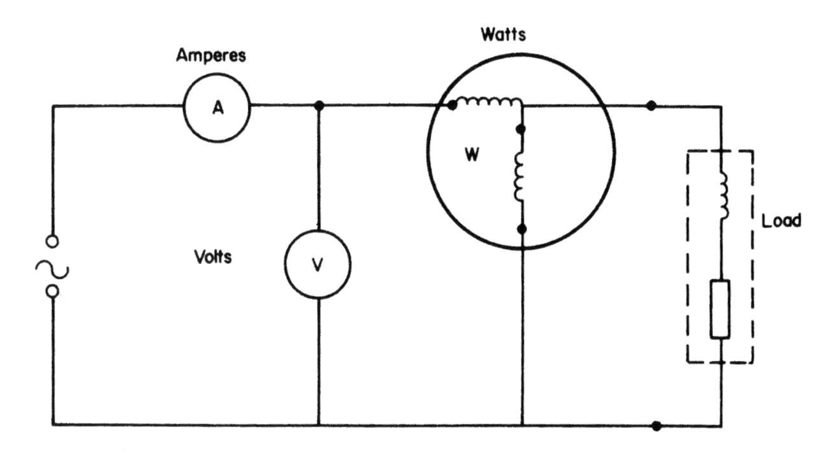

Fig. 7.10

Example

The following values were recorded from a circuit similar to that in Fig. 7.10:

Ammeter – 8 A
Voltmeter – 240 V
Wattmeter – 1.152 kW

Calculate the kVA and the power factor of the load.

$$\text{kVA} = \frac{\text{VA}}{1000}$$

$$= \frac{8 \times 240}{1000}$$

$$= 1.92\,\text{kVA}$$

$$\text{PF} = \frac{\text{kW}}{\text{kVA}}$$

$$= \frac{1.152}{1.92}$$

$$= 0.6$$

Three-phase (balanced) four-wire circuit

In this case it is necessary to measure the power only in one phase. The total power will be three times this value. Also, the PF for one phase is the overall PF.

Example

It is required to measure the power factor of a three-phase star-connected balanced inductive load. Show how the necessary instruments would be arranged, a single voltmeter being used to measure the voltage across each phase.

If the readings obtained were 20 A, 240 V and 3840 W, calculate the total power in kilowatts and the power factor (Fig. 7.11).

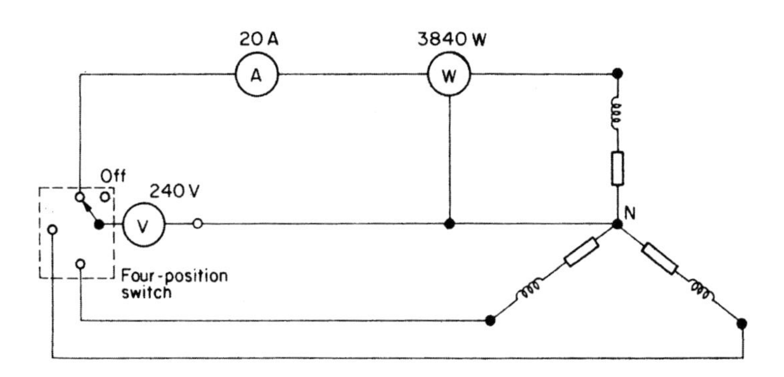

Fig. 7.11

$$\text{PF} = \frac{\text{W}}{\text{VA}}$$

$$= \frac{3840}{240 \times 20} = 0.8 \text{ lagging}$$

$$\text{Total power} = 3 \times 3840$$
$$= 11\,520$$
$$= 11.52\,\text{kW}$$

Check:

$$V_L = V_p\sqrt{3}$$
$$V_L = 240 \times \sqrt{3}$$
$$P = \sqrt{3}\,V_L \times I_L \times \text{PF}$$
$$= \sqrt{3} \times \sqrt{3} \times 240 \times 20 \times 0.8$$
$$= 11.52\,\text{kW}$$

Self-assessment questions

1 What is meant by: three-phase generation; and a four-wire system?
2 Explain with the aid of sketches the reason for the use of a neutral conductor.
3 The line current of a 415 V star-connected load is 10 A. Calculate the values of phase current and phase voltage.

4 A voltmeter, ammeter and wattmeter are arranged to measure the power in a single-phase circuit. Show how these instruments would be connected and calculate the circuit PF if the readings were 16 A, 240 V and 3600 W.

5 A three-phase star-connected load is supplied from the delta-connected secondary of a transformer. If the transformer line voltage is 190.5 V and the load phase current is 10 A, calculate the transformer phase current and the load phase voltage.

6 A 20 kW, 415 V three-phase star-connected load takes a line current of 34.8 A. Calculate the PF of the load.

7 A wattmeter, voltmeter and ammeter are arranged in a three-phase four-wire balanced system. Show how they would be connected ready to determine the total circuit power in watts and the PF. Readings obtained were 2000 W, 240 V and 10 A. Calculate the circuit power in kW and kVA.

8 A small industrial unit has the following loadings measured at the intake position:

Red phase – 88 A
Yellow phase – 72 A
Blue phase – 98 A

By means of a scaled phasor diagram determine the value of the neutral current.

8 Motors and generators

Motors play an important part in the modern domestic and industrial environment, and should therefore be of interest to the electrician.

Motors may be divided into two distinct kinds: those using direct current and those using alternating current. Each of these categories is further divided into different types.

Direct-current motors

Simple single-loop motor

The effect of the force on a conductor in a magnetic field may be used to cause the rotation of a motor armature. Fig. 8.1 illustrates a simple single-loop motor.

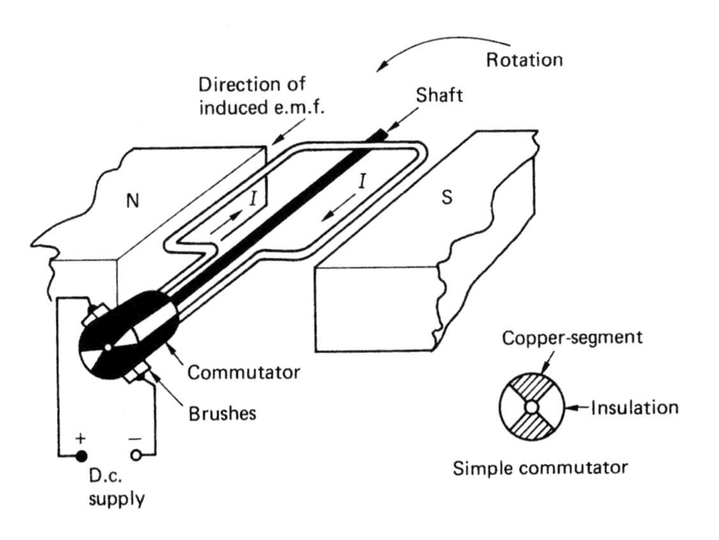

Fig. 8.1

A single loop of conductor arranged as shown in this figure has its ends connected to a simple commutator, which comprises two copper segments insulated from each other. The commutator and loop are fixed to a central shaft which enables the whole assembly to be freely rotated. Two fixed carbon brushes bear on the surface of the commutator, enabling a supply to be connected to the loop.

Fig. 8.2 shows a cross-section through the loop. The direction of movement may be determined using Fleming's left-hand rule.

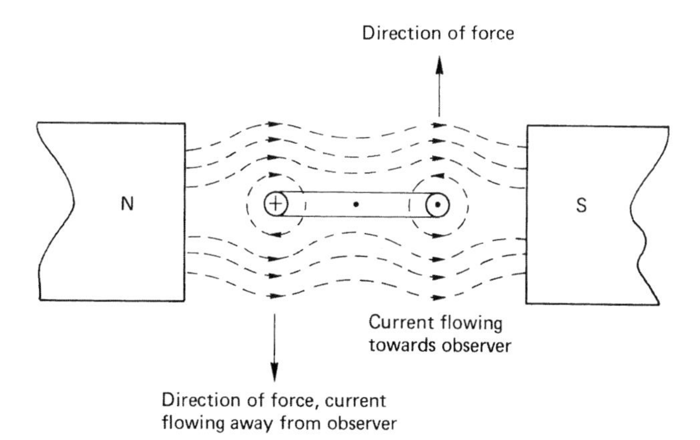

Fig. 8.2

In practice, the d.c. motor comprises an armature of many loops revolving between electromagnetic poles. Both the armature and the field are supplied from the same source. The commutator has, of course, many segments to which the ends of the armature coils are connected, and the armature core is laminated to reduce eddy currents.

Back e.m.f.: symbol E; unit, volt (V)

It is interesting to note that as the armature revolves, its coils cut across the field flux, but we know that if a conductor cuts across lines of force, an e.m.f. is induced in that conductor. This is the principle of the generator, as dealt with in this book. So, applying Fleming's right-hand rule to Fig. 8.1, we see that the induced e.m.f. is opposing the supply. This induced e.m.f. is called the 'back e.m.f.'. If the back e.m.f. were of the same magnitude as the supply voltage, no current would flow and the motor would not work. As current must flow in the armature to produce rotation, and as the armature circuit has resistance, then there must be a voltage drop in the armature circuit. This voltage drop is the product of the armature current (I_a) and the armature circuit resistance (R_a):

$$\text{Armature voltage drop} = I_a \times R_a$$

It is this voltage drop that is the difference between the supply voltage and the back e.m.f. Hence

$$E = V - (I_a \times R_a)$$

We also know (from Chapter 4) that induced e.m.f. is dependent on the flux density (B), the speed-of cutting the flux (v) and the length of the conductor (l):

$$E = B \times l \times v$$

But

$$B = \frac{\Phi}{a}\left(\frac{\text{flux}}{\text{area}}\right)$$

$$\therefore E = \frac{\Phi}{a} \times l \times v$$

Both l and a for a given conductor will be constant and v is replaced by n (revs/second) as this represents angular or rotational speed.

$$\therefore E \propto n\Phi$$

So, if the speed is changed from n_1 to n_2 and the flux from Φ_1 to Φ_2, then the e.m.f. will change from E_1 to E_2

$$\therefore \frac{E_1}{E_2} = \frac{n_1\,\Phi_1}{n_2\,\Phi_2}$$

Torque: symbol, *T*; unit, newton metre (N m)

Work = force × distance

$$\therefore \text{Turning work or torque} = \underset{(F)}{\text{force}} \times \underset{(r)}{\text{radius}}$$

We also know that

$$\text{Force} = B \times l \times I$$

$$\therefore \text{Torque } T = B \times l \times I_a \times r$$

$$\therefore T = \frac{\Phi}{a} \times l \times I_a \times r$$

Once again, for a given machine, a, l and r will all be constant.

$$\therefore T \propto \Phi \times I_a$$

Also, mechanical output power in watts is given by

$$P = 2\pi nT$$

where P is the output power in watts, n is the speed in revs/second and T is the torque in newton metres.

If we multiply $E = (V - I_aR_a)$ by I_a, we get

$$EI_a = VI_a - I_a^2R_a$$

VI_a is the power supplied to the armature and $I_a^2\,R_a$ is the power loss in the armature; therefore EI_a must be the armature power output. Hence

$$EI_a = P$$

$$\therefore EI_a = 2\pi nT$$

$$\therefore T = \frac{EI_a}{2\pi n} \text{ newton metres}$$

It is clear then that torque is directly proportional to armature current and inversely proportional to speed, i.e. if the mechanical load is lessened, the torque required is less, the armature current decreases and the motor speeds up.

Example

A 300 V d.c. motor runs at 15 revs/second and takes an armature current of 30 A. If the armature resistance is 0.5 Ω, calculate first the back e.m.f. and secondly the torque.

$$E = V - I_a R_a$$

$$= 300 - (30 \times 0.5)$$

$$= 300 - 15$$

$$E = 285\,\text{V}$$

$$T = \frac{EI_a}{2\pi n}$$

$$= \frac{285 \times 30}{2\pi \times 15}$$

$$T = 90.72\,\text{Nm}$$

Series motor (Fig. 8.3)

The series type of d.c. motor has its field windings and armature connected in series across the supply. It will be seen from this figure that the armature current I_a also supplies the field. Therefore, when I_a is large (on starting, for example), the magnetic field will be strong, and the torque will be high. As the machine accelerates, the torque, armature current and field strength will all decrease. This type of motor should never be coupled to its load by means of a belt, since if the belt breaks the required torque from the armature will be removed, the armature and field current will fall, reducing the magnetic field, and the motor will increase in speed until it disintegrates. Fig. 8.4 shows the graphs of speed and torque to a base of load current.

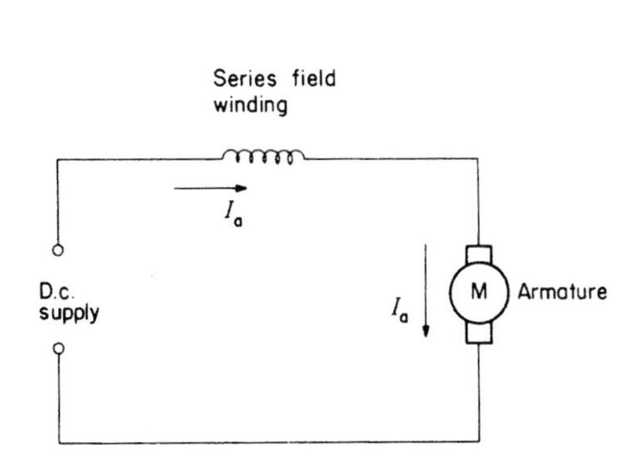

Fig. 8.3

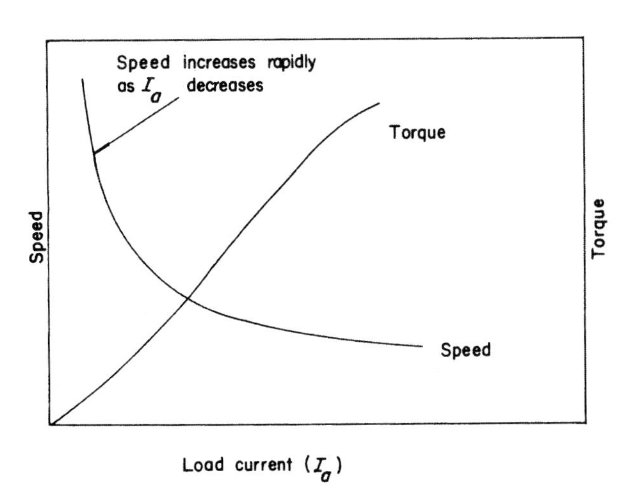

Fig. 8.4 *Load characteristics of a series motor*

Speed control

The most effective way of controlling the speed of a d.c. motor is to vary the strength of the magnetic field. On a series machine this may be achieved by diverting some of the current through a variable resistor (Fig. 8.5).

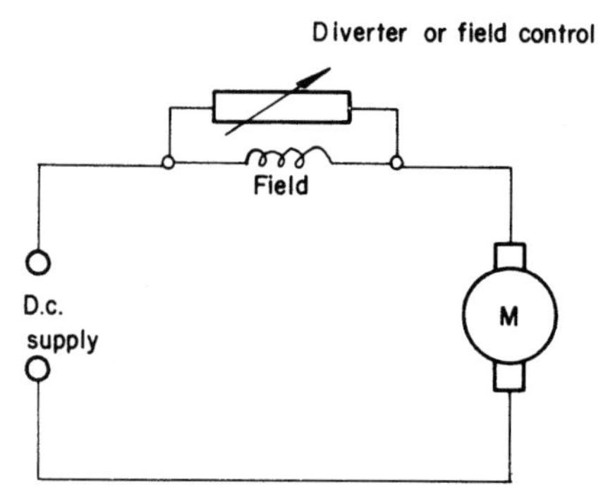

Fig. 8.5 *Speed control of a series motor*

Starting

A series motor is started by placing a variable resistor in series with the armature circuit (Fig. 8.6).

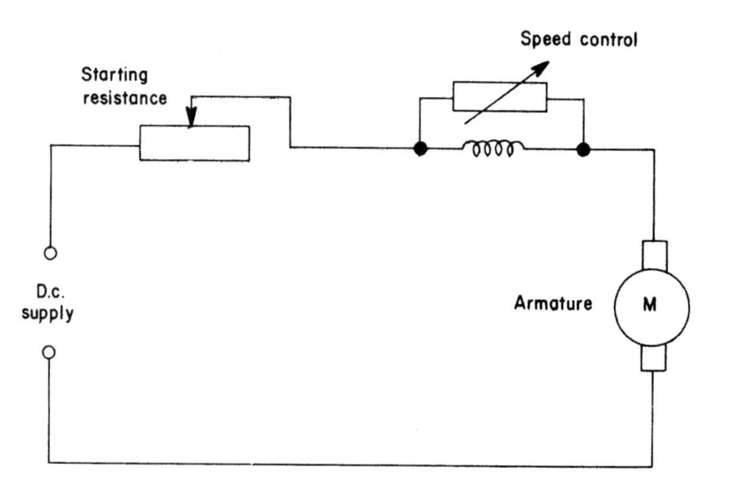

Fig. 8.6 *Starting a series motor*

Applications

The series motor is best used where heavy masses need to be accelerated from rest, such as in cranes and lifts.

Example

A 200 V series motor has a field winding resistance of 0.1 Ω and an armature resistance of 0.3 Ω. If the current taken at 5 revs/second is 30 A, calculate the torque on the armature.

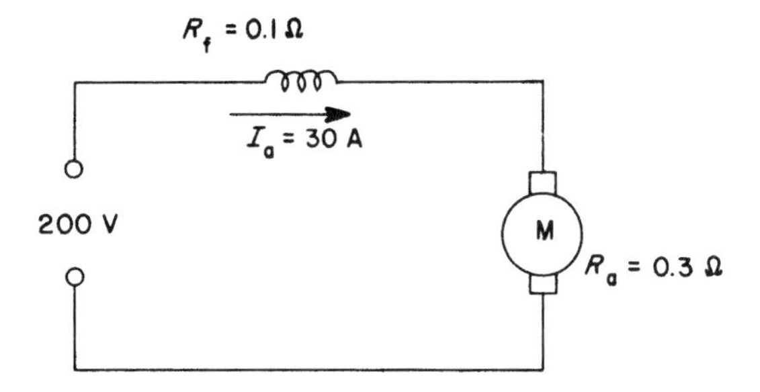

Fig. 8.7

The total armature circuit resistance is $R_a + R_f$ (Fig. 8.7) as the two resistance are in series.

$$E = V - I_a R_a$$
$$= 200 - 30(0.1 + 0.3)$$
$$= 200 - 30 \times 0.4$$
$$= 200 - 12$$
$$= 188\text{ V}$$

$$\text{Torque } T = \frac{EI_a}{2\pi n}$$
$$= \frac{188 \times 30}{2\pi \times 5}$$
$$= 180\text{ N m}$$

Shunt motor (Fig. 8.8)

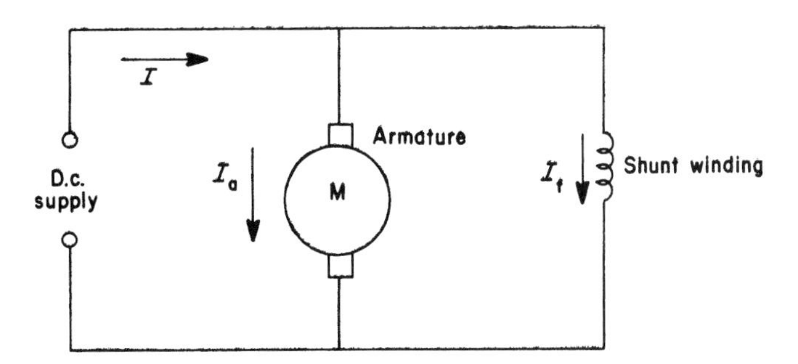

Fig. 8.8 *Shunt-wound motor*

In the case of a shunt motor, the motor winding is in parallel with the armature. It will be seen here that the supply current $I = I_a + I_f$. Unlike the series motor, if the load is removed from the motor, only the armature current will decrease, the field remaining at the same strength. The motor will therefore not continue to speed up to destruction. Fig. 8.9 shows the graphs of speed and torque to a base of load current.

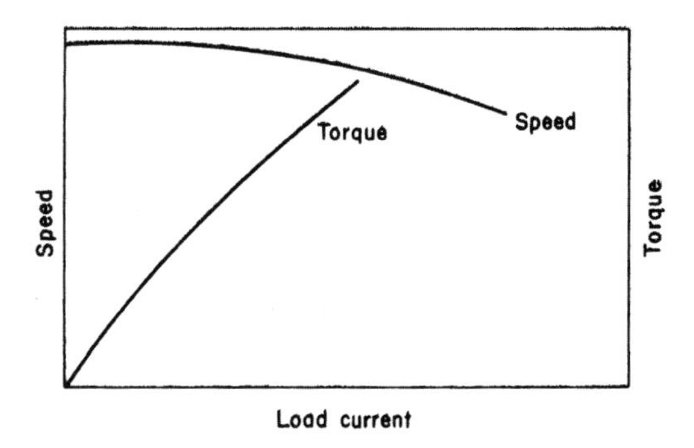

Fig. 8.9 *Load characteristics of a shunt motor*

Speed control

As in a series motor, speed control is best achieved by controlling the field strength, and in shunt motors a variable resistance is placed in series with the shunt winding (Fig. 8.10).

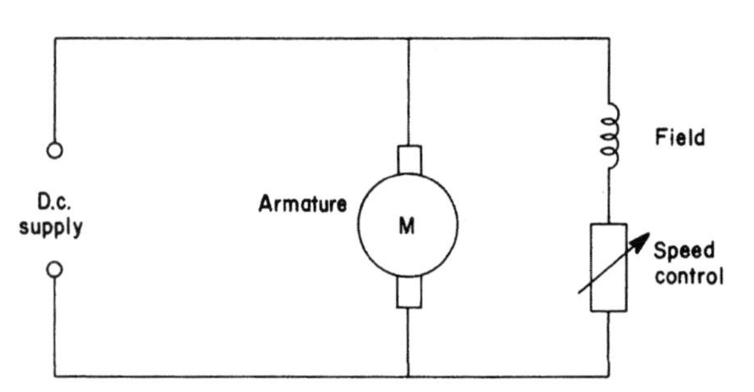

Fig. 8.10 *Speed control of a shunt motor*

Starting

Starting large d.c. shunt-wound motors is usually carried out using a d.c. face-plate starter (Fig. 8.11).

The face-plate starter comprises the following items:

(a) a series of resistances connected to brass studs;
(b) a spring-loaded handle which makes contact with two brass strips and also the brass studs;
(c) a no-volt release; and
(d) an overload release.

When the handle is located on the first stud, the field is supplied via the overload release, the top brass strip via the handle and the no-volt release (note that the field is continuously supplied in this way). The armature is supplied via the resistances.

As the handle is moved round, the resistance in the armature circuit is gradually cut out. On the final stud the handle is held in place by the no-volt release electromagnet. Should a failure in supply occur, the no-volt release will de-energize and the handle will spring back to the 'off' position. If a serious overload occurs, the overload release will energize sufficiently to attract its soft iron armature which will short out the no-volt release coil, and the handle will return to the 'off' position.

Note: The handle should be moved slowly from stud to stud.

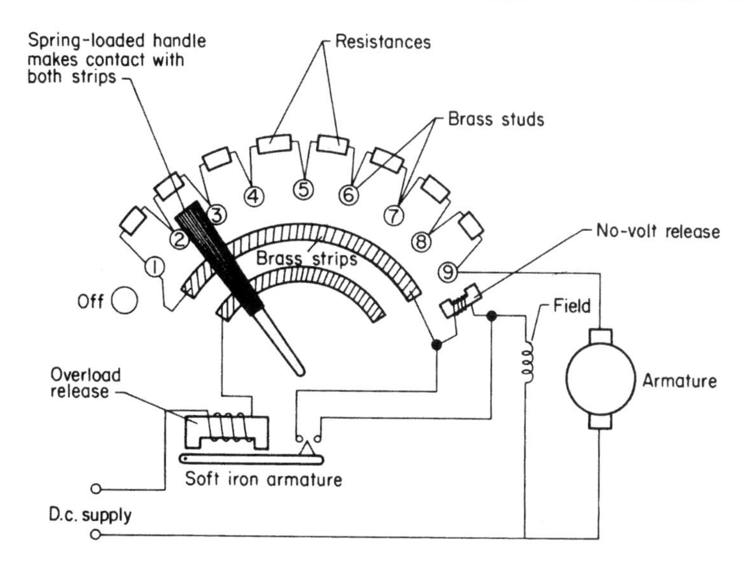

Fig. 8.11 *D.c. face-plate starter*

Applications

As the speed of a shunt motor is almost constant over a wide range of loads, it is most suitable for small machine tools.

Example

A 400 V shunt-wound motor has a field winding resistance of 200 Ω and an armature resistance of 0.5 Ω. If the current taken from the supply is 22 A, calculate the back e.m.f.

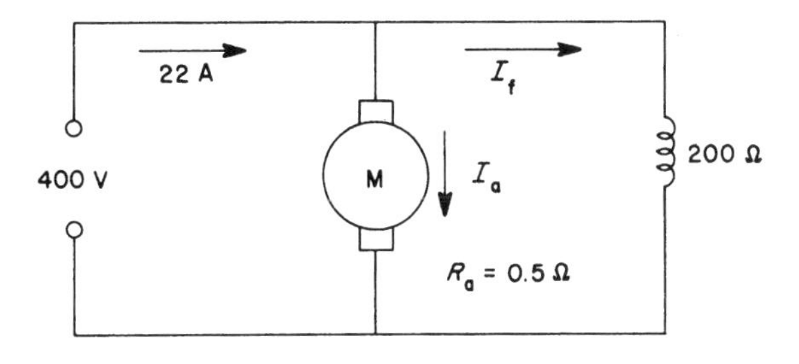

Fig. 8.12

From Fig. 8.12:

$$I = I_a + I_f$$

$$\therefore I_a = I - I_f$$

$$I_f = \frac{V}{R_f} = \frac{400}{200} = 2\text{ A}$$

$$\therefore I_a = 22 - 2 = 20\text{ A}$$

$$E = V - I_a R_a$$

$$= 400 - (20 \times 0.5)$$

$$= 400 - 10$$

$$= 390\text{ V}$$

Compound motor

A compound motor is a combination of a series and a shunt-wound motor (Fig. 8.13a and b).

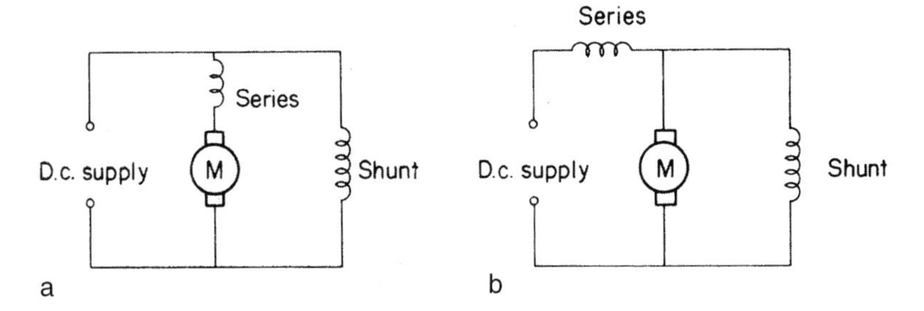

Fig. 8.13 *Compound-wound motor: (a) long shunt; (b) short shunt*

Fig. 8.14 shows the speed–torque characteristics for the compound motor.

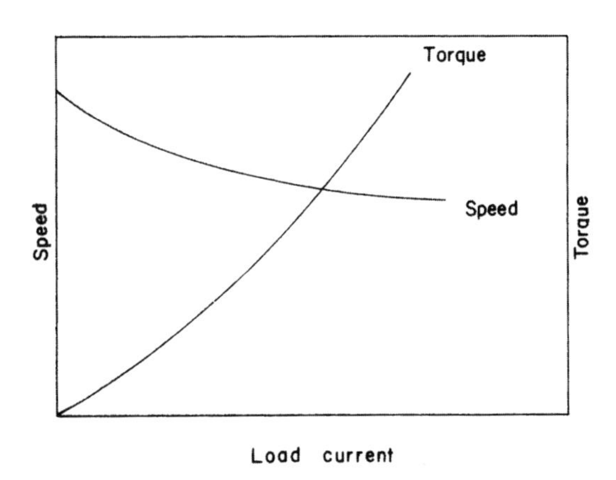

Fig. 8.14 *Speed and torque characteristics of a compound motor*

Speed control

Speed is usually controlled by variable resistors in the shunt field and armature circuit (Fig. 8.15).

The series winding may be arranged such that it aids the shunt field (*cumulative* compound) or opposes it (*differential* compound).

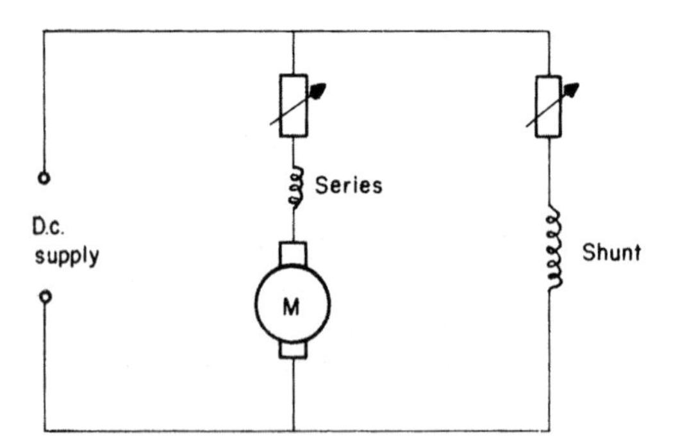

Fig. 8.15 *Speed control of a typical compound motor (long shunt)*

Cumulatively compounded motors are similar in characteristics to series motors, while differentially compounded motors are similar to shunt motors (Fig. 8.16).

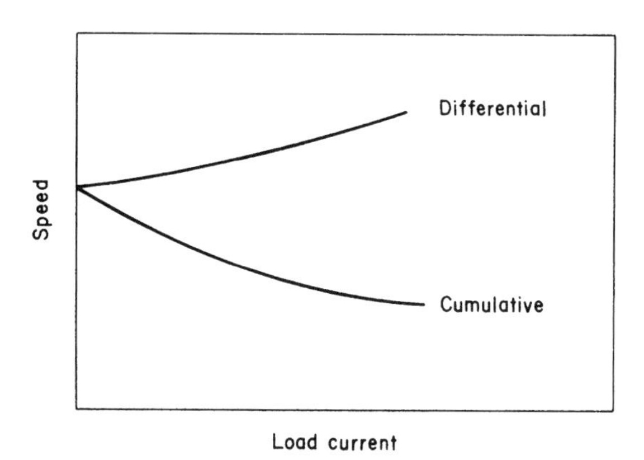

Fig. 8.16

Starting

The d.c. face-plate starter is suitable for the compound motor.

Applications

These motors can be used for applications where a wide speed range is required. However, the differentially compounded type is rarely used as it tends to be unstable. Cumulatively compounded types are suitable for heavy machine tools.

Reversing d.c. motors

D.c. motors may be reversed in direction by altering either the polarity of the field or that of the armature. This is done by reversing the connection to the armature *or* the field winding.

D.c. generators

If by some means a d.c. motor is supplied with motive power it will act as a generator. Connection of the field windings is the same as for motors; that is, either series, shunt or compound.

E.m.f. generated: symbol E; unit, volt (V)

When the armature is rotated in the field, an e.m.f. is induced or generated in the armature windings. When an external load is connected, current (I_a) will flow from the armature; this will cause a voltage drop of $I_a R_a$ where R_a is the resistance of the armature circuit. Hence the voltage available at the load is less than the generated e.m.f.

$$\therefore E = V + I_a R_a$$

Example

A 240 V shunt-wound generator has a field resistance of 120 Ω and an armature resistance of 0.4 Ω. Calculate the generated e.m.f. when it is delivering 20 A to the load.

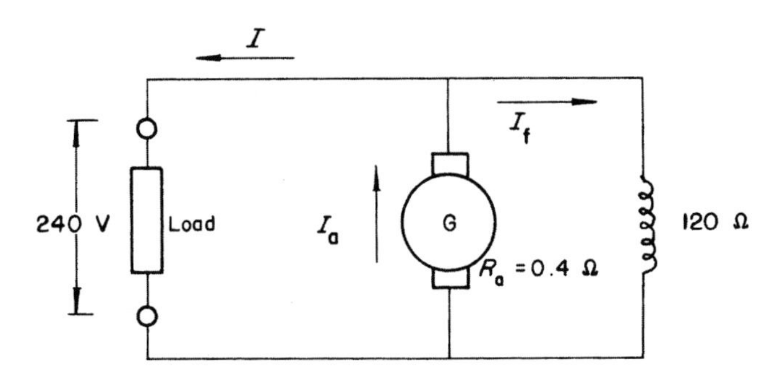

Fig. 8.17

As the generator supplies its own field, then from Fig. 8.17:

$$I_a = I + I_f$$

$$I_f = \frac{V}{R_f} = \frac{240}{120} = 2\text{ A}$$

$$\therefore I_a = 20 + 2 = 22\text{ A}$$

$$E = V + I_a R_a$$

$$= 240 + (22 \times 0.4)$$

$$= 240 + 8.8$$

$$= 248.8\text{ V}$$

Separately excited generator

The separately excited type of generator has its field supplied from a separate source (Fig. 8.18).

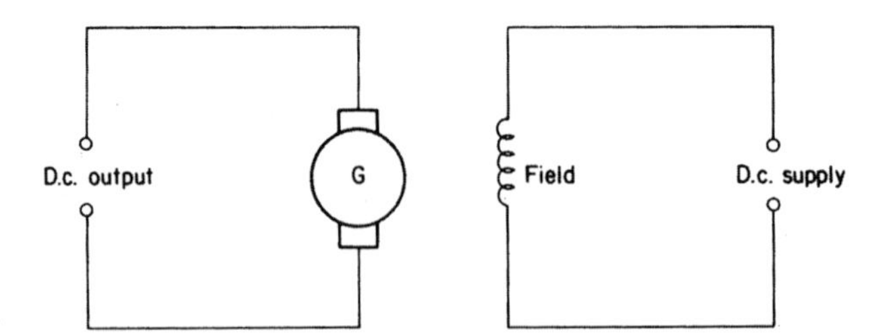

Fig. 8.18 *Separately excited generator*

Alternating-current motors

There are many different types of a.c. motor operating from either three-phase or single-phase a.c. supplies. To understand the starting problems of the single-phase types, it is best to consider three-phase motors first.

Three-phase motors

A three-phase motor depends on the rotation of a magnetic field for its movement. (Fig. 8.19) shows how this rotation is achieved.

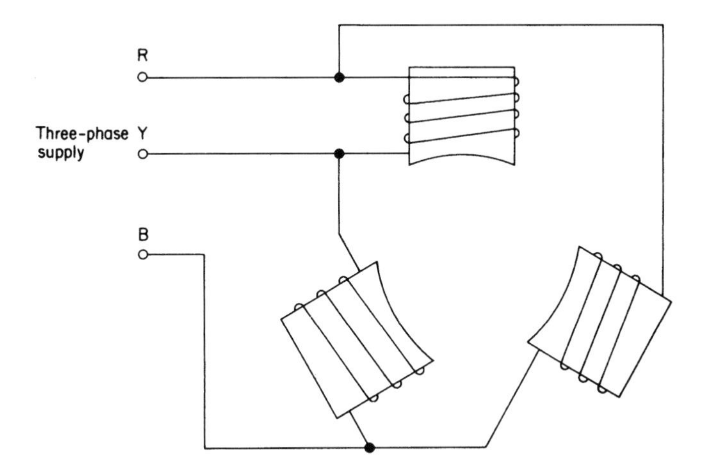

Fig. 8.19

If three iron-cored coils or poles are arranged 120° apart and connected as shown to an alternating three-phase supply, then each pole will become fully energized at a different time in relation to the others. If the poles were replaced with light bulbs, it would appear as if the light were travelling around in a circular fashion from one bulb to another.

The iron core of each coil becomes magnetized as the coil is energized, and the arrangement gives the effect of a magnetic field rotating around the coils.

The speed of rotation of the magnetic field is called the *synchronous speed* and is dependent on the frequency of the supply and the number of *pairs* of poles. Hence

$$f = n_s P$$

where f is the supply frequency in hertz, n_s is the synchronous speed in revs/second and p is the number of pairs of poles.

Example

Calculate the synchronous speed of a four-pole machine if the supply frequency is 50 Hz.

$$f = n_s p$$

$$\therefore n_s = \frac{f}{p}$$

$$= \frac{50}{2}$$

$$= 25 \text{ revs/second or } 1500 \text{ revs/min}$$

Synchronous motor

If we take a simple magnetic compass and place it in the centre of the arrangements shown in Fig. 8.19, the compass needle will rotate in the same direction as the magnetic field, because the magnetized compass needle is attracted to the field and therefore follows it. A three-phase synchronous motor is arranged in the same way as a three-phase generator (Fig. 8.20).

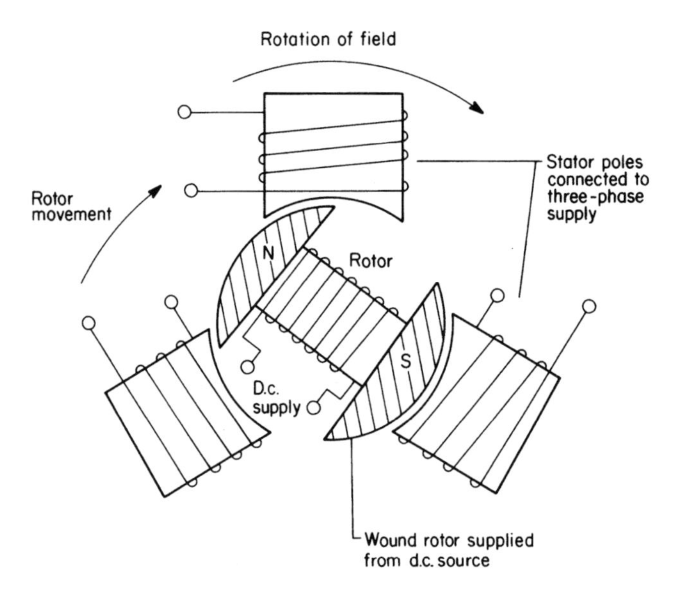

Fig. 8.20 *Synchronous motor*

This type of motor comprises the following:

(a) a stator, which supports the magnetic field poles; and
(b) the rotor, which is basically an electromagnet supplied from a d.c. source via slip rings.

The rotor will follow the rotating magnetic field at synchronous speed.

This type of motor is not self-starting and has to be brought up to or near to synchronous speed by some means, after which it will continue to rotate of its own accord. This bringing up to speed is usually achieved by providing the rotor with some of the characteristics of an induction motor rotor.

Synchronous-induction motor

The synchronous-induction type of motor is essentially an induction motor with a wound rotor. It starts as an induction motor and when its speed has almost reached synchronous speed the d.c. supply is switched on and the motor will then continue to function as a synchronous motor.

This type of motor has various applications. For example, if the d.c. supply to the rotor is increased (when it is said to be 'overexcited'), the motor can be made to run at a leading power factor. This effect may be used to correct the overall power factor of an installation.

As it is a constant-speed machine, it is often used in motor–generator sets, large industrial fans and pumps.

A great advantage of the synchronous-induction type is its ability to sustain heavy mechanical overloads. Such an overload pulls the motor out of synchronism, but it continues to run as an induction motor until the overload is removed, at which time it pulls back into synchronism again.

Three-phase induction motor

Squirrel-cage type

The squirrel-cage type of induction motor comprises a wound stator and a laminated iron rotor with copper or aluminium bars embedded in it, in the form of a cage (Fig. 8.21).

As the rotating magnetic field sweeps across the rotor, an e.m.f. is induced in the cage bars and hence a current flows. This current produces a magnetic

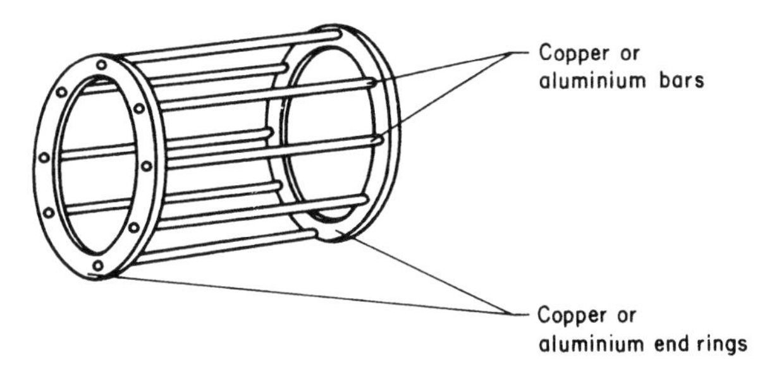

Fig. 8.21 *Cage assembly for cage rotor*

field around the conductor and the magnetic reaction between this field and the main field causes the rotor to move. Since this movement depends on the cage bars being cut by the main field flux, it is clear that the rotor cannot rotate at synchronous speed. The speed of this type of motor is constant.

Wound-rotor type (slip-ring motor)

In the wound-rotor type of motor the cage is replaced by a three-phase winding which is connected via slip rings to a starter. The starter enables the rotor currents to be controlled which in turn controls, to a degree, the speed and torque. When the machine has reached speed the rotor windings are short-circuited and the brush gear, which is no longer required, is lifted clear of the slip rings. This type of motor is capable of taking extremely high rotor currents on starting and cables must be capable of carrying such currents.

Slip

As was previously mentioned, the rotor of an induction motor cannot travel at synchronous speed, as there would be no flux cutting and the machine would not work.

The rotor is, then, said to 'slip' in speed behind the synchronous speed. Slip (S) is usually expressed as a percentage and is given by

$$\text{Slip (\%)} = \frac{(n_s - n_r)}{n_s} \times 100$$

where n_s is the synchronous speed and n_r is the rotor speed.

Example

A six-pole cage induction motor runs at 4% slip. Calculate the motor speed if the supply frequency is 50 Hz.

$$S\ (\%) = \frac{(n_s - n_r)}{n_s} \times 100$$

$$\text{Synchronous speed } n_s = \frac{f}{p}$$

$$= \frac{50}{3}$$

$$= 16.666 \text{ revs/second}$$

$$\therefore 4 = \frac{(16.66 - n_r)}{16.66} \times 100$$

$$\therefore \frac{4 \times 16.66}{100} = (16.66 - n_r)$$

$$\therefore n_r = 16.66 - \frac{(4 \times 16.66)}{100}$$

$$= 16.66 - 0.66$$

$$= 16 \text{ revs/second}$$

Example

An eight-pole induction motor runs at 12 revs/second and is supplied from a 50 Hz supply. Calculate the percentage slip.

$$n_s = \frac{f}{p}$$

$$= \frac{50}{4} = 12.5 \text{ revs/second}$$

$$S(\%) = \frac{(n_s - n_r)}{n_s} \times 100$$

$$= \frac{(12.5 - 12)}{12.5} \times 100$$

$$= \frac{0.5 \times 100}{12.5}$$

$$= \frac{50}{12.5}$$

$$= 4\%$$

Frequency of rotor currents

As the rotating field is an alternating one, the currents induced in the rotor cage bars are also alternating. These are, however, not the same frequency as the supply. The frequency of the rotor currents f_s is given by

$$f_s = \text{slip} \times \text{supply frequency}$$

$$\therefore f_s = S \times f$$

Note: S here is expressed as a per unit value; i.e. for 4% slip,

$$S = \frac{4}{100} = 0.04$$

Example

An eight-pole squirrel-cage induction motor has a synchronous speed of 12.5 revs/second and a slip of 2%. Calculate the frequency of the rotor currents.

$$f = n \times p$$

$$= 12.5 \times 4$$

$$= 50\,\text{Hz}$$

$$f_s = S \times f$$

$$= \frac{2}{100} \times 50$$

$$= 1\,\text{Hz}$$

Note: *Three-phase* motors may be reversed by changing over any *two* phases.

Single-phase induction motors

With a three-phase motor the field is displaced by 120°. In the case of a single-phase supply there is no phase displacement and hence the rotor has equal and opposing forces acting on it and there will be no movement. The motor is therefore not self-starting. However, if the rotor is initially spun mechanically it will continue to rotate in the direction in which it was turned. Of course this method of starting is out of the question with all but the very smallest motors and is therefore confined to such items as electric clocks.

The creation of an artificial phase displacement is another and more popular method of starting.

Shaded-pole induction motor

The shaded-pole type of motor has a stator with salient (projecting) poles and in each pole face is inserted a short-circuited turn of copper (Fig. 8.22).

The alternating flux in the pole face induces a current in the shading coil which in turn produces an opposing flux. This opposition causes a slight phase

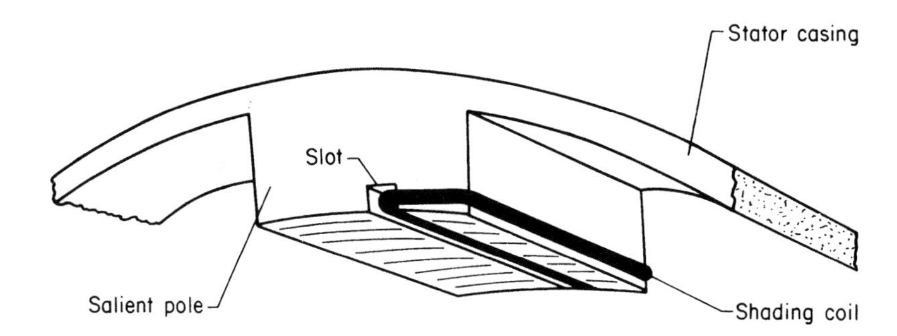

Fig. 8.22 *Shaded-pole arrangement*

displacement of the fluxes in the two parts of each pole which is enough to start the rotor turning.

As the phase displacement is very small the motor has a very small starting torque, thus limiting its use to very light loads.

Capacitor-start induction motor

With the capacitor-start induction motor the stator has a secondary winding, in series with which is a capacitor. This gives the effect of 90° phase difference and the motor will start. A second or two after starting, a centrifugal switch cuts out the secondary winding (Fig. 8.23).

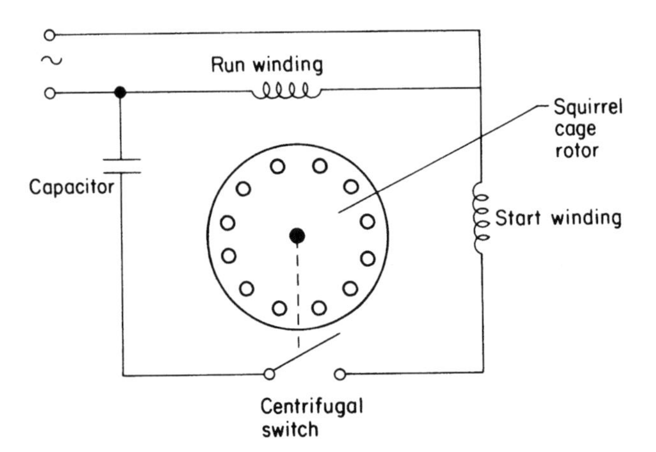

Fig. 8.23 *Capacitor-start motor*

This type of motor may be reversed in direction by reversing the connections to the start winding.

Reactance or induction-start induction motor

A phase displacement can be achieved by connecting an inductor in series with the start winding (Fig. 8.24). The centrifugal switch is as for the capacitor-start type.

Larger motors of this type take heavy starting currents, and series resistances are used to limit this.

Reversal of rotation is as for the capacitor-start type.

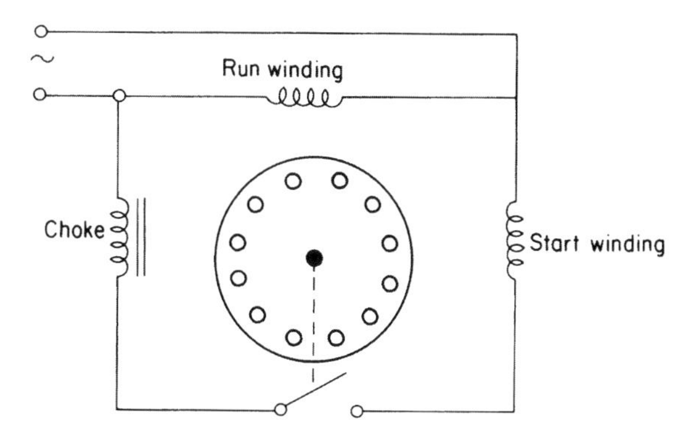

Fig. 8.24 *Reactance-start motor*

Resistance-start induction motor

In the case of a resistance-start induction motor a resistance replaces the choke or capacitor in the start winding to give a phase displacement (Fig. 8.25).

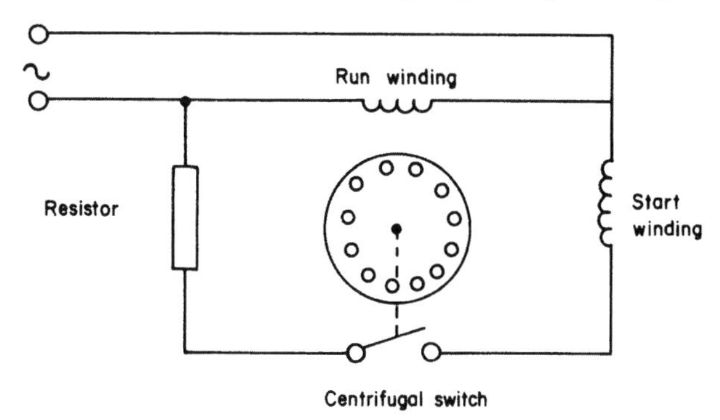

Fig. 8.25 *Resistance-start motor*

Capacitor-start capacitor-run induction motor

The most efficient of the range of single-phase induction motors is the capacitor-start capacitor-run type. The main feature is that the starting winding is not switched out but is continuously energized, the only change between starting and running being the value of capacitance. This change is achieved by using two capacitors and switching one out with the centrifugal switch (Fig. 8.26).

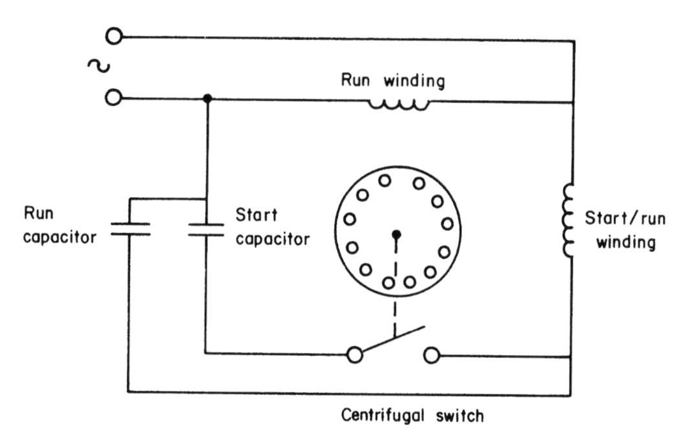

Fig. 8.26 *Capacitor-start capacitor-run motor*

Repulsion-start motor

Repulsion-start motors are of the wound-rotor type, the windings being terminated at a commutator, the brush gear of which is shorted out and arranged about 20° off centre. A transformer action takes place between the stator and rotor windings (mutual inductance) and as both windings will have the same polarity they repel or repulse each other. Speed control is effected by slight movement of the brushes around the commutator (Fig. 8.27).

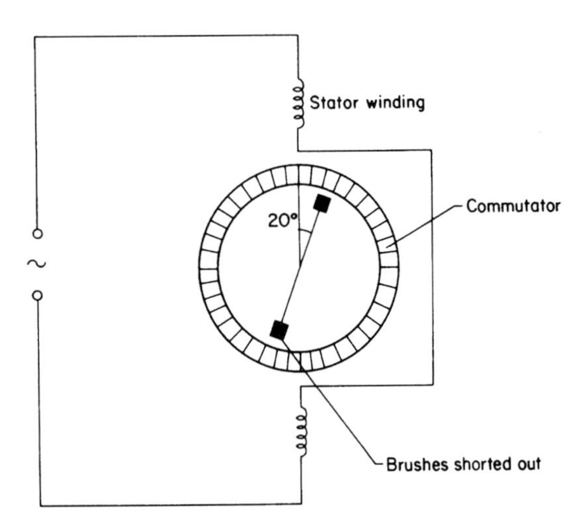

Fig. 8.27 *Repulsion motor*

A variant of this type starts as a repulsion motor; centrifugal gear then shorts out the commutator and lifts the brush gear clear, the motor then continuing to run as an induction motor.

The universal or series motor

The *universal* or *series* motor is simply a d.c.-type armature with commutator and an a.c. field. It is connected as for a d.c. series motor (Fig. 8.28).

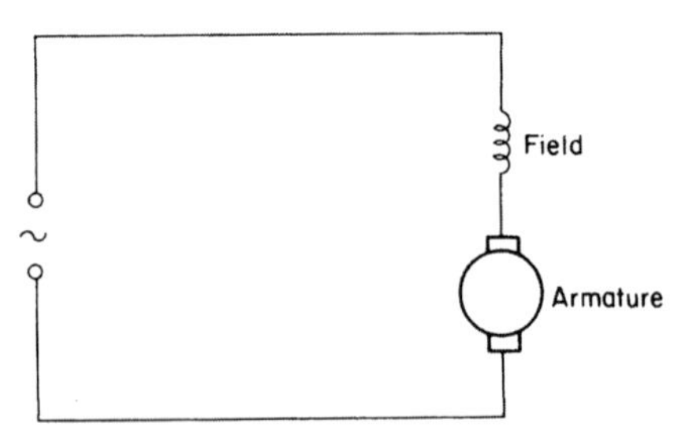

Fig. 8.28 *Single-phase series motor*

This motor will operate on an alternating current because the polarity of the a.c. supply changes on *both* field and armature; the motor will therefore rotate in one direction. Reversing is achieved by reversal of either field or armature connections.

Starters

Direct-on-line (DOL) starter

Fig. 8.29 illustrates a typical three-phase direct-on-line starter. When the start button is pressed, the 415 V contactor coil is energized and the main and auxiliary contacts close and the motor will start. The auxiliary contact in parallel with the start button holds the coil on.

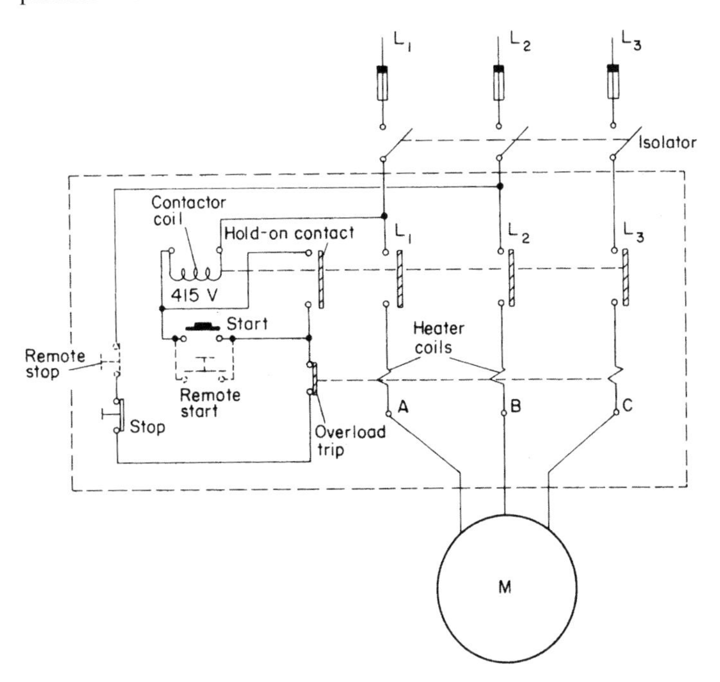

Fig. 8.29 *Three-phase direct-on-line starter*

Both single-phase and three-phase types use the same control circuit, as illustrated in Figs. 8.30 and 8.31. For Fig. 8.31 it is important to note that if a 240 V coil were to be used instead of a 415 V type, the coil connections would require a neutral conductor in the starter.

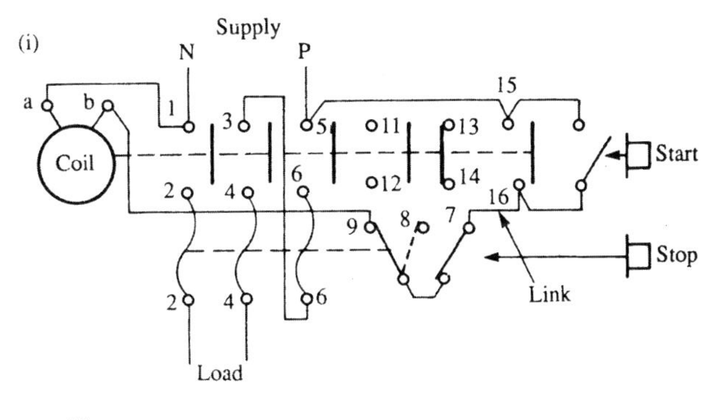

Fig. 8.30 *(i) Direct-on-line starter, single phase. (ii) Connections for remote push-button (start/stop) control; omit link and connect as shown*

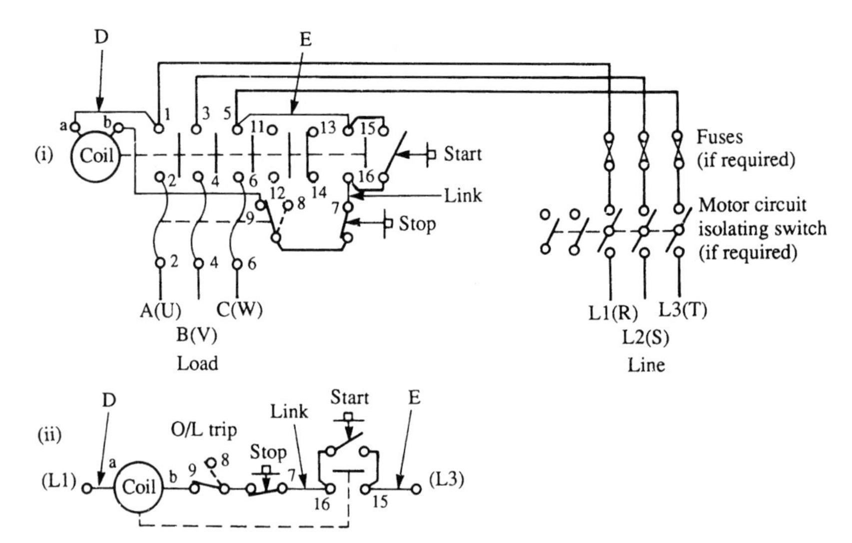

Fig. 8.31 *(i) Direct-on-line starter, three phase. (ii) Schematic diagram. Control circuit supply: for phase to phase, connect as shown; for phase to neutral, omit connection D and connect neutral to terminal a; for separate supply, omit D and E, and connect separate coil supply to terminals a and 15*

Overload or overcurrent protection is provided by either thermal or magnetic trips.

Thermal overload protection relies on the heating effect of the load current to heat the thermal coils which in turn cause movement of a bimetallic strip. This trips out a spring-loaded contact in the control circuit. The speed at which the tripping takes place is adjusted to allow for normal starting currents, which may be four or five times as large as running currents.

Magnetic protection uses the principle of the solenoid to operate the tripping mechanism. The time lag in this case is achieved by the use of an oil or air dashpot which slows down the action of the solenoid plunger (Fig. 8.32).

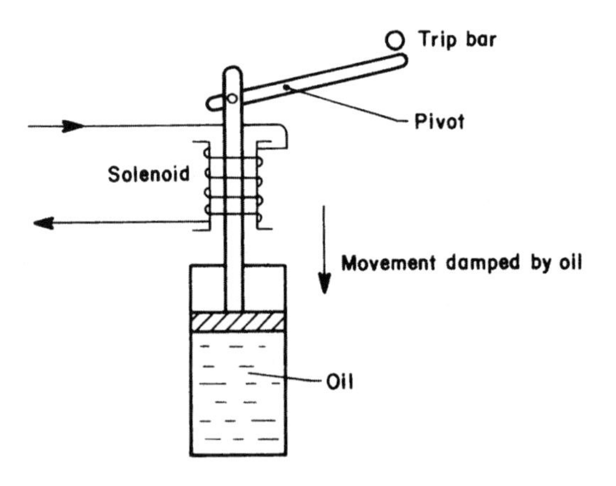

Fig. 8.32 *Oil dashpot damping*

Another form of thermal protection is given by the use of a *thermistor*, which is a temperature-sensitive semiconductor. It is embedded in the stator winding and activates a control circuit if the winding temperature becomes excessive.

Star–delta starter

If a motor's windings are connected in the star configuration, any two phases will be in series across the supply and hence the line current will be smaller (by 57.7%) than if the windings were connected in the delta arrangement. Hence larger-type motors with heavy starting currents are first connected in star, and then, when the starting currents fall, in delta. This of course means that all six of the ends of the windings must be brought to terminations outside the casing (Fig. 8.33).

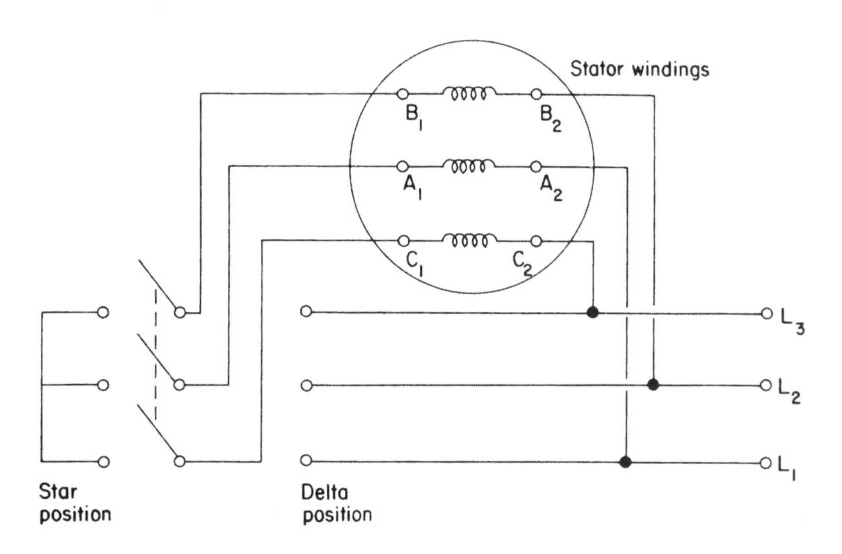

Fig. 8.33 *Basic star–delta starter*

The automatic version of this starter incorporates a timing relay which automatically changes the connections from star to delta.

Figs. 8.34 and 8.35 show the wiring and schematic diagrams for a star–delta starter. This is clearly a more complicated system than the DOL type.

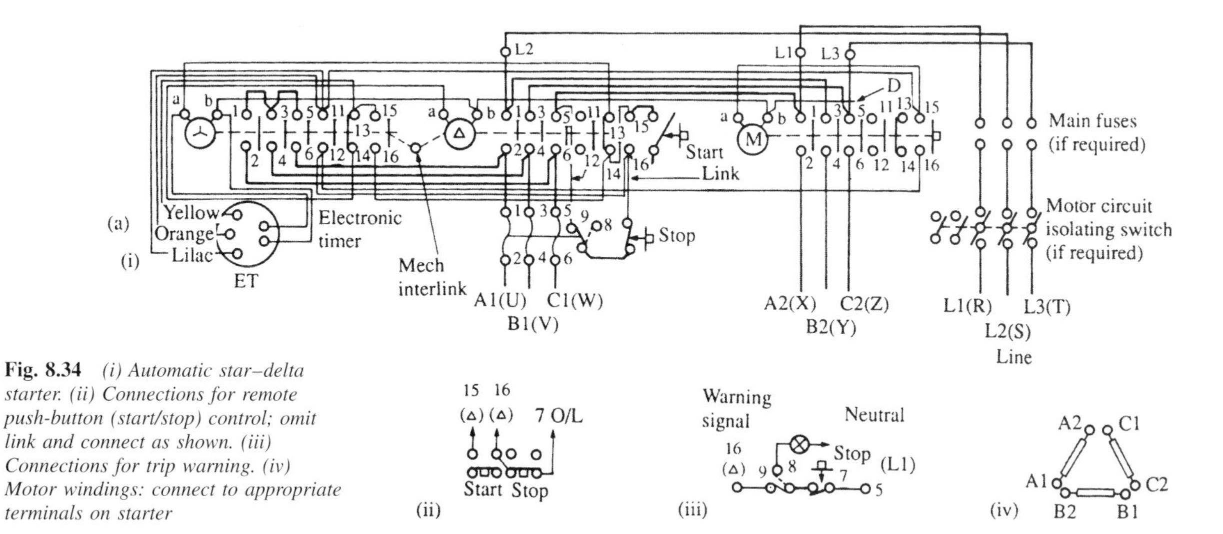

Fig. 8.34 *(i) Automatic star–delta starter. (ii) Connections for remote push-button (start/stop) control; omit link and connect as shown. (iii) Connections for trip warning. (iv) Motor windings: connect to appropriate terminals on starter*

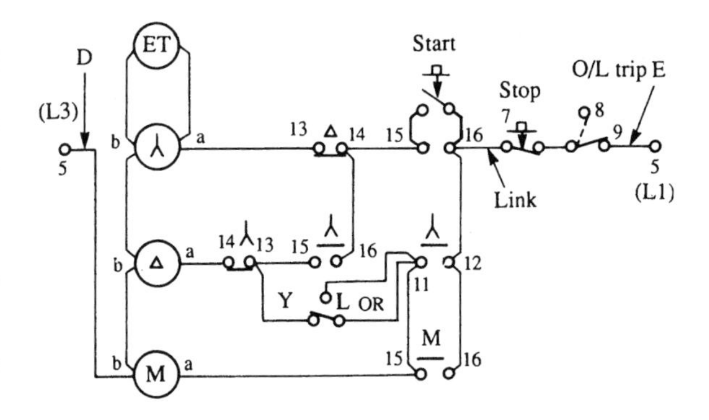

Fig. 8.35 *Schematic diagram. Control circuit supply; for phase to phase, connect as shown; for phase to neutral, omit connection D and connect neutral to terminal b; for separate supply, omit D and E, and connect separate coil supply to terminals b and 9. Connections for remote pilot switch control: remove connection 14 to 15 on delta contactor; connect between 14 and 16 on M contactor to terminal 14 on delta contactor; connect pilot switch in place of connection E*

Nevertheless, reference to the schematic Fig. 8.35 will indicate how the system functions:

1 When the start is pushed, supply is given to the star contactor (⅄) and electronic timer ET, from L1 to L3, and hence all contacts marked ⅄ will operate. Supply to (⅄) and ET is maintained, after the start is released, via ⅄ 11 and 12, time contacts OR-Y, and ⅄ 15 and 16. The main contactor Ⓜ is also energized via ⅄ 11 and 12 and M15, and thereafter maintained via its own contacts M15 and 16.
2 The motor has of course started. After a predetermined time delay ET operates and its contacts OR-Y change to OR-L. This cuts off supply to (⅄) and ET. All ⅄ contacts return to normal, and ET resets OR-L to OR-Y.
3 Supply is now given to the delta contactor (Δ) via M15 and 16, OR-Y, and ⅄ 13 and 14. Delta contacts Δ 13 and 14 open and prevent further energization of the star contactor.

The reader will notice that the line and load terminal markings in Fig. 8.34 show letters in brackets; these are the continental equivalents.

Auto-transformer starting

A star-connected auto-transformer with tappings gives lower starting currents than the star–delta type (Fig. 8.36).

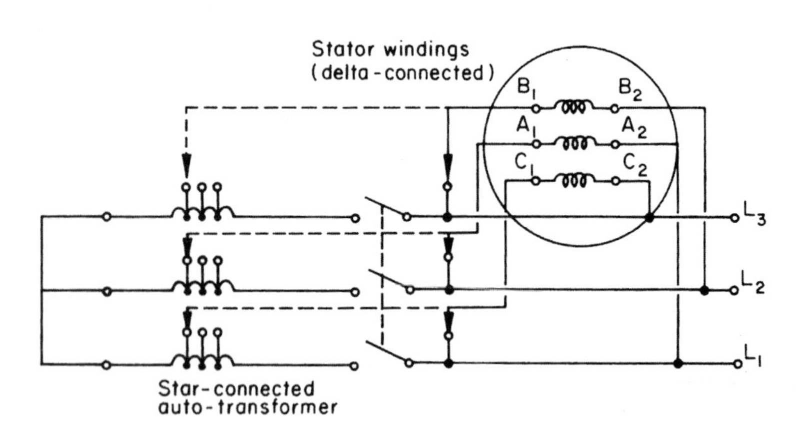

Fig. 8.36 *Basic auto-transformer starter*

Rotor-resistance starting

Fig. 8.37 shows the rotor-resistance-type starter for use with wound-rotor induction motors.

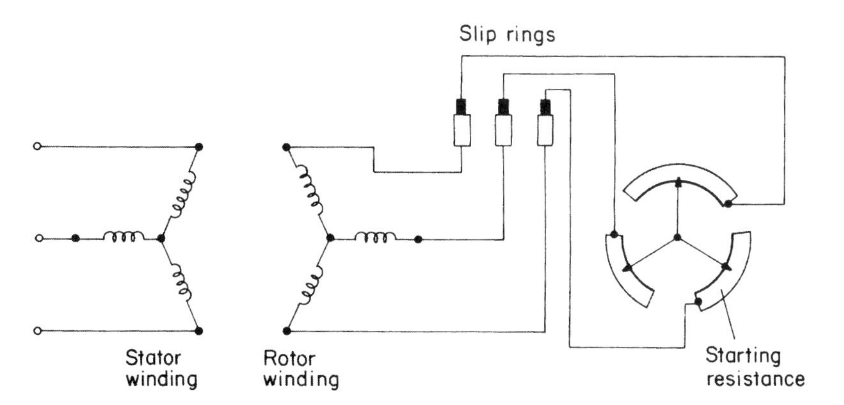

Fig. 8.37 *Rotor resistance starter*

Installing a motor

The correct handling, positioning, fixing and aligning of a motor are very important. Although it is a robust piece of machinery, great care should be taken, when transporting or positioning it, not to crack the casing or damage the feet.

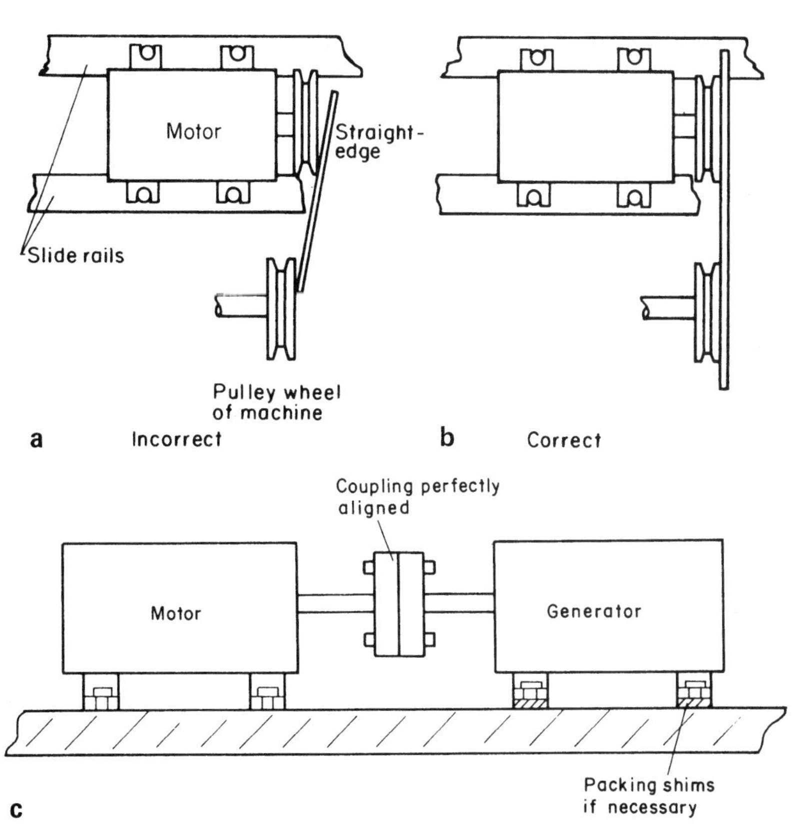

Fig. 8.38 *(a) Incorrectly and (b) correctly aligned belt-driven machine; (c) adjustment in the case of direct coupling*

Once the machine is in position it has to be fixed and aligned and this procedure will depend on the type of coupling used and the size of machine.

Very large machines are usually fixed to a concrete base or plinth. The concrete is a mixture of one part cement, two parts quartz sand and four parts of gravel. Fixing bolts should be grouted in the base with a mixture of 'one to one' washed sand and cement.

Most motors are in fact mounted on iron-slide-rails, the rails being fixed to the floor, wall or ceiling depending on the motor's use. In this way, the motor can be adjusted accurately for alignment, belt tensioning and direct coupling. Fig. 8.38a–c indicate the correct methods for alignment.

Note: Always check that the insulation resistance of the motor is satisfactory before switching on the supply; dampness may have been picked up during storage.

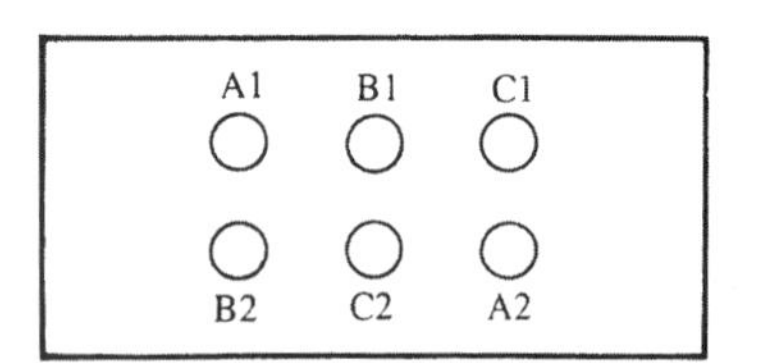

Fig. 8.39

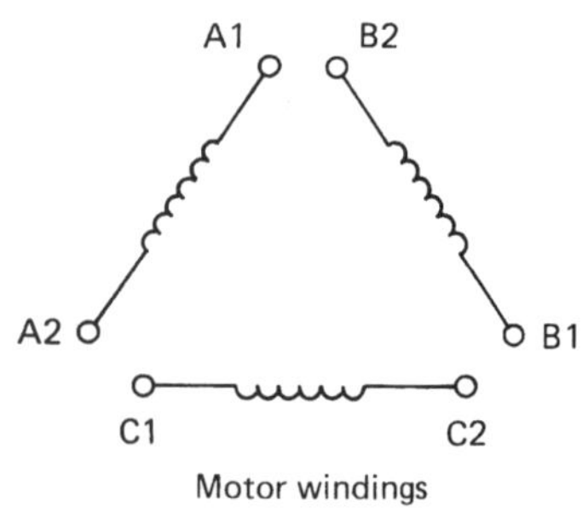

Fig. 8.40

Motor replacement

If a motor has been diagnosed as being faulty, and is either repaired or replaced, always ensure that the reconnections are correct. A reversal of the connections to the windings of a three-phase star–delta motor could have serious implications for the motor as well as for the operation of the circuit protection.

Figs. 8.39 and 8.40 show the terminal arrangements for a six-terminal motor and the correct winding connections.

If the connections to any winding are reversed, the magnetic fields will work against each other and a serious overload will occur, especially when the starter changes the motor to the delta configuration. For example, *never* allow A2 to be connected to B2 and A1 to C1.

This must not be confused with the reversal of any two *phases* of the supply when motor rotation needs to be reversed.

Motor enclosures

Once the most appropriate type of motor for a particular task has been selected, it is necessary to ensure that the motor enclosure is also suitable for its working environment. Various kinds of enclosure are listed below (Table 8.1) and their applications are summarized in Tables 8.2 and 8.3.

Table 8.1 Applications of various types of motor enclosure

Type	Applications
Screen-protected	General purposes; engineering worktops etc.
Drip-proof	Laundries, pump rooms, etc.
Pipe-ventilated	Flour mills, cement works, paper mills, etc.
Totally enclosed	Boiler houses, steelworks, outdoor winches, etc.
Flameproof	Gasworks, oil plants, chemical works, etc.

Table 8.2 Summary of the characteristics and applications of single-phase a.c. motors

Type	Main Characteristics	Applications
Universal (series)	Good starting torque, high PF	Small tools, drills, sanders, etc., vacuum cleaners
Repulsion	Good starting torque, low PF, and efficiency, speed control by brush shifting	Lifts, cranes, hoists, etc.
Repulsion-start	Good starting torque, low PF, and efficiency	Non-reversing load with heavy starting demands
Split-phase induction	Poor starting torque and PF	Used only with light starting conditions
Capacitor-start induction	Quite good starting torque and PF, quiet running	Refrigerators
Synchronous	Constant speed. Poor starting torque and PF	Clocks and timing devices

Table 8.3 Summary of the characteristics and applications of three-phase motors

Type	Main Characteristics	Applications
Small synchronous, no d.c. excitation	Constant speed, self-starting, light loads only, low PF	All drives needing synchronization
Synchronous with d.c. excitation	Constant speed, controllable PF, high efficiency with large outputs	Compressors, PF correction, ships' propulsion
Induction, squirrel-cage or wound-rotor	Starting performance good for small squirrel-cage but poor for large squirrel-cage motors good starting with all slip-ring motors	General service in engineering, pumps, machine tools, etc.

Screen-protected type

The most common enclosures in use are of the screen-protected type. The end covers are slotted and an internal fan draws cool air through the motor. Screen-protected enclosures can be used only in dust-free atmospheres.

Drip-proof type

The end plates on this type of motor are solid except for narrow slots on the underside. It can be used in damp and dust-free situations but is not waterproof.

Pipe-ventilated type

Cooling air is brought in via pipes from outside the building and circulated by an internal fan. This type is very suitable for extremely dusty environments.

Totally enclosed type

This type has no ventilation slots, its casing instead having ribs or fins to help cooling. The totally enclosed type is excellent for moist or dusty situations.

Totally enclosed flameproof type

This type is similar to the totally enclosed type but is more robustly built. It can withstand an internal explosion and prevent flames or sparks from reaching the outside of the casing.

Fault location and repairs to a.c. machines

The following tables indicate characteristics and fault diagnoses for motors.

Table 8.4 Possible faults on polyphase induction motors

Symptoms	Possible Causes	Test and/or Rectification
Fuses or over-current trips operate at start	Premature operation of protective gear Overload Reversed phase of stator winding Short circuit or earth fault on stator circuit Short circuit or earth fault on rotor circuit	
Motor will not start	Faulty supply or control gear Overload or low starting torque Open circuit in one stator phase Reversed phase of stator winding Open circuit in rotor circuit	
Overheated bearing or noisy operation	Bearing or mechanical defects	
Periodical growl	Reversed stator coil or coils	
Humming of squirrel-cage motor	Loose joints on rotor conductors	
Fluctuating stator current	Open circuit in rotor circuit	
General overheating of case	Faulty ventilation, mechanical or electrical overload Rotor core not fully in stator tunnel Open circuit in one of two parallel stator circuits	 Reassemble motor correctly
Overheating and over-labouring, two phases of star-connected stator or one phase of delta winding hotter than the rest	Single phasing owing to open-circuited supply line Open circuit in one phase of stator circuit	

Table 8.4 continued

Symptoms	Possible Causes	Test and/or Rectification
Reduced speed	Mechanical overload, low volts or low frequency Open circuit in rotor circuit	
Reduced speed of slip-ring motor	Rotor starter not fully operated	Overhaul protective gear to ensure correct operation
	Slip rings not short-circuited	Use slip-ring short-circuiting gear
	Voltage drop on cables to rotor starter	Fit rotor starter nearer motor, or use larger rotor circuit cables

Table 8.5 Possible faults on single-phase induction and capacitor types of motor

Symptoms	Possible Causes	Test and/or Rectification
Fuses or overcurrent trips operate at start	Premature operation of protective gear Overload Section of stator windings reversed Short circuit or earth fault on stator circuit Stator windings in parallel instead of series Short circuit or earth fault on slip-ring rotor circuit	
Motor will not start	Faulty supply or control circuit Overload or low starting torque Open circuit or reversed coils on stator winding Open circuit on slip-ring rotor circuit Centrifugal switch or relay sticking open	
Overheated bearing or noisy operation	Bearing or mechanical defects	
General overheating of case	Faulty ventilation, mechanical or electrical overload	
	Rotor core not fully in stator tunnel	Reassemble motor correctly
	Open circuit in one of two parallel stator circuits	
	Short circuit on auxiliary stator winding Short circuit on centrifugal switch or relay	Overhaul switch or relay and check operation
	Centrifugal switch or relay sticking closed	Overhaul switch or relay and check operation
	Reversed section of stator windings Prolonged or too frequent starting	
Reduced speed of all motors	Mechanical or electrical overload Low volts or frequency Open circuit in rotor	
Reduced speed of slip-ring motor	Rotor starter not fully operated	Overhaul protective gear to ensure correct operation
	Slip rings not short-circuited	Use slip-ring short-circuiting gear
	Voltage drop on cables to rotor starter	Fit rotor starter nearer motor, or use larger rotor circuit cables

Table 8.6 Possible faults on series a.c. and universal motors

Symptoms	Possible Causes	Test and/or Rectification
Fuses or overcurrent trips operate at start	Premature operation of protective gear	
	Overload	
	One field coil reversed	
	Short circuit or earth fault on field winding	
	Short circuit or earth fault on armature	
	Wrong brush position	
Motor will not start	Faulty supply or control circuit	
	Brushes not making good contact	
	Open circuit in field windings	
	Wrong brush position	
	Short circuit or earth fault on armature	
	Short circuit or earth fault on field windings	
	Reversed field coil	
Overheated bearing or noisy operation	Bearing or mechanical defects	
General overheating of the case	Faulty ventilation, mechanical or electrical overload	
	Short circuit, open circuit, or earth fault on armature	
	Short circuit or earth fault on field windings	
Reduced speed	Mechanical or electrical overload	
	Low voltage	
	Wrong brush position	
Increased speed	High voltage	
	Motor unloaded	It is inadvisable to run unloaded
Sparking at brushes	Faulty brushes or commutator	
	Mechanical or electrical overload	
	Wrong brush position	
	Incorrect brush spacing	
	Open circuit, short circuit or earth fault in armature	
	Reversed armature coil	

Table 8.7 Possible faults on repulsion-type motors

Symptoms	Possible Causes	Test and/or Rectification
Fuses or overcurrent trips operate at start	Premature operation of protective gear	
	Overload	
	Selection of stator windings reversed	
	Short circuit or earth fault on stator circuit	
	Stator windings in parallel instead of series	
	Short circuit or earth fault on armature	
	Wrong brush position	
	Commutator short-circuiting gear sticking in running position	Overhaul centrifugal gear and check operation

Table 8.7 continued

Symptoms	Possible Causes	Test and/or Rectification
Motor will not start	Faulty supply or control circuit Overload or low starting torque Short circuit or earth fault on stator circuit Open circuit or reversed coils on stator windings Brushes not making good contact Short circuit or earth fault on armature Commutator short circuiting gear sticking in running position Wrong brush position	Overhaul centrifugal gear and check operation
Overheated bearing or noisy operation	Bearing or mechanical defects	
General overheating of case	Faulty ventilation, mechanical or electrical overload Rotor core not fully in stator tunnel Open circuit in one of two parallel stator circuits Wrong brush position Burnt contacts on commutator short-circuiting gear	Reassemble motor correctly Overhaul short-circuiting contacts
Overheating of repulsion–induction motor	No load	Motor usually runs hotter unloaded than on full load
Reduced speed Wrong brush position Faulty commutator short-circuiting gear	Low volts or frequency Overhaul short-circuiting gear, check operation Overload Short circuit or earth fault on armature of plain repulsion motor Open circuit in squirrel-cage of repulsion–induction motor	
Hunting of repulsion-start induction motor	Faulty commutator short-circuiting gear or brushes	Overhaul short-circuiting gear and brushes
Sparking at brushes	Faulty brushes or commutator Mechanical or electrical overload Wrong brush position Incorrect brush spacing Reversed armature coil Short circuit, open circuit or earth fault on armature	

Table 8.8 Possible faults on synchronous types of motor

Symptoms	Possible Causes	Test and/or Rectification
Fuses or overcurrent trips operate at start	Premature operation of protective gear Overload Short circuit or earth fault on armature	
Motor will not start	Faulty supply or control gear Low starting voltage Overload Open circuit in one armature phase	Adjust tappings on transformer

Table 8.8 continued

Symptoms	Possible Causes	Test and/or Rectification
Motor fails to synchronize	External field resistance too high	Adjust field-regulating resistor
	Open circuit in field circuit	
	No excitation	Faulty exciter
Overheated bearing or noisy operation	Bearing or mechanical defects	
Vibration	Faulty supply	
	Open circuit in one armature phase	
General overheating	Faulty ventilation	
	Overload	
	High voltage	
	Short circuit, open circuit or earth fault on armature	
	Incorrect field strength	Adjust field-regulating resistor
	Unequal pole strength	Test field coils
	Unequal air gap	
Motor runs fast	High frequency	
Motor runs slow	Low frequency	
Motor pulls out of synchronism	Overload	
	External field resistance too high	Adjust field-regulating resistor
	Open circuit in field circuit	
	No excitation	Faulty exciter

Power factor of a.c. motors

Motors, being highly inductive pieces of equipment, have lagging power factors, some more so than others. In situations where a large number of machines are used, as in industrial premises, it is clear that some action should be taken to correct this lagging power factor. Where motors are used intermittently it is perhaps best to correct the power factor of each motor rather than the overall power factor of the installation, and capacitors connected across the terminals of each machine are used.

Example

A 240 V, 50 Hz single-phase induction motor takes a current of 13 A at a PF of 0.35 lagging. Calculate the value of capacitor required to correct the PF to 0.85 lagging.

The phasor diagram of the top branch (Fig. 8.41) is shown in Fig. 8.42, while the phasor diagram of the bottom branch is as shown in Fig. 8.43. Combining both phasor diagrams so that the resultant current is at 0.85 lagging (31.8°), we have Fig 8.44.

Hence,

$\cos \alpha = \text{PF} = 0.85$

$\therefore \alpha = 31.8°$

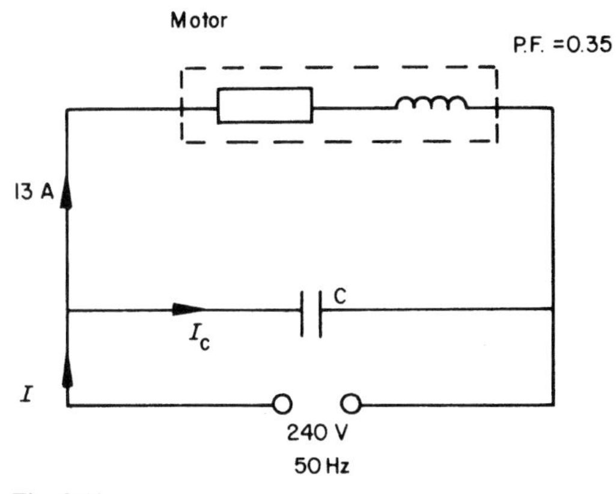

Fig. 8.41

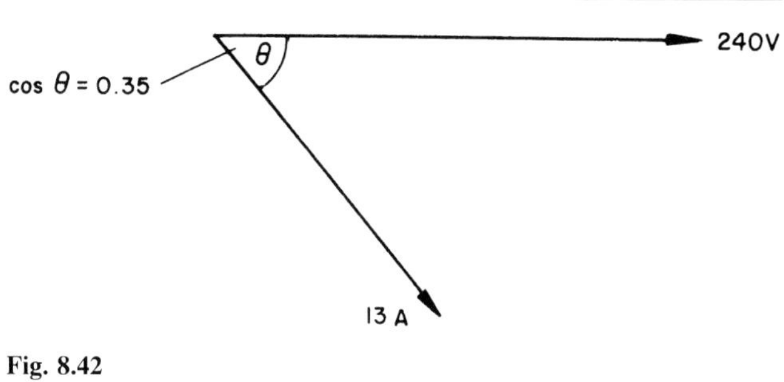

Fig. 8.42

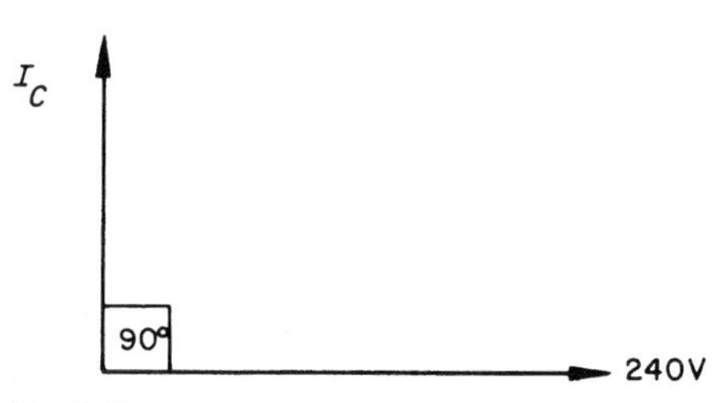

Fig. 8.43

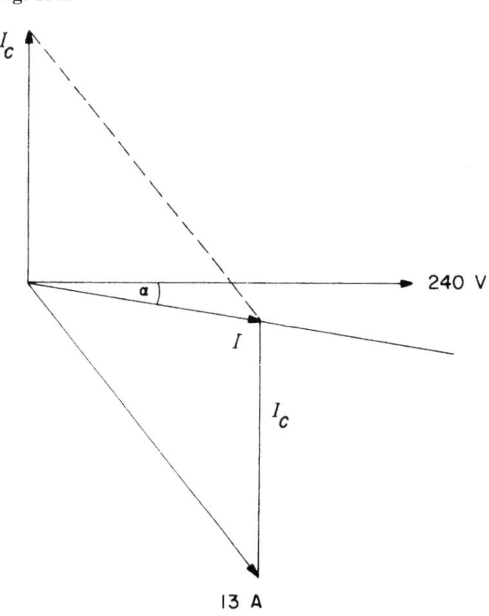

Fig. 8.44

By measurement,

$$I_C = 9.4\,\text{A}$$

$$X_C = \frac{V}{I_C}$$

$$= \frac{240}{9.4}$$

$$= 25.5\,\Omega$$

$$X_C = \frac{1}{2\pi fC}$$

$$\therefore C = \frac{1}{2\pi fX_C}$$

$$= \frac{1}{314.16 \times 25.5}$$

$$= 125\,\mu\text{F}$$

Motor ratings

As motors are a.c. plant, their electrical input is rated in kVA. The mechanical output from the machine is rated in horsepower (hp) or kilowatts (kW) where 1 hp = 746 W, and this is the rating usually displayed on the motor.

Note: Horsepower is no longer used but will still be found on older machines.

As machines have moving parts there are mechanical as well as electrical losses and they will have an efficiency given by

$$\text{Efficiency (\%)} = \frac{\text{output} \times 100}{\text{input}}$$

Example

A 5 kW, 240 V, 50 Hz induction motor has a running PF of 0.7 lagging and an efficiency of 80%. Calculate the current drawn by the motor.

$$\text{Efficiency (\%)} = \frac{\text{output} \times 100}{\text{input}}$$

$$\therefore \text{Input} = \frac{\text{output} \times 100}{\text{efficiency (\%)}}$$

$$= \frac{5 \times 100}{80}$$

$$= 6.25\,\text{kW}$$

$$\text{PF} = \frac{\text{kW}}{\text{kVA}}$$

$$\therefore \text{kVA} = \frac{\text{kW}}{\text{PF}}$$

$$= \frac{6.25}{0.7}$$

$$= 8.93\,\text{kVA}$$

$$I = \frac{\text{VA}}{V}$$

$$= \frac{8.93 \times 10^3}{240}$$

$$= 37.2\,\text{A}$$

Example

A 25 kW, 415 V, 50 Hz three-phase squirrel-cage induction motor is 87% efficient and has a PF of 0.92 lagging. Calculate the line current of the motor.

$$\text{Efficiency (\%)} = \frac{\text{output} \times 100}{\text{input}}$$

$$\therefore \text{Input} = \frac{\text{output} \times 100}{\text{efficiency (\%)}}$$

$$= \frac{25 \times 100}{87}$$

$$= 28.73\,\text{kW}$$

$$\text{Power (watts)} = \sqrt{3}\ V_L \times I_L \times \text{PF}$$

$$\therefore I_L = \frac{P}{\sqrt{3}\ V_L \times \text{PF}}$$

$$= \frac{28\,730}{\sqrt{3} \times 415 \times 0.92}$$

$$= 43.45\,\text{A}$$

Torque and output

Example

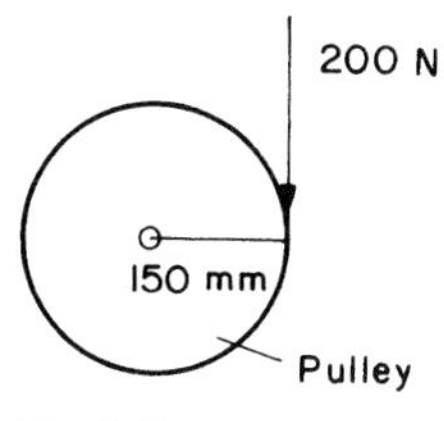

Fig. 8.45

A four-pole cage induction motor is run from a 50 Hz supply and has a slip of 3%. The rotor shaft drives a pulley wheel 300 mm in diameter, which has a tangential force of 200 N exerted upon it. Calculate the power output from the rotor in watts.

From Fig. 8.45:

$$\text{Torque} = \text{force} \times \text{radius}$$

$$= 200 \times 150 \times 10^{-3}$$

$$= 30\,\text{N m}$$

$$\text{Slip (\%)} = \frac{(n_s - n_r)\ 100}{n_s}$$

and

$$n_s = \frac{f}{p}$$

$$= \frac{50}{2}$$

$$= 25\ \text{revs/second}$$

$$\therefore 3 = \frac{(25 - n_r)\ 100}{25}$$

$$\therefore \frac{3 \times 25}{100} = 25 - n_r$$

$$\therefore n_r = 25 - \frac{(3 \times 25)}{100}$$

$$= 25 - 0.75$$

$$= 24.25 \text{ revs/second}$$

$$P = 2\pi nT$$

$$= 2\pi \times 24.25 \times 30$$

$$= 4.57\,\text{kW}$$

Points to note

1 The voltage drop between the supply intake position and the motor must not exceed 4% of the supply voltage.
2 The motor enclosure must be suitable for the environment in which it is to work; for example, a flameproof enclosure is required for explosive situations.
3 Every motor must have a means of being started and stopped, the means of stopping being situated within easy reach of the person operating the motor.
4 Every motor should have a control such that the motor cannot restart after it has stopped because of mains voltage drop or failure (i.e. undervoltage protection). This regulation may be relaxed if a dangerous situation will arise should the motor fail to restart.
5 *Every* stopping device must have to be reset before a motor can be restarted.
6 A means of isolation must be provided for every motor and its associated control gear. If this means is remote from the motor, isolation adjacent to the motor must be provided, or the remote isolator must be capable of being 'locked off'.
7 If a single motor and/or its control gear in a group of motors is to be maintained or inspected, a single means of isolation for the whole group may be installed provided the loss of supply to the whole group is acceptable.
8 Excess current protection must be provided in control gear serving motors rated above 370 watts and/or in the cables between the protection and the motor.
9 Cables carrying the starting and load currents of motors must be at least equal in rating to the full-load current rating of the motor. This includes rotor circuits of slip-ring or commutator motors.
10 The final circuit supplying a motor shall be protected by fuses or circuit breakers of rating not greater than that of the cable, unless a starter is provided that protects the cable between itself and the motor, in which case the fuses or circuit breakers may be rated up to twice the rating of the cable between the fuse and the starter (Fig. 8.46).

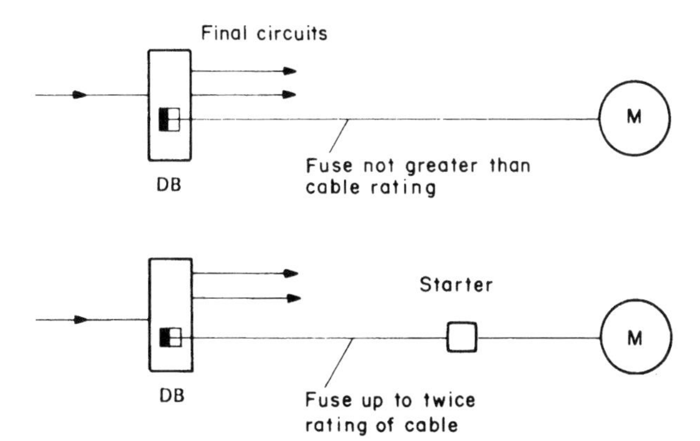

Fig. 8.46 *(DB = distribution board)*

Self-assessment questions

1 (a) Explain what is meant by the term 'back e.m.f.' in a motor.
(b) Outline the basic differences between series, shunt and compound-wound d.c. motors.

2 (a) With the aid of diagrams explain how the speed of series, shunt and compound-wound motors may be controlled.
(b) A 440 V d.c. shunt-wound motor has a field resistance of 200 Ω and an armature resistance of 0.6 Ω. When its speed is 20 revs/second the current drawn from the supply is 12.2 A. Calculate its back e.m.f. at this speed. If the speed were decreased to 19 revs/second, the field flux remaining unchanged, calculate the new back e.m.f. and the new armature current.

3 A load of 6.4 kW at 240 V is supplied from the terminals of a shunt-wound d.c. generator. The field resistance is 180 Ω. Calculate the armature current.

4 (a) Explain with the aid of a diagram how a rotating magnetic field may be obtained.
(b) What is meant by 'synchronous speed'? Calculate the synchronous speed of a 12 pole motor if the supply frequency is 50 Hz.

5 (a) Explain the action of a synchronous motor.
(b) What methods are available to start a synchronous motor? Explain with diagrams.

6 (a) Explain the action of a cage induction motor.
(b) What is meant by the term 'slip'? Calculate the percentage slip of a six-pole induction motor running at 16.2 revs/second from a 50 Hz supply.

7 Describe with sketches three different ways of starting a single-phase induction motor.

8 Explain with the aid of a sketch the action of a three-phase DOL starter. What are dashpots used for? How are remote start and stop buttons connected?

9 A 10 kW, 240 V, 50 Hz single-phase cage rotor induction motor is 85% efficient and has a PF of 0.68 lagging. Calculate the motor current and the value of capacitor required to raise the PF to 0.93 lagging.

10 Calculate the torque developed by an 18 kW four-pole induction motor run at 3.5% slip from a 50 Hz supply.

9 Cells and batteries

General background

In 1789, an Italian professor of botany, Luigi Galvani (1737–98), noticed by chance that freshly skinned frogs' legs twitched when touched by two dissimilar metals. He called this effect *animal electricity.*

It was, however, another Italian, Alessandro Volta (1745–1827), a professor of physics, who showed that the electric current which produced the muscular spasm was not due to the animal limb itself, but to the moisture in it. In 1799 he developed a simple battery comprising copper and zinc discs separated by a brine-soaked cloth. This type of assembly is known as a *voltaic pile*. From this primitive beginning have come the cells and batteries we use today. The materials used may be more refined, but the basic concept has remained unchanged.

The primary cell

If two dissimilar metals are immersed in an acid or salt solution, known as an *electrolyte*, an e.m.f. will be produced; this assembly is known as a *cell*. The e.m.f. may be used to supply a load, but will only do so for a limited time, as the chemical qualities of the electrolyte deteriorate with use. The chemicals have to be renewed to render the cell useful again.

The most common forms of primary cells in use are the Leclanché wet cell and the dry cell.

Fig. 9.1 illustrates the component parts of these two types of cell.

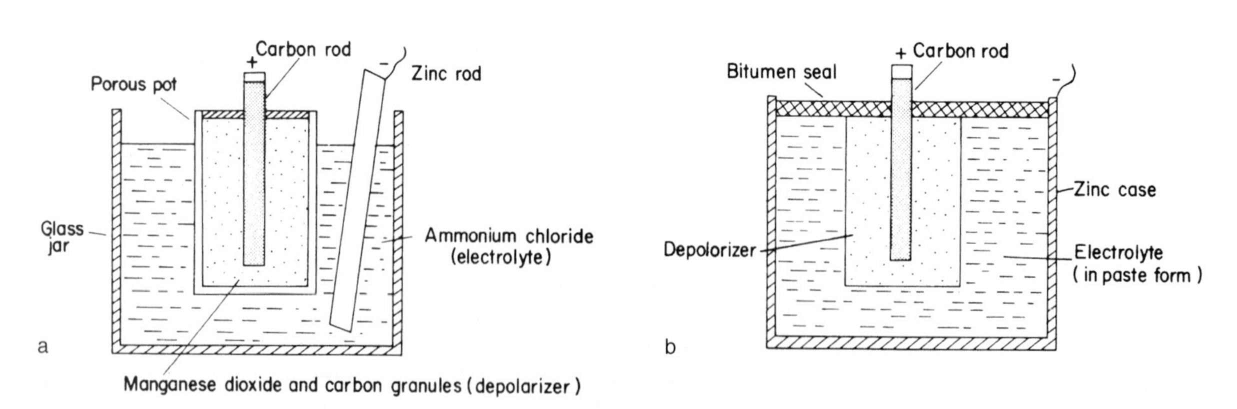

Fig. 9.1 *(a) Wet cell; (b) dry cell*

The depolarizing agent is used to remove hydrogen bubbles from around the carbon rod. These bubbles, which are formed during the chemical action, impair the performance of the cell.

Applications

The dry cell has an obvious advantage over the wet cell because it is portable and so is commonly used for appliances such as torches, door bells, etc.

The wet cell, although almost obsolete, is used in larger bell and indicator circuits and for railway signalling.

The secondary cell

Unlike the primary cell, the secondary cell can be used again after it has discharged all its electrical energy. It can be recharged by *supplying* it with electrical energy. This reverses the chemical process which took place during discharge.

There are two types of secondary cell, the lead–acid and the alkaline cell.

The lead–acid cell

This cell consists of positive and negative lead electrodes, and an electrolyte of *dilute sulphuric acid* all placed in an acid-resistant container.

The electrodes are made of several plates, the positive and negative being insulated from one another by separators of insulating material such as wood or ebonite.

The construction of the plates is of considerable importance and is discussed below.

Formed plates

Repeated charging and discharging of a cell under manufacture causes the plates to be covered in lead compounds, the negative plate with *spongy lead* and the positive plate with *lead peroxide*, these being important to the chemical action of the cell. This process is, however, both expensive and time consuming and for smaller types of cell, pasted plates are more popular.

Pasted plates

These plates are manufactured in the form of a grid, into which a compound of *sulphuric acid and red lead* is pressed; only a short initial charge is needed for the cell to be ready for use. These plates, however, disintegrate more easily than the formed type.

A combination of formed and pasted plates is used in large-capacity cells; the positive plate is formed and the negative plate is pasted.

When current is drawn from the cell the active chemicals on the positive plate expand and the plate tends to distort, especially under heavy loads. Some measure of protection against this distortion or buckling, is achieved by arranging for each positive plate to be adjacent to two negative plates (Fig. 9.2).

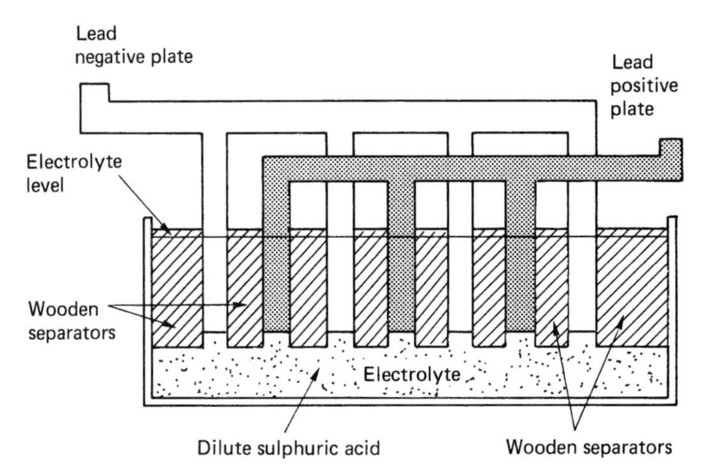

Fig. 9.2 *Lead–acid cell*

Action of lead–acid cell

When an external load is connected to the terminals of the cell, electrical energy is delivered to the load. During this *discharge* period, a chemical reaction between the plates and the electrolyte takes place and a layer of *lead sulphate* is deposited on the plates. However, this process successively weakens the electrolyte until the cell is unable to deliver any more electrical energy.

If a d.c. supply is then connected to the cell terminals, and a current is passed through it, the *lead sulphate* is converted back into *sulphuric acid* and restores the cell to its original condition. This process is known as *charging*.

Care and maintenance of lead–acid cells

Provided that a lead–acid cell is maintained regularly and is cared for, it should last for an indefinite period of time. A *weekly* check on its condition is to be recommended.

Electrolyte level

The level of the electrolyte should never be allowed to fall below the tops of the plates. Any loss of electrolyte due to evaporation may be made up by the addition of distilled water.

Specific gravity of electrolyte

As a cell discharges, the electrolyte becomes weaker and its specific gravity (SG) falls, until the cell can no longer deliver energy. The state of charge of a cell can therefore be measured by the SG of the electrolyte. A hydrometer is used for this purpose. It consists of a glass syringe containing a weighted, graduated float. The syringe has a rubber nozzle for insertion into the electrolyte, and a rubber bulb at the upper end for sucking the liquid into the syringe.

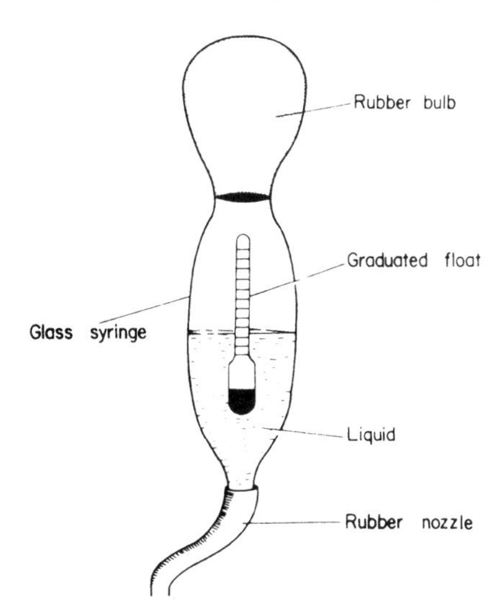

Fig. 9.3 *Hydrometer*

The nozzle is inserted in the electrolyte and a sample is drawn up into the syringe by squeezing the rubber bulb. The level of the liquid in relation to the position of the float gives a direct reading of the SG of the electrolyte. The higher the float, the higher the SG (see Fig. 9.3); the lower the float, the lower the SG.

The following table indicates typical values of the specific gravity in relation to charge:

SG	Percentage of charge
1.28	100
1.25	75
1.22	50
1.19	25
1.16	Fully discharged

A record of results obtained should be kept for each cell.

Note: Cells should not be allowed to fall below an SG of 1.18.

Terminal voltage

A check on the no-load terminal voltage should be made with a high-resistance voltmeter at the end of a discharge period. This reading should not be below 1.85 V. A fully charged cell should indicate about 2.2 V.

Plate colour

An indication of the state of charge is the colour of plates. In a healthy cell the positive plate is chocolate brown and the negative plate is slate grey.

General maintenance procedures

1 When preparing an electrolyte, *always* add acid to water, *never* water to acid.
2 Ensure that any maintenance is carried out in a well-ventilated area.
3 Do not permit the use of any naked flame near the cells.
4 Cells which are to be taken out of commission for any time should be fully charged, the electrolyte left in, and a periodic charge given to keep the cell healthy until it is needed again.
5 Never leave a cell in an uncharged state, as a layer of whitish *sulphate* will form on the plates, which will increase the internal resistance and reduce the capacity of the cell. This process is known as *sulphation of the plates*.
6 Terminals should be coated with petroleum jelly to prevent corrosion.

Application of lead–acid cells

The most common use for this type of cell is the car battery (a battery is a group of cells). Other applications include standby supplies, alarm and control circuits. The electrolyte of such a battery has a negative temperature coefficient (Chapter 2) and hence a drop in temperature causes a rise in its resistance, so less current will be delivered. We have all experienced starting problems with vehicles in very cold weather.

The nickel–alkaline cell

There are two types of alkaline cell, the nickel–iron and the nickel–cadmium.

The nickel–iron cell

Here the positive plate is made of *nickel hydroxide*, the negative plate of *iron oxide* and the electrolyte is *potassium hydroxide*.

The nickel–cadmium cell

In this cell both the positive plate and the electrolyte are the same as for the nickel–iron cell; however, the negative plate is *cadmium* mixed with a small amount of iron.

The active chemicals in the plates of alkaline cells are enclosed in thin nickel–steel grids insulated from one another by ebonite rods. The whole assembly is housed in a welded steel container.

Care and maintenance of the nickel–alkaline cell

Unlike the lead–acid cell the nickel–alkaline needs minimal attention.

The open type only needs periodic topping up to compensate for the electrolyte lost by evaporation. The totally enclosed type needs no maintenance.

Applications

It has limited use owing to its cost and is mainly used in situations where a robust construction is needed, i.e. marine work.

Capacity of a cell

If a cell delivers, say, 10 A for a period of 10 hours, it is said to have a capacity of 100 ampere hours (A h) at the 10 hour rate; taking any more than 10 A will discharge the cell in less than 10 hours.

Efficiency of a cell

The efficiency of any system is the ratio of the output to the input. The efficiency of cells is given in two forms:

$$\text{The ampere hour efficiency } \% = \frac{\text{discharge amperes} \times \text{time} \times 100}{\text{charging amperes} \times \text{time}}$$

$$\text{The watt hour efficiency } \% = \frac{\text{discharge VA} \times \text{time} \times 100}{\text{charging VA} \times \text{time}}$$

Comparison of cell characteristics

	Lead–acid	Alkaline
SG charged	1.28	1.2
SG discharge	1.18	1.2
Pd charged	2.1 V	1.3 V
Pd discharge	1.85 V	1.0 V

A h capacity

The A h obtainable from the alkaline cell at the higher discharge rates, i.e. 2 and 4 hours, is much greater than that of the lead–acid cell (Fig. 9.4).

This is because the SG of the electrolyte does not change during discharge.

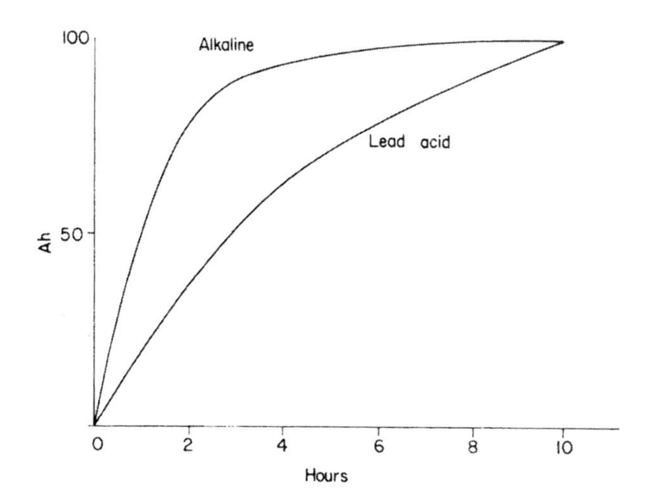

Fig. 9.4

Charge and discharge

The dips in the charging curves (Fig. 9.5) are due to the reduction of the charging current, to prevent overheating of the cell.

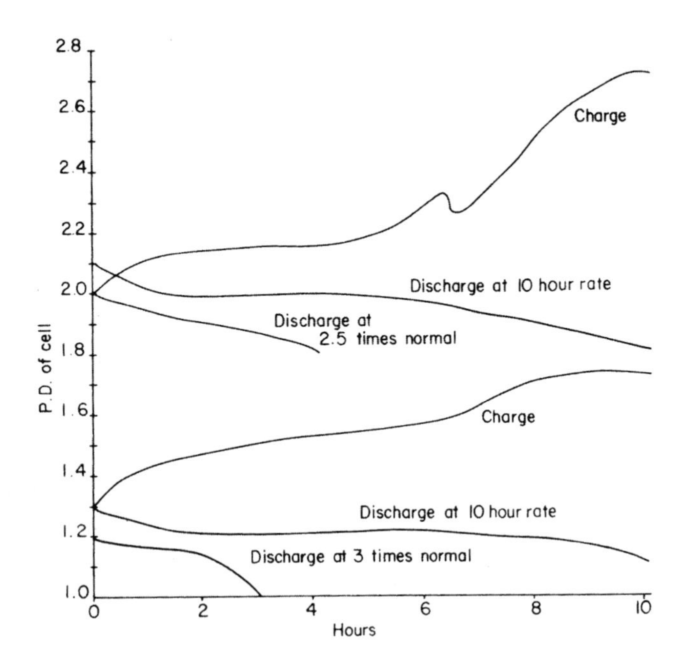

Fig. 9.5

	Advantages	Disadvantages
Lead–acid	Inexpensive High discharge voltage Uses plentiful materials	Fragile Self-discharges when not in use Requires regular maintenance
Alkaline	Very robust Retains its charge when not in use Needs little or no maintenance	Very expensive Low discharge voltage

Cell and battery circuits

E.m.f. of cell (E, volts)

This is the maximum force available (measured in volts) in a cell to produce current flow.

Internal resistance

When current flows through the cell there is some resistance to its flow (less than 1 Ω in a good cell) and hence a voltage drop across it.

P.d. of a cell or terminal (V, volts)

This is the voltage measured at the terminals of a cell, and is less than the cell e.m.f. owing to the voltage drop across the internal resistance of the cell (Fig. 9.6). E = e.m.f. of cell; r = internal resistance of cell; R = resistance of load; I = circuit current; and V = terminal voltage or p.d. across load.

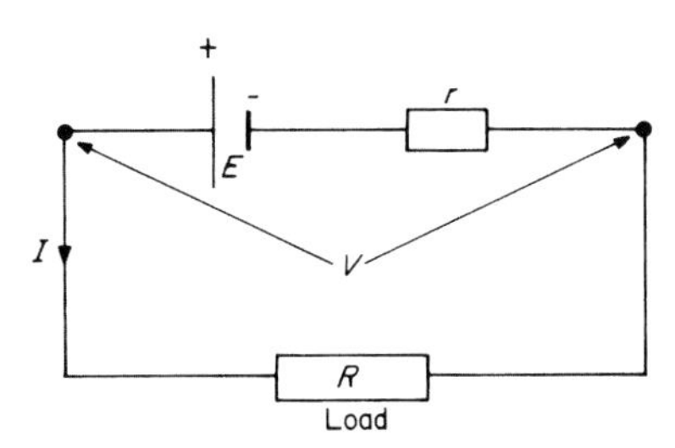

Fig. 9.6

The terminal voltage (V) available across the load is clearly the e.m.f. (E) less the voltage drop across the internal resistance (r):

$V = E - (I \times r)$

This is the same principle as the voltage available across the terminals of a load which is supplied by a long cable.

Example

If a cell with an e.m.f. of 2 V and an internal resistance of 0.2 Ω is connected across a 0.8 Ω load resistor, calculate the current that will flow (see Fig. 9.7).

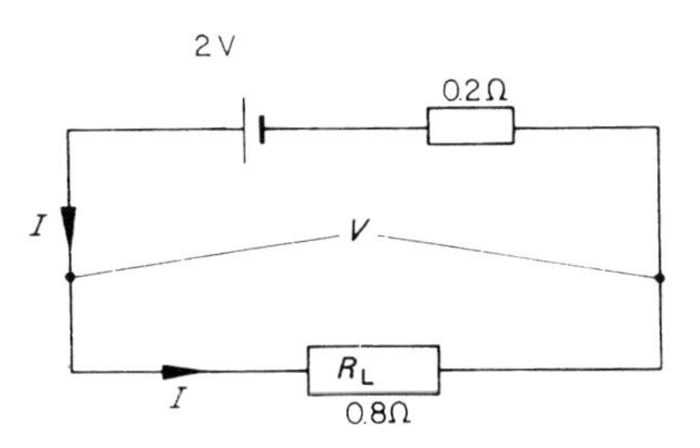

Fig. 9.7

$V = E - (I \times r)$ and $V = I \times R_L$

$$\therefore I \times R_L = E - (I \times r)$$

$$\therefore (I \times R_L) + (I \times r) = E$$

$$\therefore (I \times R_L + r) = E$$

$$\therefore I = \frac{E}{R_L + r}$$

$$= \frac{2}{(0.8 + 0.2)}$$

$$= \frac{2}{1}$$

$$= 2\,\text{A}$$

Cells in series

If a high p.d. is required, then cells are connected in series and internal resistances are added (Fig. 9.8).

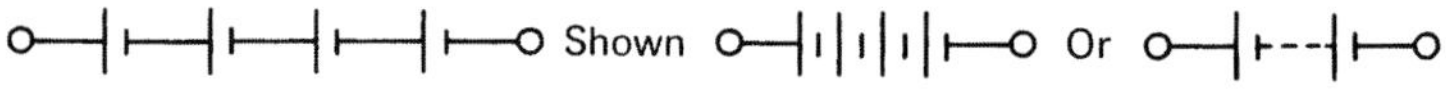

Fig. 9.8

Cells in parallel

For cells in parallel the p.d. is the same as that for one cell, but as the internal resistances are added in parallel their resultant internal resistance is less than for one cell and heavier currents can be drawn.

Battery charging

Cells and batteries are charged by connecting them to a controlled d.c. source. This source may be obtained in several ways; (1) rectified a.c.; (2) motor generator set; (3) rotary converter; and (4) d.c. mains supply.

The most commonly used method is rectified a.c. and there are two ways in which this system is used: (1) the constant-voltage method, and (2) the constant-current method.

Constant-voltage charging

In this method the d.c. charging voltage is kept constant at a value just above that of the final value of the battery e.m.f. The charging current is initially high, decreasing as the e.m.f. of the battery approaches that of the supply (Fig. 9.9).

Constant-current charging

In constant-current charging the current is kept constant by varying the d.c. input voltage as the battery e.m.f. increases (Fig. 9.10).

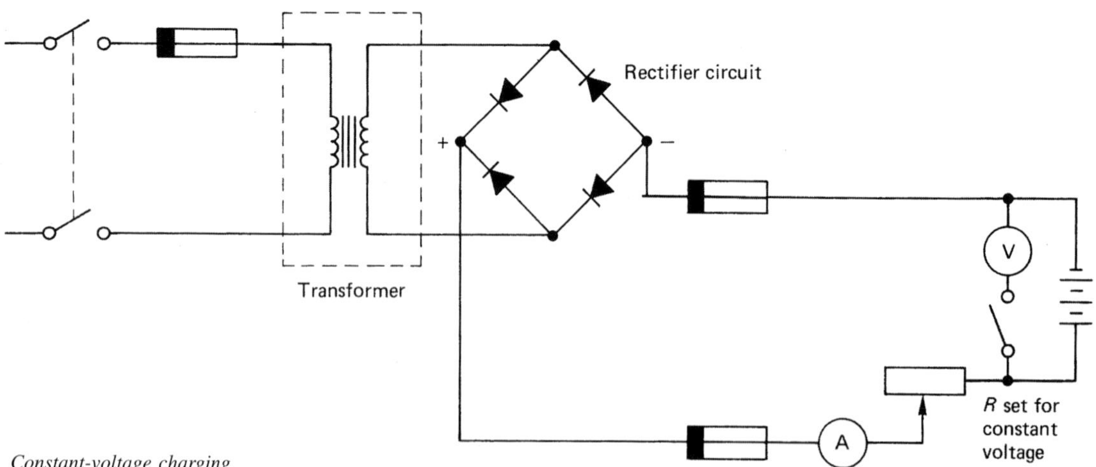

Fig. 9.9 *Constant-voltage charging*

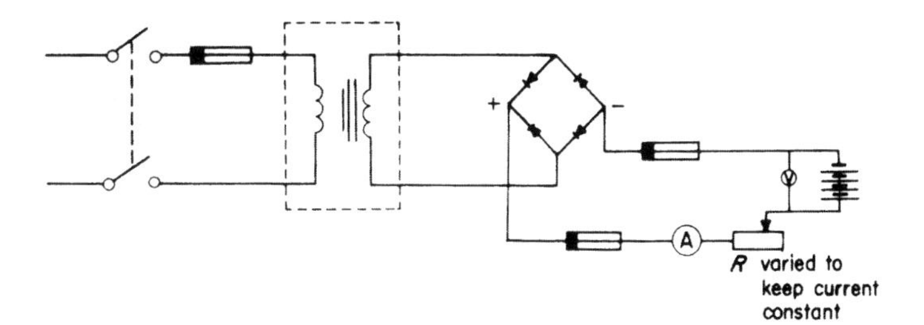

Fig. 9.10 *Constant-current charging*

The more popular method, for everyday use, is the constant-voltage method.

Cells, batteries and their associated charging equipment are frequently used in installation work such as indicator and call systems in hospitals and hotels, fire alarm and burglar alarm systems, and emergency lighting installations.

Self-assessment questions

1 Explain the difference between a primary and a secondary cell.

2 Describe two methods of testing the state of charge of a secondary cell. What figures would you expect?

3 How does an alkaline cell differ from a lead–acid cell? What effect does discharging an alkaline cell have on the SG of its electrolyte?

4 (a) What is the difference between the e.m.f. of a cell and its terminal voltage?

(b) A cell of e.m.f. 2 V and internal resistance 0.15 Ω delivers a current of 3 A to an external load. Calculate the resistance of the load and the terminal voltage of the cell.

5 State the two common methods of battery charging and explain the difference between them.

6 Why do batteries perform less well at low temperatures?

10 Illumination and ELV lighting

Light sources

The range of modern lighting fittings and lamps is so large that only the basic types will be considered here. Before we continue, however, it is perhaps wise to list the various units and quantities associated with this subject.

Luminous intensity: symbol, I; unit, candela (cd)

This is a measure of the power of a light source and is sometimes referred to as brightness.

Luminous flux: symbol, F; unit, lumen (lm)

This is a measure of the flow or amount of light emitted from a source.

Illuminance: symbol, E; unit, lux (lx) or lumen/m²

This is a measure of the amount of light falling on a surface. It is also referred to as *illumination*.

Luminous efficacy: symbol, K; unit, lumen per watt (lm/W)

This is the ratio of luminous flux to electrical power input. It could be thought of as the 'efficiency' of the light source.

Maintenance factor (MF): no units

In order to allow for the collection of dirt on a lamp and also ageing, both of which cause loss of light, a maintenance factor is used.

As an example, consider a new 80 W fluorescent lamp with a lumen output of 5700 lm. After about 3 or 4 months this output would have fallen and settled at around 5200 lm. Hence the light output has decreased by

$$\frac{5200}{5700} = 0.9$$

This value, 0.9, is the maintenance factor and should not fall below 0.8. This is ensured by regular cleaning of the lamps.

Coefficient of utilization (CU): no units

The amount of useful light reaching a working plane will depend on the lamp output, the reflectors and/or diffusers used, position of lamp, colour of walls and ceilings, etc. The lighting designer will combine all of these considerations and determine a figure to use in his or her lighting calculations.

Tungsten-filament lamp

Fig. 10.1 shows the basic components of a tungsten-filament lamp.

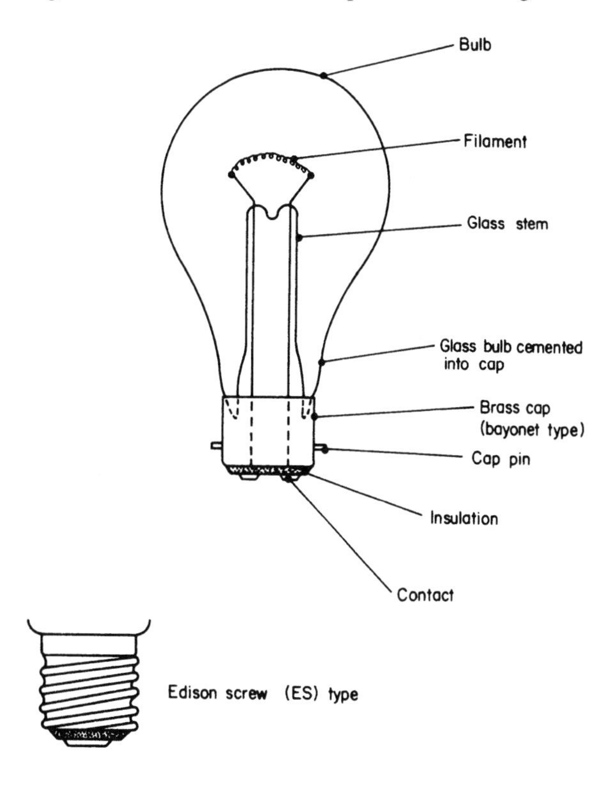

Fig. 10.1 *Tungsten-filament lamp*

The tungsten filament is either single or double coiled (coiled coil). Fig. 10.2 illustrates these types.

The efficacy of gas-filled lamps is increased by using a coiled-coil filament, as this type has in effect a thicker filament which reduces the heat loss due to convection currents in the gas.

Filament lamps are of two main types; vacuum and gas filled.

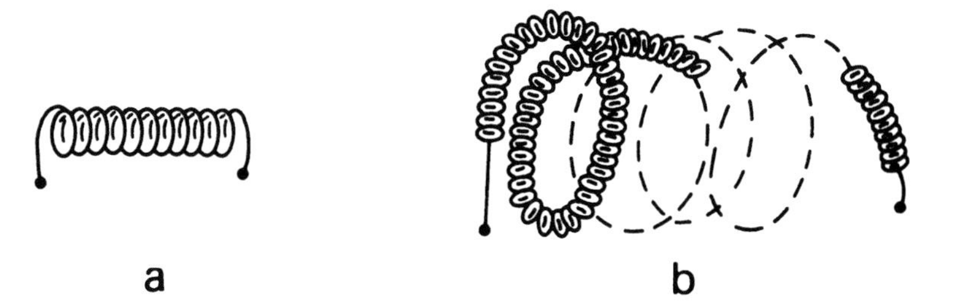

Fig. 10.2 *Filament arrangements: (a) single coil; (b) coiled coil*

Vacuum type

The filament operates in a vacuum in the glass bulb. It has a poor efficacy as it can operate only up to around 2000°C.

Gas-filled type

In this case the bulb is filled with an inert gas such as nitrogen or argon. This enables the operating temperature to reach 2500°C. The efficacy increases and the bulb is usually so bright that it is given an opaque coating internally. This type of lamp is usually called a 'pearl' lamp.

The following code refers to lamp caps;

BC – Bayonet
SBC – Small bayonet
SCC – Small centre contact
ES – Edison screw
SES – Small Edison screw
MES – Miniature Edison screw
GES – Goliath Edison screw

The efficacy of a tungsten lamp will depend on several factors, including the age of the lamp and its size, but tends to be around 12 lm/W for a 100 W lamp.

The colour of its light tends to be mostly red and yellow and in its basic form this type of lamp is used only in situations that do not require a high level of illumination.

Other lamps of the filament type include tubular strip lights, oven lamps, infrared heating lamps, spot- and floodlights, and tungsten-halogen lamps.

Discharge lighting

This type of lighting relies on the ionization of a gas to produce light. As high voltages are present in such lighting circuits, special precautions, outlined in the IEE Regulations, must be taken. Typical discharge lamps include decorative neon signs, fluorescent lighting, and mercury and sodium-vapour lamps used for street lighting.

Neon tube

In the same way that the trade name 'Hoover' is colloquially used to indicate any make of vacuum cleaner, so 'neon' tends to be used to describe any sort of gas-filled tube. There are in fact several different gases used to give different colours, including helium, nitrogen and carbon dioxide.

Fig. 10.3 shows the basic circuit for a cold-cathode neon-sign installation.

Sodium-vapour lamp

There are two types of sodium-vapour lamp available, working at high pressure and low pressure respectively.

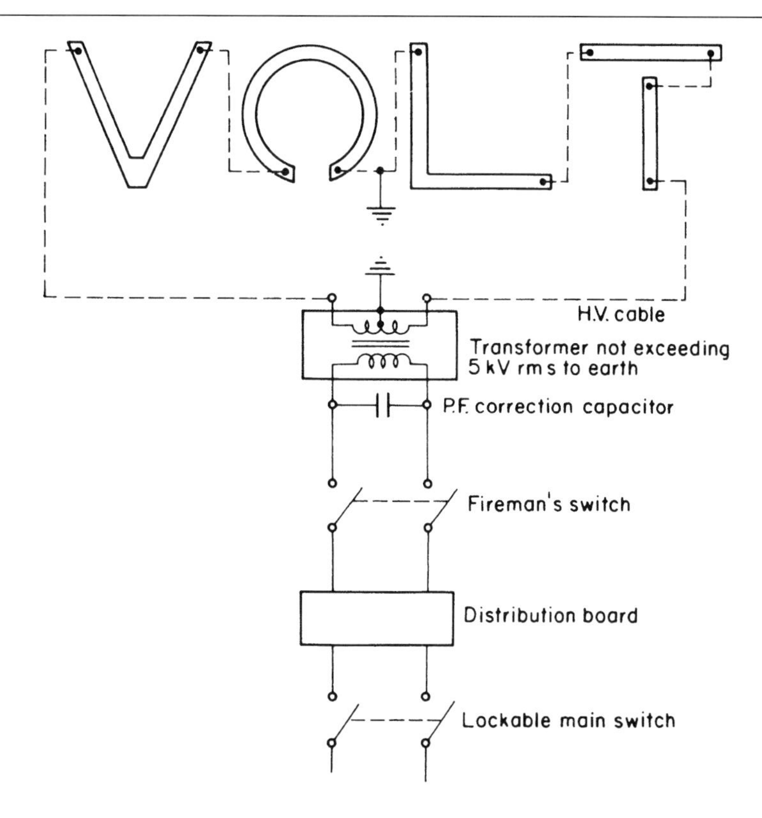

Fig. 10.3 *Single neon-sign circuit*

The low-pressure type consists of a U-shaped double-thickness glass tube, the inner wall of which is of low-silica glass which can withstand attack by hot sodium. Inside the tube is a quantity of solid sodium and a small amount of neon gas (this helps to start the discharge process). An outer glass envelope stops too much heat loss from the inner tube. Fig. 10.4 shows the components of a low-pressure sodium-vapour lamp, while Fig. 10.5 shows the control circuit for a sodium-vapour lamp.

The output from the auto-transformer is in the region of 480 V and the PF correction capacitor is important, as the PF of the lamp and transformer can be as low as 0.3 lagging.

The recommended burning position of the lamp is horizontal ± 20°, this ensures that hot sodium does not collect at one end of the tube in sufficient quantities to attack and damage it.

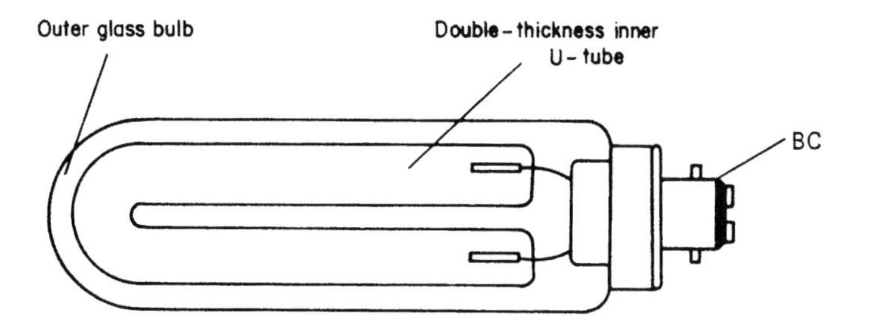

Fig. 10.4 *Low-pressure sodium-vapour lamp type SOX. BC = bayonet cap*

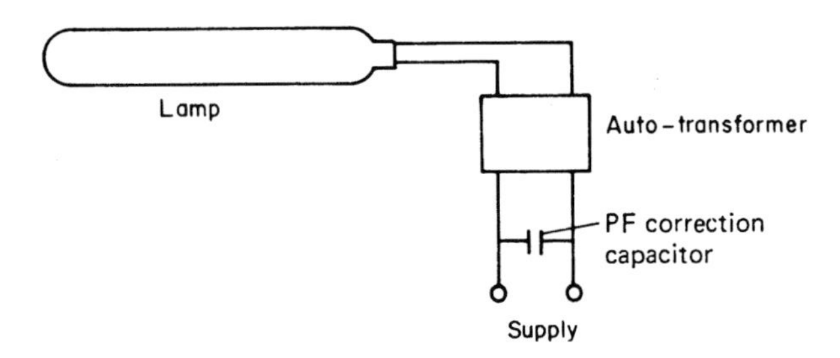

Fig. 10.5

The light output is almost pure yellow, which distorts surrounding colours, and as such is useful only for street lighting. The modern SOX type (superseding the SOH type) has a high efficacy, a 90 W lamp giving in the region of 140 lm/W. (The SOH type gives around 70 lm/W).

The high-pressure type of sodium-vapour lamp differs from other discharge lamps in that the discharge tube is made of compressed aluminium oxide, which is capable of withstanding the intense chemical activity of the sodium vapour at high temperature and pressure. The efficacy is in the region of 100 lm/W, and the lamp may be mounted in any position. The colour is a golden white and as there is little surrounding colour distortion, it is suitable for many applications including shopping centres, car parks, sports grounds and dockyards.

High-pressure mercury-vapour lamp

This type consists of a quartz tube containing mercury at high pressure and a little argon gas to assist starting. There are three electrodes, two main and one auxiliary; the latter is used for starting the discharge (Fig. 10.6).

Fig. 10.7 shows the control circuit for a high-pressure mercury-vapour lamp. The initial discharge takes place in the argon gas between the auxiliary electrode and the main electrode close to it. This causes the main electrode to heat up and the main discharge between the two main electrodes takes place.

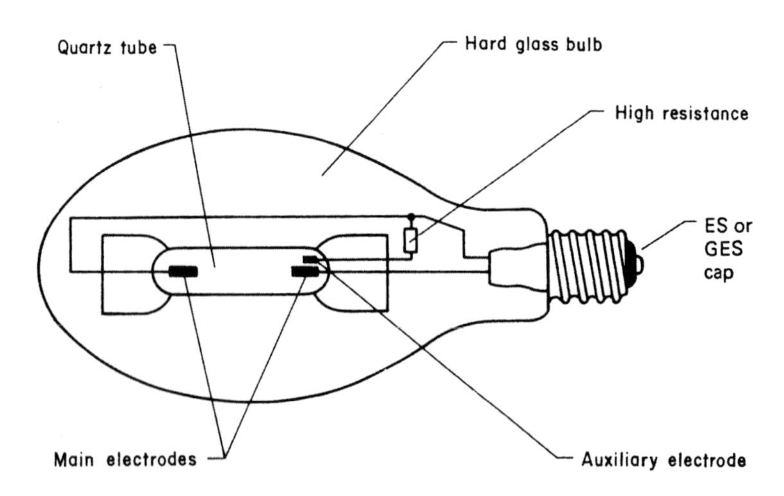

Fig. 10.6 *High-pressure mercury-vapour lamp. ES = Edison screw type; GES = Goliath Edison screw type*

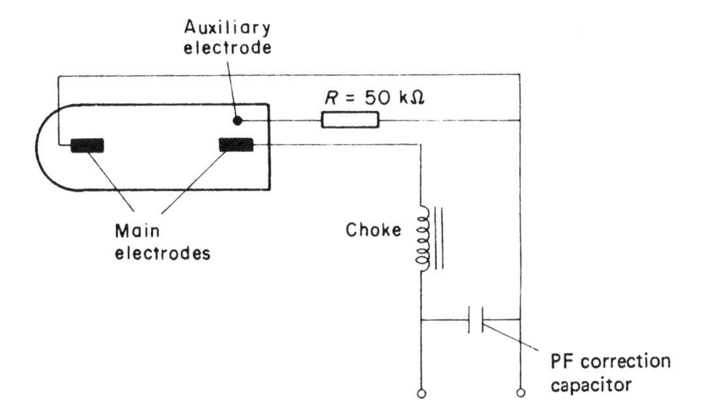

Fig. 10.7

Several types of mercury-vapour lamp are available, including the following two popular types:

MB type – Standard mercury-vapour lamp; ES or GES cap; any mounting position; efficacy around 40 lm/W. Largely superseded by the MBF type.

MBF type – Standard, but with fluorescent phosphor coating on the inside of the hard glass bulb; ES or GES cap; efficacy around 50 lm/W. Used for industrial and street lighting, commercial and display lighting. Any mounting position.

The colour given by high-pressure mercury-vapour lamps tends to be blue-green.

Low-pressure mercury-vapour lamp

A low-pressure mercury-vapour lamp, more popularly known as a *fluorescent lamp*, consists of a glass tube, the interior of which is coated in fluorescent phosphor. The tube is filled with mercury vapour at low pressure and a little argon to assist starting. At each end of the tube is situated an oxide-coated filament. Discharge takes place when a high voltage is applied across the ends of the tube. Fig. 10.8 shows the circuit diagram for a single fluorescent tube.

Practical operation

When the supply is switched on, the circuit is completed via: the choke, first lamp element, starter switch, second lamp element and the neutral. The elements, which are coated in oxide, become warm and the oxide coating emits some electrons and the gas ionizes at the ends of the tube (this helps the main ionization process). The starter contacts (usually of the bimetallic type) separate, owing to the current passing through them, and the choke is open-circuited.

As we have seen, breaking an inductive circuit causes high voltages to appear across the breaking contacts, and energy is released in the form of an

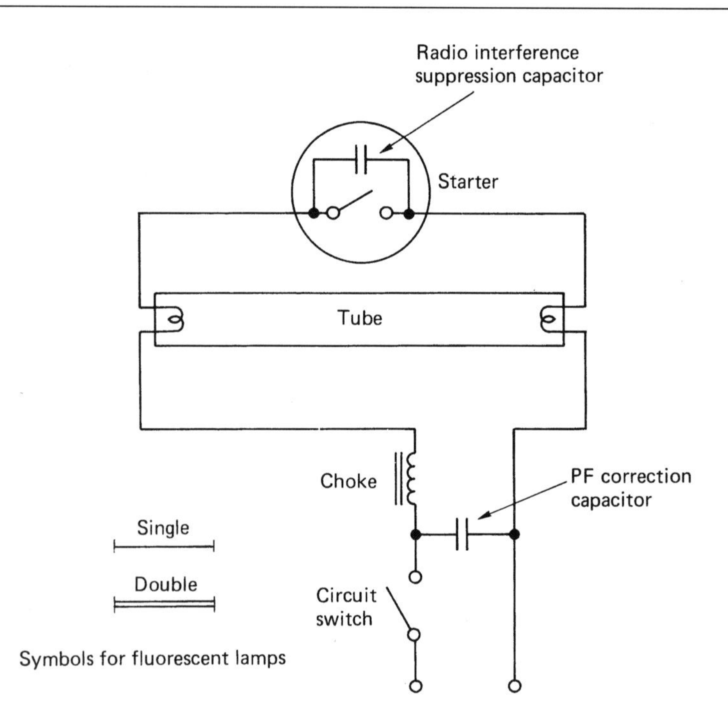

Fig. 10.8 *Basic circuit diagram for fluorescent lamp*

arc. In this case, however, there is an easier way for the energy to dissipate – via the gas, and the high voltage appears across the ends of the tube.

When the gas if fully ionized, the choke limits the current to a predetermined value, and the light emitted, which is mostly ultraviolet, is made visible by the fluorescent powder coating.

The radio interference suppression capacitor is usually located in the starter. The PF correction capacitor is part of the control circuitry common to all fluorescent lighting installations.

Starters

Three methods are commonly available for starting the discharge in a fluorescent tube: the thermal start, the glow start and the quick start.

A *thermal starter* consists of two contacts (one of which is a bimetal) and a heater. Fig. 10.9 shows how such a starter is connected.

When the supply to the lamp is switched on the heater is energized. Also, the lamp filaments are energized via the starter contact. The heater causes the contacts to part and the choke open-circuits across the tube, so that discharge takes place.

The *glow starter* is the most popular of all the means of starting the discharge. It comprises a pair of open contacts (bimetallic) enclosed in a sealed glass bulb filled with helium gas. This assembly is housed in a metal or plastic canister. Fig. 10.10 shows how this type of starter is connected.

When the supply is switched on the helium gas ionizes and heats up, causing the contacts to close, and this energizes the tube filaments. As the contacts have closed, the discharge in the helium ceases, the contacts cool and part, open-circuiting the choke across the tube, and discharge takes place.

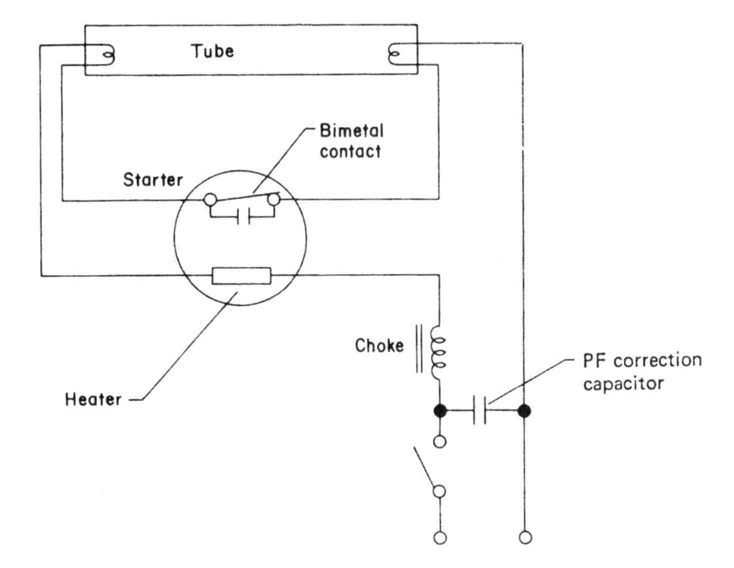

Fig. 10.9 *Thermal starter*

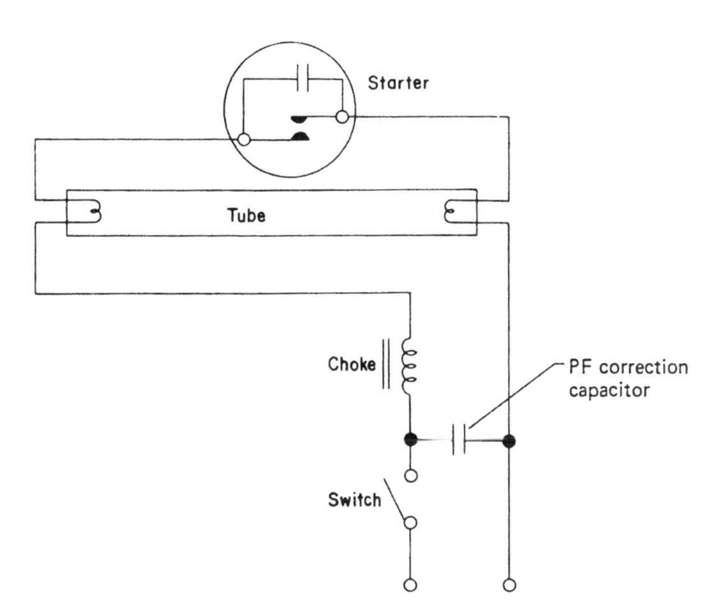

Fig. 10.10 *Glow starter*

In the case of the *quick start* or *instant starter*, starting is achieved by the use of an auto-transformer and an earthed metal strip in close proximity to the tube (Fig. 10.11).

When the supply is switched on, mains voltage appears across the ends of the tube, and the small part of the winding at each end of the transformer energizes the filaments, which heat up. The difference in potential between the electrodes and the earthed strip causes ionization, which spreads along the tube.

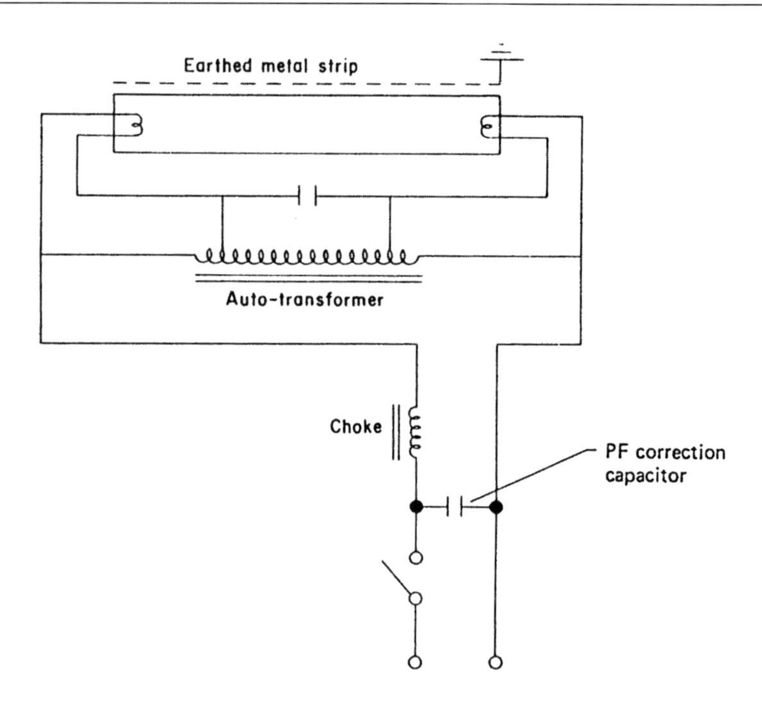

Fig. 10.11 *Quick start or instant starter*

Fluorescent tube light output

There is a wide range of fluorescent tubes for different applications as Table 10.1 indicates.

The white tube has the highest efficacy, which for a 2400 mm, 125 W tube is around 70 lm/W.

Table 10.1

Tube Colour	**Application**
White and warm white	General illumination requiring maximum efficacy, as in drawing offices
Daylight and natural	Any situation requiring artificial light to blend with natural daylight – jewellery, glassware, etc. (main shop areas)
Artificial daylight	Areas where accurate colour matching is carried out
De luxe warm white	Offices and buildings requiring a warm effect, e.g. restaurants, furniture stores
Northlight	Colour-matching areas such as in tailors and furriers
De luxe natural	Florists, fishmongers, butchers, etc.
Green, gold, blue, red and pink	For special effects

Points to note

1 If a switch, not designed to break an inductive load, is used to control discharge lighting, it must have a rating not less than twice the steady current it is required to carry, i.e. 10 A switch for a 5 A load.

2 Although a discharge lamp is rated in watts, its associated control gear is highly inductive and therefore the whole unit should have a VA rating. It is on this rating that the current rating of the circuits is calculated. If no technical information is available, a figure of 1.8 is used to calculate the VA rating. That is,

$$\text{VA rating of 80 W fitting} = 80 \times 1.8$$
$$= 144\text{ VA}$$

3 No discharge lighting circuit should use a voltage exceeding 5 kV r.m.s. to earth, measured on open circuit.

4 If a circuit exceeds low voltage and is supplied from a transformer whose rated input exceeds 500 W then the circuit must have protection such that the supply is cut off automatically if short-circuit or earth leakage currents exceed 20% of the normal circuit current.

5 All control equipment including chokes, capacitors, transformers, etc., must either be totally enclosed in an earthed metalwork container or be placed in a ventilated fireproof enclosure. Also, a notice must be placed and maintained on such a container or enclosure, reading 'DANGER – HIGH VOLTAGE'. The minimum size of letters and notice board is as shown in Fig. 10.12.

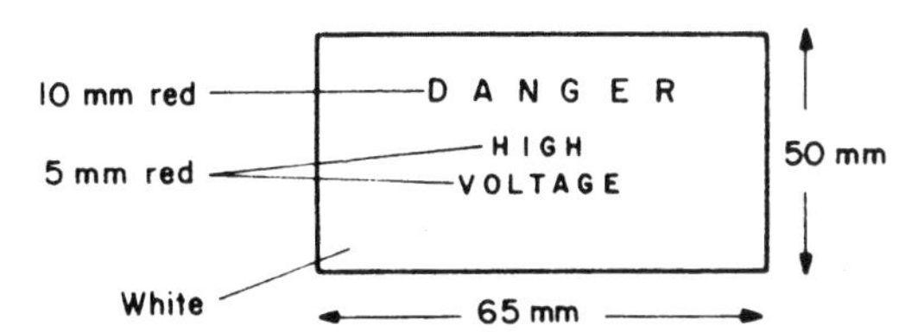

Fig. 10.12

6 Care must be taken to ensure that the only connection between discharge lamp circuits, operating at a voltage exceeding low voltage, and the mains supply is an earth conductor and/or the earthed neutral conductor of an auto-transformer having a maximum secondary voltage of 1.5 kV.

7 It is important that discharge lighting has a means of isolation from all poles of the supply. This may be achieved in one of the following ways:

   (a) An interlock device, on a self-contained discharge lighting unit, so that no live parts can be reached unless the supply is automatically disconnected (i.e. microswitch on the lid of the luminaire which will disconnect supply to a coil of a contactor when the lid is opened).

   (b) A plug and socket close to the luminaire or circuit which is additional to the normal circuit switch.

   (c) A lockable switch or one with a removable handle or a lockable distribution board. If there is more than one such switch, handles and keys must not be interchangeable.

8 Every discharge lighting installation must be controlled by a firefighter's switch which will isolate all poles of the supply (it need not isolate the neutral of a three-phase four-wire supply).

9 The firefighter's switch should be coloured red and have fixed adjacent to it a notice as shown in Fig. 10.13. The notice should also display the name of the installer and/or maintainer of the installation.

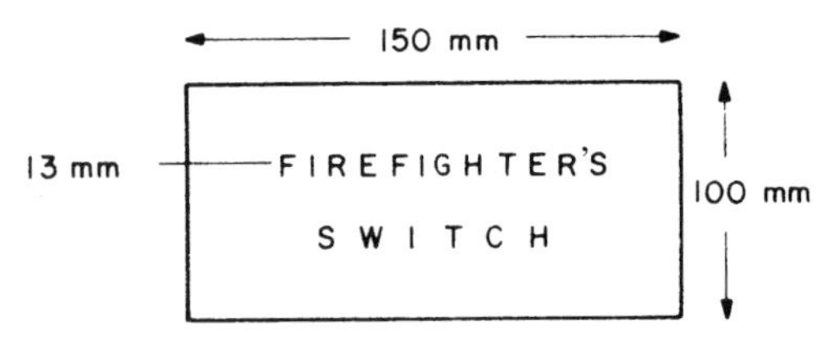

Fig. 10.13

10 The firefighter's switch shall have its ON and OFF positions clearly marked, the OFF position being at the top of the switch. The switch should be placed in a conspicuous and accessible position, no more than 2.75 m from ground level.

11 The firefighter's switch should be outside and adjacent to the installation for external installations and in the main entrance of a building for interior installations.

12 In general, cables used in discharge lighting circuits exceeding low voltage should be metal sheathed or armoured unless they are housed in a box sign or a self-contained luminaire or are not likely to suffer mechanical damage.

13 All cables should be supported and placed in accordance with the tables shown in the Regulations.

14 If it is not clear that a cable is part of a circuit operating above low voltage, it should be labelled every 1.5 m as shown in Fig. 10.14.

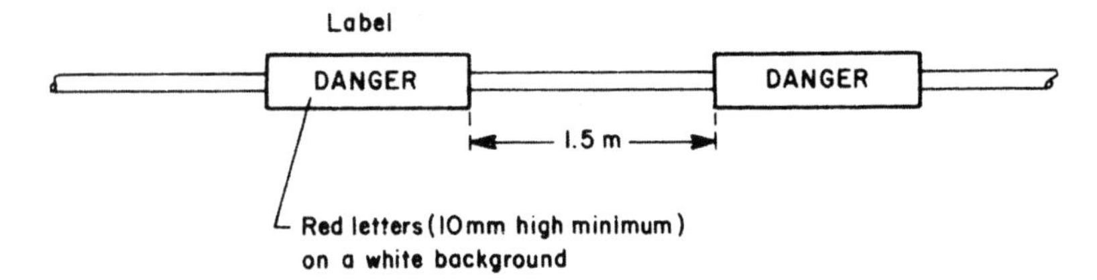

Fig. 10.14

15 ELV lighting is now popular in many kitchens and bathrooms. Such systems comprise a 240 V/12 V transformer with 12 V diachroic 12 V lamps. This lighting is often incorrectly termed low voltage.

Calculation of lighting requirements

Inverse-square law

If we were to illuminate a surface by means of a lamp positioned vertically above it, measure the illumination at the surface, and then move the lamp twice as far away, the illumination now measured would be four times less. If it were moved away three times the original distance the illumination would

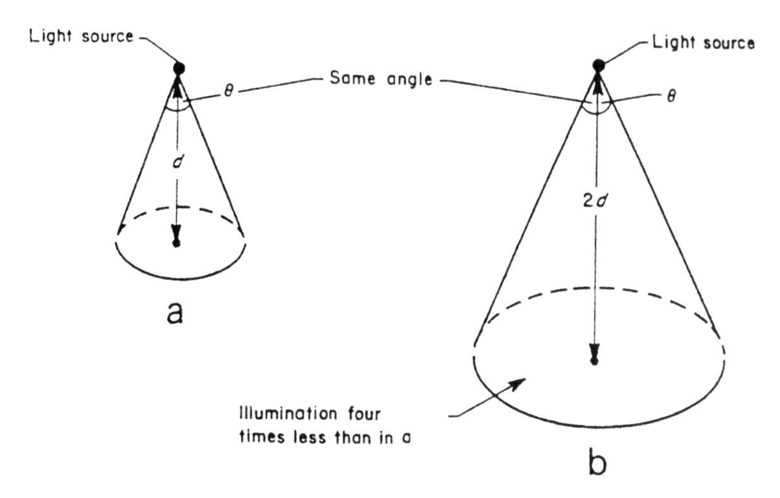

Fig. 10.15

be nine times less. Hence it will be seen that the illuminance on a surface is governed by the square of the vertical distance of the source from the surface (Fig. 10.15).

$$\therefore \text{Illuminance } E \text{ (lux)} = \frac{\text{luminous intensity (cd)}}{d^2}$$

$$E = \frac{I}{d^2}$$

$$\therefore \text{Illuminance } E \text{ (lux)} = \frac{\text{luminous intensity (cd)}}{d^2}$$

$$E = \frac{I}{d^2}$$

Example

A light source of 900 candelas is situated 3 m above a working surface. (a) Calculate the illuminance directly below the source. (b) What would be the illuminance if the lamp were moved to a position 4 m from the surface?

$$\text{(a) } E = \frac{I}{d^2}$$

$$= \frac{900}{9}$$

$$= 100\,\text{lx}$$

$$\text{(b) } E = \frac{I}{d^2}$$

$$= \frac{900}{16}$$

$$= 56.25\,\text{lx}$$

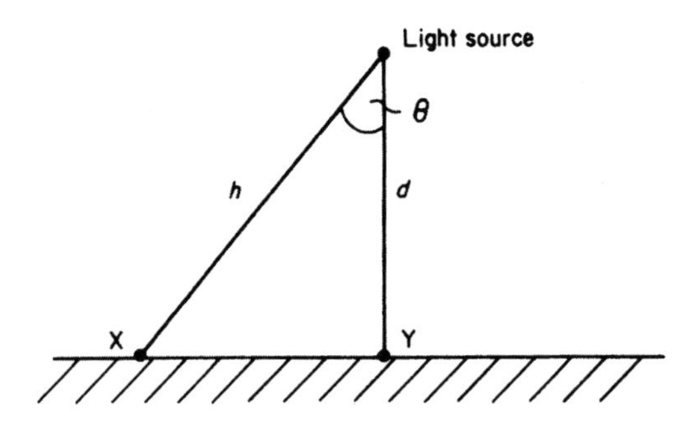

Fig. 10.16

Cosine rule

From Fig. 10.16 it will be seen that point X is further from the source than is point Y. The illuminance at this point is therefore less. In fact the illuminance at X depends on the cosine of the angle θ. Hence,

$$E_X = \frac{I \times \cos^3 \theta}{d^2}$$

Example

A 250 W sodium-vapour street lamp emits a light of 22 500 cd and is situated 5 m above the road. Calculate the illuminance (a) directly below the lamp and (b) at a horizontal distance along the road of 6 m. (Fig. 10.17).

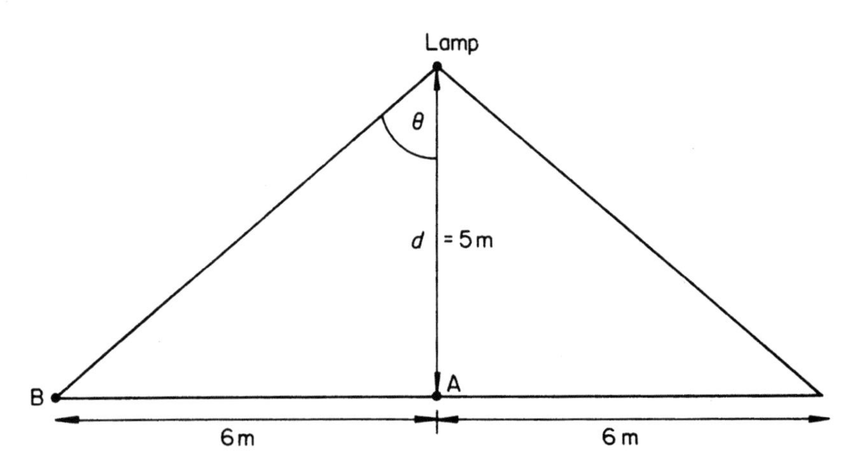

Fig. 10.17

From Fig. 10.17, it can be seen that the illuminance at A is given by

$$E_A = \frac{I}{d^2}$$

$$= \frac{22\,500}{25}$$

$$= 900\,\text{lx}$$

The illuminance at B is calculated as follows. Since the angle θ is not known, it can be found most simply by trigonometry:

$$\tan\theta = \frac{AB}{d}$$

$$= \frac{6}{5} = 1.2$$

From tangent tables,

$$\theta = 50.2°$$

and from cosine tables,

$$\cos 50.2° = 0.64$$

$$\therefore E_B = \frac{I\cos^3\theta}{d^2}$$

$$= \frac{22\,500 \times 0.64^3}{25}$$

$$= 236\,\text{lx}$$

In order to estimate the number and type of light fittings required to suit a particular environment, it is necessary to know what level of illuminance is required, the area to be illuminated, the maintenance factor and the coefficient of utilization, and the efficacy of the lamps to be used.

Example

A work area at bench level is to be illuminated to a value of 300 lx, using 85 W single fluorescent fittings having an efficacy of 80 lumens/watt.

The work area is 10 m × 8 m, the MF is 0.8 and the CU is 0.6. Calculate the number of fittings required

$$\text{Total lumens } (F) \text{ required} = \frac{E(\text{lx}) \times \text{area}}{\text{MF} \times \text{CU}}$$

$$F = \frac{300 \times 10 \times 8}{0.8 \times 0.6}$$

$$= 50\,000\,\text{lm}$$

Since the efficacy is 80 lm/W,

$$\text{Total power required} = \frac{50\,000}{80}$$

$$= 625\,\text{W}$$

As each lamp is 85 W,

$$\text{Number of lamps} = \frac{625}{85}$$

$$= 8$$

PF Correction

Example

The PF correction capacitor in a 240 V, 50 Hz fluorescent light unit has broken down and needs replacing. A test on the unit shows that, without the capacitor, the supply current is 0.86 A at a PF of 0.5 lagging. The values quoted on the original capacitor have faded and the only other information is that the working PF of the unit should be 0.95. Determine the value of the capacitor needed (Fig. 10.18).

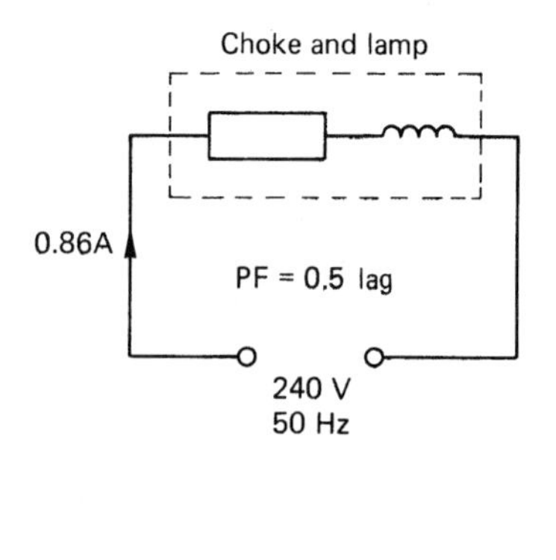

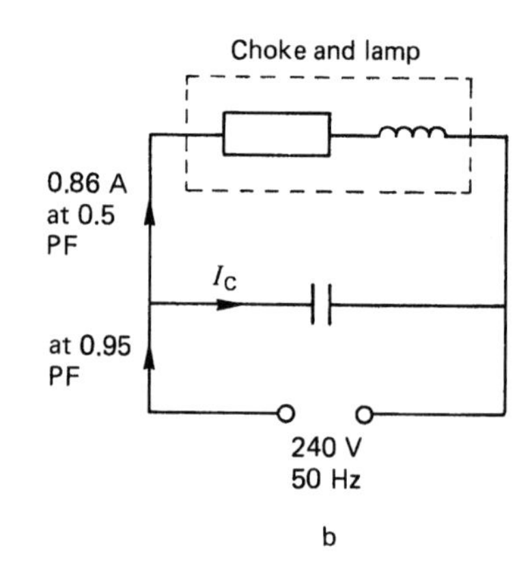

Fig. 10.18 *(a) No capacitor; (b) with capacitor*

First the phasor diagram for the unit without a capacitor is drawn (Fig. 10.19). PF = 0.5 and cos θ = 0.5. Therefore: θ = 60°.

The phasor diagram of the unit showing the supply current at working PF is drawn (Fig. 10.20).

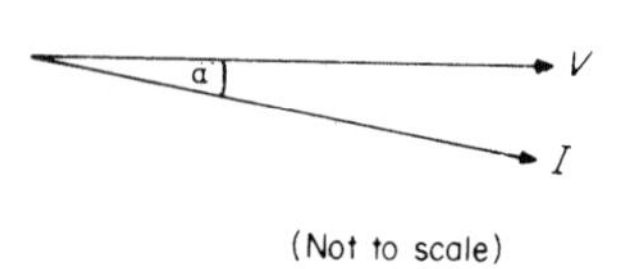

Fig. 10.19

$$PF = 0.95$$

$$\cos \alpha = 0.95$$

$$\alpha = 18.2°$$

Finally the combined phasor diagram can be drawn (Fig. 10.21).

The value of the capacitor current required to raise the PF to 0.95 lagging must be I_c, which is the same distance (ab). By measurement:

(Not to scale)

Fig. 10.20

$$I_c = 0.6 \text{ A}$$

$$\therefore X_c = \frac{V}{I_c}$$

$$= \frac{240}{0.6}$$

$$= 400\,\Omega$$

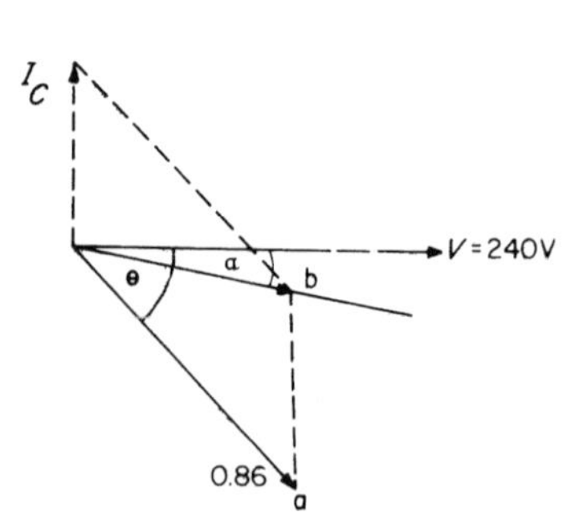

Fig. 10.21

But

$$X_c = \frac{I}{2\pi fC}$$

$$\therefore C = \frac{1}{2\pi f X_c}$$

$$= \frac{1}{2\pi \times 50 \times 400}$$

$$= 8\ \mu\text{F}$$

Note: By measurement, the supply current is 0.45 A.

Rating of fluorescent circuits

Fluorescent tubes are rated in watts, but as we have seen, the circuit of which the lamp is part is inductive, and even after improvement has a lagging PF.

We know from Chapter 6 that plant is not rated in watts, but in volt amps (VA). It is recommended that, if no other information is available, *the lamp wattage may be multiplied by 1.8 in order to determine the VA rating*. For example the VA rating of a fluorescent unit with an 80 W tube is 1.8 × 80 = 144 VA. Hence, when supplied at 240 V, the current taken would be:

$$I = \frac{VA}{V}$$

$$= \frac{144}{240}$$

$$= 0.6\ \text{A}$$

When a fluorescent lamp is switched off, the choke is again open-circuited; this time the voltage appears across the switch contacts. This can damage the switch and unless it is specially designed to break an inductive circuit, *it should have a rating of not less than twice the total steady current it is required to carry.* For example, if a fluorescent light unit draws a current of 1 A, then the switch controlling it should have a rating of at least 2 A.

Example

A consumer has a work area that he or she wishes to illuminate with single 65 W, 240 V fluorescent fittings. The existing lighting points, which are to be removed, are controlled by a single-gang 5 A switch. This switch is to remain. How many 60 W fittings may be installed?

$$\text{VA rating of fitting} = 1.8 \times 65$$

$$= 117\ \text{VA}$$

$$\therefore \text{Current rating of fitting} = \frac{117}{240}$$

$$= 0.4875\ \text{A}$$

$$\text{The 5 A switch must only carry } \frac{5}{2} = 2.5\ \text{A}$$

$$\therefore \text{Number of fittings permitted} = \frac{2.5}{0.4875}$$

$$= 5.13$$

$$= 5 \text{ fittings}$$

Stroboscopic effect

While a fluorescent lamp is in operation the light may flicker. Under some circumstances this may make it appear that rotating machinery has slowed down or even stopped. This is called the *stroboscopic effect*. This is an undesirable state of affairs which is usually remedied by one of the two following methods.

Balancing the lighting load (three phase)

If a large lighting load is installed in a three-phase installation where there is some rotating machinery, the stroboscopic effect may be overcome by connecting alternate groups of lamps to a different phase. This also has the advantage of balancing the lighting load (Fig. 10.22).

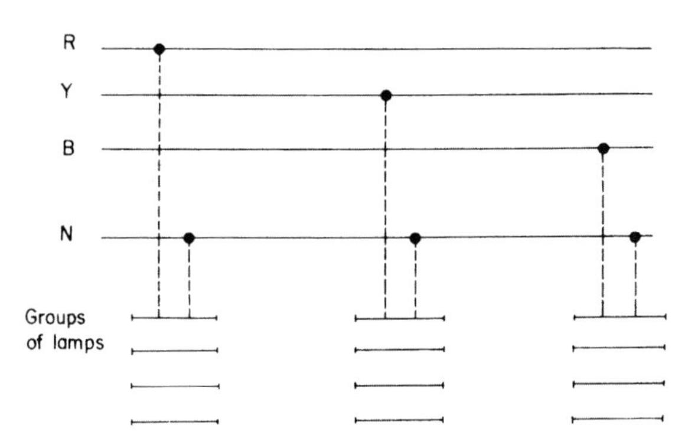

Fig. 10.22 *Lighting load*

The lead–lag circuit

In this method a capacitor is wired in series with every alternate lamp in a group. The value of the capacitor is such that the lamp unit it is fitted to has an overall *leading PF.* This means that any pair of lamps have a lagging and a leading PF. This has the effect of cancelling out the resultant flicker, in the same way as two equal but *opposing* forces cancel each other out (see Fig. 10.23).

Example

Two 240 V fluorescent lamp units A and B are arranged to overcome stroboscopic effects. Unit A has a series capacitor fitted and takes a current of 0.4 A at a PF of 0.985 leading. Unit B takes a current of 0.53 A at 0.5 PF lagging (Fig. 10.24). Draw a scaled phasor diagram showing these two currents and from it determine the total current and the overall PF. Ignore PF improvement.

Choose a suitable scale (Fig. 10.25).

$$\cos \alpha = 0.985$$

$$\therefore \alpha = 10^\circ$$

$$\cos \theta = 0.5$$

$$\therefore \theta = 60^\circ$$

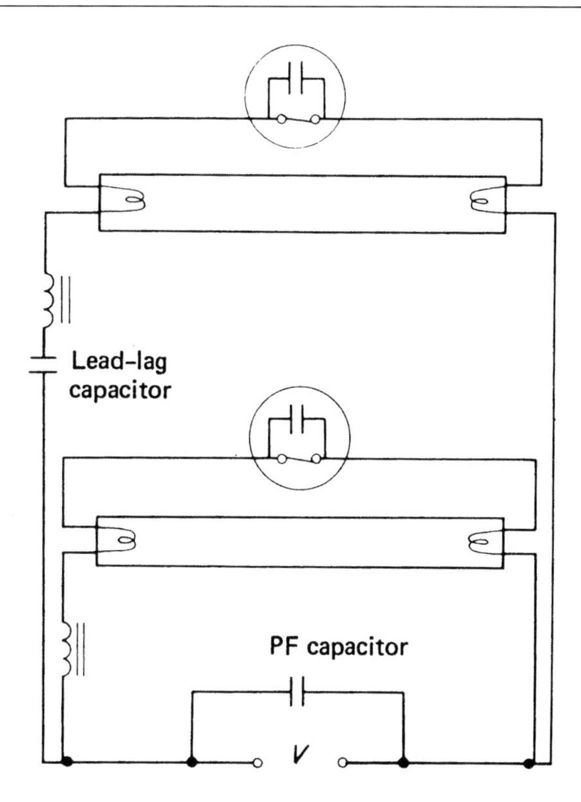

Fig. 10.23 *Lead–lag circuit*

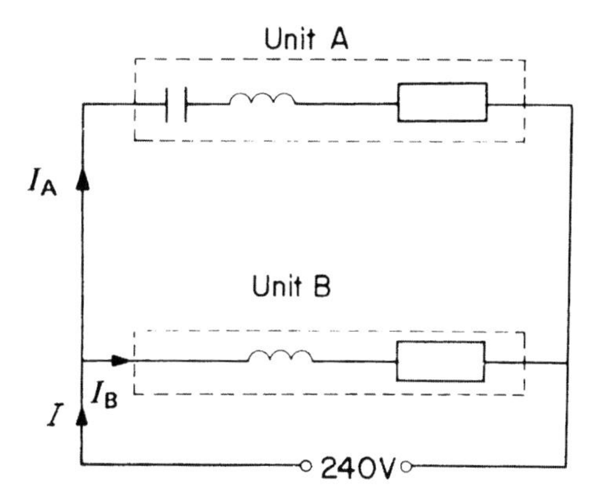

Fig. 10.24 I_A = 0.4 A at 0.985 lead; I_B = 0.53 A at 0.5 lag

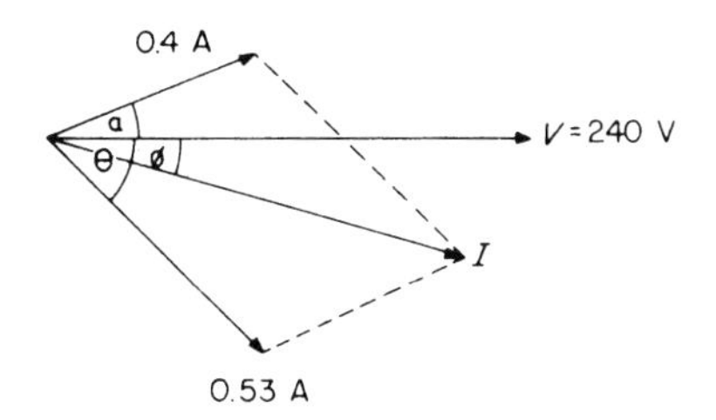

Fig. 10.25

By measurement:

$$I = 0.77$$

$$\phi = 30°$$

$$\therefore \text{PF} = \cos \phi$$

$$= 0.866$$

Example

Two 240 V fluorescent lamps are arranged to overcome the stroboscopic effect. One unit takes 0.8 A at 0.45 PF, leading, the other takes 0.7 A at 0.5 PF lagging (Fig. 10.26). Determine the total current drawn and the overall PF.

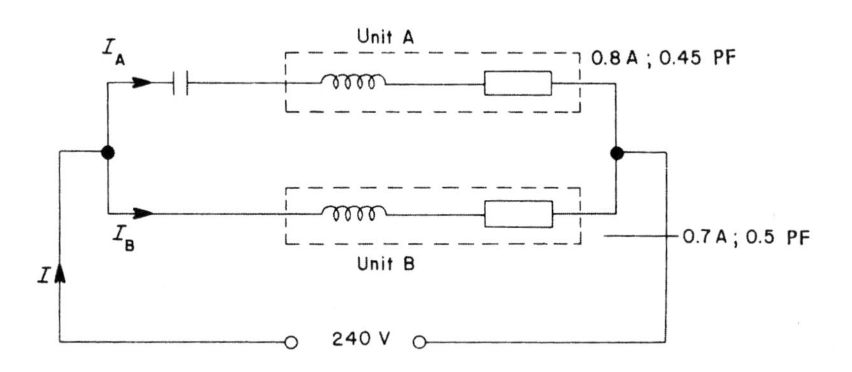

Fig. 10.26

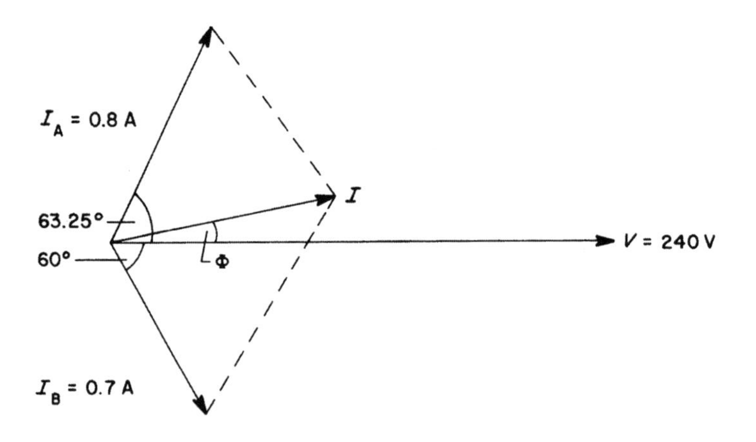

Fig. 10.27

Method 1, by phasors (Fig. 10.27)

Unit A: $\cos \theta = 0.45$

$$\therefore \theta = 63.25°$$

Unit B: $\cos \alpha = 0.5$

$$\therefore \alpha = 60°$$

By measurement,

$$I = 0.71 \text{ A}$$

and

$$\Phi = 8.5°$$

$$\therefore \cos \Phi = \text{PF} = 0.989 \text{ leading}$$

Method 2, by trigonometry

In this method, the active and reactive components of currents I_A and I_B are found (Fig. 10.28).

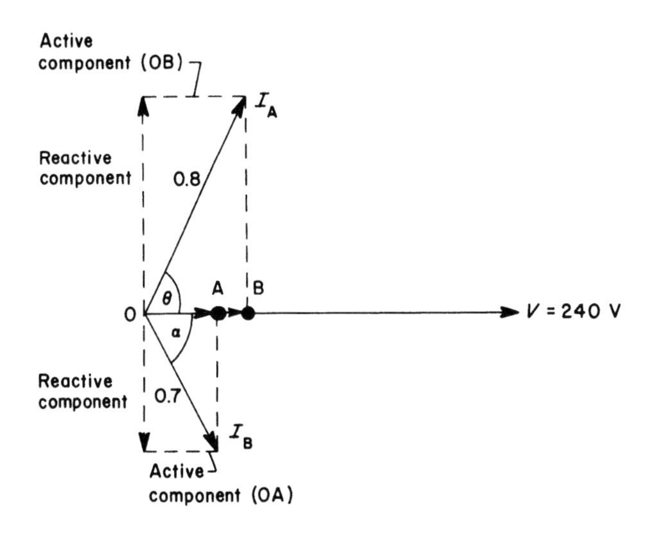

Fig. 10.28

Active or horizontal component of I_A = OB = $I_A \times \cos \theta$

$$\therefore \text{OB} = 0.8 \times \cos 63.25°$$
$$= 0.8 \times 0.45$$
$$= 0.36 \text{ A}$$

Active component of I_B = OA = $I_B \times \cos \alpha$

$$\therefore \text{OA} = 0.7 \times \cos 60°$$
$$= 0.7 \times 0.5$$
$$= 0.35 \text{ A}$$

Since all the active components are in phase they may be added. Hence:

Total of active components = 0.36 + 0.35 = 0.71 A

Reactive or vertical components of I_A = $I_A \times \sin \theta$

$$= 0.8 \times \sin 63.25°$$
$$= 0.8 \times 0.893$$
$$= 0.714 \text{ A}$$

$$\text{Reactive component of } I_B = I_B \times \sin \alpha$$
$$= 0.7 \times 60°$$
$$= 0.7 \times 0.866$$
$$= 0.606 \text{ A}$$

Since the two reactive components are opposite in phase they must be subtracted.

$$\therefore \text{Total of reactive components} = 0.714 - 0.606$$
$$= 0.108 \text{ A}$$

The resultant of the total active and reactive components will be the total current taken (Fig. 10.29):

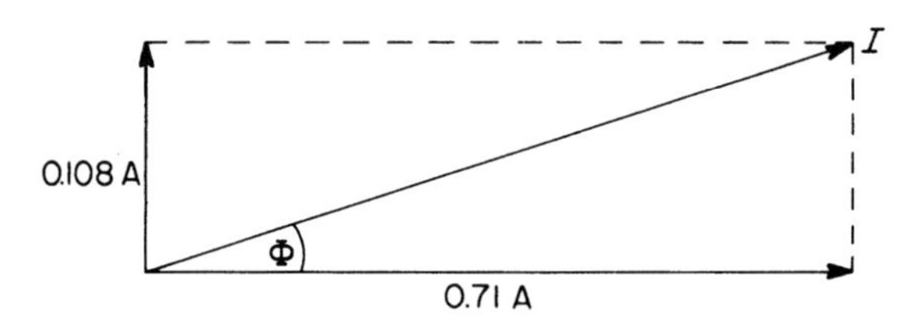

Fig. 10.29

$$\tan \Phi = \frac{\text{perpendicular}}{\text{base}}$$

$$= \frac{0.108}{0.71}$$

$$\tan \Phi = 0.152$$

$$\therefore \Phi = 8.64°$$

$$\therefore \text{PF} = \cos \Phi = 0.988 \text{ leading}$$

and

$$\cos \Phi = \frac{\text{base}}{\text{hypotenuse}} = \frac{0.71}{I}$$

$$\therefore I = \frac{0.71}{\cos \Phi} = \frac{0.71}{0.988}$$

$$I = 0.718 \text{ A}$$

Self-assessment questions

1 With the aid of circuit diagrams explain the difference between a thermal and a glow-type starter for a fluorescent lamp.

2 Show with the aid of a sketch the construction of a high-pressure mercury-vapour lamp. Clearly label all parts.

3 Why should a low-pressure sodium-vapour lamp be mounted horizontally? What is the most common application for such a lamp? Why is this?

4 Compare the colour and efficacy of tungsten-filament, low-pressure sodium-vapour and low-pressure mercury-vapour lamps.

5 What colour of fluorescent lamp should be used for the following: (a) a butcher's shop, (b) a restaurant and (c) a jeweller's?

6 Explain the purpose of the choke in a discharge lighting circuit.

7 Explain what is meant by (a) 'maintenance factor', (b) 'coefficient of utilization' and (c) 'the inverse-square law'.

8 (a) What is meant by 'illuminance'?

(b) A light source of 850 cd is situated 2.5 m above a work surface. Calculate the illuminance directly below the light and 3 m horizontally away from it (at work surface level).

9 A small workshop 27 m × 17 m requires illuminance at bench level of 130 lx. Two types of lighting are available; (a) 150 W tungsten-filament lamps at 13 lm/W, or (b) 80 W fluorescent lamps at 35 lm/W. Assuming that the maintenance factor in each case is 0.8 and that the coefficient of utilization is 0.6, calculate the number of lamps required in each case.

10 (a) A small supermarket 20 m long by 15 m wide is to be illuminated to a level of 600 lx by 2400 mm 125 W fluorescent lamps having an efficacy of 65 lm/W. The maintenance factor is 0.85 and the coefficient of utilization is 0.6. Calculate the number of fittings required and show their positions on a scale plan.

(b) Calculate the total current taken by the lighting.

11 What are the recommendations with regard to (a) rating of fluorescent lamp units, and (b) current rating of ordinary switches controlling fluorescent lighting circuits?

12 With the aid of a diagram explain the principles of operation of a fluorescent light unit.

13 What is meant by the *stroboscopic effect*, and how can it be minimized?

14 The following data relate to two 240 V fluorescent lighting units arranged to minimize stroboscopic effect:

Unit 1 – 0.75 A at 0.96 PF leading.
Unit 2 – 0.8 A at 0.6 PF lagging.

By using a scaled phasor diagram determine the value of the total current taken by the two units. Check your answer by calculation.

11 Electricity, the environment and the community

Environmental effects of the generation of electricity

In order to produce electricity by means of generators, we must provide a method of propulsion. This can be achieved in one of the following ways:

1 water power, or
2 steam power.

Water power (hydro-electric stations)

Where a natural and continuous flow of water is available (i.e. waterfalls, or fast-flowing rivers), its potential energy can be used to turn a waterwheel coupled to a generator. This method is obviously inexpensive due to the fact that no cost is involved in providing the propulsion. However, in the UK the number of sites available for such generation on any useful scale is limited, and such sites are usually to be found in mountainous areas such as Scotland and Wales.

In some instances an artificial flow is produced by siting the generating station between two reservoirs, allowing the top one to discharge via the generator to the lower one, and then pumping the water back to the top reservoir (Fig. 11.1).

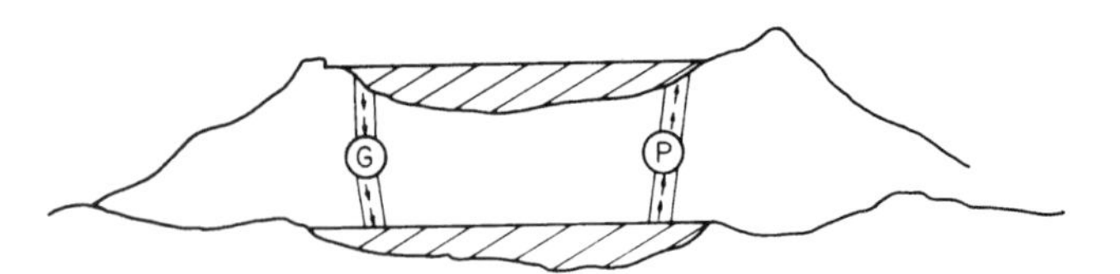

Fig. 11.1

Finding such sites, however, where two natural reservoirs are available, is almost impossible, and valleys have to be flooded to provide the right conditions. This, of course, poses environmental problems, even if every effort is made to protect the local population from loss of agricultural land and hence livelihoods, and to avoid spoiling the landscape and adversely affecting tourist trade.

Siting of power stations

In view of the previous comments, it should be clear that the siting of a power station is extremely important in order for it to generate electricity at an economic level. Fig. 11.2 shows the location of power stations mostly in England and Wales.

Steam power

Steam power involves the use of steam at very high pressure directed on to the blades of a wheel (turbine) coupled to a generator.

In order to produce this steam economically, readily available supplies of water and fuel are required.

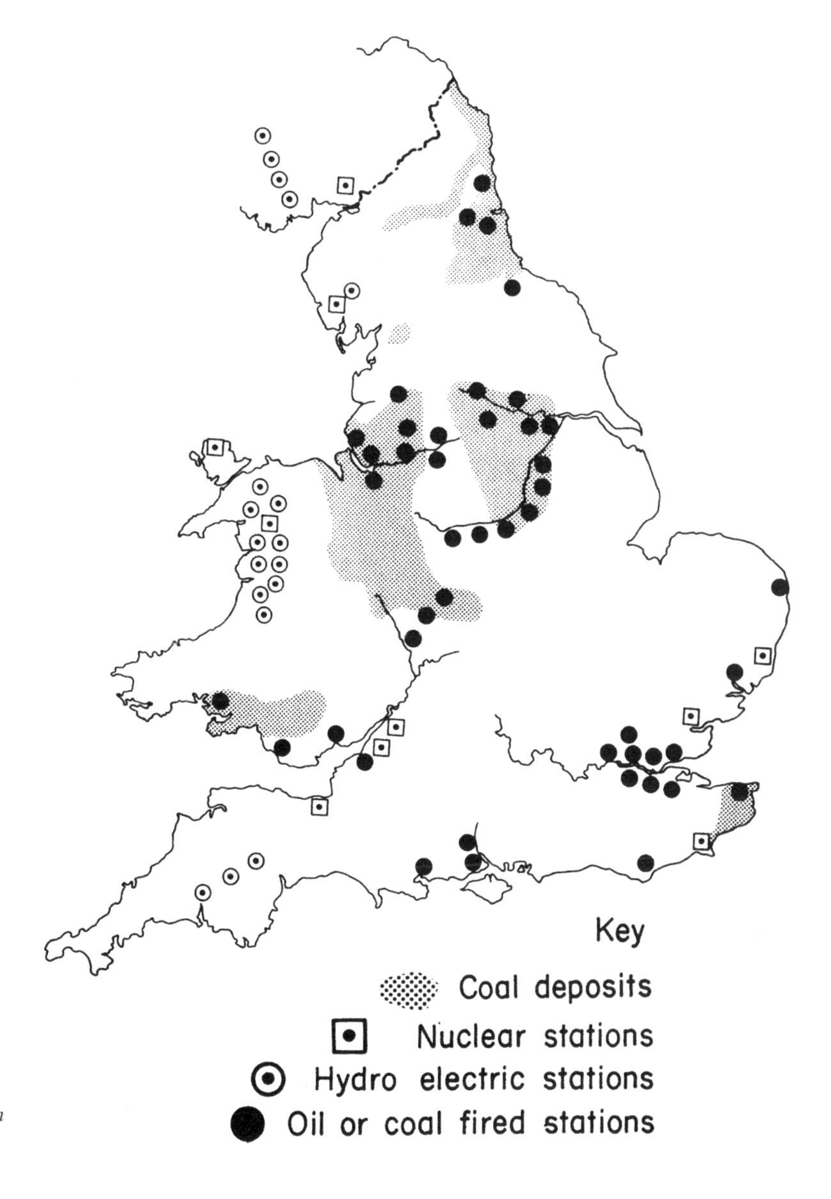

Fig. 11.2 *Location and type of British power stations*

Fuels

Fuels used in modern generating stations may be divided into two categories:

1 fossil fuels (coal, oil, gas); and
2 nuclear fuel (uranium).

All fossil fuels have to be mined and transported to the power stations, and in view of this, it is clear that a power station must be sited in an economic position. It would be pointless to locate a new station high in the Welsh mountains (conveniently out of sight) if it was impossible to transport the fuel to it. In this respect atomic or nuclear stations have an advantage in that the fuel is not bulky and transportation is relatively easy; hence it is possible to locate such stations in remote coastal areas.

Coal-fired stations

By far the greater part of electrical energy generated in Britain is provided by coal-fired stations.

The ideal situation for such a station would be adjacent to a coal mine and a large source of water (river or sea). If a mine is not to hand, the next best thing is good access for the delivery of the fuel, such as a railway or port.

Oil-fired stations

Many of the existing coal-fired stations have been converted to oil firing since the discovery of North Sea oil and the construction of overland pipelines. Another contributory factor to this changeover is the running down of the coal industry and the subsequent closure of many pits.

The oil is transported by rail, sea or by the pipelines previously mentioned.

However, it is an increasingly expensive fuel, and only a small percentage of stations are oil fired. Oil-fired stations should preferably be sited near ports.

Gas-fired stations

Gas-fired stations are rare. They are usually restricted to local generation and do not significantly contribute to the National Grid.

Nuclear stations

Although they are a subject of much controversy, new nuclear stations *are* being built, and modern technology is providing new techniques to make them more economic, relative to other types of power station.

The fuel is uranium, and relatively small quantities will produce vast amounts of heat. It is, however, a very expensive fuel, and the capital cost of building a power station to cope with all the dangers of using such fuel renders generation in this way, at present, uneconomic.

Nuclear power stations are usually sited in coastal areas.

Pollution

In any situation where a fuel is burnt, there is the attendant problem of pollution, and because of the large scale of the use of fuel in power stations, careful control of pollution is required.

Air pollution

When coal is burnt, the resulting smoke contains grit, dust and sulphur fumes, which have to be prevented from contaminating the environment. Much of the sulphur is washed out and the grit and dust remaining after treatment are prevented from reaching the ground, via low-level air currents, by ensuring that the chimney is very tall and that the discharge or emission of smoke into the air is at high velocities (15 to 25 m/s). In densely populated areas where high-rise buildings now exist, some older coal-fired stations have been converted to oil firing.

Water pollution

Pollution of sea- and especially river-water occurs when the water used in the station is recycled to its source at a higher temperature and with a considerable amount of impure solids suspended in it. The effect of this is to kill the river or marine life.

The discharge of polluted water into rivers is not illegal but the National Rivers Authority have the authority to pass by-laws controlling the amount discharged.

Control of pollution

Every month the National Grid Co. take readings from some three or four hundred different gauges at 30 or more power stations in order to monitor the amounts of pollution directly due to the generation of electricity, and carry out research into grit removal and sulphur washing from smoke and the cleansing of waste water.

Waste disposal

Another major problem which has considerable environmental implications is the safe and effective disposal of waste.

Coal-fired stations

The waste from coal-fired stations is chiefly pulverized fuel ash, of which a controlled amount is allowed to be discharged into the sea or rivers. An increasingly large amount is used in the manufacture of building materials such as breeze-block or Thermalite bricks, and the rest is deposited into old clay and gravel pits or built into artificial hills and landscaped.

Nuclear stations

Nuclear stations do not cause atmospheric pollution during normal operation, but there is always the attendant problem of the disposal of radioactive waste. A great deal of research is being carried out to investigate the possibility of using the waste, and recently some use has been found in a new type of nuclear station, the 'fast-breeder reactor'.

Disposal of radioactive waste is costly, as methods must be found to contain the material, with no risk of leakage, for the rest of its active life, which may be as long as 300 years. Lead-lined containers in steel shells and encased in concrete or glass many metres thick are effective in storing the contaminated waste; the whole is then deposited into the sea, or buried deep underground.

New developments

As our natural resources are used in greater and greater quantities, the reserves of fuel are rapidly diminishing and much research is being carried out into new methods of propelling generators. The general idea now is to use energy which costs nothing and is renewable, i.e. wind, wave and solar energy.

Wind power

Wind turbines are no more than an adaptation of the windmill principle, but as yet the amount of electrical energy that even a very large machine can produce is small.

Wave power

Research is at present being carried out into the possibility of harnessing the power of the waves by converting the motion of cam-shaped floats into circular motion to drive a generator.

Solar energy

At present, solar energy seems to be by far the best bet in terms of renewable energy. Once again, however, it is a question of developing a system which will give the required amount of electrical energy at an economic level.

The purpose and function of the National Grid

Historical background

In the early days of electricity supply, each British town or city had its own power station which supplied the needs of its particular area.

Standardization was not evident and many different voltages and frequencies were used throughout the country. By the time of the First World War (1914–18), there were some 600 independent power stations in use. However, the heavy demands made by the war industry showed the inadequacies of the system and several select committees were set up to investigate possible changes. Little was achieved until 1926 when it was suggested that 126 of the largest and most efficient power stations should be selected and connected by a grid of high-voltage transmission lines covering the whole country, and, at the same time, the frequency standardized at 50 Hz. The remaining power stations would be closed down, and local supply authorities would obtain their electricity in bulk from the grid, via suitable substations. By 1932 the system was in operation with the addition of several large, new power stations. The system voltage was 132 000 V (132 kV), and the supply frequency 50 Hz.

On 1 April 1948 the whole of the electricity supply industry was nationalized and in 1957 the 'Central Authority' responsible for the generation of electricity was renamed the 'Central Electricity Generating Board' (CEGB).

Since then, of course, the electricity industry has become privatized and the CEGB has been replaced by the National Grid Co. who buy, at the lowest price, generated electricity from such companies as National Power, PowerGen, Nuclear Electric, French Electric, Scottish Hydro Electric, etc.

The purpose of the grid system

The purpose of the grid system is to maintain a secure supply of electricity at a standard voltage and frequency to consumers throughout the country. Having stated its purpose, we can now list several advantages that have resulted from its introduction:

1. security of supplies;
2. standardization of frequency and voltages;
3. economy;
4. the ability to transmit very large loads for considerable distance without loss; and
5. the ability to transfer electricity to and from different parts of the country.

The function of the grid

In order to fulfil its purpose, the grid system must function in the following way.

The National Grid Control Centre in London, in association with the various grid control centres around the country, estimates the load required in different areas each day. This information is then used to arrange to purchase the country's power depending on the demand. In this way stations are used to their maximum efficiency, which in turn reduces the cost of generation. Due to the fact that the system is interconnected, bulk supply points can be fed from other areas, should a failure of the usual supply occur.

Generation, transmission and distribution systems

The very nature of the grid system is such that power has to be transmitted over large distances. This immediately creates a problem of voltage drop. To overcome this problem, a high voltage is used for transmission (400 kV or 132 kV), the 400 kV system being known as the *Super Grid*. We cannot, however, generate at such high voltages (the maximum in modern generators is 25 kV), and transformers are used to step up the generated voltage to the transmission voltage. At the end of a transmission line is a grid substation where the requirements of the grid system in that area can be controlled, and where the transmission voltage is stepped down via a transformer to 132 kV.

It is at this voltage that the different Area Boards transmit the power required by their consumers around that particular area. The system voltage is then further reduced at substations to 33 000 V, 11 000 V and 415/240 V.

The declared voltage at consumer terminals is 400 V three-phase/230 V single-phase. However, the measured voltage is still likely to be 415/240 V for many years.

Overhead lines and underground cables

The only economic method of transmitting power at grid voltages is by means of overhead lines; cables are used only in very short lengths. However, the cost of producing cable rated at the lower voltages 33 kV, 11 kV and 415 V enables its use to be widespread. The topic of lines and cables is dealt with in greater detail later.

System layout

Fig. 11.3 illustrates how electricity is made and conveyed to the consumer.

The aesthetic effects of the siting of generation and transmission plant

An unfortunate situation has arisen in modern times from the public's conflicting desires for:

1 an ever-increasing supply of electrical energy; and
2 the beauty of the countryside to be preserved.

The task of completely satisfying both of these requirements will remain impossible until man can find a way of generating and transmitting the

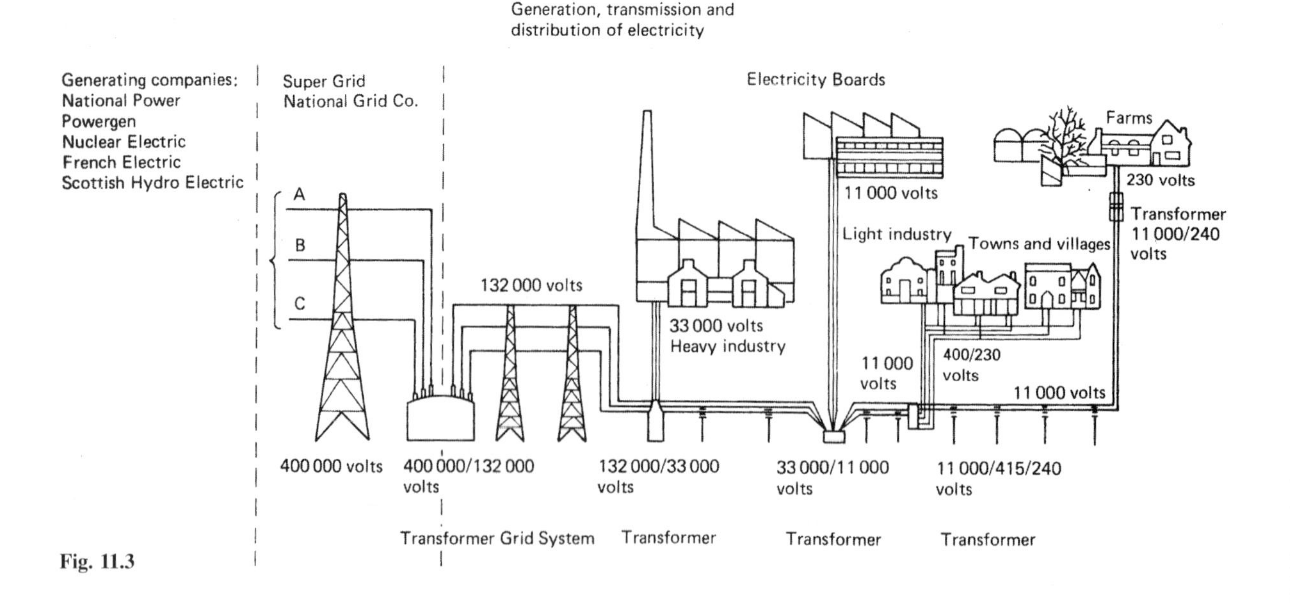

Fig. 11.3

country's energy needs without the use of power stations and overhead lines.

Contrary to popular opinion, however, the National Grid Co. and Electricity Boards *cannot* build power stations and erect lines wherever they please. A large body of organizations oppose or approve decisions made by the Boards until a satisfactory compromise can be reached.

Siting of power stations

The siting of power stations in relation to fuel and water resources was discussed on p. 196, and, of course, these are primary considerations in deciding where a new station should be located. However, the aesthetic aspects must also be considered.

How can something as large as a modern power station be hidden?

Obviously it cannot, but could the design be made to blend with the surrounding countryside? Once again this would be almost impossible to achieve, as the efficient working of such a building depends on a very definite design that does not lend itself to the irregular contours of the landscape.

It only remains, therefore, to try to make the best of a difficult job. Within the design limits, the buildings are made streamlined using modern materials and as much glass as possible, together with an off-white finish to the main structures. This, at least, is more acceptable to the majority of people than was the older type of brick-built station.

Transmission lines

Opinion is divided as to the beauty of transmission lines; some think pylons graceful, others think them ugly. Once again, within the design limits, an attempt has been made to make them as graceful as possible. Why, the public asks, cannot the power be transmitted by underground cables? The simple answer is, economy.

Enormous costs would be involved in manufacturing a cable whose insulation was capable of withstanding 400 kV or 132 kV between phases. The total cost of manufacturing and installation of 400 kV oil-cooled cable, together with all the necessary oil pressure equipment along the route, and compensation to farmers for destruction of agricultural land, is 12 times as much per mile as for transmission lines.

If this information is coupled with the fact that locating and repairing a fault on underground cable would involve many times the cost compared with that for a fault on an overhead line, it will be seen that overhead transmission lines are the only economic proposition.

12 Health and safety

Whilst enjoying leisure pursuits, especially those of a sporting nature, most of us wear the accepted clothing suited to the sport in question, such as cricket and football pads, knee and elbow protectors and crash-helmets for skateboarding, and padded jackets and visors for fencing. However, in the hazardous and sometimes potentially lethal environment of our place of work, many of us choose to ignore the warnings and carry out our duties without the use of the recommended protective clothing. It is this careless attitude that accounts for thousands of lost man-hours in industry every year, not to mention the pain and suffering of the casualty.

By the same token many employers are seriously at fault in not providing safe working conditions, and it is up to us as potential casualties to bring the lack of safety to our employers' notice. Remember accidents don't just happen, they are caused, either by our own carelessness or by that of others.

It is for these reasons that over the years a number of safety regulations have come into being, designed to protect us from injury at work. The following list indicates such regulations and, whilst all are important, those that are mostly relevant to electrical installation work are the H&SWA, EAWR, PPE, CDM and COSHH.

Health and Safety at Work Act 1974 ***(H&SWA)***
Management of Health and Safety at Work Regulations 1999
Workplace (Health, Safety and Welfare) Regulations 1992
Provision and Use of Work Equipment Regulations 1998
Personal Protective Equipment Regulations 1992 ***(PPE)***
Manual Handling Operations Regulations 1992
Health and Safety (First Aid) Regulations 1981
Reporting of Injuries, Diseases and Dangerous Occurrences Regulations 1995
Ionising Radiation Regulations 1999
Noise at Work Regulations 1989
Control of Substances Hazardous to Health Regulations 2002 ***(COSHH)***
Electricity at Work Regulations 1989 ***(EAWR)***
Construction (Design and Management) Regulations 1994 ***(CDM)***
The Building Regulations

These are known as statutory regulations, which means they have the force of law, and non-compliance with them can result in a fine, imprisonment, or, in some cases, both.

The regulations issued by the Institution of Electrical Engineers (IEE Wiring Regulations), which are not statutory, are designed to ensure minimum safety standards in the installation of electrical equipment in buildings, and are based on an internationally agreed plan.

There are other statutory Acts which cover situations such as railways, mines and quarries and, for the time being, all these separate Acts will remain in force, even though the Health and Safety at Work Act 1974 deals with the welfare of virtually every person at work.

This act is extremely important and the next few sections give a general outline of its requirements.

The Health and Safety at Work Act 1974

The act outlined

The purpose of the Health and Safety at Work Act 1974 is to provide the legislative framework to promote, stimulate and encourage high standards of health and safety at work.

The minister primarily concerned with the Act is the Secretary of State for Employment, and he makes most of the Regulations in consultation with other appropriate ministers.

For the time being the existing Acts, such as the Factories, Offices, Shops and Railway Premises and the Mines and Quarries Acts, remain in force, and the Health and Safety at Work Act makes one comprehensive and integrated system of law to deal with the health and safety of virtually all people at work, and the protection of the public where they may be affected by the activities of people at work. The Health and Safety Commission and the Health and Safety Executive administer the legislation and are a focus of initiative for all matters relating to health and safety at work.

All 'persons at work', whether employees or self-employed, are covered with the exception of domestic servants in a private household. About 5 million people, such as those employed in education, medicine, leisure industries and in some parts of the transport industry who have not previously been covered by safety legislation, are protected for the first time.

Duties of employers

Employers must safeguard, as far as is reasonably practicable, the health, safety and welfare of all the people who work for them. This applies in particular to the provision and maintenance of safe plant and systems of work, and covers all machinery, equipment and appliances used.

Some examples of the matters which many employers need to consider are:

1 Is all plant up to the necessary standards with respect to safety and risk to health?
2 When new plant is installed, is latest good practice taken into account?
3 Are systems of work safe? Thorough checks of all operations, especially those operations carried out infrequently, will ensure that danger of injury or to health is minimised. This may require special safety systems, such as 'permits to work'.

4 Is the work environment regularly monitored to ensure that, where known toxic contaminants are present, protection conforms to current hygiene standards?
5 Is monitoring also carried out to check the adequacy of control measures?
6 Is safety equipment regularly inspected? All equipment and appliances for safety and health, such as personal protective equipment, dust and fume extraction, guards, safe access arrangements, monitoring and testing devices, need regular inspection (Sections 2(1) and 2(2) of the Act).

No charge may be levied on any employee for anything done or provided to meet any specific requirement for health and safety at work (Section 9).

Risks to health from the use, storage, or transport of 'articles' and 'substances' must be minimised. The term *substance* is defined as 'any natural or artificial substance whether in solid or liquid form or in the form of gas or vapour' (Section 53(1)).

To meet these aims, all reasonably practicable precautions must be taken in the handling of any substance likely to cause a risk to health. Expert advice can be sought on the correct labelling of substances, and the suitability of containers and handling devices. All storage and transport arrangements should be kept under review.

Safety information and training

It is now the duty of employers to provide any necessary information and training in safe practices, including information on legal requirements.

Duties to others

Employers must also have regard for the health and safety of the self-employed or contractors' employees who may be working close to their own employees; and for the health and safety of the public who may be affected by their firm's activities.

Similar responsibilities apply to self-employed persons, manufacturers and suppliers.

Duties of employees

Employees have a duty under the Act to take reasonable care to avoid injury to themselves or to others by their work activities, and to co-operate with employers and others in meeting statutory requirements. The Act also requires employees not to interfere with or misuse anything provided to protect their health, safety or welfare in compliance with the Act.

Enforcement

If an inspector discovers a contravention of one of the provisions of the existing Acts (Factory Acts, etc.), he or she can:

1 Issue a prohibition notice to stop the activity.
2 Issue an improvement notice to remedy the fault within a specified time.
3 Prosecute any person instead of or in addition to serving a notice.

Contravention of some of the Regulations can lead to prosecution summarily in a Magistrates' Court. The maximum fine is £20,000. Imprisonment for up to 2 years can be imposed for certain offences. In addition to any other penalty, the Court can make an order requiring the cause of the offence to be remedied.

The Health and Safety Commission

The Health and Safety Commission consists of representative of both sides of industry and the local authorities. It takes over from government departments the responsibility for developing policies in the health and safety field.

The Health and Safety Executive

This is a separate statutory body appointed by the Commision which works in accordance with directions and guidance given by the Commission. The Executive enforce legal requirements, as well as provide an advisory service to both sides of industry.

Electricity at Work Regulations 1989

Persons on whom duties are imposed by these Regulations

(1) Except where otherwise expressly provided in these Regulations, it shall be the duty of every:
  (a) employer and self-employed person to comply with the provisions of these Regulations in so far as they relate to matters which are within his control; and
  (b) manager of a mine or quarry (within in either case the meaning of section 180 of the Mines and Quarries Act 1954) to ensure that all requirements or prohibitions imposed by or under these Regulations are complied with in so far as they relate to the mine quarry or part of a quarry of which he is the manager and to matters which are within his control.

(2) It shall be the duty of every employee while at work:
  (a) to co-operate with his employer so far as is necessary to enable any duty placed on that employer by the provisions of these Regulations to be complied with; and
  (b) to comply with the provisions of these Regulations in so far as they relate to matters which are within his control.

Employer

1 For the purposes of the Regulations, an employer is any person or body who (a) employs one or more individuals under a contract of employment or apprenticeship; or (b) provides training under the schemes to which the HSW Act applies through the Health and Safety (Training for Employment) Regulations 1988 (Statutory Instrument No. 1988/1222).

Self-employed

2 A self-employed person is an individual who works for gain or reward otherwise than under a contract of employment whether or not he employs others.

Employee

3 Regulation 3(2)(a) reiterates the duty placed on employees by section 7(b) of the HSW Act.
4 Regulation 3(2)(b) places duties on employees equivalent to those placed on employers and self-employed persons where these are matters within their control. This will include those trainees who will be considered as employees under the Regulations described in paragraph 1.
5 This arrangement recognises the level of responsibility which many employees in the electrical trades and professions are expected to take on as part of their job. The 'control' which they exercise over the electrical safety in any particular circumstances will determine to what extent they hold responsibilities under the Regulations to ensure that the Regulations are complied with.
6 A person may find himself responsible for causing danger to arise elsewhere in an electrical system, at a point beyond his own installation. This situation may arise, for example, due to unauthorized or unscheduled back feeding from his installation on to the system, or to raising the fault power level on the system above rated and agreed maximum levels due to connecting extra generation capacity, etc. Because such circumstances are 'within his control', the effect of regulation 3 is to bring responsibilities for compliance with the rest of the regulations to that person, thus making him a duty holder.

Absolute/reasonably practicable

7 Duties in some of the regulations are subject to the qualifying term 'reasonably practicable'. Where qualifying terms are absent the requirement in the regulation is said to be absolute. The meaning of reasonably practicable has been well established in law. The interpretations below are given only as a guide to duty holders.

Absolute

8 If the requirement in a regulation is 'absolute', for example if the requirement is not qualified by the words 'so far as is reasonably practicable', the requirement must be met regardless of cost or any other consideration. Certain of the regulations making such absolute requirements are subject to the Defence provision of regulation 29.

Reasonably practicable

9 Someone who is required to do something 'so far as is reasonably practicable' must assess, on the one hand, the magnitude of the risks of a particular work activity or environment and, on the other hand, the costs in terms of the physical difficulty, time, trouble and expense which would be

involved in taking steps to eliminate or minimize those risks. If, for example, the risks to health and safety of a particular work process are very low, and the cost or technical difficulties of taking certain steps to prevent those risks are very high, it might not be reasonably practicable to take those steps. The greater the degree of risk, the more weight that must be given to the cost of measures needed to prevent that risk.

10 In the context of the Regulations, where the risk is very often that of death, e.g. from electrocution, and where the nature of the precautions which can be taken are so often very simple and cheap, e.g. insulation, the level of duty to prevent that danger approaches that of an absolute duty.

11 The comparison does not include the financial standing of the duty holder. Furthermore, where someone is prosecuted for failing to comply with a duty 'so far as is reasonably practicable', it would be for the accused to show the court that it was not reasonably practicable for him to do more than he had in fact done to comply with the duty (section 40 of the HSW Act).

Personal Protective Equipment Regulations (PPE)

These regulations detail the requirements for safety regarding protective clothing, tools, etc., for example hard hats, protective footwear, rubber mats and insulated tools. Signs are usually posted on site to indicate the need for the use of such equipment.

Construction (Design and Management) Regulations (CDM)

Whilst these regulations are important to all operatives 'on-site', they are the result of risk assessment made by architects, designers, planners etc., at the initial design stage of a project. They have to ensure that all safety matters are addressed and followed for the duration of any contract.

Control of Substances Hazardous to Health Regulations (COSHH)

In this case these regulations deal with the use, storage, transportation and disposal of any substance which may be dangerous to the health of persons and/or livestock. These may include bottled gases, flammable liquids, corrosive materials: for example propane or oxyo-acetalyne for plumbing or welding; petrol or paraffin for motors or burners; caustic soda for cleaning; paint strippers; and the powder in fluorescent tubes.

Clearly such substances must be stored somewhere on site and provision must be made to ensure that such storage is safe. In the case of the disposal of used substances, once again such materials will need to be collected and stored ready for collection by an appropriate disposal company. In any event, different substances should not be stored in the same area (e.g. toxic and explosive materials). All dangerous materials should be housed and locked in a secure area away from normal work activities. Records should be kept of the movement and storage of materials and only authorised personnel should have access.

The Building Regulations

Clearly these Regulations encompass all aspects of building construction and are known as Approved documents (Ads) or Parts. There are some Ads that are relevant to electrical installations. These are as follows:

Ad Part A. This, overall, deals with the building structure and, to some extent, on how the fabric of the building may be violated to accommodate electrical systems via chases, notching, drilling, etc. See pages 248–266 regarding joists.
Ad Part B. This concerns fire safety. All electrical systems should be constructed and installed to prevent the start of and the spread of fire. Such considerations would include: heat resistant sleeving on conductors, hoods over downlighters, fire barriers in trunking and sealing of holes made in the building fabric during the installation process. Added to this there is a requirement to provide smoke and fire alarm detection systems.
Ad Part F. Ventilation is important, especially in kitchens and bathrooms, and in consequence particular attention should be paid to the positioning and effectiveness of extract fans and the number of appliances installed.
Ad Part L. This is about the efficient use of fuel and energy. Generally in domestic dwellings, lighting is probably most abused and in consequence the types and control of luminaries have to be carefully considered.
Ad Part M. Disabled persons need access to buildings and facilities and hence the position of sockets and switches are important. The Part M document indicates reasonable heights of such accessories where there is access/use by the disabled.
Ad Part P. There is now a requirement to inform the Local Authority Building Control (LABC) of any electrical work carried out in a domestic dwelling. This is generally confined to kitchens, bathrooms and gardens, where the risk of shock is considerably higher. This does not extend to changing old for new fittings or accessories. Those who carry out electrical work in such areas must be competent and registered with an authoritative body before certification can legitimately be issued.

General safety

None of us enjoys the discomfort of injury. It is, therefore, clearly sensible to avoid such distress by working in a safe manner and ensuring that our colleagues and employers do likewise.

Working on 'site', even if the site is a domestic premises where rewiring is taking place, may involve considerable risk to oneself and/or any occupants or other trades. Hence the type and size of the site is of no real consequence, danger in a working environment is always present in one form or another.

All employees and employers should be aware of the risks and should be alert to danger at all times. No situation, however seemingly innocent, should be ignored as it may develop into the cause of a serious accident.

If the situation or environment requires it, the PPE must be used at all times. On 'bona-fide' construction sites, signs and notices are displayed and all personnel are required to comply. Failure to do so will result in exclusion from the site.

The CDM requirements for a site will include all the procedures for dealing with emergency situations such as fire, explosion and leakage of toxic materials, and should be available for all to aquaint themselves with. Everyone should know the basic common-sense actions to take together with the specific requirements for each particular site.

Basically, in the event of discovering a fire:

1 raise the alarm by operating relevant call points and/or shouting **fire**;
2 call emergency services;
3 make sure no-one ignores the alert and evacuation is in process;
4 only if safe to do so and the fire is relatively small, attempt to control using an appropriate extinguisher.

When an explosion has occurred:

1 alert all personnel to evacuate the area;
2 call emergency services;
3 do **not** attempt to approach the area as other explosions may occur.

The same procedure should be adopted if there is a leakage of toxic material. Only trained personnel should deal with such situations. If in any doubt, **All work should be stopped and the area evacuated**. A good example of this is the cutting, drilling and disposal of any substance that looks like, but has not been cleared of containing, asbestos. In no circumstances should work continue until experts have investigated.

Listed below are some important DOs and DON'Ts associated with working conditions.

DO know the site procedures for evacuation in the event of an emergency.

DO know where any emergency exits are located.

DO know where telephones are located in order to contact emergency services.

DO wash hands after accidental contact with any corrosive or irritant substance.

DO always ensure that there is enough working space and adequate levels of lighting when working, especially where electrical equipment is present: the EAWR requires this. Items of electrical equipment should be sited such that work can be carried out without difficulty and in such a location to afford easy access. Hence distribution boards should not be installed in areas used for storage of non-electrical equipment (consumer units in understairs cupboards!). Controls for washing machines, dishwashers etc. should be visible and accessible. Switch rooms housing electrical equipment should have good lighting, preferably of the maintained emergency variety with switches inside the room.

DO wear the correct protective clothing for the job, e.g. goggles when using a grindstone, safety helmet when working on a building site.

DO adopt the correct posture when lifting.

DO use the correct access equipment, e.g. ladders and scaffolding.
DO ensure that any machinery is mechanically and electrically safe to work on or with.
DO report any unsafe situation that you personally cannot remedy.
DO ensure that when a job of work is completed, everything is put back to normal, e.g. fixing floorboards and tacking down carpets, and replacing trunking covers, etc.
DO know where first aid is available.

DON'T leave pools of liquid on floors, or objects lying in access ways.
DON'T wear loose clothing that may catch or snag, especially when using rotating machinery.
DON'T work in an untidy fashion.
DON'T misuse tools and equipment or use damaged tools.
DON'T attempt to lift and carry objects that are too heavy, or that obstruct vision.
DON'T take short-cuts or take risks.
DON'T ignore warning signs or alarm bells.
DON'T play the fool in a working environment.

In order to carry out some of the DOs and DON'Ts it is necessary to have a greater understanding of the subject we are dealing with. Others, of course, are common sense, like wearing goggles to protect the eyes.

The mechanics of lifting and handling

Simple levers

Figs. 12.1a–c illustrate three forms of the simple lever and how the principles of leverage are used in practice. We can see from the diagrams that the load is as close to the fulcrum as possible, and the effort is as far from it as possible. In this way heavy loads can be lifted with little effort. Fig. 12.1a and c are first-order levers, 12.1b is a second-order lever.

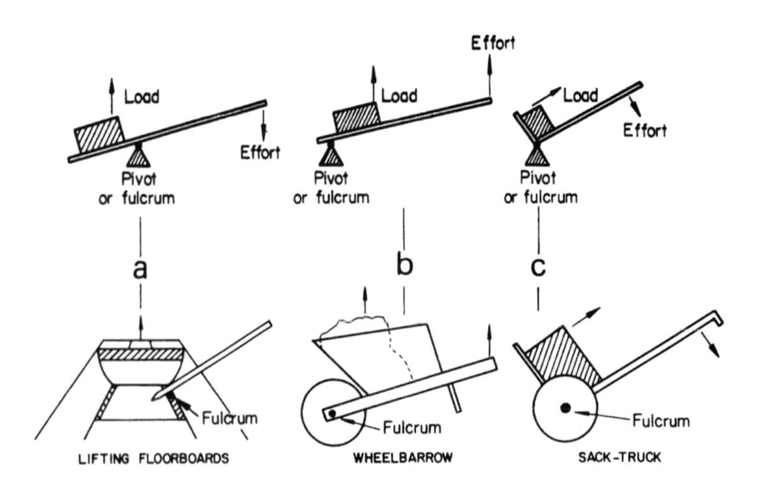

Fig. 12.1

The inclined plane

The inclined-plane method of moving a load is best illustrated as in Fig. 12.2a and b. Most of us at some time have had to push or help push a car up a gentle slope. Imagine, then, trying to get a car from the road on to the pavement. Would it be possible for us physically to lift it from one level to another? Doubtful. More than likely we would arrange two wooden planks from the pavement to the road, and then, without too much effort, push the car up.

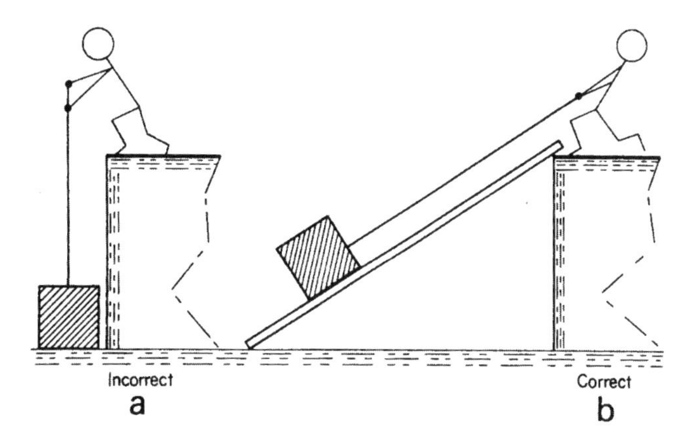

Fig. 12.2

Variations of the inclined plane

The wedge

From Fig. 12.2b it can be seen that the arrangement of an inclined plane has a wedge shape.

Fig. 12.3a and b show the use of wedges.

Simple screw jack

As a screw thread is at an angle, the screw jack is a variation of the inclined plane (Fig. 12.4a). Another application of this tool is a bearing extractor (Fig. 12.4b).

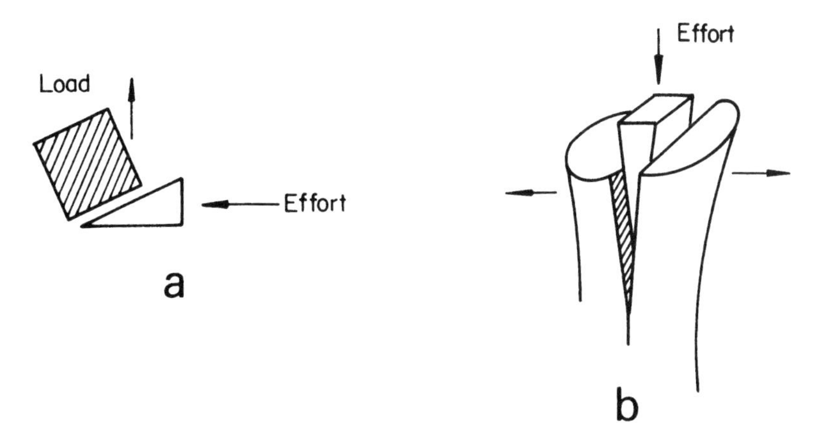

Fig. 12.3 *(a) Single inclined plane; (b) double inclined plane*

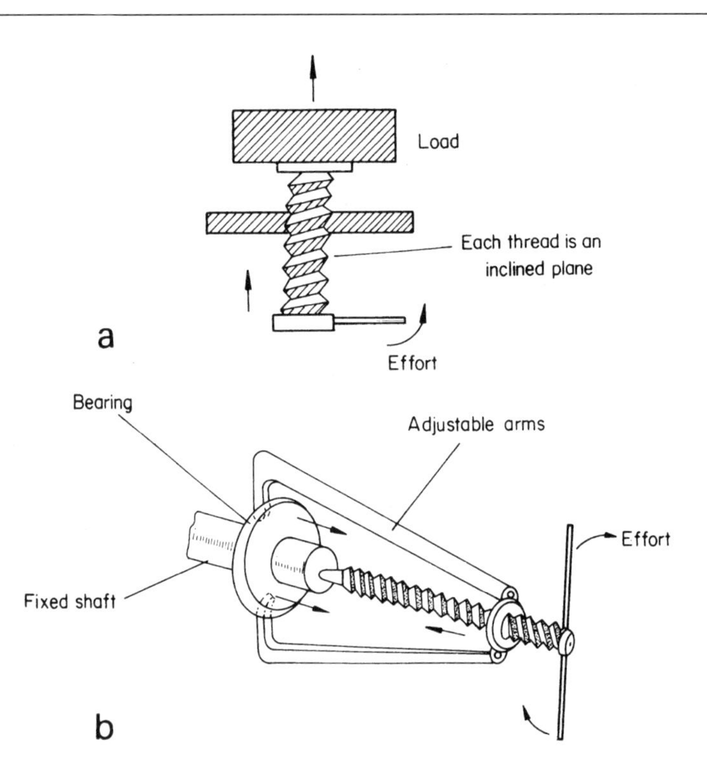

Fig. 12.4 *(a) Screw jack; (b) bearing extractor*

Pulleys

Fig. 12.5 below shows some simple pulley systems. The figure shows multiple sheave blocks, the most common form of rope and pulley system. In practice, all the wheels in a pulley system are the same size.

Manual lifting and handling

Many injuries in industry and in the home are the direct result of incorrect lifting or handling. Fig. 12.6 illustrates some of the DOs and DON'Ts of lifting and handling.

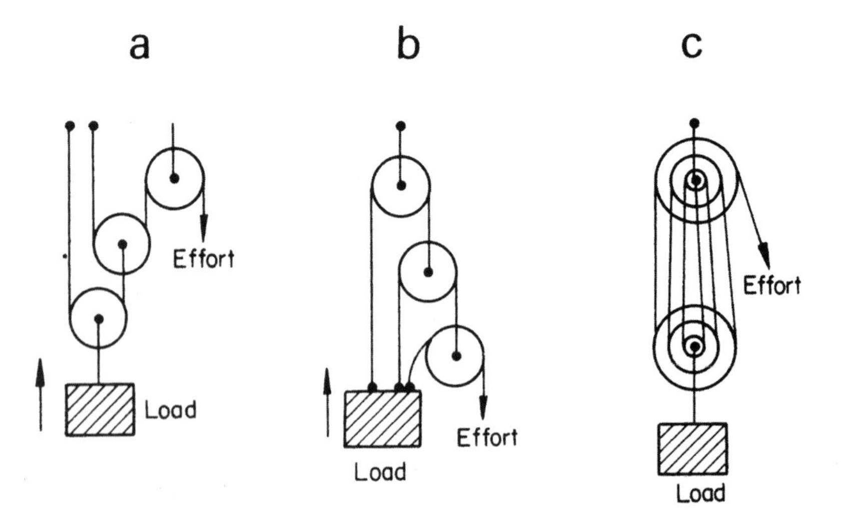

Fig. 12.5

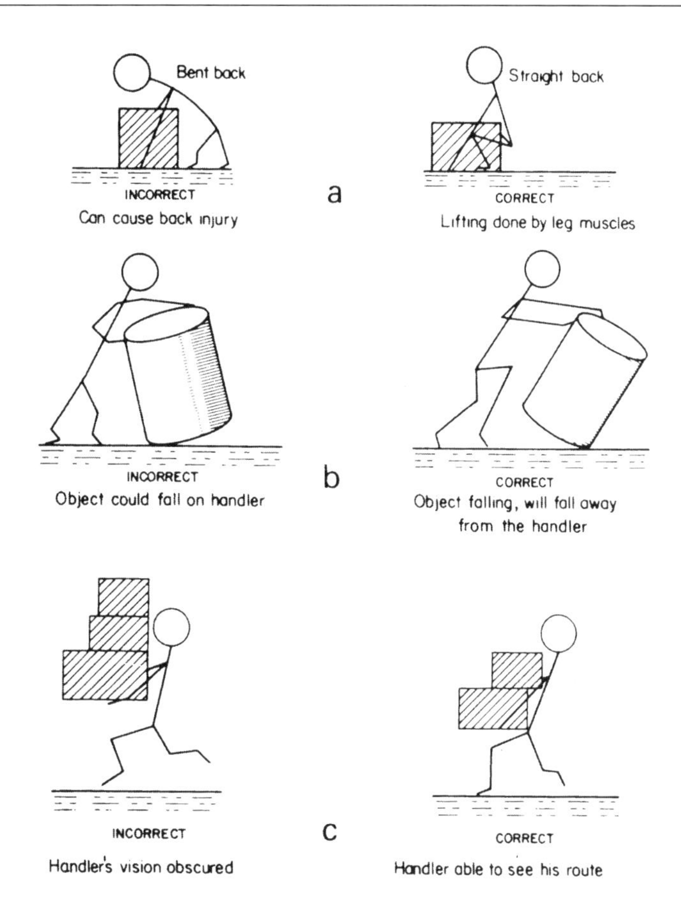

Fig. 12.6

Apart from considering one's own welfare when lifting and handling, the safety of others must be taken into account.

Always ensure that others are in no danger whilst you are lifting, lowering or moving items of equipment especially full lengths of conduit, tray or trunking. Never throw items to fellow workers, always be **alert**.

Work, load and effort

Work: symbol, *W*; unit, joule (J)

If an object is moved from one place to another, work is done. The heavier the object and the greater distance it is moved, the more work is done. Therefore work is a product of the amount of force or effort used to move the object, and the distance it is moved:

Work = force × distance

$$W = F \times l$$

If we ignore friction and any other losses, then the force we exert is equal to the force the load exerts against us.

Load force: symbol, F; unit, newton (N)

The mass of a load is a measure of the amount of material making up the load and the units we use are kilograms (kg). The weight or force that a mass exerts on the surface of the Earth is measured in newtons (N), and mass in kilograms can be converted into load force by multiplying by 9.81.

Load force (N) = load mass (kg) $\times$ 9.81

Effort

Effort is the force we require to move a load and as we have already seen, it is equal to the load force or weight (ignoring friction).

Example

A concrete block has a mass of 100 kg. Ignoring friction, calculate the work done in moving the block 8 m.

Work (J) = force (N) $\times$ distance (m)

First we must convert the load force into newtons.

Load force = mass $\times$ 9.81

= 100 $\times$ 9.81

= 981 N

$\therefore$ Work = 981 $\times$ 8

= 7848 J

or 7.848 kj (kilojoules)

Calculations involving simple levers

The effort required to lift a load is calculated as shown in Fig. 12.7.

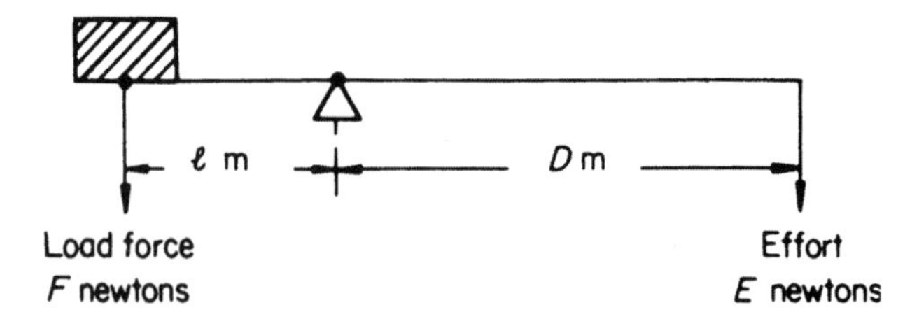

Fig. 12.7 *($E \times D = F \times l$)*

Example

Calculate the effort required, using a simple first-order lever, to lift a load of mass 500 kg if the effort is exerted 3.5 m from the fulcrum and the load is 0.5 m from the fulcrum (Fig. 12.8).

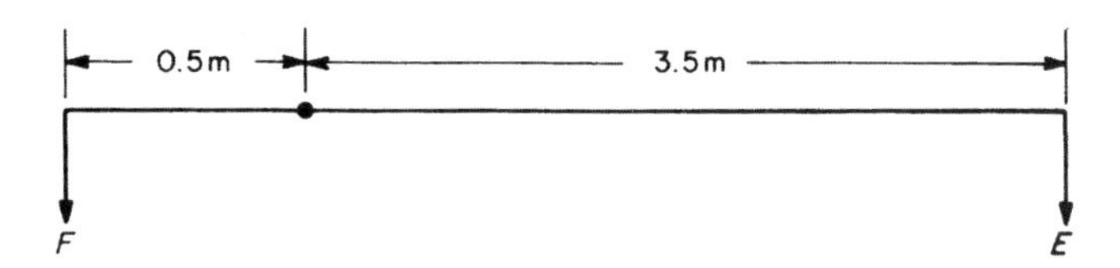

Fig. 12.8

First we must convert the load mass into load force;

$$F = 500 \times 9.81$$
$$= 4905 \text{ newtons}$$

Now

$$E \times D = F \times l$$

$$\therefore E = \frac{E \times l}{D}$$

$$= \frac{4905 \times 0.5}{3.5}$$

$$= 700.7 \text{ N}$$

Access equipment

In order for us to carry out certain tasks in installation work, it is often necessary to work above ground level (installing conduit and trunking, etc.). It is obvious, then, that a safe means of access must be used. The following items are typical of access equipment available:

Ladders	No-bolt scaffolding
Swingback steps (step-ladders)	Bolted tower scaffold
Ladder and step-ladder scaffold	Standard independent scaffold
Trestle scaffold	Putlog scaffold

Ladders

There are various types of ladder: wooden or aluminium; single, double and triple extension; rope-operated extension.

Safe use of ladders

1 Ladders must not be used with:
   broken, missing or makeshift rungs;
   broken, weakened or repaired stiles;
   broken or defective ropes and fittings.
2 Rungs must be clean and free from grease.
3 Ladders must never be painted as this may hide defects.
4 Aluminium ladders must not be used near low- or high-voltage cables.
5 Ladders over 3 m long must be secured at the top or bottom, or a second person must 'foot' the ladder on the bottom rung.
6 Ladders should be erected at an angle of 75° to the ground.
7 Ladders used for access to working platforms must extend five rungs or 1 m beyond the working surface (Fig. 12.9).

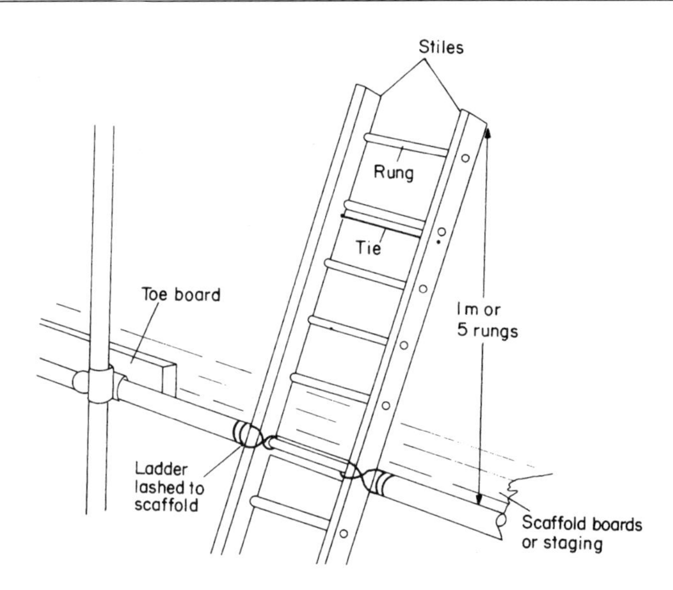

Fig. 12.9

Swingback steps (step-ladders)

The safe use of wooden or aluminium step-ladders is as follows:

1 As for items 1 to 4 for ladders.
2 Hinges must be secure.

Ladder or step-ladder scaffold

Two ladders may be used to support scaffold boards (Fig. 12.10)

1.5 m span – use 38 mm scaffold board.
2.5 span – use 50 mm scaffold board.
Above 2.4 m – use two scaffold boards.

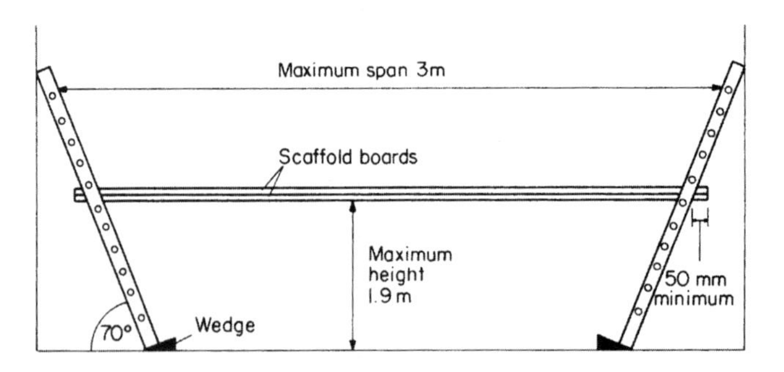

Fig. 12.10 *Ladder scaffold*

Maximum span is 3 m.
Maximum height is 1.9 m.
Maximum overhang of scaffold boards is four times the thickness of board.
Minimum overhang of scaffold boards is 50 mm.
Access to the platform should be by a third ladder.

The same span distances, etc, apply to scaffold boards used with step-ladders.

Trestles (wooden or aluminium)

Trestles are used to support scaffold boards (Fig. 12.11) and should not be used in the closed position as a ladder.

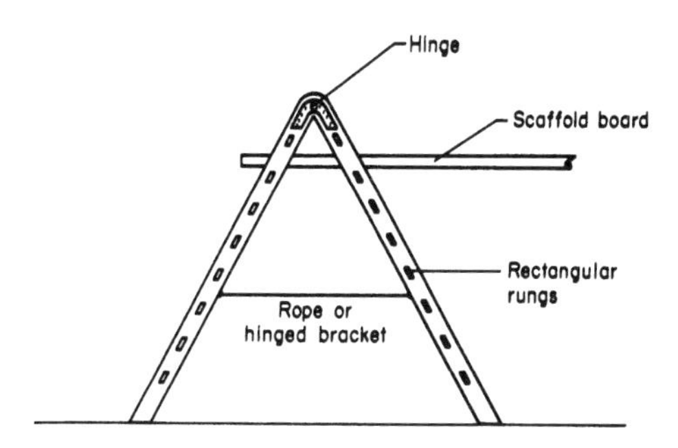

Fig. 12.11 *Trestle*

Span lengths and overhangs as for ladders.
Safety precautions as for ladders.
Maximum platform height 4.57 m.

Scaffold boards

Scaffold boards (Fig. 12.12) are made to the British Standard BS 2482/70 and are the only boards that should be used. Their maximum length is usually no greater than 4 m; beyond this, special staging is used.

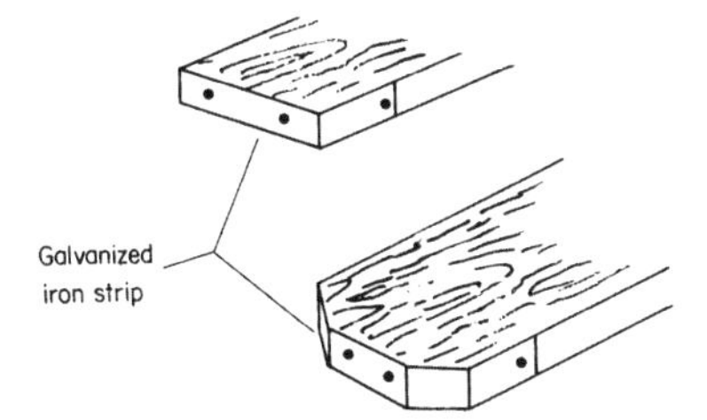

Fig. 12.12 *Typical scaffold board ends*

Scaffold boards should be:

1 clean and straight;
2 free from decay or damage; and
3 free from paint, dirt or grease.

Tower scaffolding

Tower scaffolding can have either the modern no-bolt interlocking type of construction or the conventional bolted type.

Safety precautions are the same for both types:

1 Beyond a height of 6.4 m, the tower must be tied to the building or have an outrigger fitted.
2 Maximum working platform height should not exceed:
 3 times the width of the narrow side out of doors;
 $3\frac{1}{2}$ times the width of the narrow side indoors.
3 Working platforms above height of 2 m must have a toe board and hand rail (Fig. 12.13).

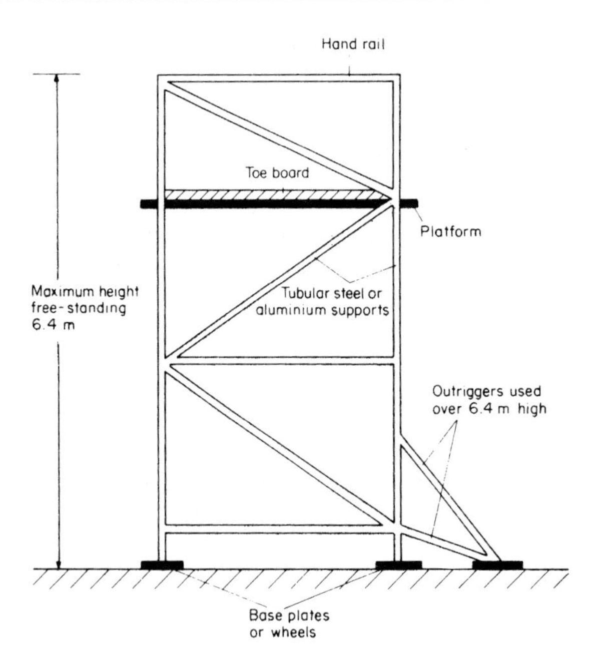

Fig. 12.13 *Tower scaffold*

Note: For any kind of scaffolding:

1. Toe boards must be at least 150 mm high.
2. Hand rails must be between 920 mm and 1150 mm above the working platform.
3. The distance between boards must not exceed 25 mm.
4. Mobile towers must have locking devices fitted to the wheels.

Overlaps and spans of scaffold boards are as for ladders.

Independent scaffolding

Independent scaffolding (Fig. 12.14) is one of the typical systems used on building sites.

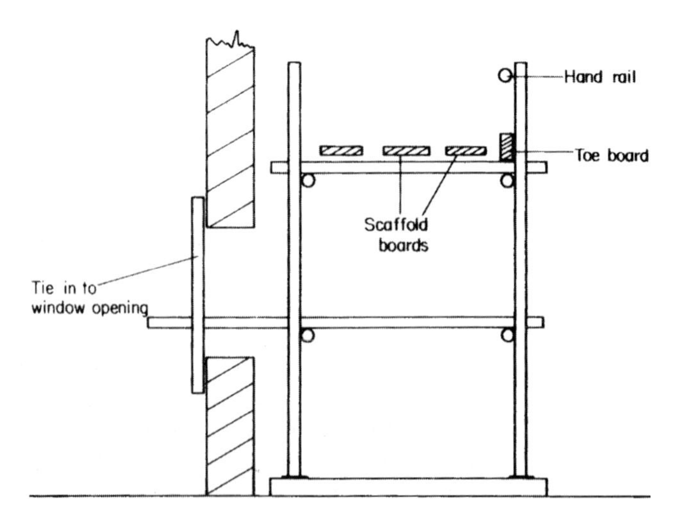

Fig. 12.14 *Independent scaffolding*

Putlog scaffolding

The maximum safe height of putlog scaffolding (*see* Fig. 12.15) is 45.5 m.

The distance between supports for platforms on both putlog and independent scaffolding should be as follows:

990 mm for 32 mm board thickness
1520 mm for 38 mm board thickness
2600 mm for 50 mm board thickness.

All scaffolding should be inspected once every 7 days and after bad weather.

The joining of materials

It is often required to bond together two materials together, for example cables or metal sheets. Three typical methods used in the electrical industry are soldering, riveting and crimping.

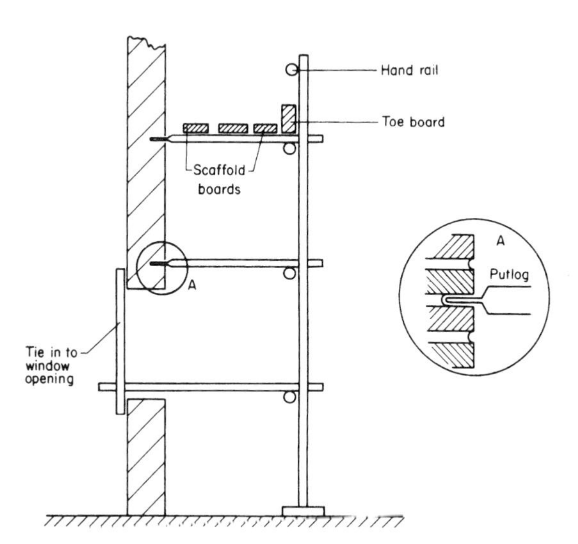

Fig. 12.15 *Putlog scaffolding*

Soldering

Care must be taken when soldering to avoid the risk of burns or starting a fire, or inhaling toxic fumes.

When using molten solder to pour over joints or into lugs, wear protective gloves and goggles.

When using soldering irons, always use a fireproof stand to hold the iron. Never flick surplus solder from the iron; wipe it off.

Riveting

There are many different designs of riveting tool available, the most potentially dangerous being the powered types which rely on compressed air or an explosive charge. These must *never* be operated incorrectly. Follow the written operating procedure and if this is not available, seek instruction from a person competent in the use of the tool.

Crimping

Crimping is a method used to join cables together. The ends of the conductors to be connected are placed in a sleeve which is then crushed to provide a mechanical weld between sleeve and conductors. Once again, the most hazardous of the tools available are the powered types and the same safety precautions as for riveting should be observed.

Fire safety

Fire prevention

It is obviously more sensible to prevent a fire than to be forced to put one out. Most establishments carry notices prohibiting the use of naked flames or smoking in certain areas, and provided that we comply with these instructions, and also make sure that all appliances are switched off after use and that all highly combustible materials are kept away from any source of heat, then the risk of fire will be minimised.

Fire detection and control

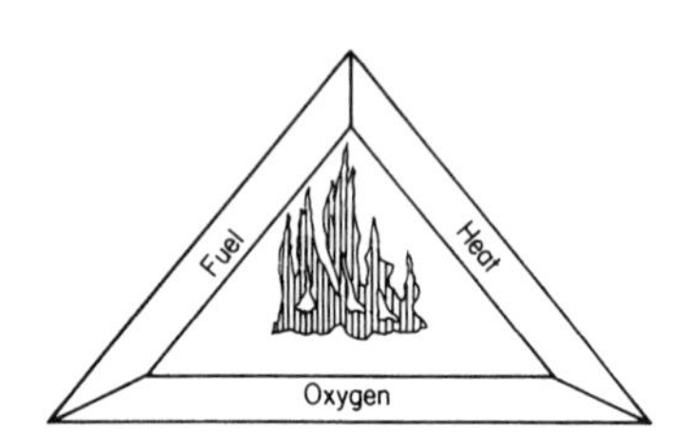

Fig. 12.16

In order for fire to exist, three properties must be present (Fig. 12.16). These are fuel (oil, paper, wood, etc.), oxygen and heat (naked flame, spark, etc.).

Detection of the outbreak of fire is normally confined to three methods: human detection, smoke detection and heat detection.

Human detection is limited to an individual noticing the fire and raising the alarm manually at a break-glass call point. Both smoke and heat detection rely on automatic means of sensing the fire and operating the alarms.

The control of fire is limited to two methods, one manual (hand-held extinguishers) and the other automatic (sprinkler system). Both systems, however, use the same substances (water, foam, etc.), depending on the type of fire.

Classes of fire

Fires may be divided into three classes depending on the location and the fuel:

Class	Types of fire
A	Solids – wood, paper, textiles, etc.
B	Flammable liquids – oil, petrol, paint, etc.
C	Flammable gases – acetylene, butane, methane, etc.
D	Metals – magnesium, sodium, etc.
E	Cooking oils

Types of extinguisher

It is extremely important to understand that **not all fires can be extinguished by the same kind of extinguishing media**. It would be dangerous to use water on a liquid fire as the burning fluid would only be spread on the surface

Table 12.1 Fire extinguishers

Type	Most suited for	Class	Colour
Water	Solids	A	Red
Carbon dioxide (CO_2)	Liquids, electrical	B	Black
Foam (AFFF)	Solids, liquids	A, B	Cream
Powder	Solids, liquids, gases	A, B, C	Blue

of the water. It would also be hazardous to use water on an electrical fire as water is a conductor of electricity. Table 12.1 shows the correct kind of extinguisher, its class and colour for a particular type of fire.

Current legislation requires that **all** fire extinguishers are coloured red but with a block or band of colour denoting the type.

Remember, some kinds of materials such as plastics and certain chemicals give off highly toxic fumes when ignited. Under these circumstances, in order to minimise the danger to the fire fighter, special breathing apparatus would be needed.

Electrical safety

Electricity, with the exception of extra-low voltage, is potentially lethal. It must be treated with respect at all times and used only in approved ways. Many people die every year as a result of the misuse of electricity. They die as a direct result of electric shock, or as an indirect result of faulty or misused electrical equipment that causes fire.

Listed below are some important points to remember when working with electricity.

1 Never work on live equipment (unless a special live test is required, for which you will need to be an experienced and competent person).
2 Always ensure, by using approved test instruments, that equipment is dead.
3 Never accept another person's word that a circuit is safe to work on; always check.
4 Ensure that all supplies to equipment to be worked on are isolated at the appropriate places and locked 'OFF' if possible, and that all supply fuses are removed and retained in a safe place. Place 'CAUTION MEN AT WORK' notices at positions of isolation.
5 If work is to be carried out on dead equipment which is adjacent to live supplies, ensure that barriers are used to define safe areas, or 'DANGER LIVE APPARATUS' notices are placed on all adjacent live equipment.

Fig. 12.17a shows the ways in which a test lamp is used to indicate the circuit condition, and Fig. 12.17b and c show how the human body can become part of a circuit. Either position in Fig. 12.17a will indicate whether the live terminals is live or dead. In effect the body replaces the test lamp of Fig. 12.17a. At the normal domestic voltage of 230 V, it needs only 0.05 ampere (50 mA) of current flowing through the body to cause death.

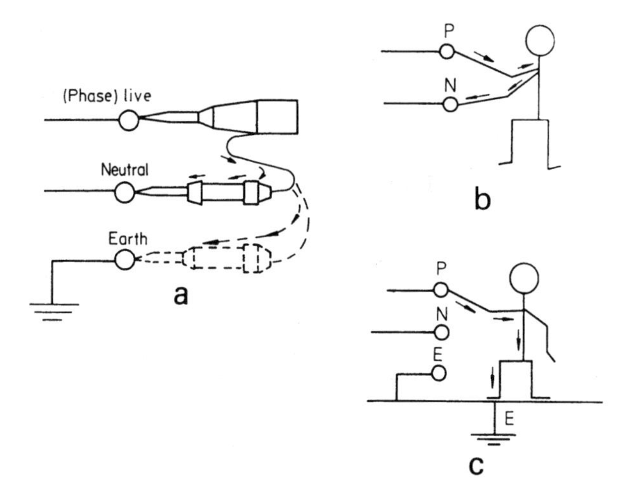

Fig. 12.17

Safe isolation of supplies

The Electricity at Work Regulations require that **no live** working is undertaken unless it is impracticable to work **dead**.

All electrical circuits, apparatus etc. must, therefore, be switched off, isolated (these can sometimes be achieved by one action, e.g. switching of a circuit breaker also isolates the circuit) and, whenever possible, locked in the off position.

In order to prove the system dead and safe to work on, an approved voltage indicator should be used (this is usually a test lamp, see p. 366). Neon screwdrivers, 'volt sticks' and the like should **not** be used. The following steps should be observed:

1 Identify the circuit/equipment to be worked on.
2 Ensure that it is safe/convenient to switch off.
3 Isolate the supply.
4 Using an approved voltage indicator:
   - check the indicator works, on a known supply
   - then use it to test to check the system is dead
   - then re-check the indicator is still working, on the known supply again.
5 Lock off the isolator or remove the fuse and keep in a safe place. If no locking facilities are available, post warning signs or notices.

Safe use of electrical equipment

When one is using electrical equipment such as drills, saws, sanders, etc. on site or in a workshop, great care must be taken to ensure that the tools are in good condition and that the cables supplying them are not damaged in any way and are adequate for the job they have to do.

Any connections of cables must be carried out by a competent person using approved tools and equipment.

For work on building sites, tools using a voltage lower than usual (110 V instead of 230 V) are recommended.

All current-using and current-carrying apparatus used on sites must be inspected and checked at regular intervals. A 3-month period is recommended, but the user should always check before use that all electrical apparatus is in good condition.

Ensure that all cables exposed to mechanical damage are well protected.

First aid

No matter how careful we are, there are times when we or our workmates have an accident. Prompt action can relieve unnecessary suffering, prevent permanent disability and, in extreme cases, save life. This action is *first aid*. Table 12.2 gives the first-aid action recommended in some common cases; electric shock is dealt with separately.

Table 12.2 Recommended first-aid action

Symptoms	Action
Shock (suffered by most casualties)	Move casualty as little as possible. Loosen all tight clothing (belts and ties, etc.). If possible lay him/her down and raise legs over head level. Keep casualty warm with blankets or coats. Reassure casualty. Send for medical help
Burns	Immerse burnt areas in cold water. Treat for shock. Do not remove burnt clothing. Do not apply oils or grease to burns. Cover burns with clean dry cloth. Arrange for medical help if burns are severe
Cuts and grazes	Small cuts and grazes should be washed in running water, and a dressing applied. More severe bleeding of deep cuts should be stopped by applying pressure to the wound. Treat for shock. Call for medical assistance
Bruising	Apply an ice bag or damp cloth until the pain stops
Falling	Make casualty lie still. Check for obvious injuries. If necessary treat for shock. If in doubt call for medical help
Contact with chemicals	Wash affected area with water, cover with dry dressing. Get casualty to hospital
Contact with toxic fumes	Remove casualty from gas-laden or oxygen-deficient area. Loosen tight clothing. Apply resuscitation if breathing has stopped

Electric shock

An electric shock is experienced when a current passes through the body. Not everyone would have the same level of shock from the same source; it would depend on variables such as the individual's body resistance, his or her health, etc. However, it is generally accepted that 50 mA (0.05 ampere) is the lethal level. Below this level, contact with a live source throws us away from the source. Above 50 mA the muscles contract or freeze and we are unable to break contact. Also, interruption of the heart's rhythm takes place and its beating may stop altogether (ventricular fibrillation). Burns to the parts of the body in contact can occur, together with burning of internal organs and loss of breathing (see p. 28).

It is clearly essential that prompt action be taken in the case of severe electric shock. The following procedure must be adopted:

1 **Do not touch the casualty with the hands**.
2 Switch off the supply.
3 If this is not possible, pull the casualty away from contact using **insulating material**, i.e. a scarf, piece of wood, newspaper, etc.
4 If heart or breathing or both have stopped, apply resuscitation and cardiac massage until the casualty recovers.
5 Treat for burns.
6 Check for other injuries; treat as necessary.
7 Treat for shock.
8 Call for medical assistance.

Note: If possible get a third person to call for medical help as soon as the casualty is discovered.

Method of resuscitation (Fig. 12.18)

Cardiopulmonary resuscitation (CPR)

Place victim flat on back on a hard surface and check the mouth for foreign bodies, false teeth, etc. and remove any loose objects.
A. Open the airway.

If unconscious and not breathing tilt head back and support jaw.

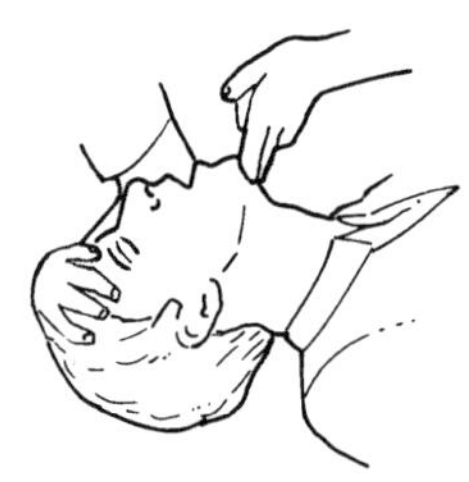

B. Breathing.
Look, listen and feel for breathing. If not breathing give mouth to mouth resuscitation.

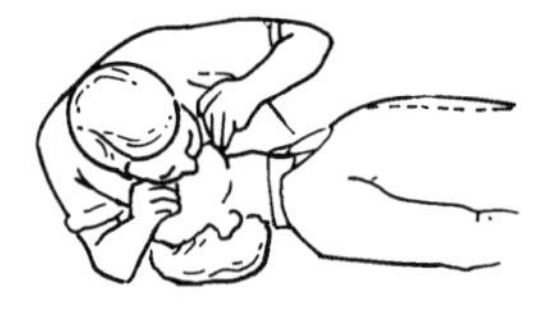

Close nostrils with fingers and blow into mouth.
Make sure there is no air leak and the chest expands.
Give 2 normal breaths.

Feel for carotid pulse in neck.
If no pulse:

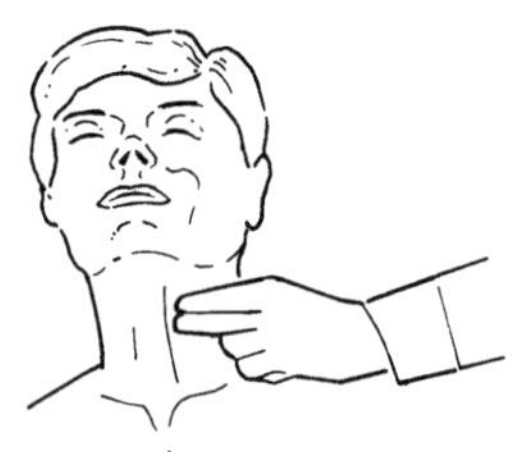

C. Perform chest compressions.

Place both hands on the lowest third of breastbone.
Depress sternum to one-third of the depth of the patient's chest at a rate of 100 compressions per minute. Alternate 2 breaths with 15 compressions.
Continue cycle until the victim shows signs of recovery by coughing, vomiting or making signs to move.

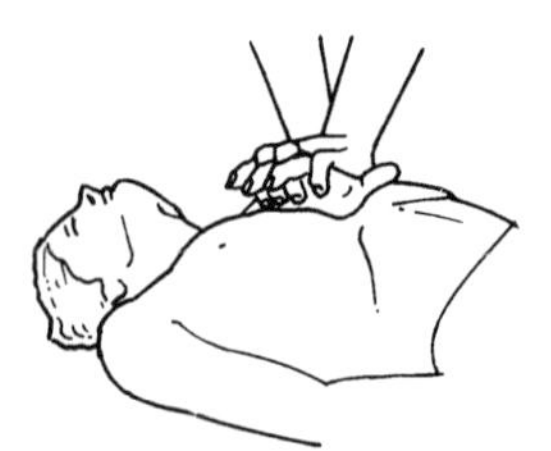

Fig. 12.18

13 The electrical contracting industry

Let us imagine that we are about to start an electrical installation company employing half a dozen or so staff. We will assume that initial finance is no problem and that suitable premises are available. What, then, are the organizations, rules and regulations, and standards that could affect the way in which we will be working?

Just about all working processes are governed by one or more sets of regulations. The use of electricity is no exception to this, the main body of regulations being as follows:

Electricity at Work Regulations 1989
Electricity Supply Regulations 1988
IEE (Institution of Electrical Engineers) Wiring Regulations 16th edition
British Standards Specification (BS)
British Standards Codes of Practice (CP)

Electricity at Work Regulations 1989

These are statutory regulations (i.e. they may be enforced by law) and are the basis of the IEE Regulations. They lay down, in broad terms, the measures to be taken to ensure the safe installation and use of electrical equipment.

Electricity Supply Regulations 1988

Once again these are statutory regulations. In this case they ensure the safety and welfare of the public and that a proper and sufficient supply of electricity is provided.

IEE Wiring Regulations 16th edition

The IEE Regulations are not statutory, but are based on the statutory Acts and internationally agreed codes of safety. They are also a British Standard, BS 7671. They lay down recommendations for the safe installation and use of electrical equipment in buildings, and are therefore of great importance to members of the electrical contracting industry.

British Standards Specifications

British Standards (abbreviated to BS) ensure a national uniformity in the quality, performance, dimensions and listing of materials. Many items of equipment have a BS EN number which signifies European acceptance.

British Standards Codes of Practice

Codes of Practice (CPs) are issued by the British Standards Institution and recommend standards of good practice. In the case of electrical installations, they follow, in general, the recommendations of the IEE Regulations. In some cases, when a choice of methods is available, they select one as preferred practice.

All sources of information mentioned are important to the installation electrician and he or she should be familiar with them all.

Also important is a knowledge of the many different Boards, Associations, unions and schemes which influence the electrician's work.

Joint Industrial Board

The Joint Industrial Board (JIB) is a national organization which basically acts as a means of liaison between unions and employers in the subject of grading and rates of pay for employees. Hence a JIB-graded electrician will have attained a certain academic and practical standard and will receive a set basic wage wherever he or she is working.

Trade unions

There are several unions to which the electrician can belong. Whichever he or she chooses, if indeed there is a choice, he or she should remember that the unions' basic role is to ensure that employees enjoy satisfactory working conditions and rates of pay. They are the employees' voice with which to speak to management, and in the event of any serious dispute with the employer, it is best to let the union deal with the matter in the correct way, using the approved disputes procedure and conciliation machinery.

Electrical Contractors' Association

The Electrical Contractors' Association (ECA) was established in 1901 and requires its members to have a high standard of workmanship.

Anyone employing the services of an ECA-registered firm can expect top-quality work carried out in an efficient manner. Contractors can become members only after their work has been carefully examined by ECA experts.

A customer has a certain comfort in employing an ECA company, as, if the company were to cease trading in the middle of a contract, the ECA would ensure that the work was completed by another ECA firm at no extra cost.

National Inspection Council for Electrical Installation Contractors

The NICEIC concerns itself solely with standards of workmanship and contractors can apply to be placed on the NICEIC role as an approved contractor. Regular inspections of the contractors' work are undertaken to ensure that high standards are maintained.

Contract and tenders

The majority of electricians are employed by a contracting firm and therefore are not likely to become involved with the administrative side of the business. However, it is important that the electrician is aware of the procedures involved in obtaining the work he or she is to carry out.

Usually, the first step in obtaining a contract to carry out an installation is to tender a price for the work.

Tenders

A tender is, by dictionary definition, an offer to supply goods and/or services at a fixed rate. In many cases this is a simple procedure but on larger jobs the tender can become complicated, and considerable experience is necessary to complete such a tender correctly.

Contracts

The law relating to contracts is extremely complicated and involved, and hence only the most basic concepts will be considered.

In simple terms, for any job there is a *main contractor*, which can be an electrical installation firm, building firm or a decorating firm, etc., depending on the work to be done. This main contractor is responsible to the client (that is, the person ordering the work to be done), either directly or via an agent such as an architect.

Should the main contractor employ the services of another firm, this firm is called the *subcontractor* and is responsible to the main contractor.

A typical sequence of events is as follows:

1 The client approaches an architect with a view to having, say, a hotel designed and built.
2 The architect designs the building and the design is approved by the client.
3 A specification and a bill of quantities are prepared.
4 The work is put out to tender, and eventually one is selected – not necessarily the lowest priced.
5 The architect may nominate the subcontractors, i.e. painters, electricians, plumbers, etc., or leave it to the main contractor. In any event the subcontracts will go to tender. This is where the ability to read and interpret drawings, bills of quantities and specifications is so important.

Specifications

The specification indicates the quantity of sockets, lighting points and other fixed accessories required for each room, and sometimes the preferred manufacturer. It may also indicate the type and size of cable to be used (*see* sample specification).

The architect's drawings normally show elevations and plan views, with details of recommended positions for all accessories.

It is from these two sources of information, specification and drawing, that a design for the electrical installation can be prepared and a competitive tender submitted.

Bills of quantities

Usually prepared by a quantity surveyor, bills of quantities indicate, for each trade concerned with the work, the quantity and sometimes cost of the materials to be used.

Variation order

In the case of electrical installation work, there is every chance, on a big job, that some variation from the programme will occur. It is very important that the site electrician notifies his or her superiors *immediately* of any change. A *variation order* can then be made out which will enable the new work to be carried out without breaking any of the terms of the contract.

Daywork

It may be necessary to carry out work in addition to that referred to in the contract and this work will be the subject of daywork.

Daywork is normally charged at a higher rate than the work tendered for on the main contract, and the charges are usually quoted on the initial tender.

Typical additional charges are as follows:

120% – labour
20% – materials
5% – plant
} added to normal rates

It is important that the members of the installation team on site, record on 'daywork forms' all extra time, plant and materials used.

It must always be remembered that employer and employees are both essential for a business to exist. For it to succeed, the two sides must work in harmony. From the point of view of the electrician on site, this involves – apart from doing a good wiring job – accurate recording of time on time sheets and daywork sheets, recording and checking deliveries of materials on site, ensuring that all materials stored on site are safe, and keeping in constant contact with the employers, the main contractor and other subcontractors.

Sample specification

Phone: Waterlooville 2583

F. A. Smith (Waterlooville) Ltd.

Building Contractor

Registered Office

88 Jubilee Road, Waterlooville, Hants. PO7 7RE

Company No 662680 Registered in England

Your Ref:

Our Ref: FB/HI

3 May 1979

M Jones & Co
34 Queensway
Anytown
England

Dear Sirs

Re: Electrical Installation at Hunters Meadow Estate

We should be pleased to receive by return your lowest price for materials/subcontract work as detailed below, delivered to/executed at the above project.

Workmanship and materials must comply with the appropriate clauses of the current edition of the registered House-Builder's Handbook Part II: Technical.

As the main developers for this project we have the order to place for these items.

Yours faithfully
F A Smith (Waterlooville) Limited

Fred Bloggs

Fred Bloggs
Buyer

E/O for connecting central heating.

Wiring To be carried out in PVC cable; lighting wired in 1.0 MM: ring main in 2.5 MM

Sheathing PVC sheathing to be fixed on walls under plaster.

Boxes Metal boxes to be used for all fittings on plastered walls.

Fittings To be of cream flush Crabtree or equivalent; 13 mm socket outlets for ring main.

HUNTERS MEADOW ESTATE, ACACIA AVENUE, ANYTOWN

TYPE 'B' 3 BED LINKED

This spacious Link/Detached three-bedroomed house with attractive elevations has a larger than average lounge with separate dining area, good-sized bedrooms and downstairs cloakroom making the ideal family home.

Heating is by means of gas-fired central heating and an electric fire in hardwood surround to lounge. Thermoplastic floor tiles to choice from our standard range for the whole of the ground floor.

ACCOMMODATION

Lounge	19'0" × 13'0"
Dining room	9'0" × 10'9"
Kitchen	9'9" × 10'9"
Hall	5'0" × 5'0"
Cloaks	5'0" × 3'9"
Bedroom 1	12'0" × 12'0"
Bedroom 2	9'0" × 11'9"
Bedroom 3	9'0" × 7'0"
Bathroom	6'6" × 6'9"
Airing cupboard	3'0" × 2'0"

EXTERNAL DOORS Are softwood glazed in two squares to match windows

GARAGE DOORS Metal up and over

INTERNAL DOORS Sapele hardwood flush

CEILINGS Artex stipple finish

DOOR AND WINDOW FURNITURE In satin anodized aluminium

CURTAIN BATTENS to all windows

LOUNGE

2 Ceiling light points
3 - 13 amp socket outlets
1 T.V. aerial socket outlet with down lead from roof space
1 Point for electric fire
2 Radiators

DINING ROOM

1 Ceiling light point
Twin socket outlet
1 Radiator

KITCHEN

1 Fluorescent ceiling light
3 - 13 amp socket outlets
1 Cooker control with 1 additional socket incorporated
1 Immersion heater switch
1 - 13 amp point for gas boiler
Wall-fitted gas boiler for central heating
3 Gas points

TYPE 'B' 3 BED LINKED (Continued)

KITCHEN (Continued)

Double drainer stainless steel sink unit mixer taps and cupboard under
1 Large base unit
1 Broom cupboard
1 High level cupboard

HALL

1 Ceiling light point

CLOAKROOM

1 Ceiling light point
1 Low-level W.C. suite
1 Corner hand basin

GARAGE

Metal up and over door
1 Ceiling light point
1 - 13 amp socket outlet
Electric and gas meters

LANDING

1 Ceiling light point
1 - 13 amp socket outlet
1 Full height airing cupboard with slatted shelves

BEDROOM 1

1 Ceiling light point
2 - 13 amp socket outlets
1 Radiator

BEDROOM 2

1 Ceiling light point
2 - 13 amp socket outlets
1 Radiator

BEDROOM 3

1 Ceiling light point
2 - 13 amp socket outlets
1 Radiator

BATHROOM

1 Ceiling light point
Coloured bathroom suite comprising low-level suite, pedestal basin and 5' 6" bath
1 Chromium-plated towel rail (electric)
Half-tiled walls to match suite colour
Bathroom cabinet with mirror front
1 Shaver point

TYPE 'B' 3 BED LINKED (Continued)

EXTERNAL

Tarmac drive to garage
Paved concrete paths to front and back entrance 2' 6" wide
Turfed open-plan front garden
Rear and side boundaries are defined by various means to suit the development as a whole, e.g. screen walls, chain link fences, etc.
Interwoven fencing panels erected adjacent to the property

EXTERNAL COLOUR SCHEME

Woodwork is generally white but details are featured in colours which are designed for the development as a whole.

INTERNAL COLOUR SCHEME

A limited choice of internal decoration is available from our standard range.

VARIATIONS

The house has been designed as an integrated unit and as a general rule it is not possible to accept variations.

THIS INFORMATION IS GIVEN AS A GUIDE TO PROSPECTIVE PURCHASERS, THE DETAILS MAY VARY FROM ONE PLOT TO ANOTHER AND FULL INFORMATION CAN BE OBTAINED FROM OUR SALES PERSONNEL. THIS SPECIFICATION DOES NOT FORM PART OF ANY CONTRACT AND MAY BE REVISED WITHOUT PRIOR NOTICE.

The design team

This may consist of one or more people, depending on the size of the contracting firm and the complexity of the installation. The example in question (a domestic dwelling) would require only one designer. This designer will need to be experienced, not only in electrical work, but in other trades, and be able to interpret the architect's drawings. He or she will also, in the absence of any protective information, provisionally locate all electrical accessories using symbols to BS 3939. These symbols are extremely important, especially to the installation team.

The installation team

Once again, the team will comprise one or more people, depending on the contract. For the work in this example, two people would normally be employed: an approved electrician and perhaps an apprentice.

A good electrician will have a reasonable knowledge of the basic principles of the other trades involved in the building and servicing of the house, and in this respect, the electrician will often need to discuss, competently, problems which may occur due to the location of his or her materials in relation to those of different trades. The electrician must, of course, be familiar with the symbols in BS 3939, and be able to communicate intelligently with the client or their representative.

The client

The client, in this case usually the builder, has to rely on the expertise of the design and installation teams; it is therefore important that a good liaison be maintained throughout the duration of the contract.

It is the satisfied client who places further contracts with the installation firm.

Working relationships

Working with the building trade on site can present its own problems not found in, say, private rewiring. Good working relationships with other trades go a long way to overcoming any problems. The phrase 'good working relationships' does not mean just having a pleasant attitude to other workers; it involves liaising with them either directly or via a site foreman, and endeavouring not to hinder their work progress. A schedule of intended work with proposed dates can help a great deal. There is nothing more damaging to working relationships than, for example, a plastering team arriving on site to find that the cable drops are not complete and the electrician is not on site.

Customer relationships

In a private dwelling requiring a rewire or simply additional lighting or power points, the electrician(s) must have the correct attitude to the customer and his or her property; after all, it is the customer who is paying for the work and who, justifiably, expects the best possible service. The following are some examples of conduct when working on other people's property:

DO be polite under *all* circumstances.
DO be presentable in dress and manner.
DO be tidy, clear away all unnecessary debris, replace furniture to its original position.
DO consult the customer if positions of accessories are not clearly defined.
DON'T use bad language.
DON'T place tools on furniture.
DON'T use furniture as steps.
DON'T leave without informing customer (unless he or she is out).
DON'T leave floorboards and/or carpets unsecured when work is completed.

Requisitions and estimates

Before an estimate can be prepared a requisition or materials list must be compiled. This is where experience is invaluable in choosing the correct, and also the most economic, materials for the job. The dwelling we are considering could be wired in m.i.c.c.–(mineral-insulated, copper-clad) cable, but although perfectly acceptable, this would involve the client in very high costs, so much so, that the tender for the work would stand little chance of being accepted.

The choice of cable and accessories is in fact quite easy in this case, but in more complex contracts the designer may be faced with extreme conditions, and will be involved in a considerable number of calculations for voltage drop and current rating before a choice of materials can be made.

Cost of materials and systems

Materials

The type of dwelling in the specification would only require PVC twin with CPC cable, with drops to switches etc. run in an oval PVC conduit and buried in the plaster, and run unenclosed under the floor. To run the wiring in single-core PVC cable enclosed in a conduit would be pointless and rather expensive.

Systems

The choice of a wiring system is just as important as the choice of material. The system dictates the material quantity. For example, the specification might call for ten socket outlets (downstairs). Wired as a ring system requires only one 30 A/32 A way in the consumer unit. Wiring on a radial system would require a calculation of floor area and an increase in cable size to $4.0\,\text{mm}^2$.

In many instances switch-gear and accessories are denoted not by name but by symbols on a drawing. All such symbols should be to BS EN 60617. Figure 13.1 illustrates some of those symbols commonly used.

Materials list

Once a system has been decided on and all the calculations have been completed, a list or requisition of materials can be compiled. The example shown of such a list (Table 13.1) is typical for the dwelling we are concerned with.

To complete the quantity column, information must be obtained from the specification and plan, the cost being obtained from the wholesalers, or manufacturer's catalogue.

Equipment on site

Let us assume that the tender has now been accepted, and work has to commence on site. In the case of a large site, a contracting firm may well have a site hut erected for the storing of materials; for smaller contracts, equipment would be stored at base and transported as required.

In either event, materials have to be delivered and an accurate check on goods delivered must be made and records must be kept. Starting a job only to find that items have been incorrectly sent is totally inefficient and does not make for good customer relationships.

Part 11 Architectural and topographical installation plans and diagrams

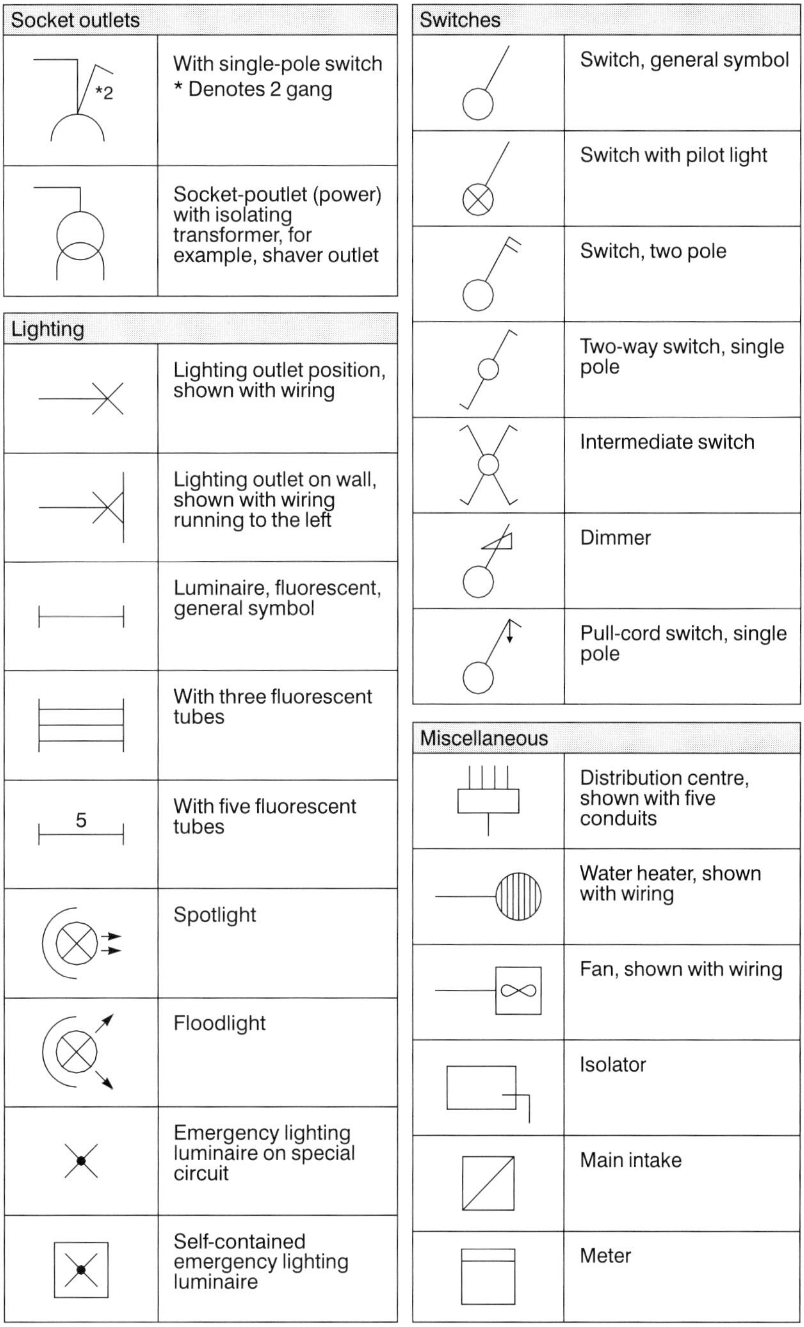

Socket outlets	
*2	With single-pole switch * Denotes 2 gang
	Socket-poutlet (power) with isolating transformer, for example, shaver outlet

Lighting	
	Lighting outlet position, shown with wiring
	Lighting outlet on wall, shown with wiring running to the left
	Luminaire, fluorescent, general symbol
	With three fluorescent tubes
5	With five fluorescent tubes
	Spotlight
	Floodlight
	Emergency lighting luminaire on special circuit
	Self-contained emergency lighting luminaire

Switches	
	Switch, general symbol
	Switch with pilot light
	Switch, two pole
	Two-way switch, single pole
	Intermediate switch
	Dimmer
	Pull-cord switch, single pole

Miscellaneous	
	Distribution centre, shown with five conduits
	Water heater, shown with wiring
	Fan, shown with wiring
	Isolator
	Main intake
	Meter

Fig. 13.1

The correct procedure is as follows:

1 Keep a copy of the original material order.

2 Delivered goods should be accompanied by an *advice note*. Ensure that there is one.

Table 13.1

Description	Manuftr	Code No.	Price Ea.	Qty.	Total cost
Cable					
1.0 mm^2					
1.0 mm^2 3 core					
2.5 mm^2					
6.0 mm^2					
16 mm^2 Tails					
6.0 mm^2 Bonding					
0.85 mm^2 Bell wire					
0.5 mm^2 Lighting flex					
1.5 mm^2 Butyl					
2.5 mm^2					
Low-loss TV co-axial					
Oval conduit					
12 mm					
20 mm					
25 mm					
Metal boxes (KO)					
Plaster depth					
25 mm deep 1 gang					
25 mm deep 2 gang					
35 mm deep 2 gang					
Cable clips					
1.0 mm					
2.5 mm					
6.0 mm					
Power					
Switched socket outlet (D)					
Switched socket outlet (S)					
Switched fused spur box					
Cooker control box					
Lighting					
1-gang 1-way plate switch					
1-gang 2-way plate switch					
2-gang 2-way plate switch					
3-gang 2-way plate switch					
Pull cord switch					
Ceiling rose					
Lampholder					
Battenholder					

Table 13.1 Continued

Description	Manuftr	Code No.	Price Ea.	Qty.	Total cost
Other					
Shaver point					
Heated towel rail					
Immersion heater					
Thermostat					
8-way consumer unit					
5 A fuse and base (or m.c.b.s)					
15 A fuse and base (or m.c.b.s)					
30 A fuse and base (or m.c.b.s)					
Earth clamp					
Bell					
Bell transformer					
Bell push					
4 ft fluorescent					
Fitting					
TV outlet					
Surface boxes (S/O)					
Surface boxes (switch)					
Earth sleeving					

3 Then, while the delivery driver is with you, check that the goods correspond with those stated on the advice note.
4 Check the advice note against the original order.

Security

This presents a major problem on the larger site, where much equipment may be stored. There is not always a nightwatchman, and security patrols cannot maintain a constant vigil.

The only answer is to have good padlocks and an alarm system. Apart from the risk of bulk theft of materials, there is always a danger of smaller amounts being stolen, which have been left at the point of work. The simple remedy for this is: do not leave any tools, materials or other equipment lying around after work. Lock everything away.

Protection of materials

All materials used in installation work should be in perfect condition. This cannot be achieved if the materials are carelessly stored or handled on site. All equipment should be kept away from damp or corrosive conditions and equipment involving delicate mechanisms, i.e. thermostats, contactors, relays, etc., should be stored or handled so as to prevent mechanical damage.

If these recommendations are not observed, it is likely that it will be necessary to return to the installation, after completion, to carry out repairs or replacements.

Disposition of equipment on site

In order that a job runs smoothly and efficiently, the positioning of materials on site must be considered. The correct procedure may be summed up as: *Always ensure that all the relevant tools and materials are taken to the place of work.* In this way needless journeys to and from the site hut or base can be eliminated and the correct amount of time can be spent on the installation.

Bar charts

In order that a job may be carried out in the most efficient manner, some job programmers use a bar chart. This is simply a method of showing graphically each stage of work to be completed on a job. This is best illustrated by an example, though it should be borne in mind that there are several ways of drawing a bar chart, this being only one of the possible methods.

Example

An electrical contracting firm has the job of rewiring an old three-storey dwelling, each floor of which is to be converted into a self-contained flat. Two pairs of men will do the work and the estimated time for each stage per pair is as follows:

Removal of old wiring per floor – 3 days
Installing new wiring per floor – 1 week
Testing and inspection per floor – 1 day

By using a bar chart (Fig. 13.2) estimate the least time in which the work may be completed.

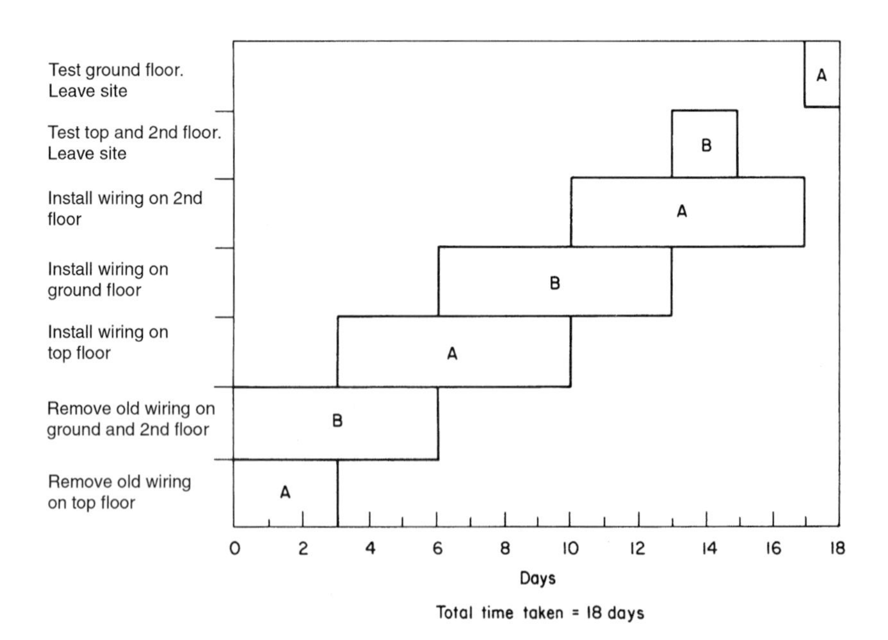

Fig. 13.2

14 Installation materials and tools

Before any wiring system is installed, account must be taken of the environment in which the system is to operate. There are many influences that may contribute to the final design of the installation. The IEE regulations give details of such influences which include, for example, ambient temperature, water, dust, vibration and corrosion.

The IEE Regulations employ an alphanumeric system with three categories: environment (A), utilisation of the environment (B) and the building structure (C). Within these categories there are subdivisions of A, B and C denoting specific conditions, and then numbers to indicate the severity of the influence.

So, for example, in the environment category A, the second letter, say, D, denotes a water influence, and the associated number the severity of the influence. For example, an AD4 code indicates an environment that is subject to splashes of water, whereas an AD8 environment is total submersion in water. In the same way, the second letter E denotes the presence of dust. Hence AE1 is negligible dust and AE6 is heavy dust

The same system is used for utilization B and building structure C.

The most likely environmental conditions that may be encountered in installation work are:

AA	Ambient temperature
AD	Water
AE	Foreign bodies and dust
AF	Corrosion
AG	Impact
AF	Vibration
AK and AL	Flora and fauna
AN	Solar

With regard to enclosures used to protect against, in particular, moisture and dust, the IP code provides the degree of protection needed. Hence an AD8 environment would require an IPX8 rated enclosure, and an AE6 environment would need an IP6X enclosure (see IP codes, page 325).

Cables

A great number of types of cable are available, ranging from the very smallest single-core wire used in electronic circuits, to the huge oil- and gas-filled cables used in high-voltage transmission systems. In this book we are concerned only with cables used in low-voltage systems (50–1000 V).

A cable comprises two parts: the conductor or conductors and the sheathing and insulation.

Conductors

A conductor may be defined as the conducting portion of a cable, which consists of a single wire or group of wires in contact with each other.

As we have seen in previous chapters, the ability of a material to be a good or bad conductor of electricity depends on the composition of that material, i.e. its resistivity. The table indicates the resistivity of some common conducting materials:

Material	Resistivity at 20°C (μΩ cm)
Silver	1.58
Copper	1.72
Gold	2.36
Aluminium	2.6
Tungsten	5.6

Silver is clearly the best conductor in the list shown, but its cost prohibits its use as a conductor material on any large scale. Gold also is a material too expensive for use in the construction of conductors.

It can be seen that copper, quite a plentiful mineral, has a low enough resistivity to make it suitable as a conductor material, and, in fact, its use in the manufacture of cable is widespread.

Aluminium, although cheap and with a relatively low resistivity, is not as suitable as copper. It has to have a large cross-sectional area to pass the same current and is mechanically inferior to copper.

Tungsten, because of its high resistivity, is used mainly in heating elements and light-bulb filaments.

Conductor construction

Conductors may be divided into two groups:

1 solid conductors; and
2 stranded conductors.

Solid conductors are either circular or rectangular in cross-section and are used for fixed wiring. Circular conductors are restricted mainly to cable cores up to 2.5 mm^2, although cross-sectional areas of up to 25 mm^2 are sometimes used in trunking, from the ground floor to the top floor of a block of flats, to provide a supply point for each floor. These conductors are called *risers*. Rectangular conductors (usually called *bus-bars*) are used in distribution

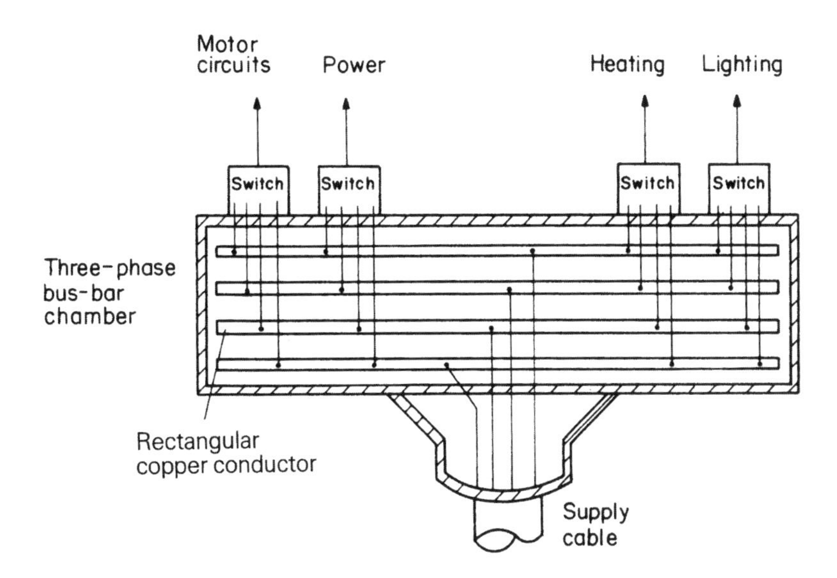

Fig. 14.1 *Bus-bar chamber*

boards or specially constructed bus-bar chambers designed to allow many different circuits to be 'tapped' off (Fig. 14.1)). Risers are a version of this type.

Stranded conductors are used in both fixed wiring cable and flexible cords, the latter being flexible cables not exceeding 4 mm^2 in cross-sectional area (c.s.a.)

Conductors for fixed wiring up to 25 mm^2 have seven strands; for example, a 6 mm^2 conductor has seven strands each of 1.04 mm diameter (7/1.04). Conductors of c.s.a. above 25 mm^2 have more strands depending on size. Flexible cords have conductors comprising a great many fine strands. This type of construction gives the conductor its flexible quality.

Sheathing and insulation

With the exception of bare conductors, i.e. bus-bars and bare risers, all conductors have some sort of insulation and/or sheathing.

Cables for fixing wiring

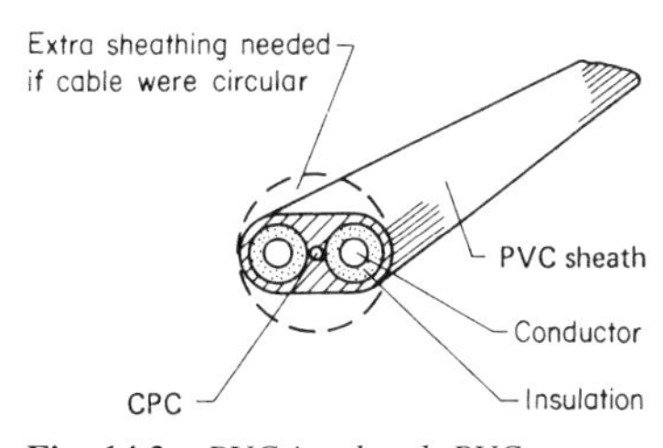

Fig. 14.2 *PVC-insulated, PVC-sheathed flat twin with earth. Colour code: red for* ***phase****, black for* ***neutral****, bare copper for* ***earth***

In many existing installations, old cables can still be found insulated with rubber with an outer sheathing of lead, cotton or rubber. This type of cable for fixing wiring is no longer manufactured. One modern insulating material for cables is PVC (polyvinyl chloride). Some cables also have a PVC sheath. (Fig. 14.2).

PVC-insulated single-core cables ('singles') are used when the installation is to be run in conduit or trunking.

Another type of fixed wiring cable is mineral-insulated metal-sheathed (m.i.m.s.) cable. The construction of this type of cable is shown in Fig. 14.3.

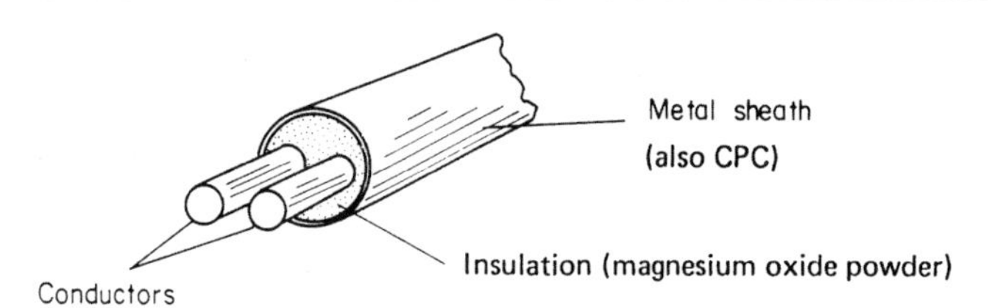

Fig. 14.3 *Twin-core m.i.m.s. cable*

The most popular version of this cable has copper conductors and a copper sheath (m.i.c.s.). It is also available with an overall PVC covering. This is a very strong and long-lasting type of cable.

Armoured cable

Armoured cable is an extension of the type of cable used in fixing wiring. Here, the inner PVC sheath is in turn sheathed in strands of steel wire and an overall PVC sheath is fitted (Fig. 14.4).

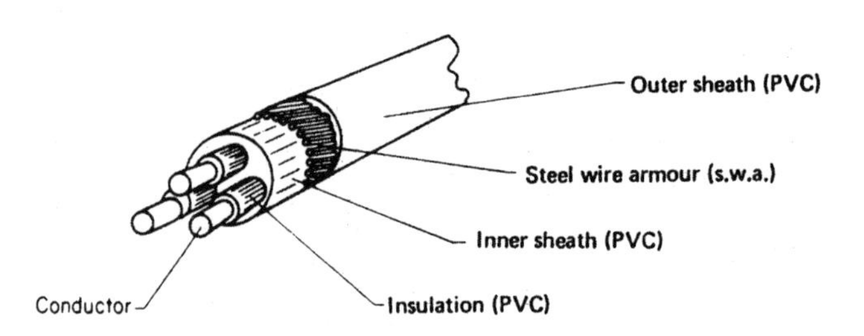

Fig. 14.4 *Three-core armoured cable*

Cables for flexible cords

In this case rubber and cotton are still used as well as PVC for insulation and sheathing. Applications include high-temperature PVC for drops to lamp-holders; butyl rubber for supply to water heaters: cotton-covered rubbers for leads to irons.

Note: **The colour code for flexible cords is as follows**:

Brown – phase
Blue – neutral
Green and yellow – earth

Reasons for sheathing and armouring

The insulation immediately surrounding a cable conductor is designed to withstand the cable's working voltage in order to prevent danger. The additional sheathing and/or armouring is added to protect the insulated conductors from mechanical damage. Some environments are more hazardous than others and cables must be chosen carefully to suit those environments: for example, a PVC-sheathed steel wire armoured (s.w.a.) cable would be used for running underground for house services.

Jointing and terminations

Wherever conductors are to be joined together, or to accessories, or to bus-bars, a safe and effective termination or joint must be made.

Jointing

There are many different ways of joining two conductors together. Here we will discuss some of the more popular methods.

The screw connector

The screw connector (Fig. 14.5) is probably the simplest method of joining conductors. Because of their simplicity these devices are often used incorrectly. Connectors of this type should always be fixed to a base, allowing the conductor no movement. They should *not* be used to connect two flexible cords together.

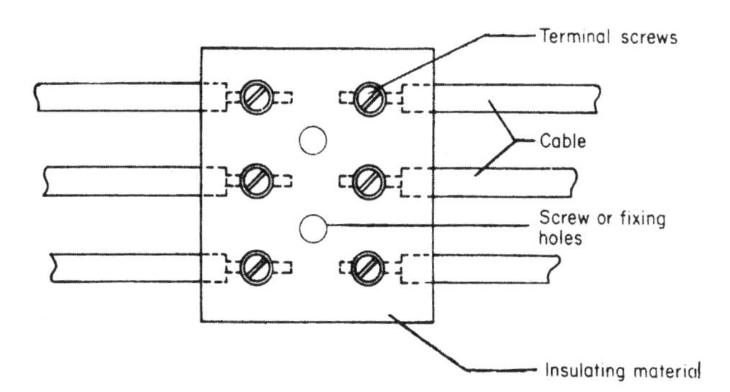

Fig. 14.5 *Simple connector*

Junction box

A junction box (Fig. 14.6) is an extension of the connector block type of joint. It is enclosed with a lid, and screwed to a base (joint, wall, etc.) and is designed for fixed wiring systems.

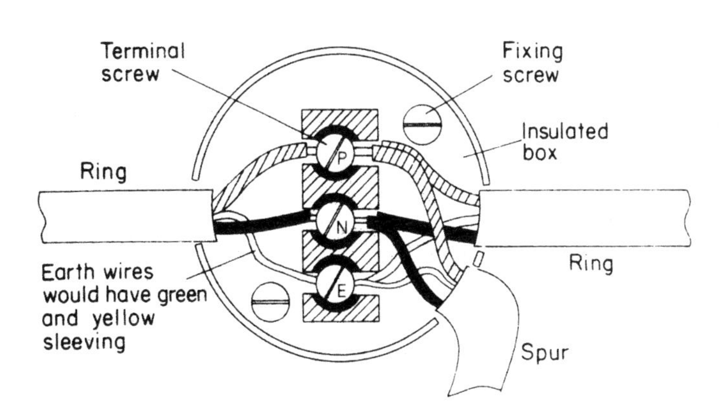

Fig. 14.6 *Tee joint or spur using a ring main junction box*

Soldered joints

This type of joint may be achieved in several ways:

1 By interlacing (like the fingers of the hands) the cable conductors (stranded) and running solder over the joint. This is called a *married joint* ('straight-through joint').
2 By a tee joint, as shown in Fig. 14.7a and b.
3 By a soldered straight-through joint using a split ferrule (Fig. 14.8).
4 By a crimped joint (Fig. 14.9a and b).

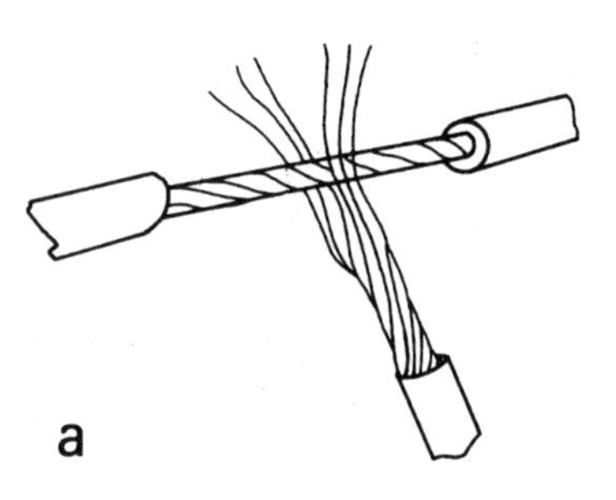

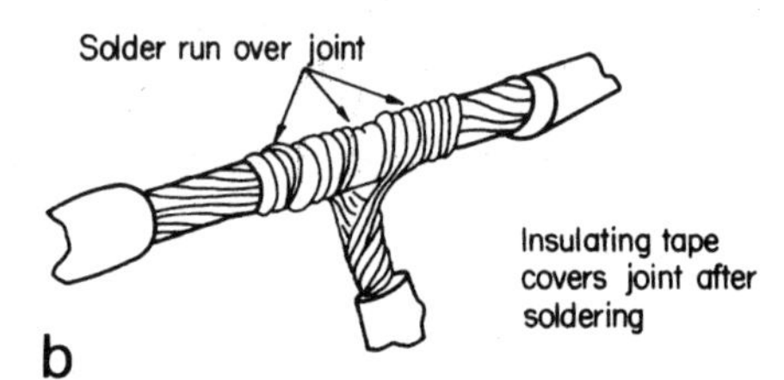

Fig. 14.7 *Soldered tee joint*

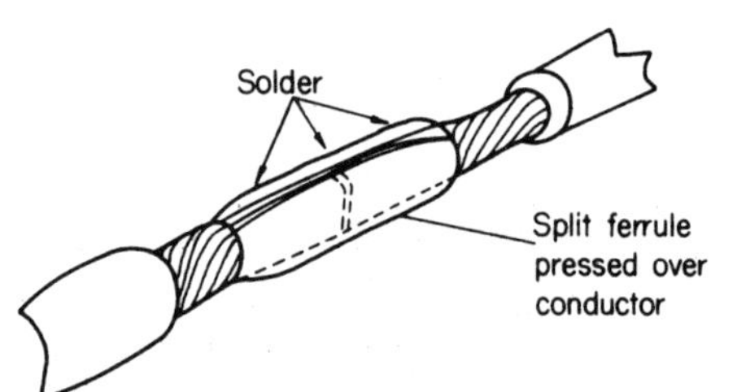

Fig. 14.8 *Joint using split ferrule*

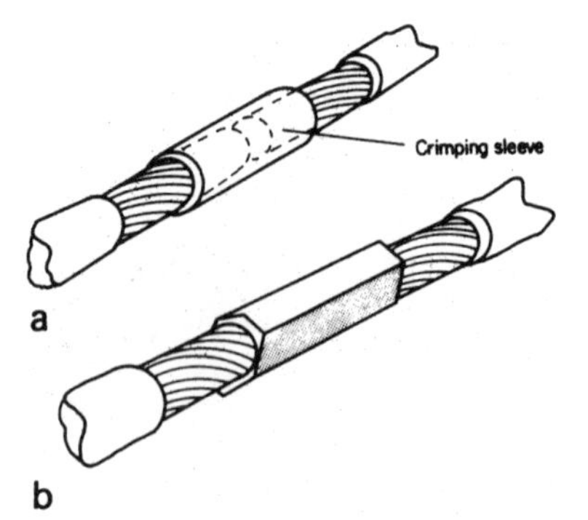

Fig. 14.9 *Crimped joint: (a) before crimping; (b) after crimping*

Terminations

When a cable conductor is finally connected to the apparatus it is supplying, a safe and effective termination of that conductor must be made.

PVC singles into screw terminals

The insulation should be removed *only* far enough to allow the conductor to enter the terminal. **Do not leave bare conductor showing outside the terminal** (Fig. 14.10a and b). Take care *not* to score the conductor surface when removing the insulation, as this may cause the conductor to break if moved.

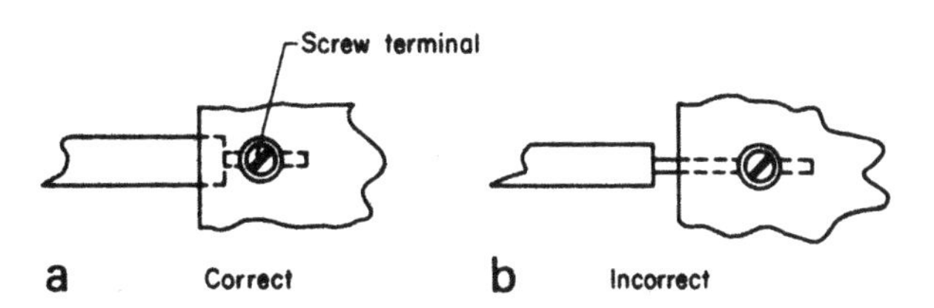

Fig. 14.10

PVC twin and three-core with earth

In this case, although the final connection into screw terminals is the same as for singles, care must be taken not to damage the conductor insulation when removing the outer sheath. Two methods used are as follows:

1 Using a sharp knife, slice the cable lengthwise, open the cable up and cut off the sheath (Fig. 14.11).

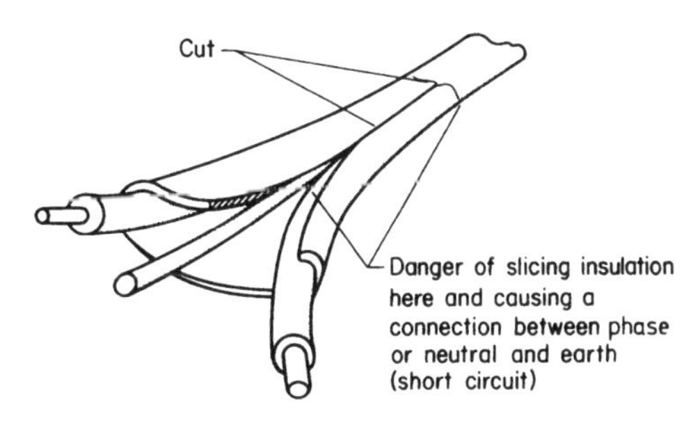

Fig. 14.11

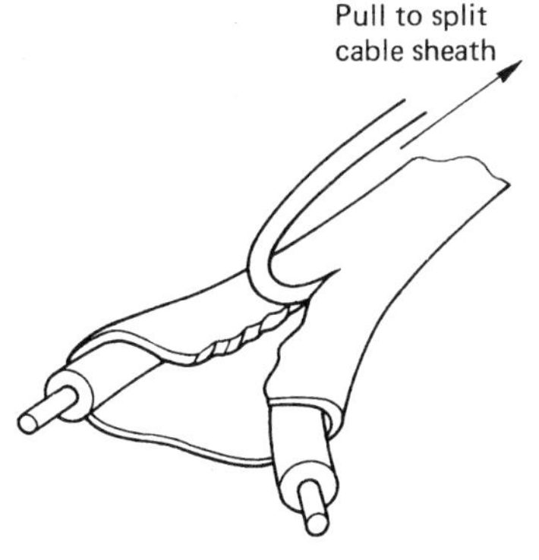

Fig. 14.12

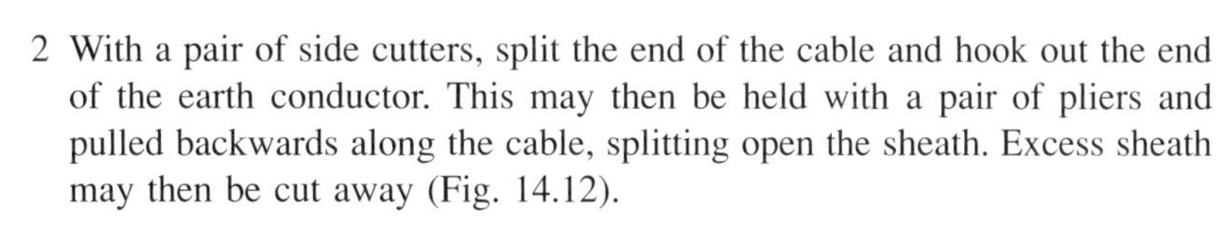
2 With a pair of side cutters, split the end of the cable and hook out the end of the earth conductor. This may then be held with a pair of pliers and pulled backwards along the cable, splitting open the sheath. Excess sheath may then be cut away (Fig. 14.12).

Mineral-insulated metal-sheathed cable

As the magnesium oxide insulation is absorbent, the termination of m.i.m.s. cable has to be watertight. The main method of terminating this type of cable is by using a 'screw-on' seal. The preparation of the cable is as follows.

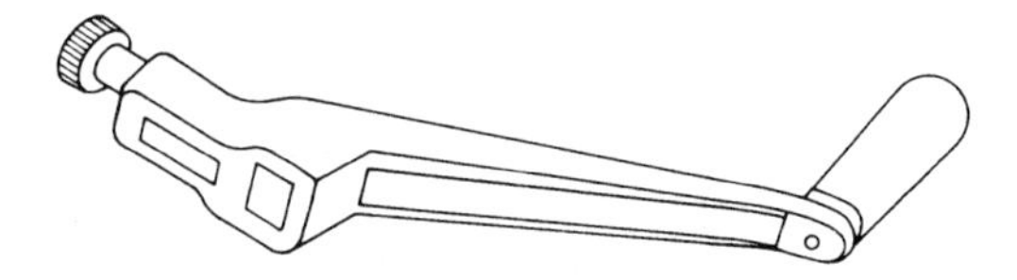

Fig. 14.13 *Stripping tool*

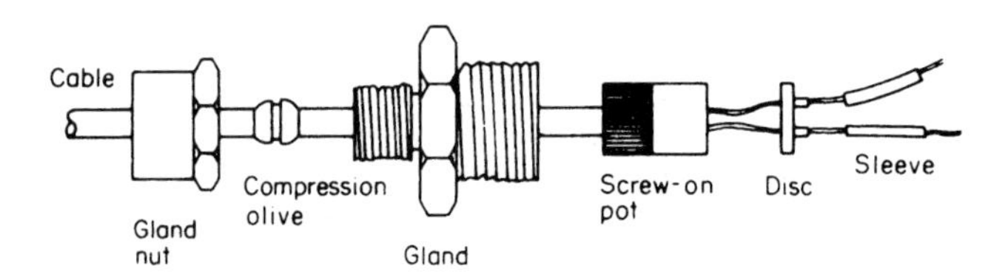

Fig. 14.14 *Screw-seal termination*

1 The sheath is stripped using a stripping tool (Fig. 14.13).
2 Loose powder is removed by tapping cable.
3 The gland nut is slipped over the cable.

After this the remainder of the termination is as shown in Fig. 14.14.

Armoured cable (steel wire armour)

The method of terminating s.w.a. cable is best illustrated by Fig. 14.15.

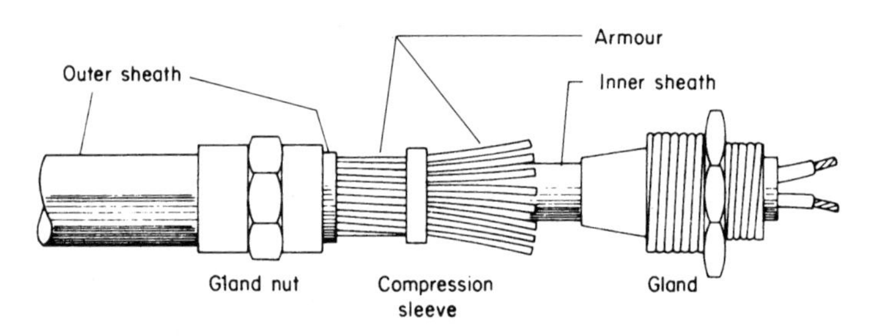

Fig. 14.15 *Armoured cable*

Lug terminations

There are two types of lug: crimped lugs and soldered lugs. A crimped lug is fastened to the end of the conductor using the same method as shown in Fig. 14.9. The soldered lug is filled with molten solder and the conductor is pushed into it. Lug terminations are frequently used for connecting a conductor to a bus-bar (Fig. 14.16).

Soldering

For soldering lugs and joints, a tinman's solder is used in conjunction with a flux. The flux keeps the work clean and prevents it from being oxidized during soldering.

For smaller jobs on fine cable, a solder with a flux incorporated in it is used.

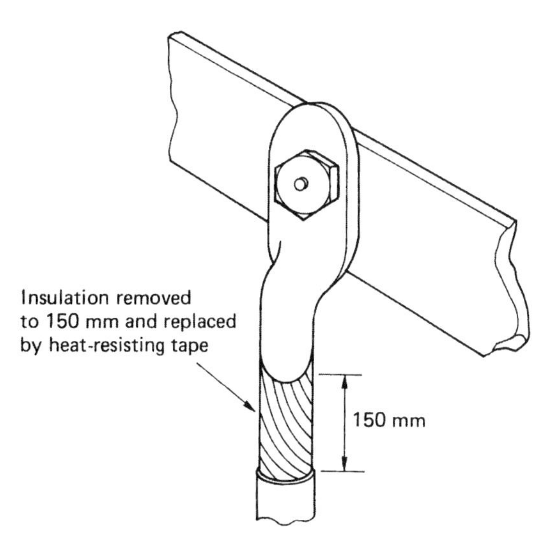

Fig. 14.16 *Lug termination*

Termination into appliances

The methods used to terminate conductors have already been discussed. However, one important aspect of termination of cable into appliances, especially the portable type, is the securing of the cable at or near the point of entry (Fig. 14.17). A good example of this is the cable clamp in a plug top.

It is important to note that in any termination and any run of conductor or cable, measures must be taken to avoid any undue strain on that conductor or cable.

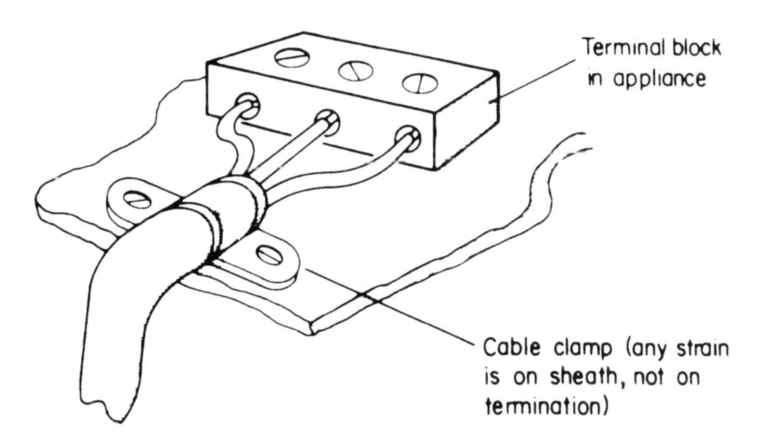

Fig. 14.17

Points to note

1. The type of insulation and sheathing or covering of every cable must be selected so that the cable or conductor is protected against heat, corrosion or mechanical damage, depending on the situation.
2. Cables and conductors must be chosen to suit the voltage and current rating of the circuit.

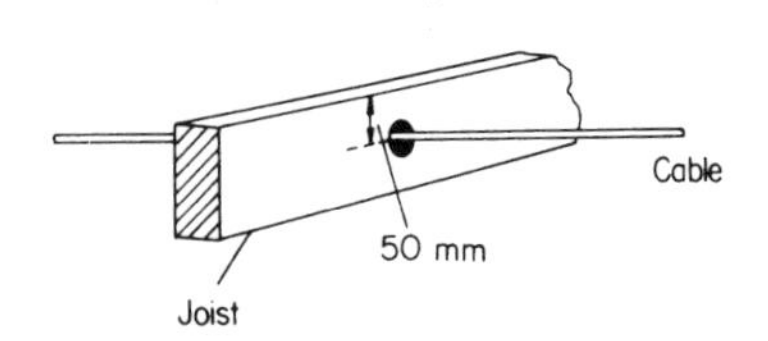

Fig. 14.18

3 Cables and conductors must be chosen such that the voltage drop between the intake position and any point in the installation does not exceed 4% of the supply voltage, i.e. for a 230 V installation, the voltage drop must not exceed 4% of 230 V = 9.2 V.
4 Cables passing through wooden joists should be 50 mm below the top or above the bottom of the joist (Fig. 14.18).
5 Where cables pass through holes in metalwork, the hole should be bushed to prevent any abrasion of the cable.
6 Cables should not be installed in a lift shaft unless they are part of the lift installation.
7 Cables – sheathed and unsheathed and/or armoured – must be supported according to the requirements shown in the appropriate IEE Regulations on-site-guide tables.
8 Every cable must be selected such that it can withstand the normal ambient temperature of its surroundings.
9 Where cables are to be connected to bus-bars, 150 mm of the insulation should be removed and replaced with heat-resisting insulation. This applies only if the original insulation is of general-purpose rubber or PVC.
10 Care must be taken to select the correct type of cable for use in flammable or explosive situations.
11 In order to ensure that fire cannot spread, any holes through which cables pass must be made good to the full thickness of the floor, wall, ceiling, etc.
12 Cables and conductors must be chosen carefully in order to avoid damage by corrosion.
13 Every core of a cable used for fixed wiring should preferably be identified throughout its length, or if this is not possible, by sleeves or discs at its terminations.
14 All terminations of cable conductors must be mechanically and electrically sound.
15 Care must be taken to avoid contact between aluminum and copper conductors, unless adequate precautions have been taken to prevent corrosion.

Plastics

The uses of plastics materials (polymers) in electrical engineering are widespread, the most common being for cable insulation. Other applications include plug tops, socket outlets, motor and transformer winding insulation.

Plastics technology is a vast and complicated subject and hence only the most basic concepts will be discussed here.

Thermoplastic polymers

Thermoplastic polymers soften on heating and solidify to their original state on cooling. Repeated heating and cooling causes no damage.

Thermosetting polymers

Thermosetting polymers become fluid when heated and change permanently to a solid state when cooled. Further heating may cause the polymer to disintegrate.

Polyvinyl chloride (PVC)

General

Rigid PVC as used in conduits etc. is a thermoplastic polymer and has the following properties:

1 high tensile strength;
2 it can be bent by hand if warmed;
3 it has high electrical resistance;
4 it is weather resistant;
5 it does not crack under stress at normal temperatures;
6 it has a low flammability;
7 it is self-extinguishing when the source of heat is removed;
8 it must be used with special saddles and expansion couplers when used in fluctuating temperatures, as its expansion is five times that of steel.

Flexible PVC (used in cable insulation):

1 is weather resistant;
2 has high electrical resistance;
3 should be kept clear of other plastics to avoid migration of plasticizer.

The effects of environmental conditions on PVS are discussed at greater length in Chapter 17.

Phenol-formaldehyde

Phenol-formaldehydes are thermosetting polymers, and are used with other compounds to manufacture plug tops, socket outlets, etc., and thin insulation for all types of winding. Such equipment can safely be used in temperatures up to 100°C.

Conduit

A *conduit* is a tube or pipe in which conductors are run. In effect, the conduit replaces the PVC sheathing of a cable, providing mechanical protection for the insulated conductors.

There are three types of conduit: metal, flexible (metal), and non-metallic (PVC).

Metal conduit

Most metal conduit used nowadays in low-voltage installations is either 'heavy-gauge welded' or 'solid-drawn'. Heavy-gauge is made from a single sheet of steel and welded along the seam, while solid-drawn is produced in tubular form and is therefore seamless. This type of conduit is used only for flameproof installations (owing to its expense).

Metal conduit may be threaded and bent, making it a versatile system. It is available covered with black enamel paint or galvanized, the choice depending on the situation in which it is to be used. It is supplied with all the fittings necessary to make it a complete installation system.

When one is preparing conduit, care must be taken to use the correct tools and preparation methods. For example, a proper conduit bending machine and accessories are essential. The machine will have several different sizes of bending wheel and come complete with a pipe vice. The principle on which it is based is that of a simple second-order lever (Fig. 14.9a–c).

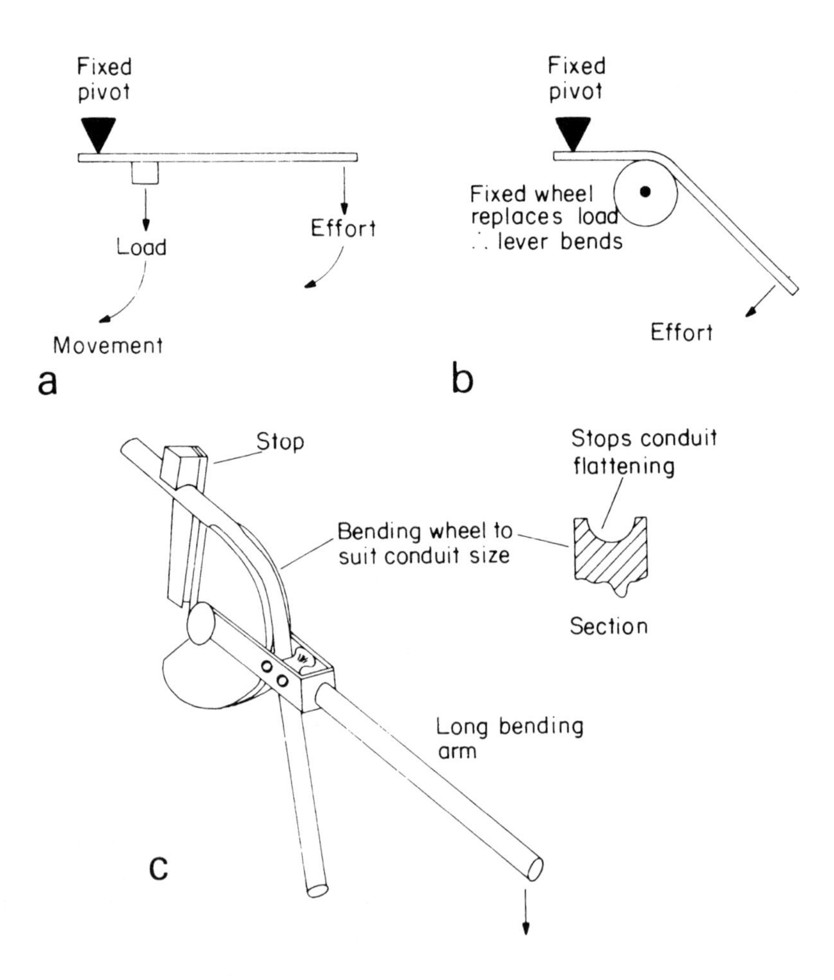

Fig. 14.19

Some typical bends in conduit are illustrated in Fig. 14.20. The *set* is frequently used in order to terminate the conduit at an outlet box (Fig. 14.21).

There are several ways of joining and terminating conduit as shown in Fig. 14.22. Fig. 14.22c shows a 'running coupler'. The conduits to be joined are

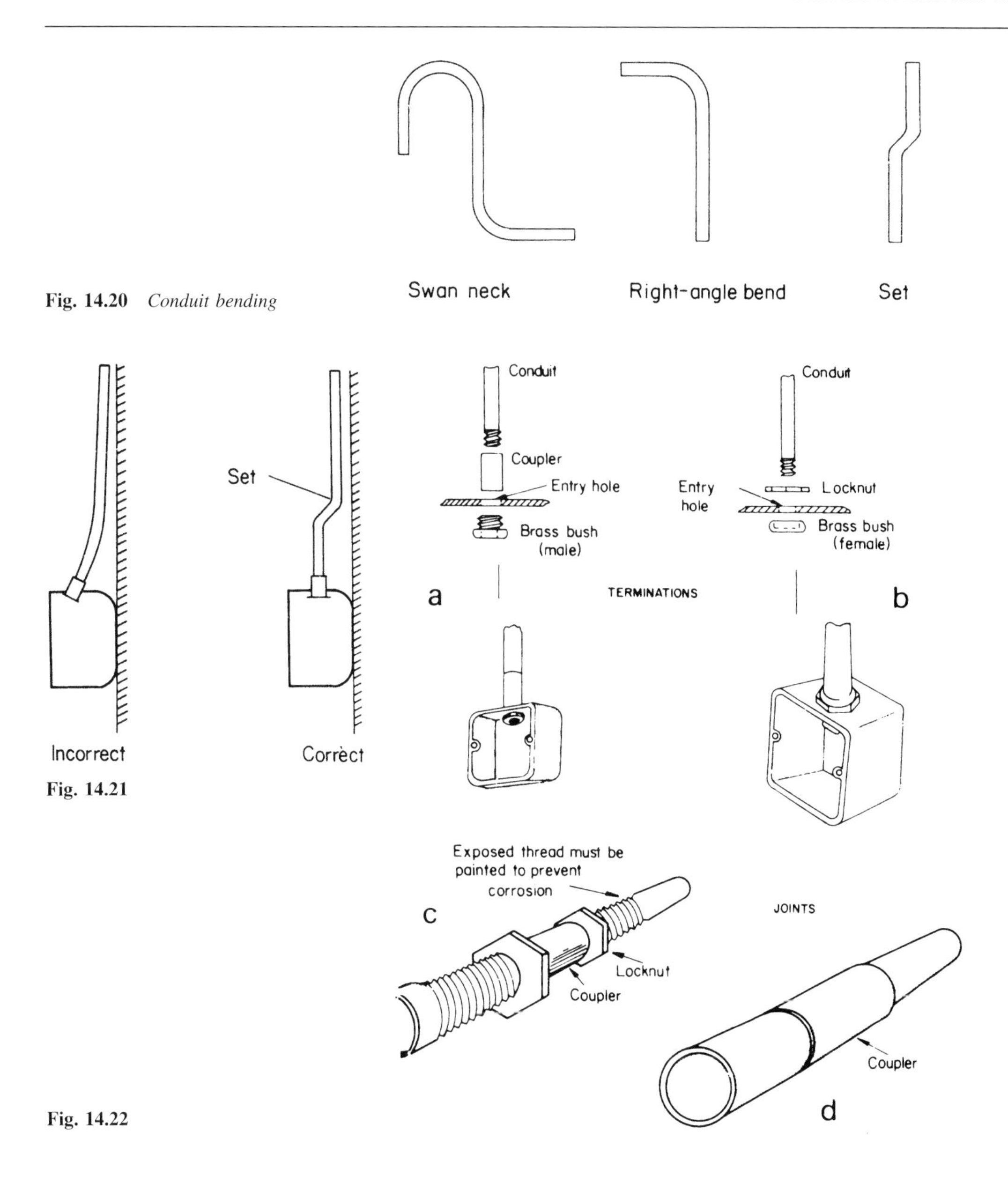

Fig. 14.20 *Conduit bending*

Fig. 14.21

Fig. 14.22

threaded to a distance greater than the length of the couplet. The coupler is then screwed right on to one thread, the other conduit is butted to the first conduit and the coupler screwed back over the second thread. This method enables two conduits to be joined without actually turning the conduits themselves. The locknuts are provided to prevent the coupler from moving.

Fixing conduit

Fig. 14.23 shows the common methods used for securing or fixing conduit.

Crampets are used for securing conduit in place prior to covering with plaster.

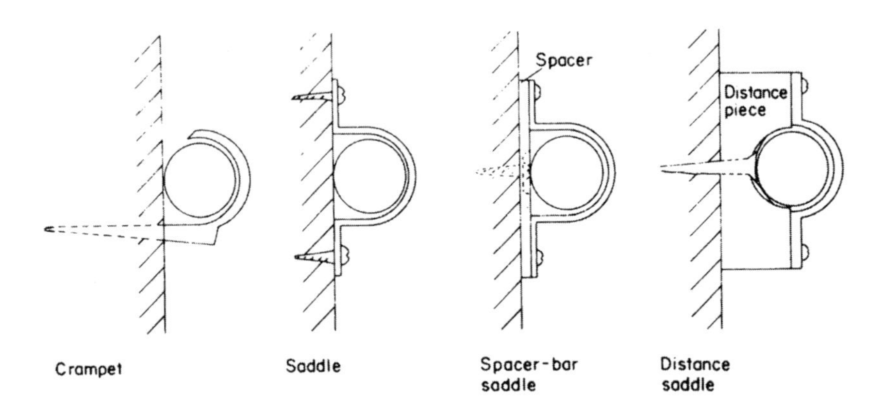

Fig. 14.23

Spacer-bar types are often used to space the conduit out from a wall in order to avoid the need for 'sets'.

Distance saddles space the conduit far enough from the fixing surface to prevent dust from collecting between the conduit and the surface. They are often used in hospitals.

Other accessories for use with conduit include elbows (sharp 90° bends), bends (gentle 90° bends) and tees. All of these are available in the inspection and non-inspection varieties. There are also circular boxes with removable lids, to permit ease of wiring a conduit system. Some of these accessories are shown in Fig. 14.24.

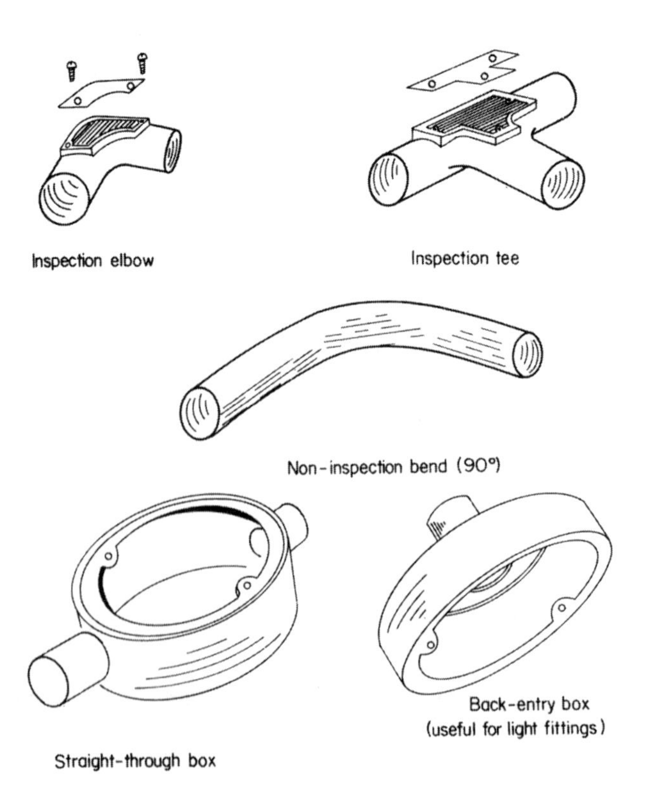

Fig. 14.24

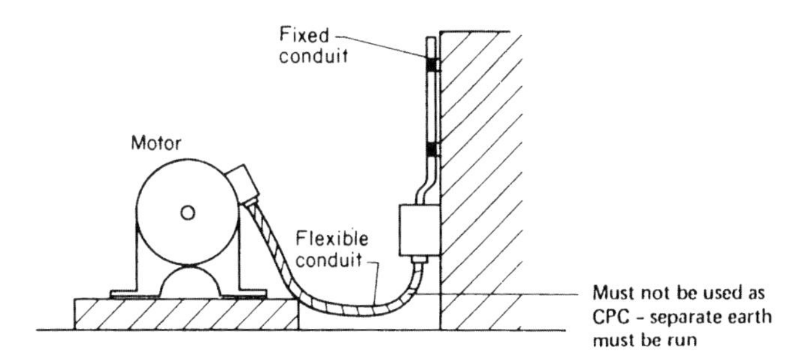

Fig. 14.25

Flexible conduit

Flexible conduit is of great use when a rigid conduit system is supplying machinery (Fig. 14.25). The use of a short connecting length of flexible conduit ensures that the vibration of the motor does not affect the rigid conduit, and enables the motor easily to be aligned, or removed for repair.

Non-metallic (PVC) conduit

PVC conduit is especially suitable for installation systems in light-industrial premises or offices where surface wiring is required. The fittings are identical to those used for metal conduit with the exception that the system is connected not by screwing but by the use of an adhesive. PVC conduit may be bent by hand using a bending spring. This spring, the same diameter as the inside of the conduit, is pushed inside the conduit. The conduit may now be bent by hand, the spring ensuring that the conduit keeps its shape. In cold weather a little warmth may need to be applied to achieve a successful bend.

Threading metal conduit

The following procedure should be adopted for the successful threading of metal conduit.

1 Using a pipe vice to hold the conduit, cut the conduit end square (with a hacksaw); if it is not square, then file it until it is correct.
2 Slightly chamfer the edge of the conduit (see Fig. 14.26). This helps to start the thread cutting (see Fig. 14.27).

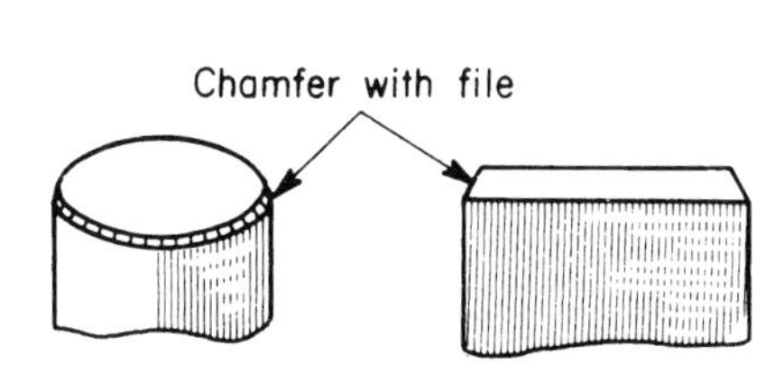

Fig. 14.26

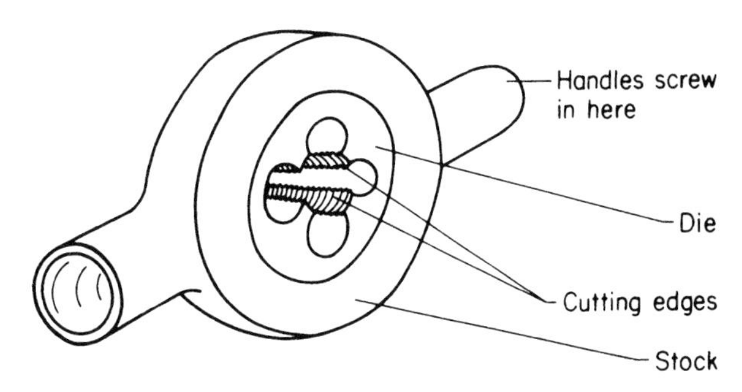

Fig. 14.27

3 Using a stock and die (see Fig. 14.27), making sure that it is offered square to the end of the conduit, begin cutting the thread. This should not be difficult if the die is sharp, and sufficient 'tallow' (lubricating agent) has been applied.
4 As the thread is cut, a small curl of metal is produced. Do not continue cutting in the same direction all the time, but make two or three turns and then reverse the motion; this will break off the curl of metal and keep the thread clean and even.
5 When the thread is complete, clean away any metal filings and ream out the inside edge to remove any burrs. A little tallow applied to the finished thread will help to screw it into a fitting.

Never cut more thread than is necessary; the finished work will be unsightly, and the exposed thread will need painting to protect it from corrosion.

Pulling a small piece of rag through the conduit with a wire will clean out any particles of metal that have accumulated during the threading process. This ensures, as does reaming (see p. 263), that cable insulation is not damaged when conductors are pulled into the conduit.

Never paint a thread before it is used; the steel conduit itself is usually used as the circuit protective conductor and it is essential that a good metallic connection is made at every join and termination.

Points to note

1 The conduit system for each circuit of an installation must be completely erected before any cables are drawn in.
2 Inspection-type fittings must always be accessible for the removal or addition of cables.
3 The number of cables drawn into a conduit may be achieved by selecting the size of conduit using the tables shown in the guidance notes to the Regulations. The system used applies to trunking as well as conduit. The IEE On-site Guide gives various tables to enable the designer to establish the number of conductors that may be run in conduit or trunking, taking into consideration space factor and the ease of drawing cable in. This system of tables is based on the requirement that, in a conduit run, there should be 40% of cable and 60% of air. In trunking the cable should only occupy 45% of the available space. So, for example, if 8 2.5 sq. mm stranded conductors were to be installed in a short straight length of conduit, the factor for the cable (43) would be multiplied by the number of conductors (8) to give an overall cable factor of $43 \times 8 = 344$. This overall factor is now used to find a suitable conduit size by comparing it with the conduit factors and selecting a size. In this instance the nearest conduit factor given is (460), giving a conduit size of 20 mm. This method is used for longer lengths with bends but a different set of tables and factors are used. The same process is also used for trunking capacity.
4 When conduit is bent, the radius of any bend must not be less than 2.5 times the outside diameter of the conduit (Fig. 14.28).
5 If non-inspection elbows or tees are used, they must be installed only at the end of a conduit immediately behind a luminaire (light fitting) or outlet

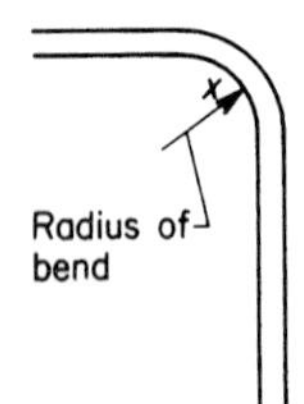

Fig. 14.28

box etc. of the inspection type. Alternatively *one* solid elbow may be located not more than 500 mm from a readily accessible outlet box in a conduit not exceeding 10 m between two outlet points providing that all other bends do not add up to more than 90°C (Fig. 14.29).

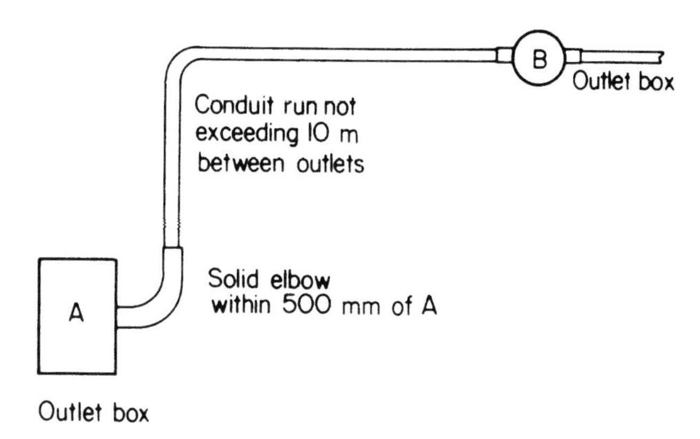

Fig. 14.29

6 The ends of conduit must be reamed and entries to boxes and accessories must be bushed in order to prevent damage to cables.
7 As we have already seen, a conductor carrying a current has a magnetic field around it and as a.c. causes the field to change direction continually the field will cut across any metal close to it and in so doing induce e.m.f.s and eddy currents (heating) in that metal. Running another conductor close to the first with its current flowing in the opposite direction will cancel out the fields. Hence cables of a.c. circuits installed in steel conduit should have all phases and neutral (if any) bunched in the same conduit.
8 Flexible metal conduit must not be used as a circuit protective conductor; a separate such conductor must be run.
9 Non-metallic conduits must not be installed where extremes of ambient temperature are likely to occur.
10 The supports for rigid PVC conduit must allow for expansion and contraction of the conduit.
11 Drainage points should be provided at the lowest point of a metallic conduit installation to allow moisture due to condensation to drain away.

Trunking

Trunking may be thought of as simply a larger and more accessible conduit system. It is available in two ranges, ordinary wiring trunking and bus-bar trunking. The ordinary system is further available in either metal or PVC.

PVC trunking

The main areas in which PVC trunking is used are domestic and office premises. It is perhaps better described as 'channelling' rather than 'trunking'. All necessary fitting are available to complete any shape of run with little difficulty. Joins in PVC trunking are usually made with adhesive.

Metal trunking

Metal trunking is used extensively in engineering premises. It allows a large number of cables to be drawn into one system with relative ease and because of its accessibility enables changes or additions to circuits to be made with the minimum of effort. As with conduit, it is available either painted or galvanized. A whole range of fittings is available, enabling the most difficult runs to be constructed.

It is available in many different sizes and designs. For example, compartment trunking allows the segregation of certain currents, and support of cables in vertical runs. Fig. 14.30 show several different types of trunking.

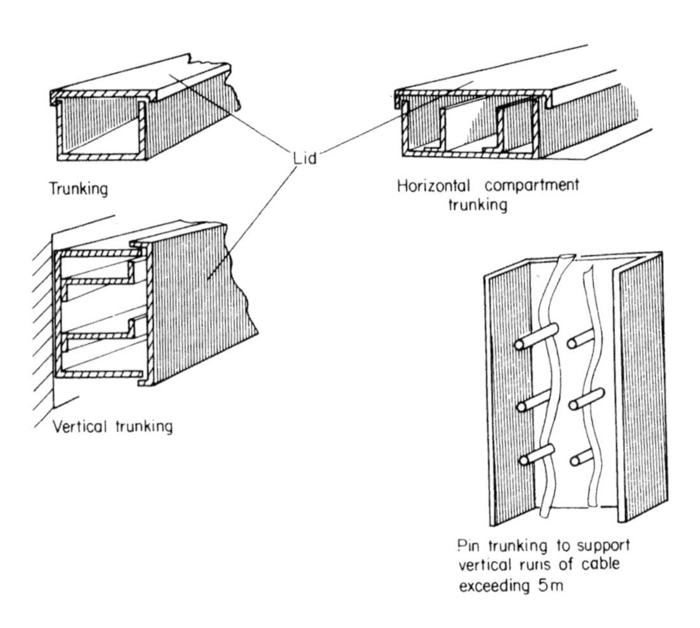

Fig. 14.30

Cutting and drilling trunking

It is often necessary to cut or drill trunking. After cutting, all edges should be filed to remove sharp metal. Burrs left after the drilling of a hole should also be filed away. When high-speed twist drills are being used, a lubricant should be applied; this aids the drilling process and prolongs the life of the drill. A lard oil such as tallow is recommended for use with steel. (No lubricant is needed when drilling PVC.)

It is often necessary to cut large holes in order to connect conduit into the trunking system; this may be achieved by the use of a circular hole saw or a pressure-type hole cutter (Fig. 14.31).

After cutting, all hole edges must be filed smooth and conduit entries bushed.

Supports for trunking

Trunking may be secured direct to a surface or suspended by means of brackets.

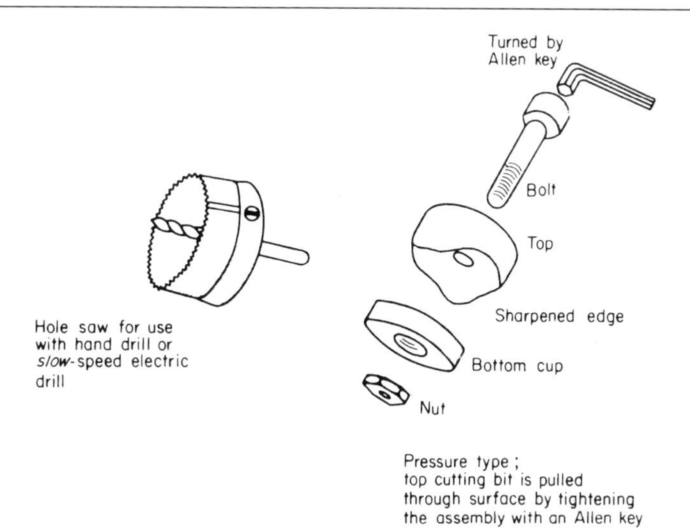

Fig. 14.31

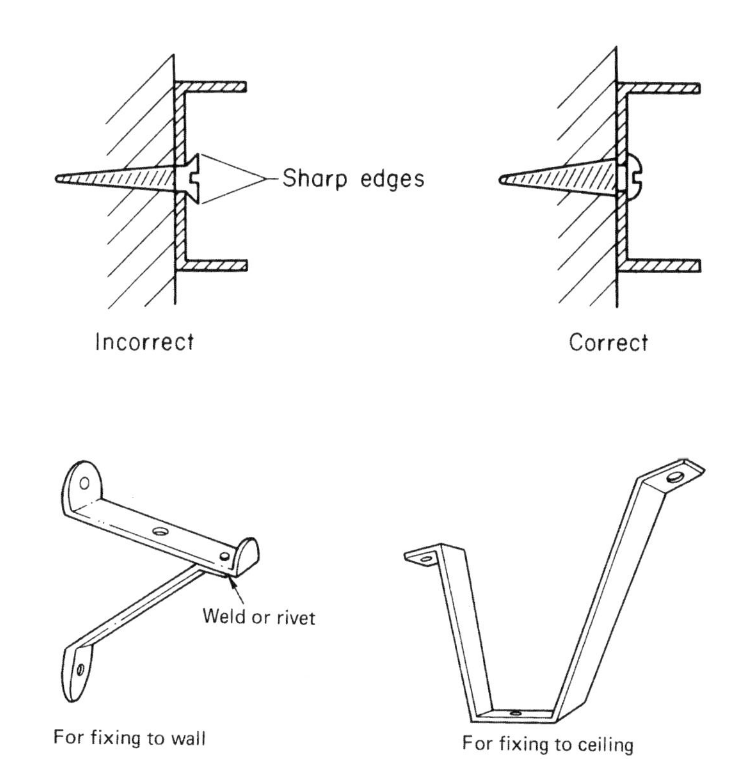

Fig. 14.32

Fig. 14.33

As trunking material is thin, there is no room for countersunk holes. Roundhead screws (Fig. 14.32) are suggested, unless of course the fixing surface requires bolts.

There are occasions when the trunking has to be suspended by means of brackets. Fig. 14.33 shows two typical brackets.

Bus-bar trunking

There are two main types of bus-bar trunking, the overhead type of bus-bar used for distribution in industrial premises, and rising-main trunking (Figs 14.34 and 14.35).

Overhead bus-bar trunking is ideal for distribution in factories. It is run at high level and the tap-off boxes enable machinery to be moved easily. When connected in the form of a ring and incorporating section switches, this system enables parts of the trunking to be isolated when necessary, without all machines losing supply.

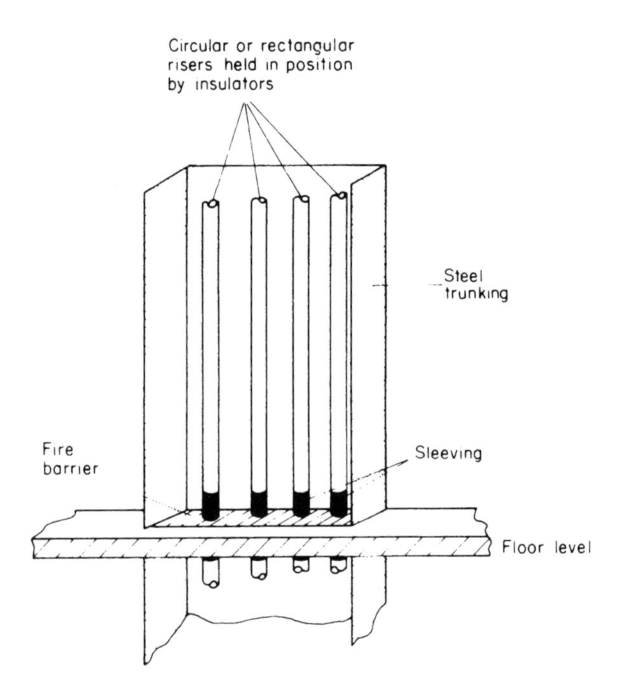

Fig. 14.34 *Bus-bar trunking*

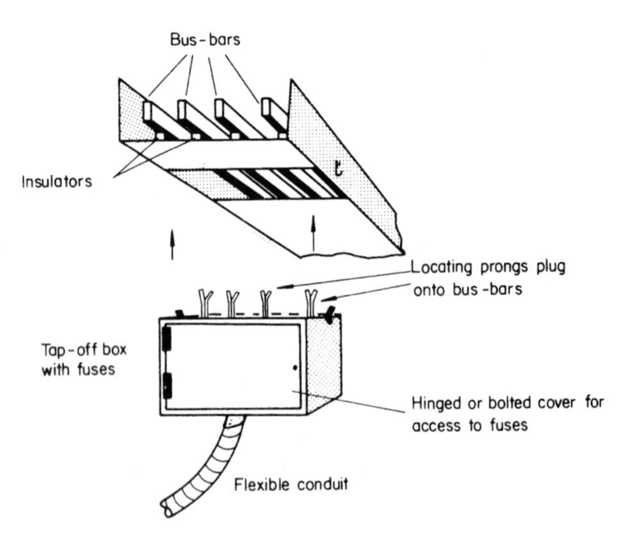

Fig. 14.35 *Overhead bus-bar trunking*

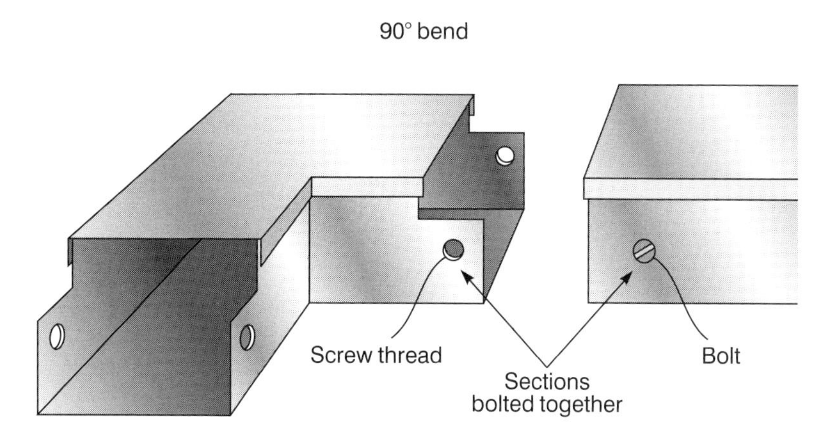

Fig. 14.36

Trunking, and in particular steel trunking, can be fabricated into various shapes to allow for bends and tees, etc. This process is, however, laborious and there are many different manufactured shapes available which simply bolt onto straight sections to allow versatility in installation. Figure 14.36 illustrates how such sections are joined together.

Points to note

1 Where trunking passes through floors, walls or ceilings, a fire barrier must be provided.
2 Many installations contain circuits of different types. These circuits are divided into two voltage bands:
   Band 1 Extra low voltage (ELV) circuits, including bell and call systems, telecommunication and alarm systems.
   Band 2 Low voltage circuits (LV), i.e. typical domestic, commercial and industrial arrangements.
   In trunking systems these bands may be enclosed together provided that either:
   (a) band 1 cables are insulated to the highest voltage present, or
   (b) band 1 and 2 circuits are in separate compartments in the trunking.
3 All conductors of bus-bar trunking must be identifiable.
4 Bus-bar trunking must be installed so that the conductors are inaccessible to unauthorized persons.
5 Bus-bars must be free to expand and contract with variations in temperature (Fig. 14.37).
6 All phases and neutral (if any) of a.c. circuits must be contained in the same metal trunking.
7 Entry to trunking must be so placed as to prevent water from entering.
8 The size of trunking may be determined from the guidance notes to the Regulations.
9 When metal trunking is used as a circuit protective conductor, copper straps should be fitted across joints to ensure electrical continuity.

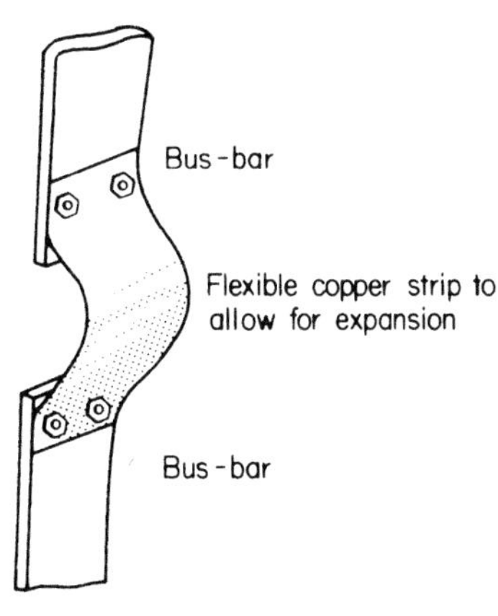

Fig. 14.37

Traywork

Traywork is another method of supporting several cables along a run. It is simply a perforated metal tray on to which cables can be tied (Fig. 14.38). Its use is restricted normally to situations such as switch rooms, boilerhouses and large cable ducts, i.e. places not normally occupied by personnel.

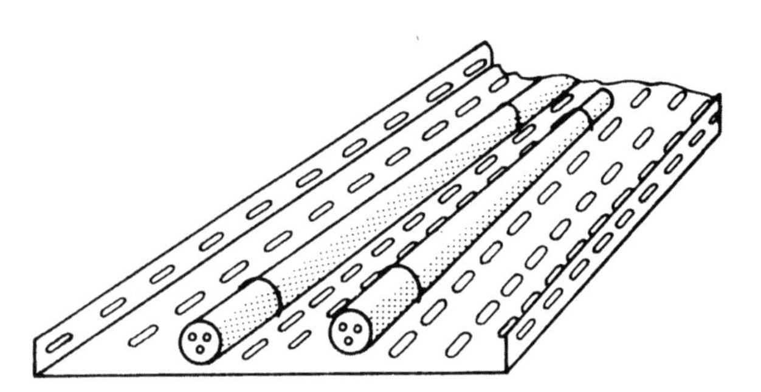

Fig. 14.38 *Cable tray*

The construction of a cable tray is such that is easily cut, shaped and bolted, to suit all situations. Support brackets are as for trunking.

Fixing and tools

One important aspect of installation work is the fixing of accessories to various surfaces.

Fig. 14.39 shows several different methods of fixing.

The main disadvantage of gravity and spring toggles is that it is not possible to remove the bolt without losing the toggle.

Tools

The following is a list of tools used by the electrician; a complete basic tool kit will have all these items.

First fix

These tools are used for chasing walls, lifting floorboards, etc.

Club hammer. This is for use with cold chisels and bolsters. Ensure that the handle is in good condition. Do not use the hammer if the head is loose.

Bolster (large and small). This tool is used for chasing walls and brickwork and for making holes for metal boxes. Ensure no mushrooming (see p. 262).

Floorboard chisel. This is similar to a bolster but with a longer handle. The same precautions should be taken as for bolsters.

Scutch hammer and combs. These tools are used for chasing brickwork.

Cold chisels (a selection of various sizes). These are for use on brickwork and concrete. The same precautions should be taken as for bolsters.

Tenon saw and padsaw. For cutting floorboards etc. (special floorboard cutters are available). They must be kept sharp, and greased when not in use.

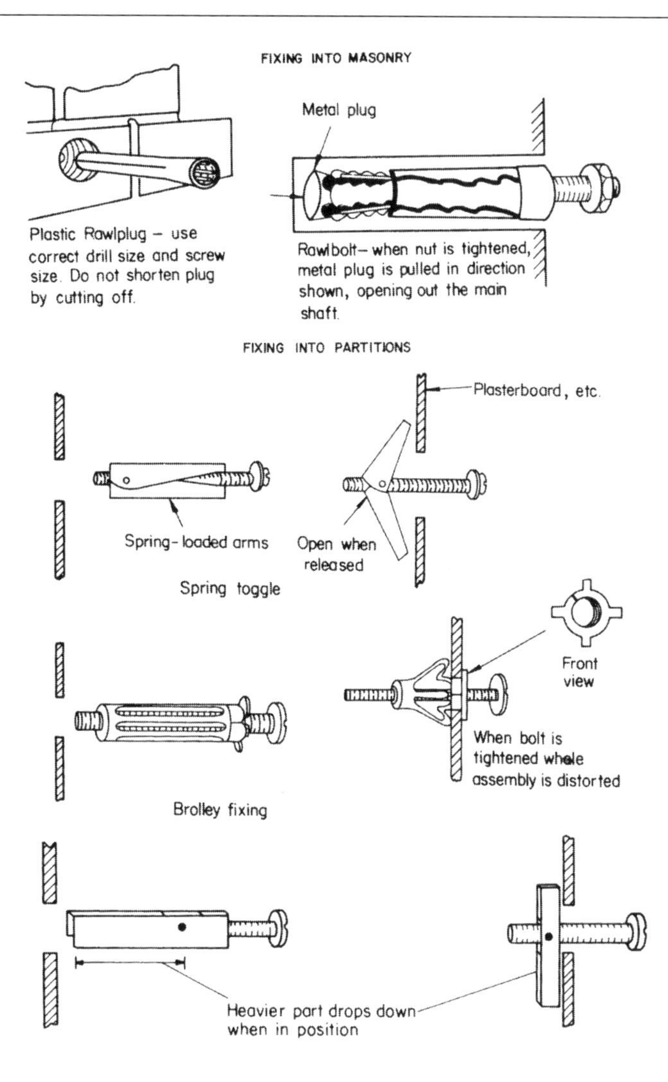

Fig. 14.39

Brace and wood bits. These are used for drilling joists etc. Wood bits should be kept sharp, and ratchets oiled.

Claw hammer. For de-nailing joists and floorboards. The precautions to be taken are the same as those for a club hammer.

Rawlplug tool and bits. These are for making holes in brickwork and concrete in order to insert Rawlplugs. Watch for mushrooming of the head.

Hand drill and electric drill. These are for use with metal drills and masonry drills. An electric drill should preferably have two speeds and a hammer action.

Second fix

These tools are used for fitting accessories such as sockets, etc.

Large cabinet-maker's screwdriver. This tool is for use in fixing screws.

Large and small electrician's screwdriver. Used for most screws in accessories. The handles of screwdrivers must be in perfect condition and the

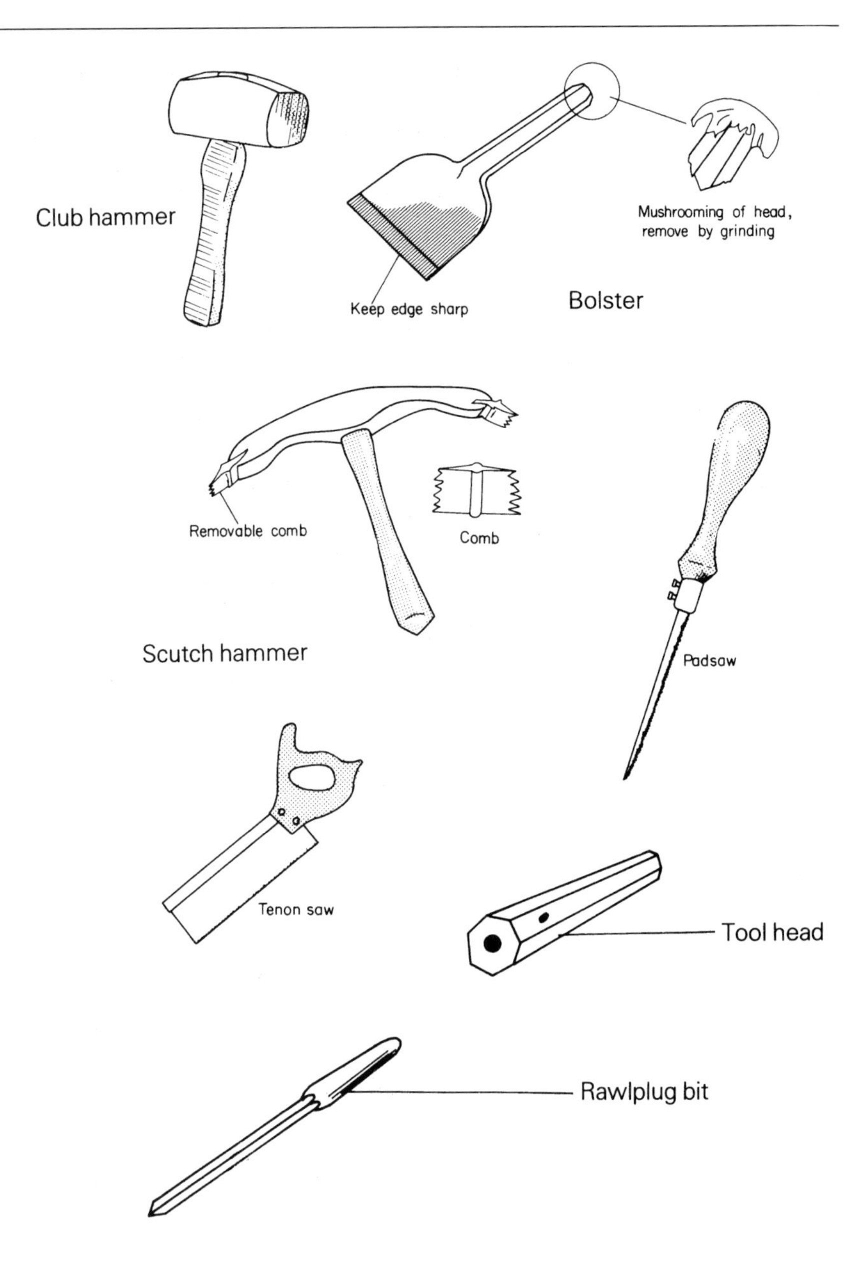

Fig. 14.40

tips must be kept at the correct size. Only the correct size of tip should be used on screws. Do *not* use screwdrivers as chisels.

Wire cutters and strippers. Used for preparing cable ends. They should be kept clean and sharp.

Pliers and sidecutters, insulated; junior hacksaw and blades; Stanley knife and blades. These are used for general work.

Set of tools for termination of m.i.m.s. cable. Cutting tools should be kept sharp and all surplus compound removed.

Tools for use on conduit and trunking. These include a bending machine with a pipe vice, designed for use with most sizes of conduit. They should be

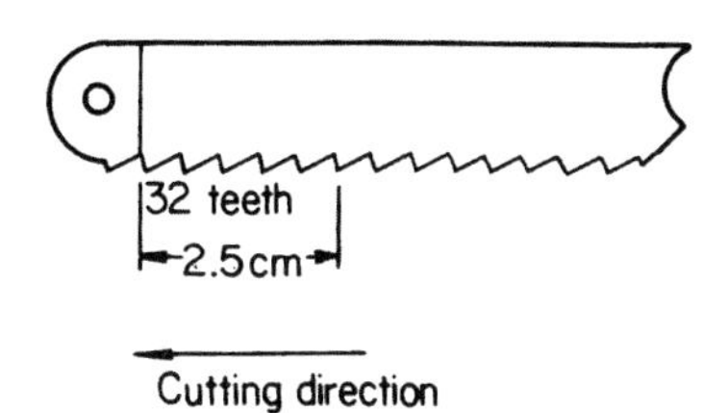

Fig. 14.41

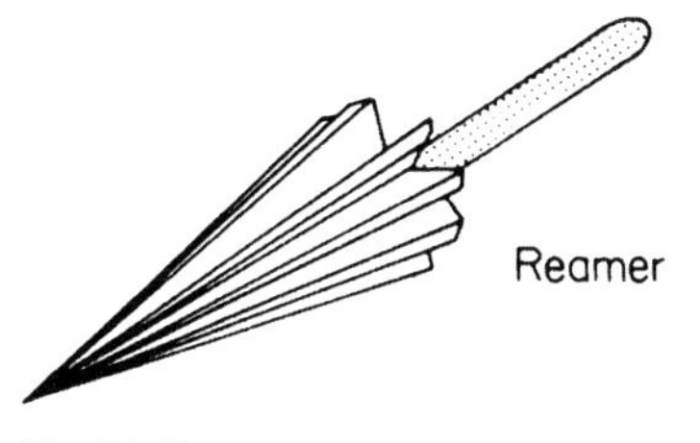

Fig. 14.42

kept clean and free from metal particles. Another tool for use on conduit is a hacksaw frame taking hand blades with 32 teeth per 2.5 cm (Fig. 14.41), the blade must be used in the correct direction.

A selection of files. Files are used for removing sharp edges and burrs. *Always* use a handle on a file.

Reaming tool (Fig. 14.42). For removing burrs from the inside edge of the conduit.

Stock and dies. These tools are used for cutting threads. A recommended lubricant such as tallow should be used. Dies must be kept clean; do not allow a build-up of tallow and metal.

Adjustable pipe wrench and Stillson (Fig. 14.43). Used for tightening and untightening conduit.

Bush spanner (Fig. 14.44). This tool is for use on inaccessible bush nuts.

Draw tape. Used for pulling cables through conduit.

Measuring tape. Steel tapes must *not* be used near live equipment.

Set of twist drills. For use with a hand or electric drill. Cutting edges should be kept sharp. A hot drill must never be quenched in water.

Selection of spanners. For general use.

Spirit level.

Set-square. For use in marking out trunking.

Scriber. Used for marking metal. The points should be kept sharp.

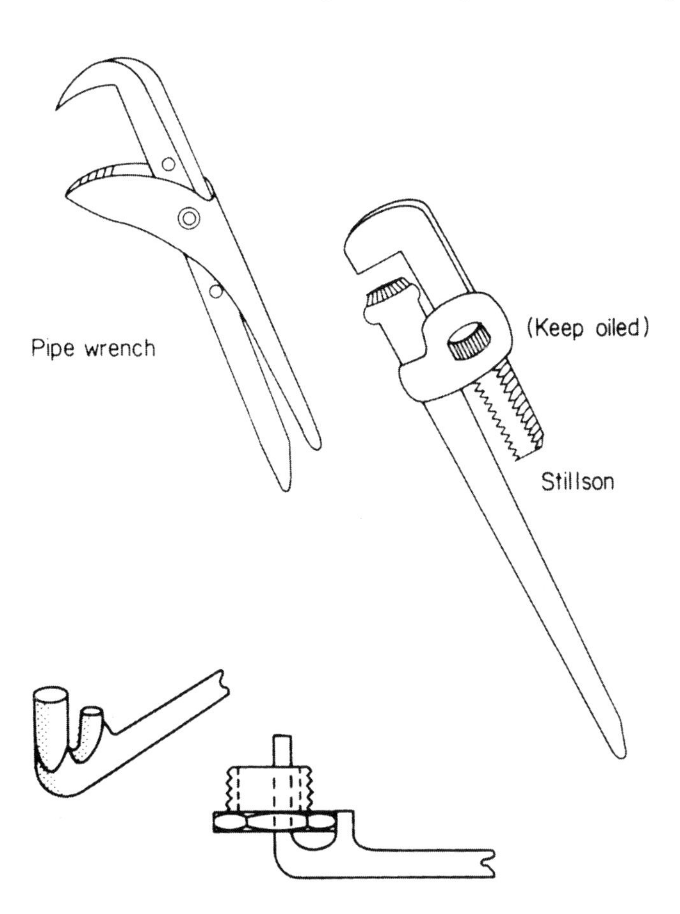

Fig. 14.43

Fig. 14.44 *Bush spanner*

Centre punch. For marking metal prior to drilling.

Hole saw. This type of saw is used for cutting holes in trunking. It should be used only with a hand drill or a low-speed electric drill.

Putty knife and pointing trowel. For making good the plasterwork, etc.

Bending springs. Used for PVC conduit.

Solder, soldering iron, blowlamp. These are used for all general electrical soldering.

Approved test lamp. Used on live circuits. Moisture must be kept away.

First-aid kit. This must be kept stocked at all times.

As tools play an important part in installation work, it is perhaps sensible to understand some of the processes involved in their manufacture. This applies particularly to hardened, tipped tools such as chisels, drills, scribes, etc.

Heat treatment

Most good-quality tools are manufactured from carbon steel. When carbon steel is heated, its physical properties change, and these changes may be used to advantage. On being heated, carbon steel increases in temperature until it reaches a point called the *lower critical limit*, when the steel starts to change. It remains at this temperature for a short while before its temperature increases again to the *upper critical limit*, where the change is complete. On slow cooling the reverse process takes place and the steel returns to normal.

Hardening

If a sample of carbon steel is heated to its upper critical limit and then cooled rapidly by plunging it into cold water, it will become very hard. It is usually too brittle for most purposes, however, and has to be tempered.

Tempering

Tempering is carried out by heating the hardened steel to just below its lower critical limit and quenching in water. The steel then loses its brittleness but remains very hard.

Annealing

In order to cold-work to shape and/or machine a sample of steel before it is hardened and tempered, it must be as soft and ductile (able to be permanently deformed without damage) as possible. This is achieved by heating the sample to its hardening temperature and then cooling it very slowly over a long period of time.

Work hardening

Excessive hammering, rolling or bending, etc., of a metal causes it to harden and become brittle. It may be returned to its original condition by annealing.

Comparison of systems

PVC-insulated PVC-sheathed cable

PVC-insulated PVC-sheathed cable is suitable for all types of domestic and commercial wiring installation where there is little risk of mechanical damage, extremes of temperature or corrosion. It is inexpensive and easy to handle, and no specialized tools are needed for working with it.

Mineral-insulated metal-sheathed cable

M.i.m.s. cable is suitable for various applications in commerce and light industry. It has good mechanical strength and can withstand a considerable amount of crushing before it breaks down electrically. It has a high degree of resistance to heat, which makes it useful for fire-alarm circuits. The sheath makes a good circuit protective conductor (CPC). M.i.m.s. cable is not to be used in damp or corrosive situations without an overall PVC covering. It is expensive, although long lasting; it needs specialized tools and skilled craftsmen to install it.

Fire-retardant cable

For fire alarm circuits and areas of high combustibility a fire retardant cable such as Fire-Tuf or FP200 is used. It is a metal (usually aluminium) sheathed cable with high heat-resistant insulation and does not require the same degree of skill for terminations.

Conduit

Steel conduit is used extensively in industrial premises. It has good mechanical strength and may be used as a CPC. Specialized tools are needed to install it, and steel conduit is not easily added to; it is subject to corrosion and is relatively expensive.

PVC conduit is used in commercial and light-industrial premises. It has quite high mechanical strength, and is inexpensive compared with steel. It does not corrode, is easily erected and no specialized tools are needed. It cannot stand extremes of temperature and distorts owing to its high degree of expansion unless this is compensated for; it cannot be used as a CPC.

Other types of conduit, including copper and aluminium, are available. these are termed *non-ferrous conduits* (non-magnetic) and are used in specialized environments only. Aluminium conduit is susceptible to a high degree of corrosion.

Trunking

Wirable trunking is used extensively in industry. It is a versatile system allowing the easy addition of further sections. A greater number of cables can be run than in conduit. No specialized tools are needed and there is a range of sizes. It allows different circuits to be segregated. It is, however, more expensive than conduit.

The bus-bar type of trunking is used for rising mains, distribution boards, and overhead supplies to machines. The tap-off box system is popular as it gives a measure of control, and selective isolation.

System installation

Once a wiring system has been chosen, decisions must be made as to how the system is to be installed and such decisions will need to take account of the fabric and structure of the building and the aesthetics. For example, PVC flat cables used to rewire a very old dwelling where the walls are 'lath and plaster' (slats of wood covered in a soft lime and sand plaster) are not suitable for chasing out or surface clipping. In this case the best solution is to use an adhesive-backed mini-trunking to enclose the cables.

Another example is that of a traditional brick construction premises with wooden joists, some of which may be load-bearing, i.e. walls are resting on them. These timbers should not be penetrated by drilling or notching in such a way as to weaken them. The Building Regulations indicate where a joist may be penetrated, as shown in Figure 14.45.

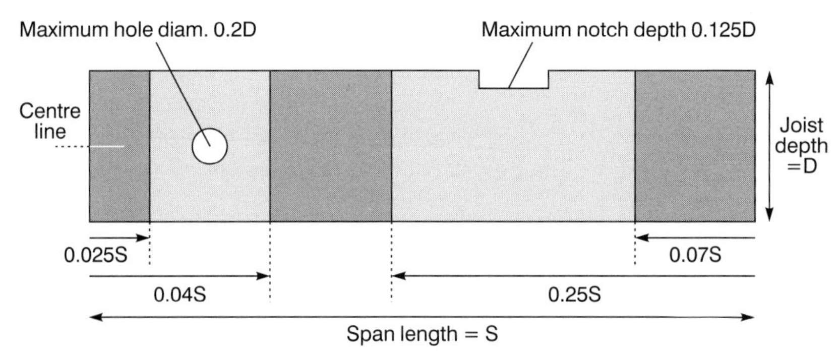

Fig. 14.45

Clearly, all exposed wiring systems will need supporting either vertically or horizontally to ensure that they are not damaged by undue strain. The IEE Regulations 'On-site Guide' includes various tables indicating distances between supports for cables, conduit and trunking systems, together with spans and heights of overhead cables, etc.

Self-assessment questions

1 Explain the difference between cable insulation and cable sheathing. What is a flexible cord and when is it used?

2 (a) What precautions should be taken when stripping cable sheaths and insulation and why?

(b) What are the requirements of the IEE with respect to identification of conductors?

3 What is a bus-bar and where is it used? Illustrate your answer with sketches.

4 (a) What is m.i.m.s. cable?
   (b) Describe with the aid of sketches one type of termination for m.i.m.s. cable.

5 (a) Describe the correct method of cutting a thread on a conduit end.
   (b) What is an inspection fitting and what are the restrictions on the use of solid (non-inspection) elbows and tees?

6 What is the meaning of the term 'space factor'?

7 Explain, with the aid of sketches, why all phases and neutral (if any) of a.c. circuits must be drawn into the same metallic conduit or trunking.

8 What is meant by 'segregation' of circuits and how is it achieved in trunking?

9 Comment on the advantages and disadvantages of conduit and trunking systems. What special precautions must be taken when using PVC conduit?

15 Installation circuits and systems

Throughout this chapter there are many references to diagrams/drawings and these may be categorised as follows:

Block diagrams	These show, using squares, rectangles etc., the sequence of a system without too much technical detail e.g. Figures 15.17 and 15.26.
Layout diagrams	These are very similar to block diagrams, but they indicate more technical detail and tend to show items in their correct geographical location. (e.g. Figures 15.18 and 15.19).
Circuit/schematic	These show how a circuit functions and takes no account of exact locations of terminals or equipment (e.g. Figures 15.28, 15.29, 15.32, 15.34 etc.).
Wiring diagram	These indicate how a circuit or system is physically wired (e.g. Figures 15.2, 15.4, 15.6 etc.).

Lighting circuits

One-way switching

Fig. 15.1 is a circuit diagram showing how the light or lights are controlled, while Fig. 15.2 shows how the point would be wired in practice.

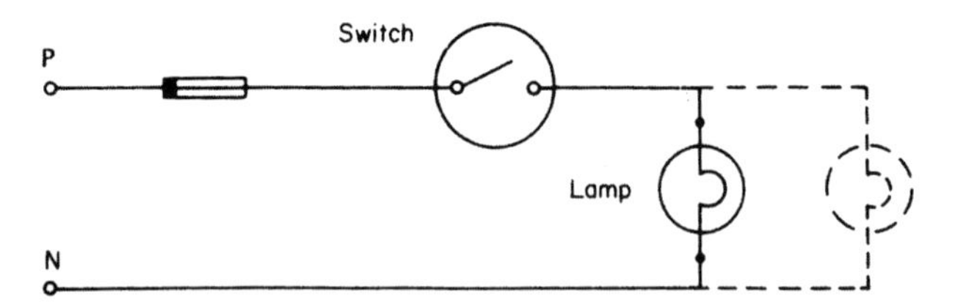

Fig. 15.1 *Single switch controlling one or more lamps. Note: switch in phase conductor*

Points to note

1 The ends of black, blue or yellow switch wires have red sleeving to denote phase conductor. (This is not required for conduit wiring as the cable will be red.)
2 The earth wire terminations have green and yellow sleeving. (This is not required for conduit.)

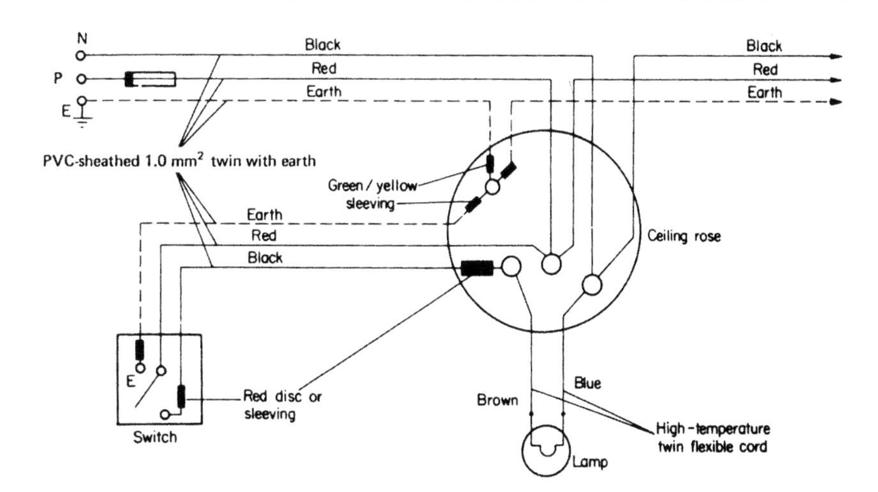

Fig. 15.2 *Wiring diagram*

3 The light-switch point has an earth terminal.
4 The ceiling rose has an earth terminal.
5 The flexible cord from the rose to the lampholder is capable of withstanding the maximum likely temperature.
6 If a batten holder is used instead of a ceiling rose, the cable entries should be sleeved with heat-resistant sleeving.
7 The maximum mass suspended by flexible cord shall not exceed:

   2 kg for a 0.5 mm^2 cord;
   3 kg for a 0.75 mm^2 cord;
   5 kg for a 1.0 mm^2 cord.

8 The phase terminal in a ceiling rose must be shrouded.
9 For the purpose of calculating the cable size supplying a lighting circuit, each lighting point must be rated at a minimum of 100 W.
10 A ceiling rose, unless otherwise designed, must accommodate only one flexible cord.

Two-way switching (Figs 15.3 and 15.4)

Points 1 to 10 are as for Fig. 15.2. A typical application is for stairway lighting.

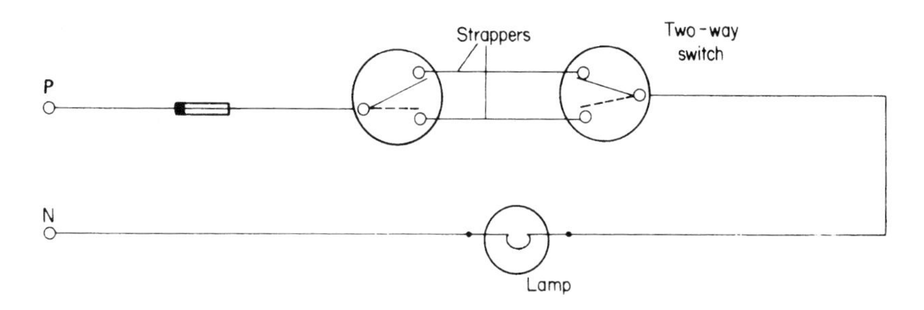

Fig. 15.3 *Circuit diagram – two-way switching*

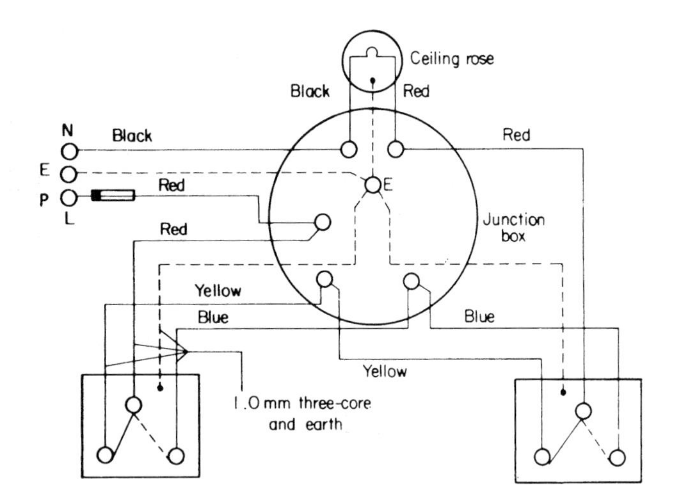

Fig. 15.4 *Wiring diagram – two-way switching*

Two-way and intermediate switching

The circuit diagram for two-way and intermediate switching is shown in Fig. 15.5. *Note*: The earth is omitted from the diagram, for clarity.

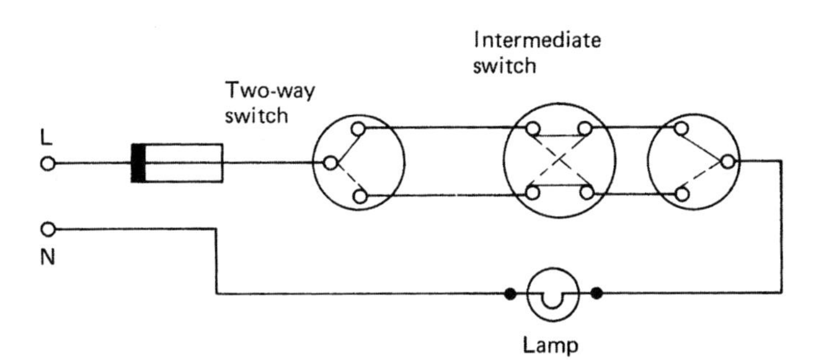

Fig. 15.5 *Circuit diagram – two-way and intermediate switching*

Red sleeving or discs are used on yellow, black and blue cables from two-way and intermediate switches to the junction box in order to denote live cables (Fig. 15.6).

The applications of this type of switching are for stairs and landings and in long corridors.

Lighting layouts

There are two main methods of wiring a lighting installation.

1 Each ceiling rose or junction box is fed from the previous one in the form of a chain.
2 The main feed is brought into a central junction box and each point is fed from it (like the spokes of a wheel).

There are of course variations involving combinations of these two methods depending on the shape and size of the installation.

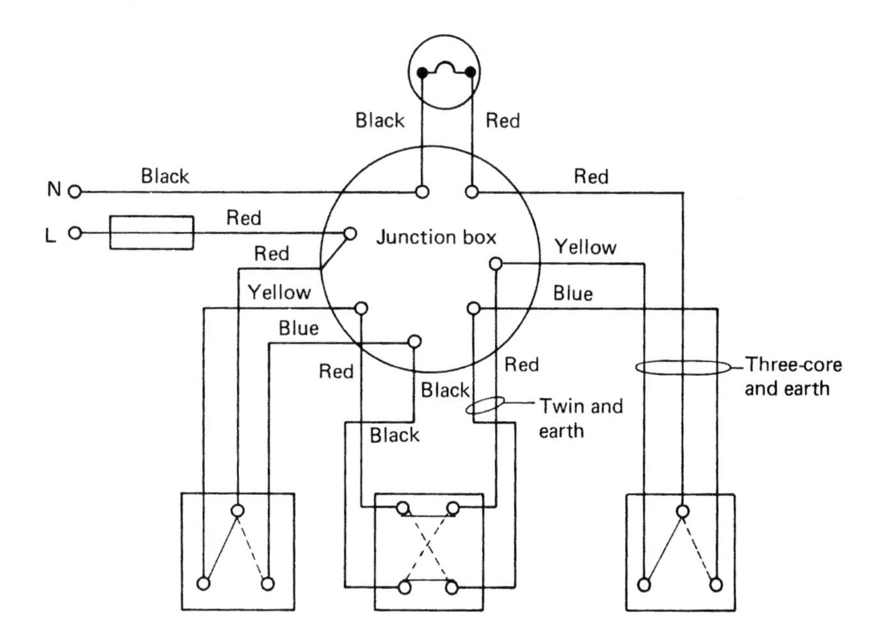

Fig. 15.6 *Wiring diagram (PVC sheathed)*

Power circuits

Radial circuits

Radial circuits are arranged in the same way as item 1, above, in lighting layouts, in that each socket outlet is supplied via the previous one (Fig. 15.7).

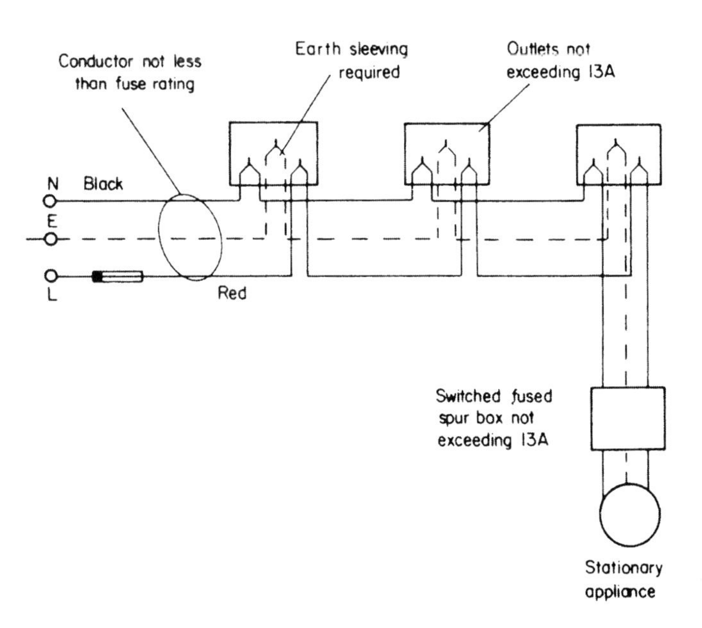

Fig. 15.7 *Radial circuit (PVC insulated)*

Points to note

1 The number of socket outlets, complying with BS 1363, is unlimited for a floor area of up to 50 m^2 if the circuit protection is a 30 A or 32 A cartridge fuse or circuit breaker, the cable being 4 mm PVC copper or 2.5 mm^2 mineral insulated (m.i.).
2 The number of socket outlets is unlimited for a floor area up to 20 m^2 with any type of circuit protection of 20 A using 2.5 mm^2 PVC copper cable or 1.5 mm^2 m.i.
3 The total number of fused spurs is unlimited.

Ring final circuits

These circuits are the same as radial circuits except that the final socket outlet is wired back to the supply position. In effect any outlet is supplied from two directions (Fig. 15.2).

Points to note

1 Every twin-socket outlet counts as two single-socket outlets.
2 The number of non-fused spurs must not exceed the total number of points on the ring. Fig. 15.8 shows seven points on the ring and five spurs.
3 The fuse rating of a fused spur box must not exceed 13 A and the current rating of all points supplied by the fused spur must not exceed 13 A.
4 Non-fused spurs must supply *no more* than one single or one double socket or one stationary appliance.

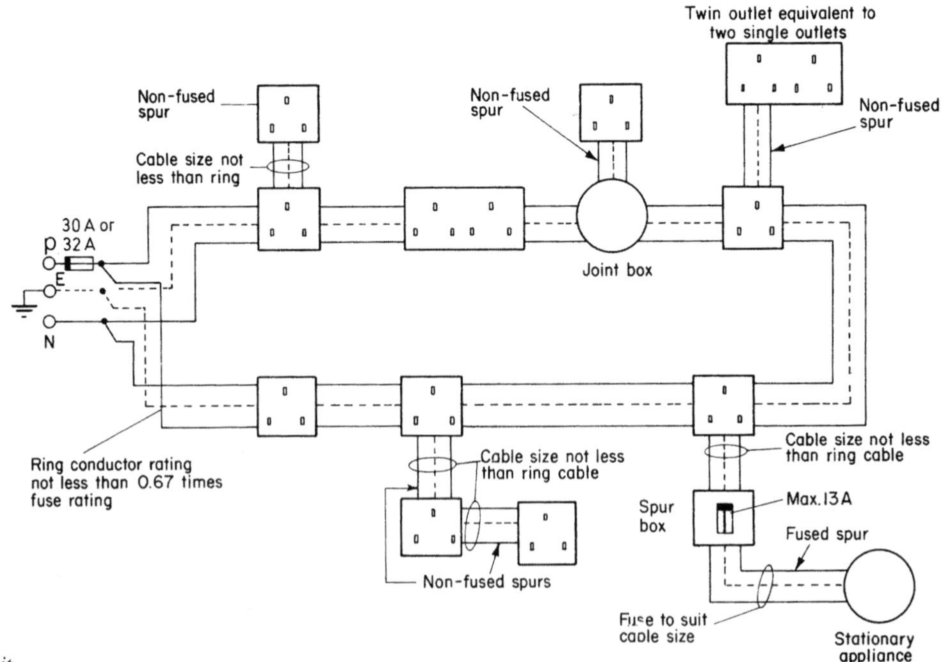

Fig. 15.8 *Ring final circuit*

5 For domestic premises of area less than 100 m^2, a ring circuit may serve an unlimited number of points.
6 Cable sizes for ring circuits using copper conductors are:

2.5 mm^2 PVC insulated or
1.5 mm^2 m.i.

Note: When using some 2.5 mm^2 flat twin with earth cable, a 1.0 mm^2 circuit protective conductor (CPC) is too small to comply with the regulations unless a circuit breaker or high breaking capacity (HBC) fuse is used.

Water heater circuits

Water heaters of the immersion type (i.e. having an uninsulated element completely immersed in water) are available in several different forms, two of the more popular being the large 140 litre storage type, and the smaller open-outlet type (10–12 litres) used for small quantities of instant hot water.

Both types are stationary appliances and could therefore be connected into a ring circuit. However, it is recommended that heaters over 15 litres have their own circuit.

Points to note

1 The heater shall be effectively connected to earth, and the inlet and outlet pipes must be made of metal. An earthing terminal must be provided.
2 The heater must be supplied from a double-pole linked switch (*not* a plug and socket).

Cooker circuits

A cooker exceeding 3 kW should be supplied on its own circuit. As it is rare to have every heating element working at once, *diversity* is applied to calculate the assumed current demand as follows.

The first 10 A of the total rated current of the connected cooking appliance, plus 30% of the remainder of the total rated current of the connected appliance, plus 5 A if there is a socket outlet in the control unit.

Example

The full-load rating of a 230 V cooker is 11.5 kW. Calculate, using diversity, the assumed current demand. The cooker control unit has a socket outlet.

Actual full load demand:

$$P = I \times V$$

$$I = \frac{P}{V}$$

$$= \frac{11\,500}{230}$$

$$I = 50\text{ A}$$

Assumed demand using diversity:

$$I = 10 + \frac{30 \times (50-10)}{100} + 5$$

$$= 10 + \frac{30 \times 40}{100} + 5$$

$$= 10 + 12 + 5$$

$$I = 27 \text{ A}$$

This means that a 27 A cable could be used safely rather than a 50 A cable. But, the use of diversity is mainly to size the main incoming tails, by adding together all the other assumed current demands of other circuits.

Points to note

1 If the assumed current demand of a cooker circuit exceeds 15 A but does not exceed 50 A, two or more cooking appliances may be fed, if they are in the same room.
2 Every cooker must have a control switch within 2 m. If two cookers are installed, one switch may be used, provided that neither cooker is more than 2 m from the switch.

Space heating systems

There are three methods of transferring heat from one place to another: conduction, convection and radiation.

Conduction of heat occurs when the source of heat and the object it is heating are in direct contact; for example, a metal rod held in a flame will become hot at the holder's end as the heat is conducted along the rod. Conduction of heat does not really concern us in the subject of space heating.

Convection heating is a method by which cold air or liquid is heated, rises, cools and falls, and is reheated again, the operation continuing until the required temperature is reached.

Radiation transfers the heat from the source directly to the object to be heated through the surrounding air.

Radiant or direct heating

Direct-heating appliances include all apparatus that gives a person or persons warmth by the direct transfer of heat from the source to the person, e.g. an electric fire (Fig. 15.9).

The infrared type of heater found in bathrooms also gives out its heat by radiation. This type is permitted in a room containing a fixed bath or shower as it has an element totally enclosed in a silicon glass tube, its control being by means of a pull-cord switch.

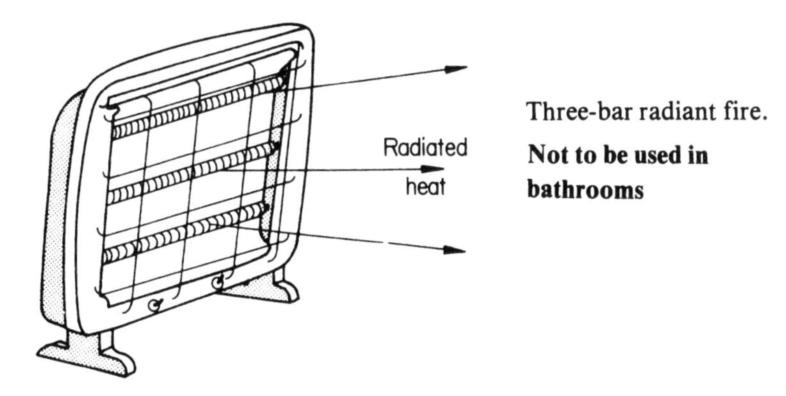

Fig. 15.9 *Three-bar radiant fires. Not to be used in bathrooms*

Convection heating

Convectors work on the principle of circulating warm air in a space (Fig. 15.10). Fig. 15.11 shows a typical domestic convector heater.

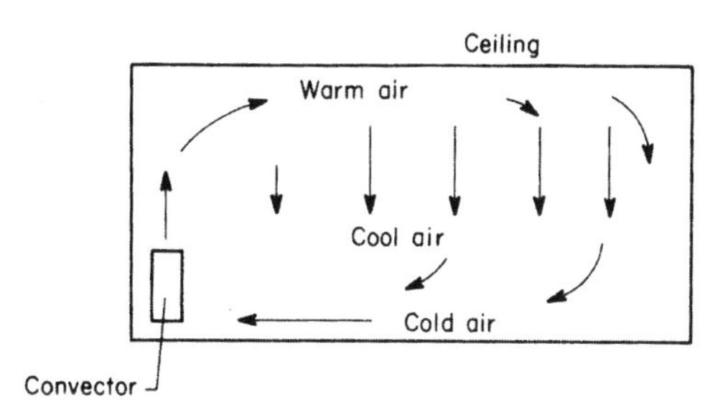

Fig. 15.10

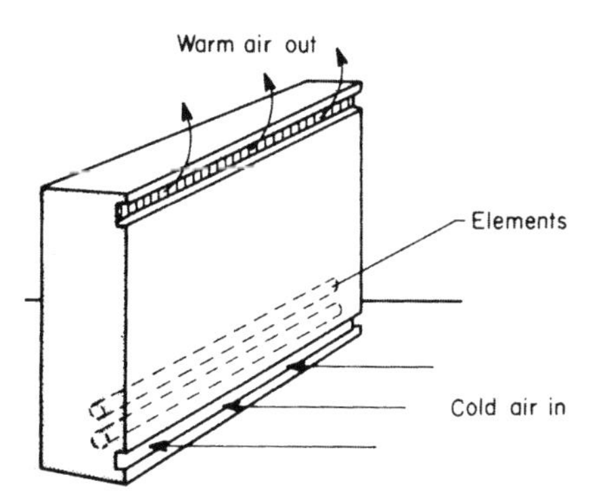

Fig. 15.11 *Floor or wall-mounted convector*

Storage heaters

Another type of heating appliance which uses the convection principle is the block storage heater.

In this case the elements are embedded in special fireproof heat-retaining bricks. The bricks collect the heat from the elements while they are energized and slowly give off the heat to the surrounding air over a period of time. These heaters are designed to be used on a special 'off-peak' tariff, switching on at

11.00 p.m. and switching off at 7.00 a.m. During the day the stored heaters are sometimes installed to give a boost to the room temperature. The fan blows hot air from around the bricks into the room.

Other variations of this system are underfloor heating installations and ducted warm-air systems.

Floor warming

Mineral-insulated single-core heating cable can be laid in the solid floor of a dwelling for heating purposes.

The floor composition is usually such that it readily retains heat, giving off the required warmth during the day.

Ducted warm-air heating

There are many different types of ducted warm-air heating systems on the market. All, however, have the same basic approach, i.e. a large, centrally located storage heater, with ducts leading from it to separate rooms. Air is forced from the unit to the rooms by means of one or more fans and released into the rooms via adjustable grills.

Thermostats

In order to control a heating appliance automatically, a method of detecting the temperature of the air, or water, or element being heated, must be found. Such a device is a *thermostat*. It detects changes in temperature and switches the heating appliance either on or off. The different types of thermostat are as follows.

The bimetallic strip

Different metals expand at different rates when heated, and this is the basis of the bimetallic-strip type of thermostat (Fig. 15.12a and b). Metal A expands faster than B when subjected to a rise in temperature, causing the strip to distort (Fig. 15.12b) and make the contact.

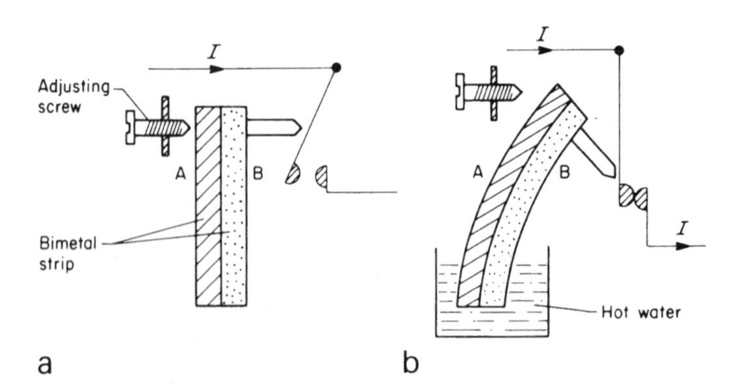

Fig. 15.12 *Bimetal strip*

The Invar rod

In a variation of the bimetallic-strip type of thermostat, a length of nickel-steel alloy, Invar, is secured inside a copper tube. With a rise in temperature the copper expands but the Invar rod does not (Fig. 15.13a and b).

Both of these types of thermostat, the Invar rod and the bimetal strip, are used with water heaters. The bimetal type is used for the control of irons.

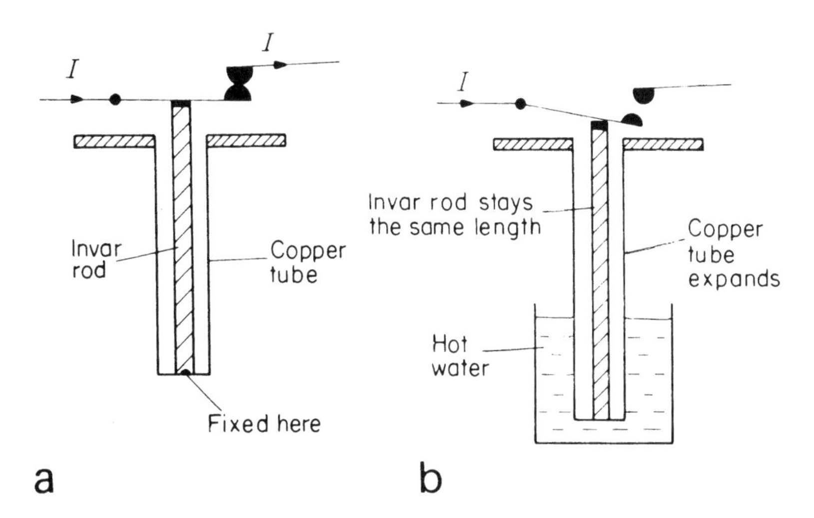

Fig. 15.13 *Invar rod*

Air thermostats

Air thermostats are commonly used to control central heating systems. Fig. 15.14a and b illustrate the two different types used.

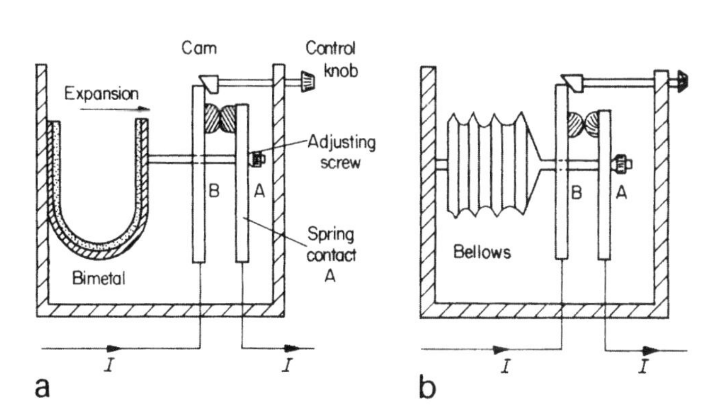

Fig. 15.14 *Air thermostat: (a) bimetal type: (b) bellows type*

In Fig. 15.14a increase in air temperature causes the bimetal to expand and the spring contact A to move away from B, opening the circuit. Heat settings can be achieved by the control knob, which alters the distance between A and B.

Fig. 15.14b shows a bellows type. In this case the air inside the bellows expands and contracts with a change in temperature, so causing contacts A and B to open or close.

Simmerstat

A simmerstat is used mainly for heat control on cooker plates and is shown in Fig. 15.15. Once the control knob has been selected to a particular value, the bimetal at A is the only part that can move, point B being fixed by the position of the cam.

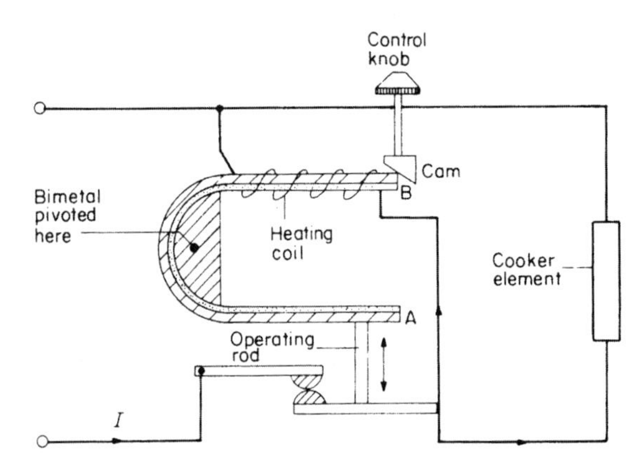

Fig. 15.15 *Simmerstat*

Ovenstat

This type of thermostat works on the bellows principle. In this case, however, the bellows are expanded by the pressure from a bulb and tube filled with liquid, which expands as the temperature rises (Fig. 15.16).

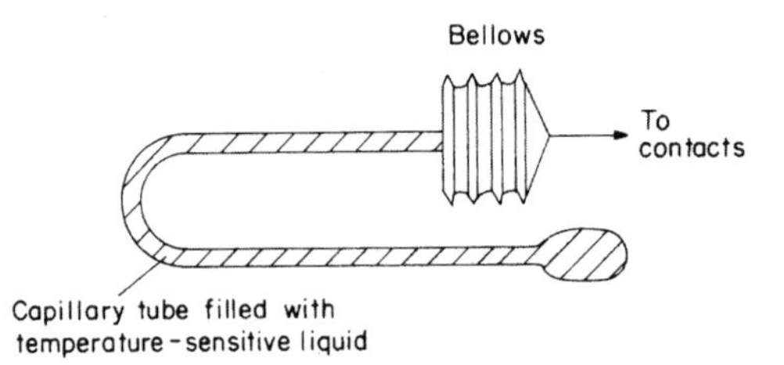

Fig. 15.16 *Ovenstat*

Installation systems

The IEE Regulations recommend that every consumer's installation should have a means of isolation, a means of overcurrent protection and a means of earth leakage protection. This recommendation applies whatever the size or type of installation, and the sequence of this equipment will be as shown in Fig. 15.17a and b.

Industrial installations

Industrial installations differ basically from domestic and commercial ones only in the size and type of equipment used. The supplies are three-phase

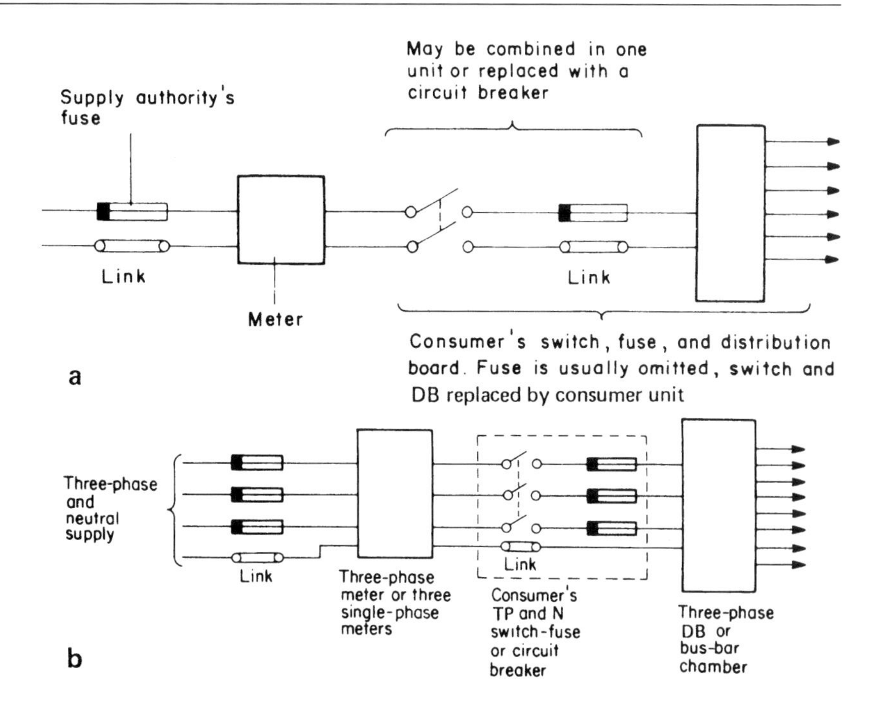

Fig. 15.17 *(a) Single-phase control: (b) three-phase control. TP and N = triple pole and neutral*

four-wire, and switchgear is usually metal clad. For extremely heavy loads, switch-fuses are replaced by circuit breakers, and much use is made of overhead bus-bar trunking systems. Figs 15.18 and 15.19 show typical layouts.

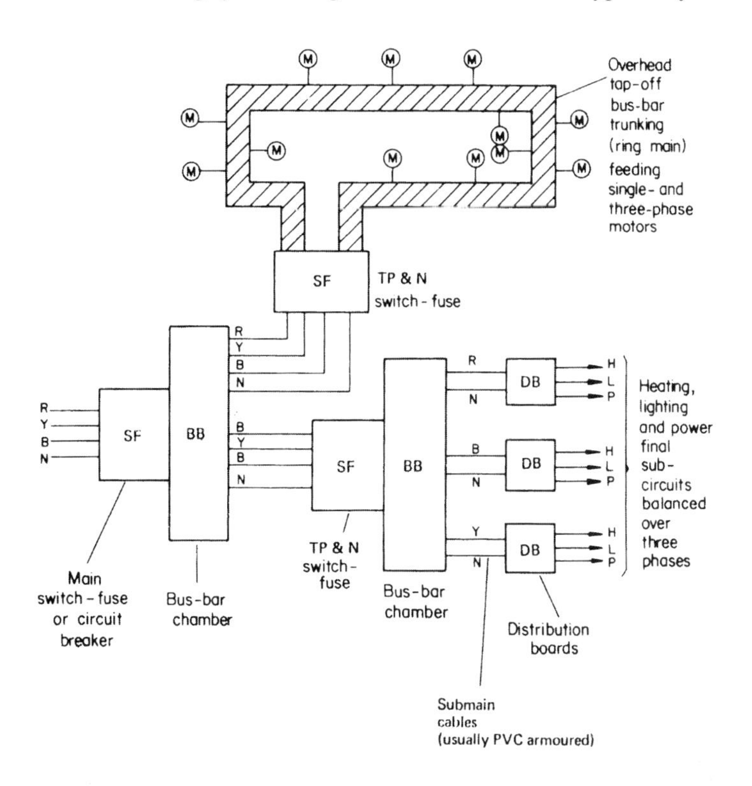

Fig. 15.18 *Layout of industrial installation*

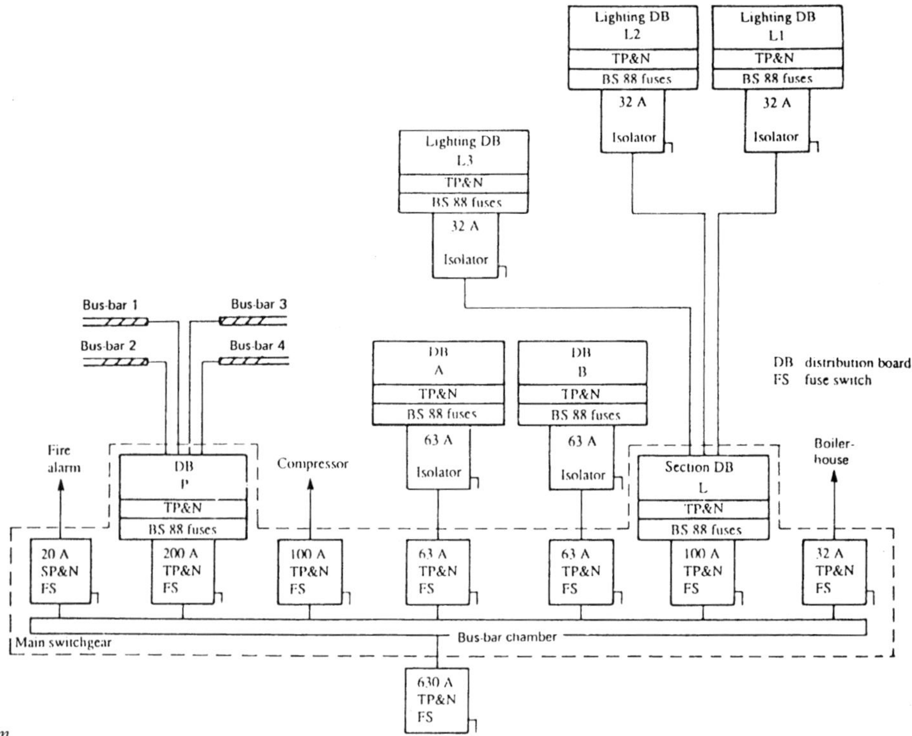

Fig. 15.19 *Distribution system*

With the larger types of installation, an alphanumeric system is very useful for cross-reference between block diagrams and floor plans showing architectural symbols.

Figs 15.20–15.22 illustrate a simple but complete scheme for a small garage/workshop. Fig. 15.20 is an isometric drawing of the garage and the installation, from which direct measurements for materials may be taken.

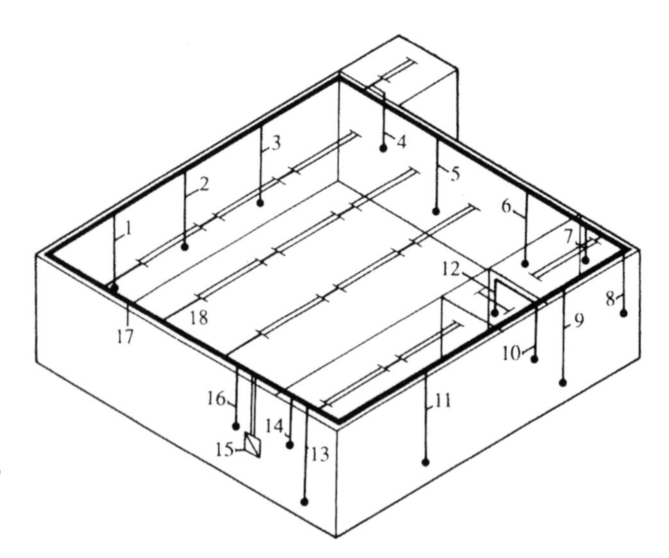

Fig. 15.20 *Isometric drawing for garage/workshop. 1, three-phase supply to ramp: 20 mm² conduit; 2, single-phase supply to double sockets: 20 mm² conduit; also 3, 5, 6, 9, 11, 13; 4, single-phase supply to light switch in store: 20 mm² conduit; 7, single-phase supply to light switch in compressor: 20 mm² conduit; 8. three-phase supply to compressor: 20 mm² conduit; 10, single-phase supply to heater in WC: 20 mm² conduit; 12, single-phase supply to light switch in WC: 20 mm² conduit; 14, single-phase supply to light switch in office: 20 mm² conduit; 15, main intake position; 16, single-phase supplies to switches for workshop lights: 20 mm² conduit; 17, 50 mm × 50 mm steel trunking; 18, supplies to fluorescent fittings: 20 mm² conduit*

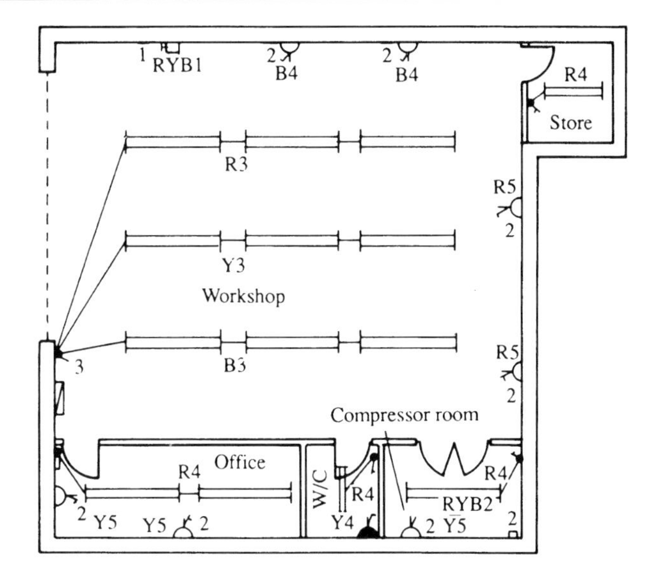

Fig. 15.21 *Floor plan for garage/ workshop*

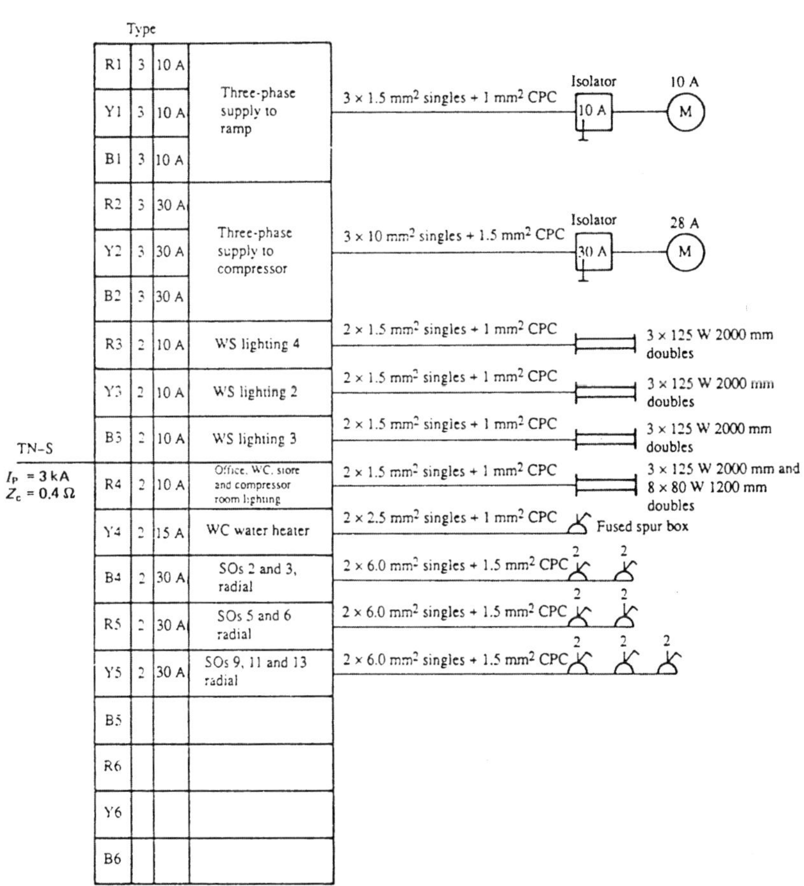

Fig. 15.22 *Details of connection diagram for garage/workshop*

Fig. 15.21 is the associated floor plan, which cross-references with the DB schedule and interconnection details shown on Fig. 15.22.

Multi-storey commercial or domestic installations

In order to supply each floor or individual flat in a block, it is necessary to run cables from the main intake position. These cables are called *risers*, and the submain cables which run from these to each individual supply point are called *laterals*.

The majority of rising mains are in the form of bus-bar trunking with either rectangular or circular conductors; this enables easy tapping off of submain cables. Fig. 15.23 shows a typical system.

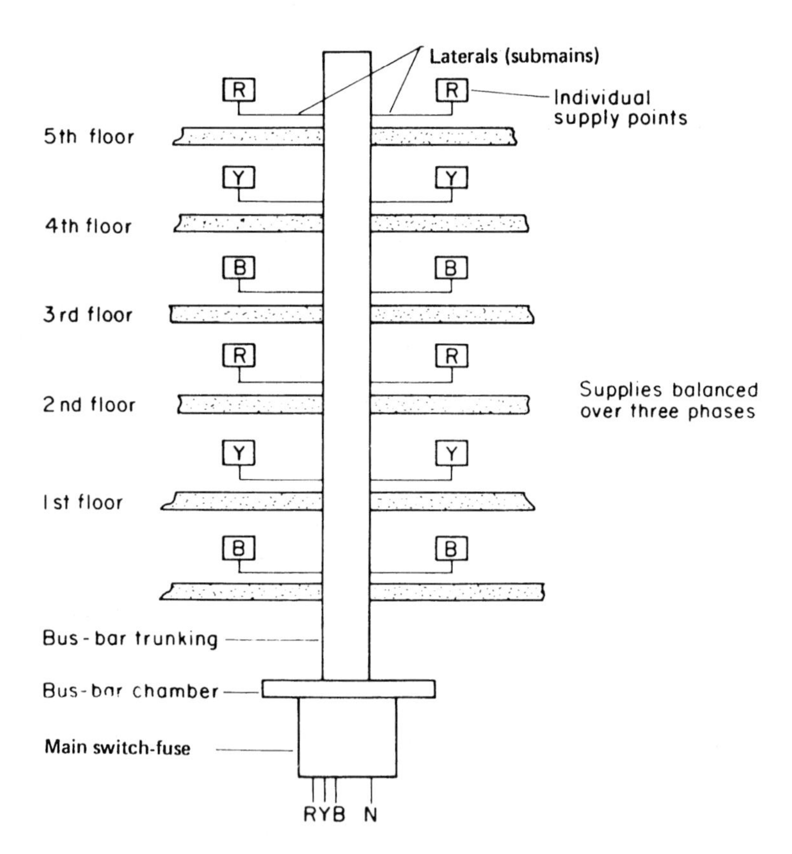

Fig. 15.23 *Rising main in a block of flats*

The rising-main system is similar to the ordinary radial circuit in that one cable run supplies several points. Hence the current flowing in the cable at the far end will be less than that at the supply end and the voltage drop will be greater at the far end with all loads connected (Fig. 15.24 and 15.25).

Hence the currents may be found in any part of a radial distribution cable. Also, if the resistance per metre of the cable is available and the position of the loads along the cable is known, the voltage drop at points along the cable may be calculated.

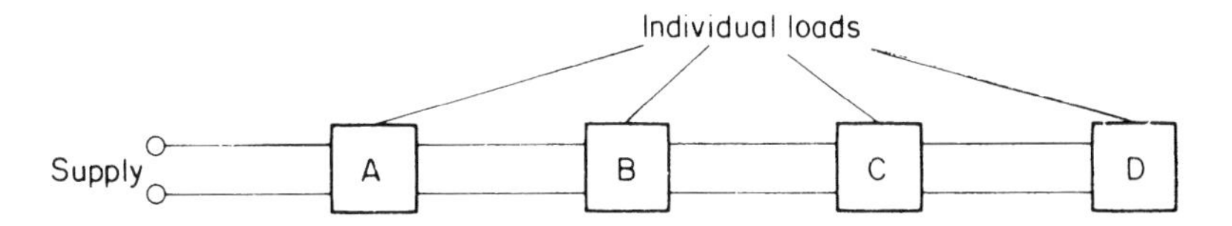

Fig. 15.24 *Radial circuit*

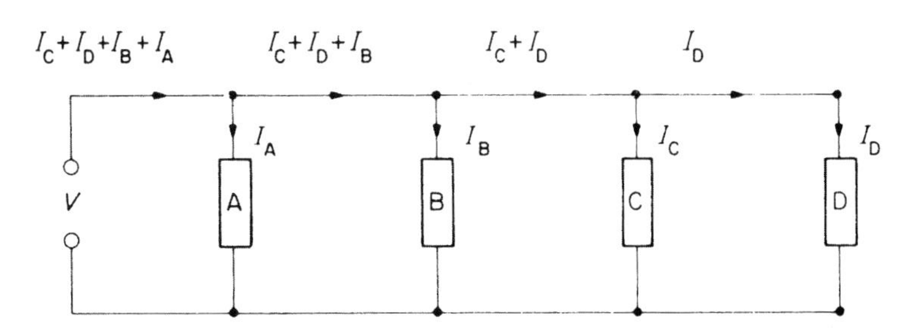

Fig. 15.25 *Circuit equivalent to that shown in Fig. 15.24*

Example

A 240 V radial distributor is 70 m long and has a resistance of 0.0008 Ω per metre supply and return. Four loads A, B, C and D rated at 30 A, 45 A, 60 A and 80 A are fed from the cable at distances of 20 m, 10 m, 15 m and 25 m respectively. Calculate the total current drawn from the supply, the current in the cable between each of the loads, and the voltage at load D if all the loads are connected.

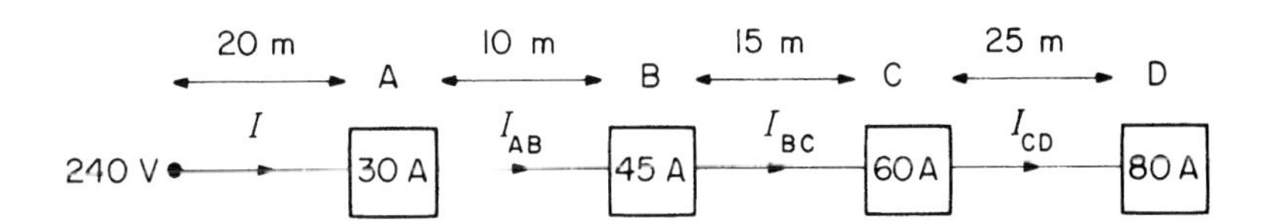

$$\text{Total load } I = 30 + 45 + 60 + 80$$
$$= 215 \text{ A}$$
$$I_{CD} = 80 \text{ A}$$
$$I_{BC} = 60 + 80$$
$$= 140 \text{ A}$$
$$A_{AB} = 45 + 60 + 80$$
$$= 185 \text{ A}$$

$$\text{Resistance between supply and A} = 20 \times 0.0008$$
$$= 0.016\Omega$$

$$\therefore \text{Voltage drop between supply and A} = 0.016 \times I$$
$$= 0.016 \times 215$$
$$= 3.44\ \text{V}$$
$$\text{Resistance between A and B} = 10 \times 0.0008$$
$$= 0.008\ \Omega$$
$$\therefore \text{Voltage drop between A and B} = 0.008 \times I_{AB}$$
$$= 0.008 \times 185$$
$$= 1.48\ \text{V}$$
$$\text{Resistance between B and C} = 15 \times 0.0008$$
$$= 0.012\ \Omega$$
$$\therefore \text{Voltage drop between B and C} = 0.012 \times I_{BC}$$
$$= 0.012 \times 140$$
$$= 1.68\ \text{V}$$
$$\text{Resistance between C and D} = 25 \times 0.0008$$
$$= 0.02\ \Omega$$
$$\therefore \text{Voltage drop between C and D} = 0.02 \times I_{CD}$$
$$= 0.02 \times 80$$
$$= 1.6\ \text{V}$$
$$\text{Total voltage drop} = 3.44 + 1.48 + 1.68 + 1.6$$
$$= 8.2\ \text{V}$$
$$\therefore \text{Voltage at load D} = 240 - 8.2$$
$$= 231.8\ \text{V}$$

Off-peak supplies

As the name implies, off-peak electricity is supplied to the consumer at a time, usually between 11 p.m. and 7 a.m., when demand is not at a peak. This ensures a greater economy in the use of generators and hence the cost per unit to the consumer is low.

These supplies are used mainly for space and water heating; however, some intake arrangements allow all energy-using devices to be used on off-peak.

Standard off-peak arrangement

Fig. 15.26 shows the arrangement at the supply intake position. In this arrangement the time clock controls the contractor coil. In installations with only a light off-peak load, the time clock contacts are able to control the load directly.

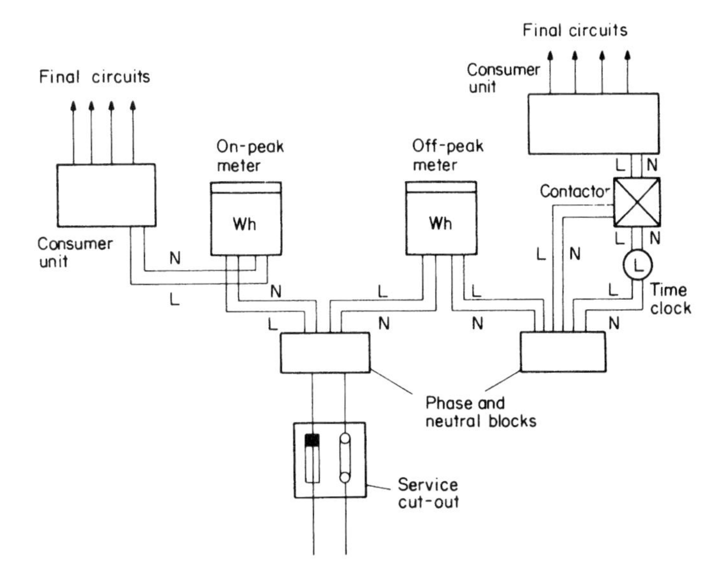

Fig. 15.26 *Arrangement for off-peak supplies (earths omitted for clarity)*

The white meter

A white meter has two recording dials, one for normal supplies, the other for off-peak. Two time clocks and a contactor are used to change over to off-peak. With this system the whole installation can be run during off-peak hours. Fig. 15.27 illustrates this system.

During normal hours, only the normal-rate dial will record. The consumer can in fact use off-peak appliances during this period, by overriding the time clock. Of course, any energy used by these appliances is charged at the normal rate.

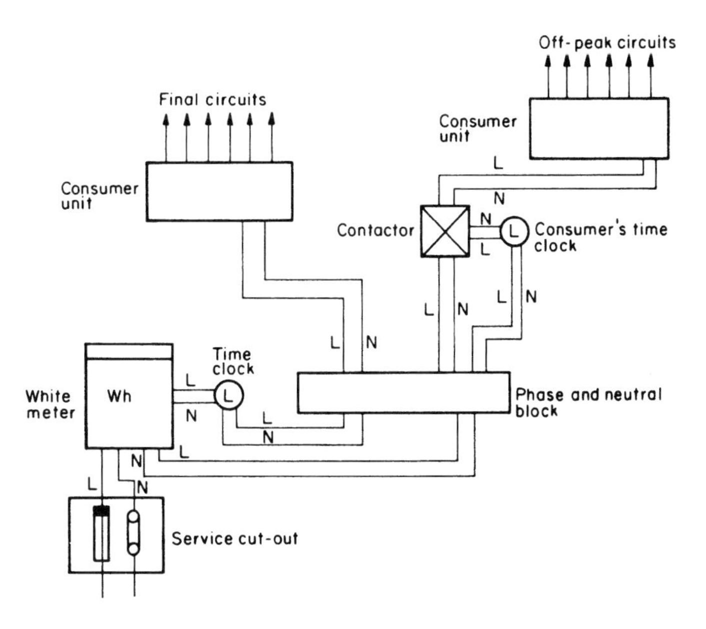

Fig. 15.27 *White meter*

At a pre-set time, say 11 p.m., the authority's time clock automatically changes the connections to the dials in the meter, and energy used by any appliance will be metered at off-peak rates.

Alarm and emergency systems

These types of systems are usually supplied by an extra-low voltage (up to 50 V) although the operating voltage may be supplied via a transformer, whose primary may be at a low voltage.

Before each of these systems is discussed, it is perhaps best to establish a convention with regard to relays. A *relay* is an electromagnet which causes pairs of contacts to make or break, when it is energized. All diagrams should show the relay *de-energized*; the contacts are then said to be in their *normal* position (Fig. 15.28).

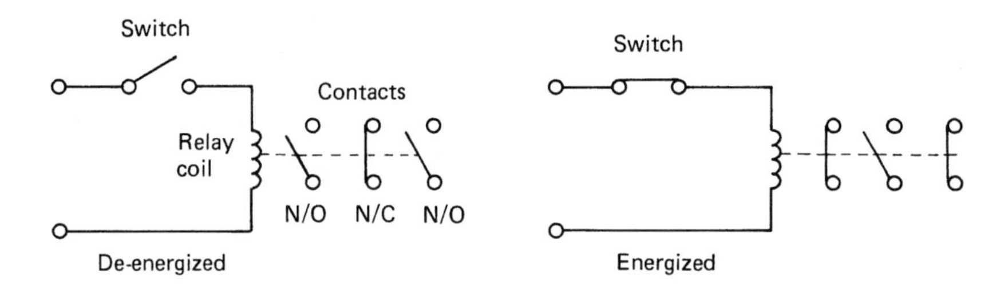

Fig. 15.28 *(a) De-energized; (b) energized*

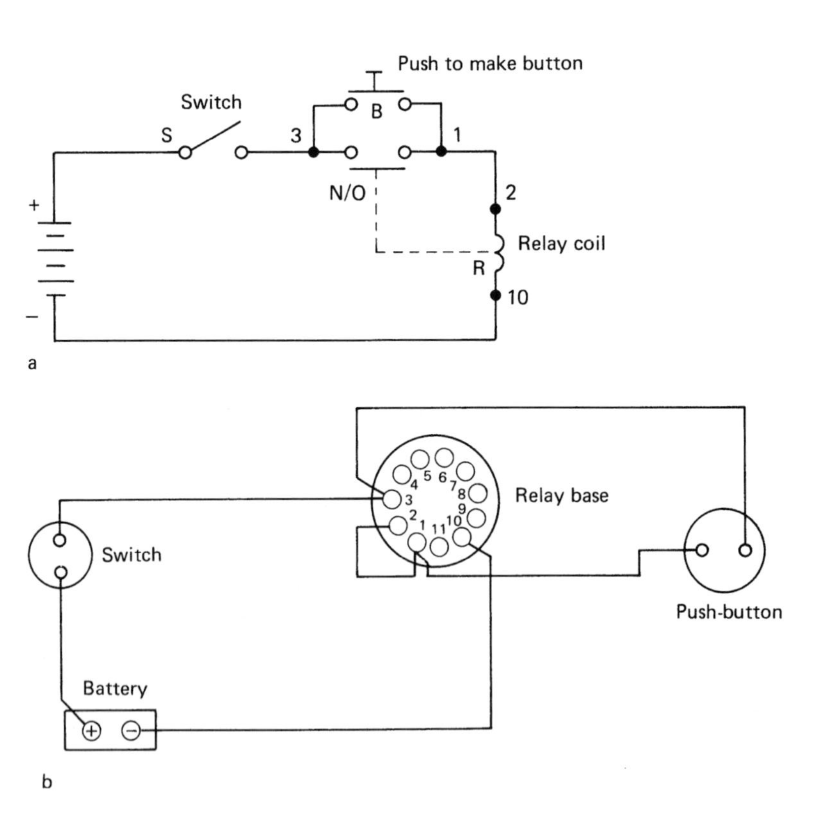

Fig. 15.29 *(a) Simple 'hold-on' circuit; (b) wiring diagram*

Another important point to note is the difference between a circuit or schematic diagram and a wiring diagram. A circuit diagram shows how the system *functions*, and relay contacts, switches, and accessories are shown on a diagram in a position most convenient for drawing and understanding. A wiring diagram shows how the system is to be *wired*, and all components of the circuit should be shown in their correct places (Fig. 15.29a and b).

The 'hold-on' circuit shown in Fig. 15.29a is most important especially in fire- and burglar-alarm systems. It operates as follows. With switch S closed R is still not energized. By pushing the button B coil R is energized and the normally open contact C will close giving another route for the supply to reach the relay coil. When the button is released the relay will 'hold-on' through its own contact. If switch S were opened, the relay would de-energize and contact C would open, but closing S again would not re-energize the relay. Button B would need to be pushed again to re-energize the relay.

Both security and fire-alarm systems are basically the same in that various sensors are wired to a control panel, which in turn will activate an alarm in the event of sensor detection (Figs 15.30 and 15.31). Some modern panels have the facility for incorporating both systems in the same enclosure.

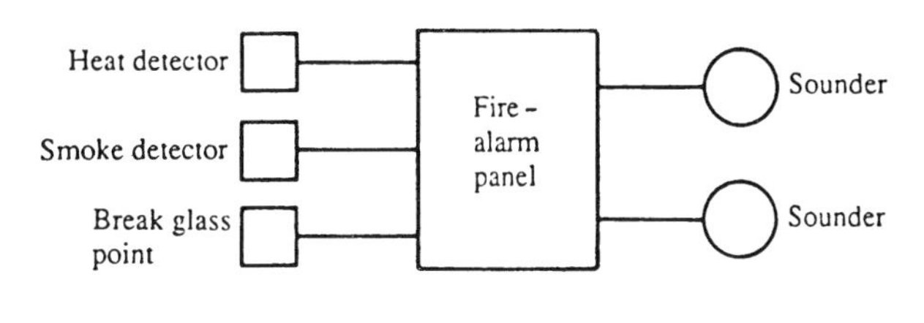

Fig. 15.30 *Block diagram for fire-alarm system*

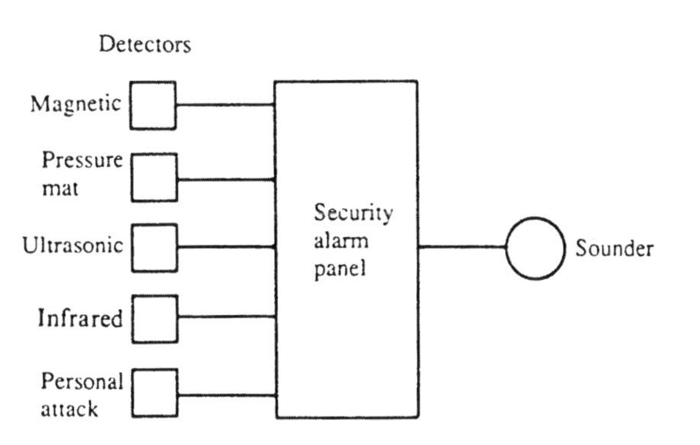

Fig. 15.31 *Block diagram for security alarm system*

Open-circuit system

In this system the call points (sensors, detectors, etc.) are wired in parallel such that the operation of any one will give supply to the relay RA and the sounder via the reset button. N/O contacts RA1 will then close, holding on the relay and keeping to the sounder via these contacts. This hold-on facility is most important as it ensures that the sounder is not interrupted if an attempt is made to return the activated call point to its original off position.

Fire-alarm systems are usually wired on an open-circuit basis, with a two-wire system looped from one detector to the next, terminating across an end-

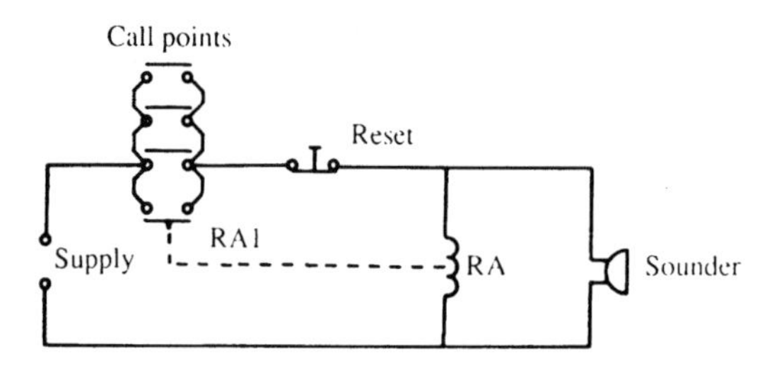

Fig. 15.32 *Open circuit*

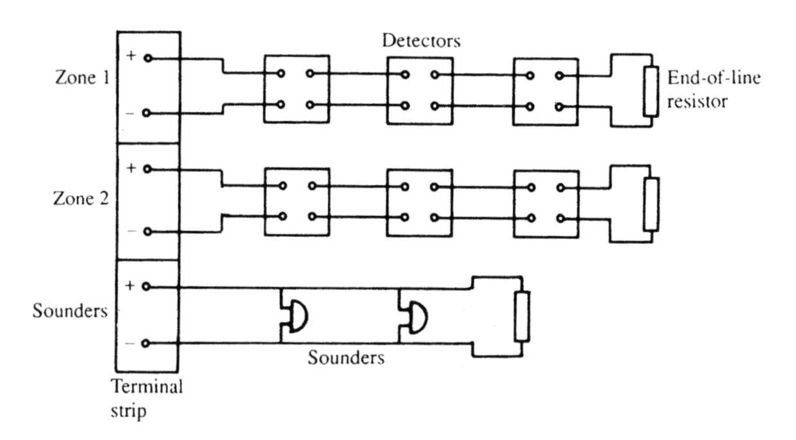

Fig. 15.33 *Fire-alarm system*

of-line resistor (EOLR). This provides a circuit cable monitoring facility; the EOLR is of sufficiently high value to prevent operation of the alarm. Fig. 15.33 shows a typical connection diagram.

Closed-circuit system

This system has the call points wired in series, and the operation of the reset button energizes relay RA. N/O contacts RA1 close and N/C contacts RA2 open, the relay RA remaining energized via contacts RA1 when the reset button is released. The alarm sounder mute switch is then closed, and the whole system is now set up.

An interruption of the supply to the relay RA, by operation of any call point, will de-energize the relay, open RA1 and close RA2, thus actuating the alarm sounder. The system can only be cancelled and reset by use of the reset button.

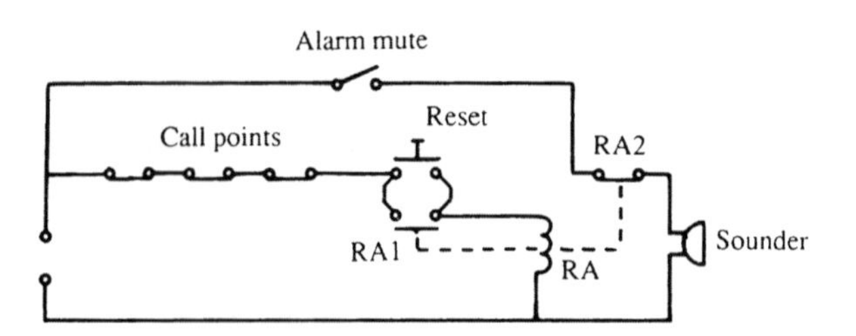

Fig. 15.34 *Closed circuit*

The closed-circuit system is quite popular, as it is self-monitoring in that any malfunction of the relay or break in the call point wiring will cause operation of the system as if a call point had been activated.

Intruder alarm systems tend, in the main, to be based on the closed circuit type. Fig. 15.35 shows the connection diagram for a simple two-zone system with tamper loop and personal attack button.

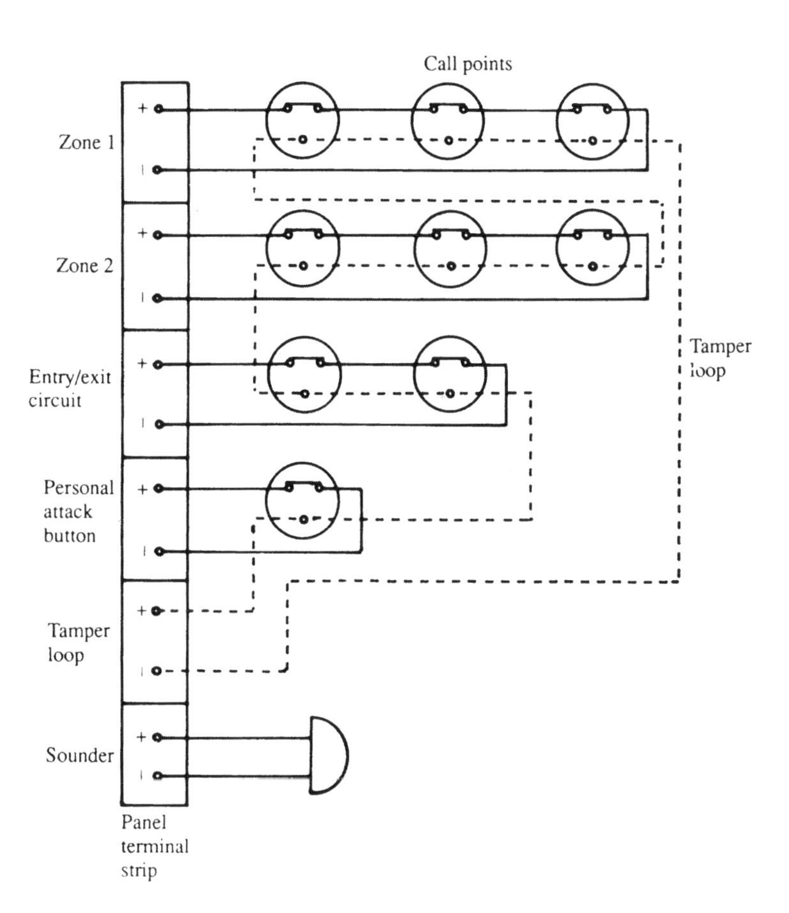

Fig. 15.35 *Security alarm system*

The tamper loop is simply a continuous conductor wired to a terminal in each detector in the system. It is continuously monitored irrespective of whether the alarm system is switched on or off, and if interrupted will cause immediate operation of the alarm.

The entry/exit circuit is usually confined to the front and/or back doors. The facility exists to alter the time delay between setting the system and exiting, and between entering and switching the system off. This adjustment is made inside the control panel.

All security and fire-alarm systems should have battery back-up with charging facilities.

Call systems

Once again these fall into different categories, such as telephone systems and page and bleeper systems. However, the nurse-call variety which uses push-buttons and lamp indication is probably the most popular.

With this type, each room is equipped with a call button of some description, a patient's reassurance light, and a cancel button. Outside each room is an indicator light, and at strategic points in the building there are zone buzzers. Centrally located is a display panel which incorporates a buzzer and an indication of which room is calling.

Fig. 15.36 illustrates, in a simple form, the principle of operation of such a system. This system should by now be quite familiar to the reader; it is simply another variation of the hold-on circuit. Any patient pushing a call button energizes his or her corresponding relay in the main control panel, which is held on by a pair of N/O contacts. At the same time the reassurance, the room and the panel lights 1, 2 and 3 are all illuminated. The zone and panel buzzers are energized via the relay's other pair of N/O contacts.

It is usual to locate the cancel button only in the patient's room, as this ensures that staff visit the patient in question.

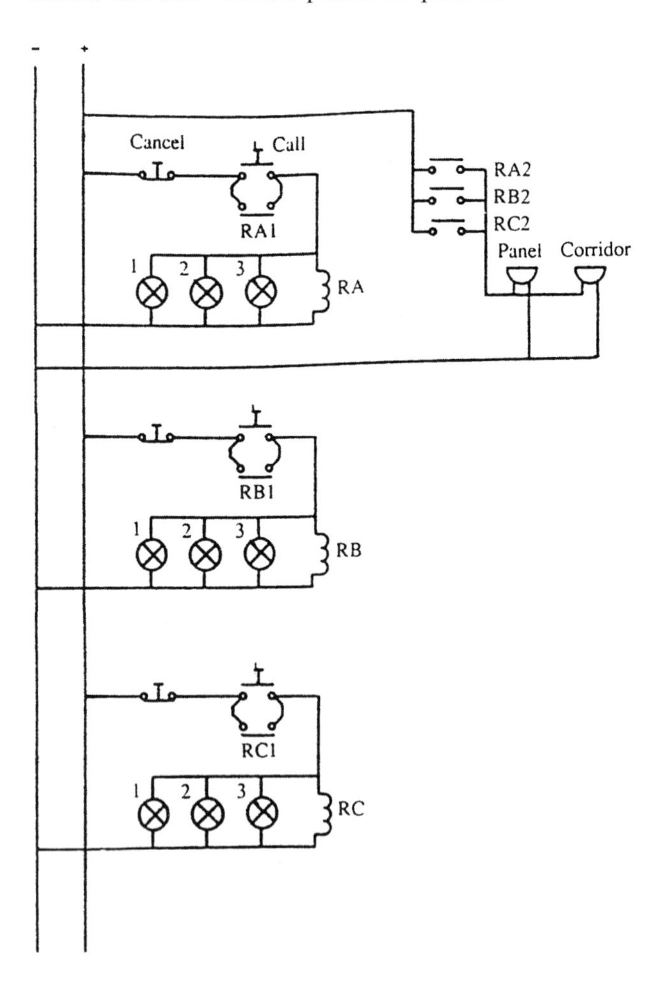

Fig. 15.36 *Nurse-call system*

Emergency lighting systems

These fall into two categories: maintained and non-maintained. Both of these systems may be utilized by individual units or by a centralized source.

Maintained system

In this system the emergency lighting unit is energized continuously via a step-down transformer, and in the event of a mains failure it remains illuminated via a battery (Fig. 15.37).

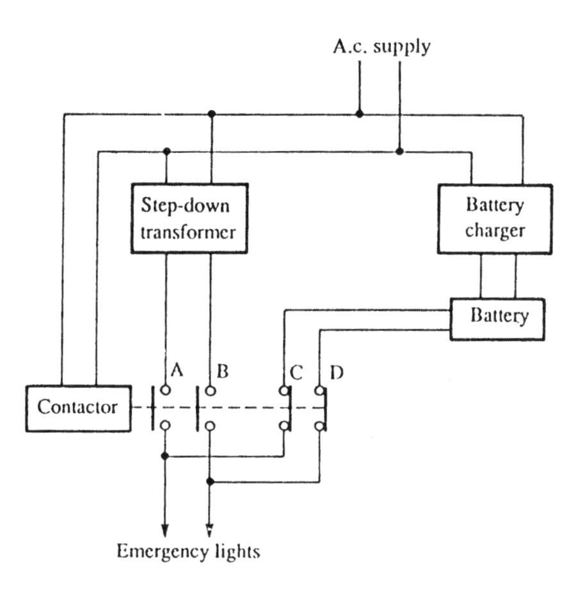

Fig. 15.37 *Maintained system*

Non-maintained system

Here the lighting units remain de-energized until a mains failure occurs, at which time they are illuminated by a battery supply (Fig. 15.38).

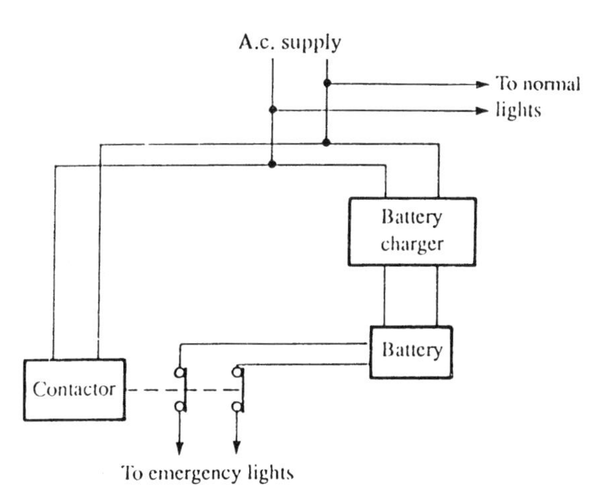

Fig. 15.38 *Non-maintained system*

It should be noted that modern systems use electronic means to provide the changeover from mains to battery supply. The contactor method, however, serves to illustrate the principle of operation.

Central heating systems

Let us take a look at the two most basic arrangements: the pumped central heating (CH) and gravity-fed hot-water (HW) system, and the fully pumped system with mid-position valve. It must be remembered that, whatever the system, it is imperative that the wiring installer has a knowledge of the function of the system in order to do a competent job.

Pumped CH and gravity HW

This system comprises a boiler with its own thermostat to regulate the water temperature, a pump, a hot-water storage tank, a room thermostat, and some form of timed programmer. The water for the HW, i.e. the taps etc., is separate from the CH water, but the boiler heats both systems.

Fig. 15.39 shows such a system. From the diagram it might appear that when the requirement for HW is switched off at the programmer, the CH cannot be called for as the boiler has lost its feed. In fact, such programmers have a mechanical linkage between the switches: HW is allowed without CH, but selection of CH automatically selects HW also.

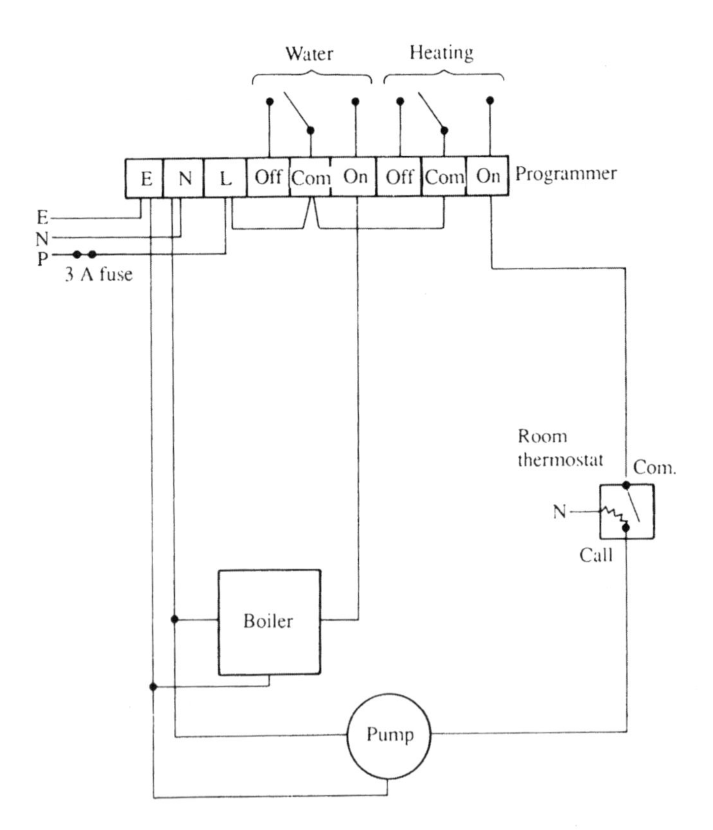

Fig. 15.39 *Gravity primary and pumped heating*

Note the little heating element in the room thermostat; this is known as an accelerator. Its purpose is to increase the sensitivity of the thermostat; manufacturers claim that it increases the accuracy of the unit to within one degree Celsius. The inclusion of an accelerator (if required) does mean an extra conductor for connection to neutral.

Fully pumped system

Two additional items are required for this system: a cylinder thermostat and a mid-position valve. In this system HW and CH can be selected independently. The mid-position valve has three ports: a motor will drive the valve to either HW only, CH only, or HW and CH combined. With this system the boiler and pump always work together. Fig. 15.40 illustrates the system, and Fig. 15.41 shows the internal connections of a mid-position valve.

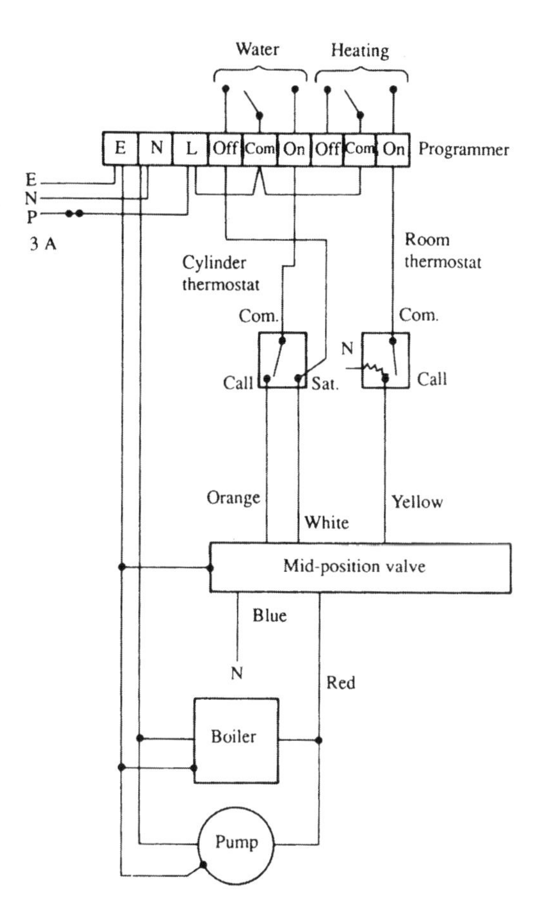

Fig. 15.40 *Fully pumped system*

Some difficulties may be experienced in wiring when the component parts of the system are produced by different manufacturers. In this case it is probably best to draw one's own wiring diagram from the various details available.

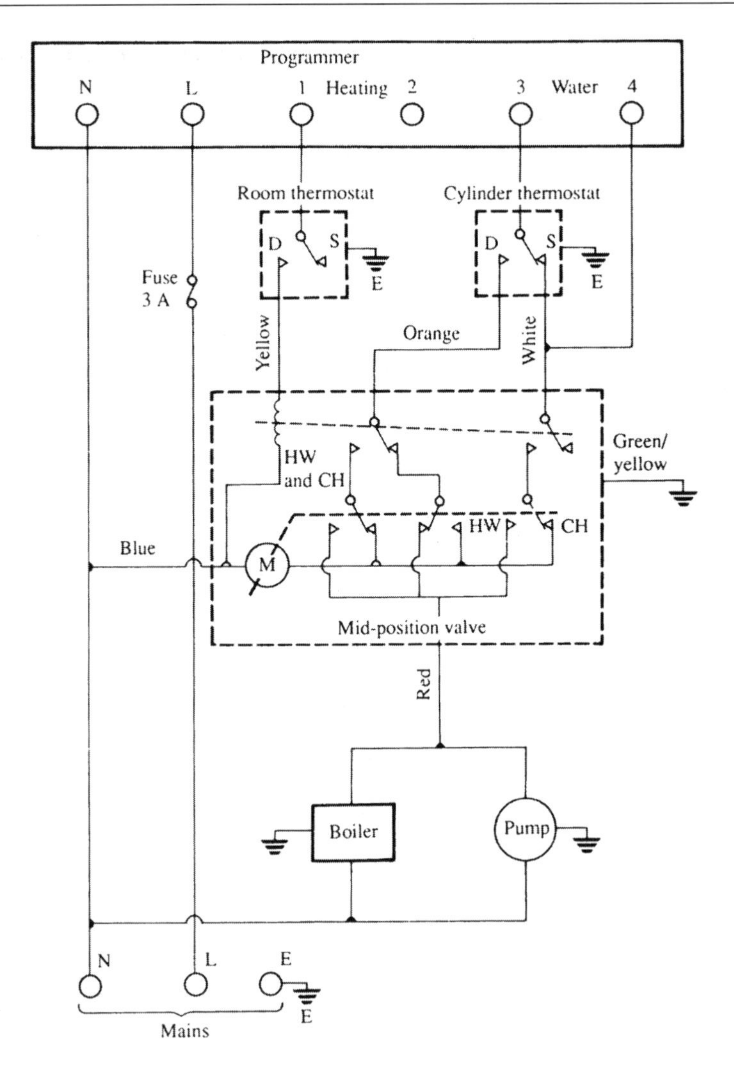

Fig. 15.41 *Internal connection of mid-position valve*

Extra-low-voltage lighting

These systems, incorrectly referred to as low-voltage lighting (low voltage is 50 V to 1000 V a.c.), operate at 12 V a.c. They employ tungsten–halogen dichroic lamps, which have a very high performance in comparison with 240 V halogen lamps. For example, a 50 W dichroic lamp has the same light intensity as a 150 W PAR LAMP.

Extra-low-voltage (ELV) lighting is becoming very popular, especially for display purposes. There is very little heat emission, the colour rendering is excellent, and energy consumption is very low.

The 12 V a.c. to supply the lamps is derived from a 240 V/12 V transformer specially designed to cater for the high starting surges, and only these types should be used. The voltage at each lamp is critical: 0.7 V overvoltage can cause premature ageing of the lamp, and 0.7 V undervoltage will reduce the light output by 30%. Hence variation in voltage must be avoided.

To achieve this, leads and cables must be kept as short as possible, and the correct size used to avoid excessive voltage drop. When several lamps are to be run from one transformer, it is advisable to use a fused splitter unit rather than to wire them in a parallel chain (Fig. 15.42).

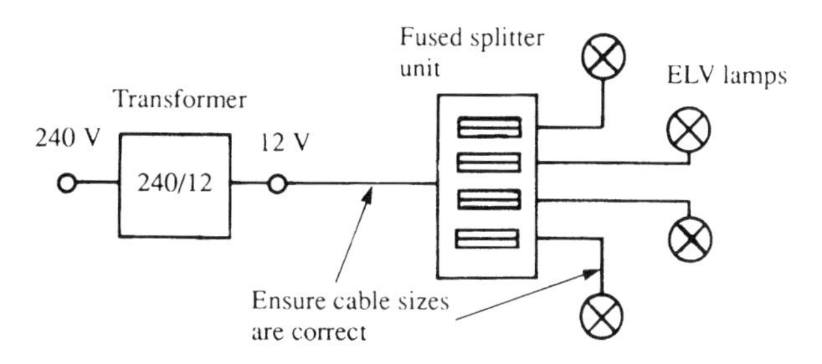

Fig. 15.42

It is important to remember that, for example, a 50 W ELV lamp will draw 4.17 A from the 12 V secondary of the transformer ($I = P/V$). Although a 1.0 mm cable will carry the current, the voltage drop for only 3 m of this cable will be 0.55 V.

Choice of system

The choice of any particular wiring system and its accessories will depend on the environment in which it is to be installed. Under normal conditions, typical wiring systems would include:

PVC-insulated PVC-sheathed	– domestic premises; small shops and offices, etc.
PVC conduit or trunking	– offices; light industry
Metal conduit; trunking or armoured cable	– any situation where there is a serious risk of mechanical damage
M.i.m.s.	– fire-alarm systems; boilerhouses; earthed concentric wiring, etc.

There are, however, certain environments which need special attention.

Circuits in bathrooms/shower rooms

Water is not a good conductor of electricity, but nevertheless it is a conductor and therefore any situation that uses electrical equipment in close proximity to water and steam should have extra safeguards.

The latest IEE Regulations suggest that such areas are divided into zones, these being zones 0, 1, 2 & 3.

Zone 0

This is the interior of the bath tub or shower basin or, in the case of a shower area without a tray, it is the space having a depth of 50 mm above the floor out to a radius of 600 mm from a fixed shower head or 1200 mm radius for a demountable head.

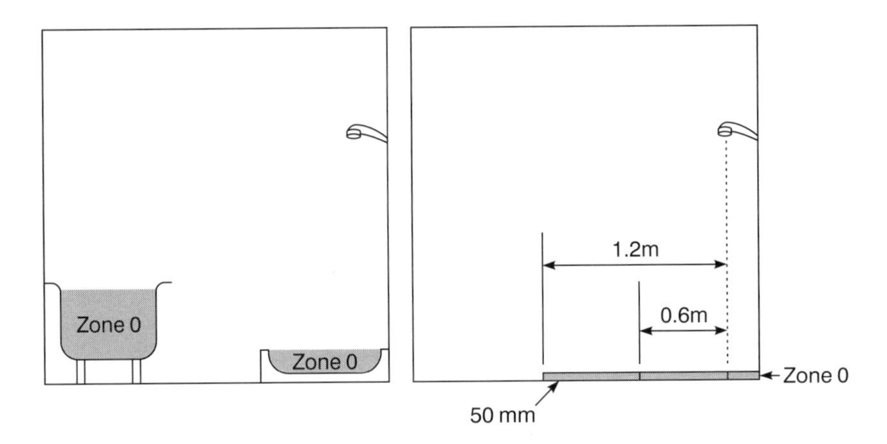

Fig. 15.43

Points to note

- Only SELV (12 V) may be used as a measure against electric shock, the safety source being outside zones 0, 1 and 2.
- Other than current using equipment specifically designed for use in this zone, **no** switchgear or accessories are permitted.
- Equipment designed for use in this zone must be to at least IP X7.
- Only wiring associated with equipment in this zone may be installed.

Zone 1

This extends above zone 0 around the perimeter of the bath or shower basin to 2.25 m above the floor level, and includes any space below the bath or basin that is accessible without the use of a key or tool. For showers without basins, zone 1 extends out to a radius of 600 mm from a fixed shower head or 1200 mm radius for a demountable head.

Points to note

- Only SELV may be used as a measure against electric shock, the safety source being outside zones 0, 1 and 2.

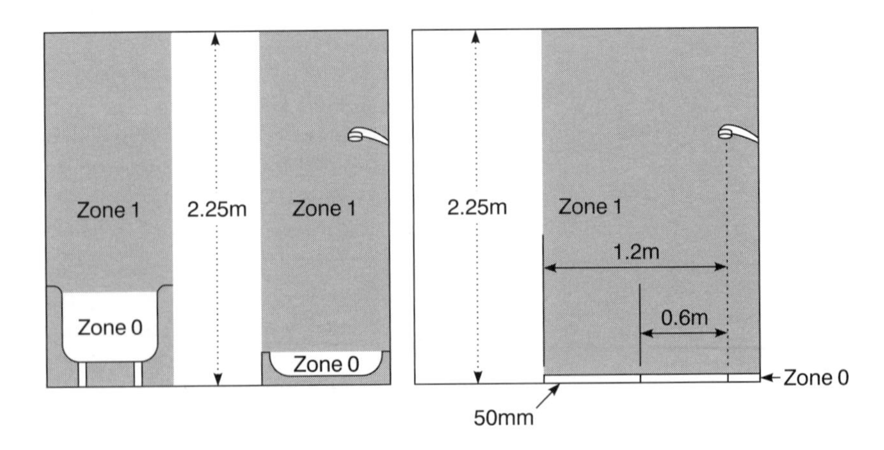

Fig. 15.44

- Other than switches and controls of equipment specifically designed for use in this zone, and cord operated switches, only SELV switches are permitted.
- Provided they are suitable, fixed items of current using equipment that may be installed in this zone are water heaters and shower pumps.
- Other equipment that can reasonably be only installed in this zone must be suitable, and be protected by an r.c.d. of rated tripping current 30 mA or less.
- Equipment designed for use in this zone must be to at least IP X4, or IP X5 where water jets are likely to be used for cleaning purposes.
- Only wiring associated with equipment in this zone and zone 0 may be installed.

Zone 2

This extends 600 mm beyond zone 1 and to a height of 2.25 above floor level and to the space above zone 1 between 2.25 m and 3 m above floor level.

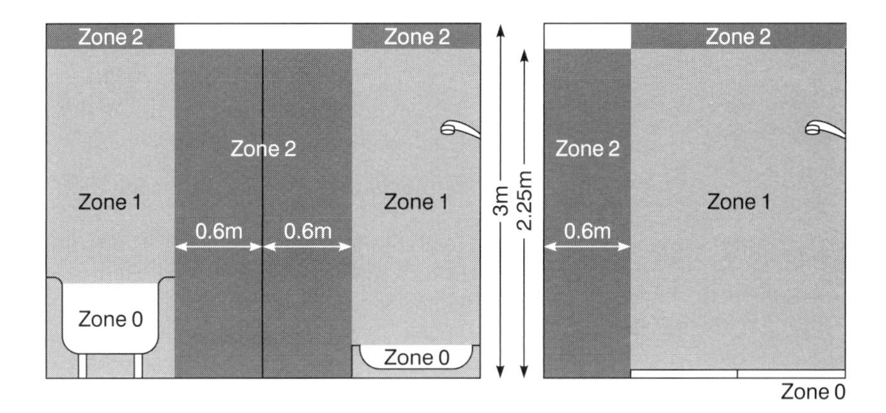

Fig. 15.45

Points to note

- Only SELV (12 V) may be used as a measure against electric shock, the safety source being outside zones 0, 1 and 2.
- Other than current using equipment specifically designed for use in this zone, only switches and socket outlets of SELV circuits and shaver units to BS EN 60742 may be installed.
- Provided they are suitable for use in this zone, water heaters, shower pumps, luminaires, fans etc. may be installed.
- Equipment designed for use in this zone must be to at least IP X4, or IP X5 where water jets are likely to be used for cleaning purposes.
- Only wiring associated with equipment in this zone and zones 0 and 1 may be installed.

Zone 3

This extends 2.4 m beyond zone 2 and to a height of 2.25 m above floor level and to the space above zone 2 between 2.25 m and 3 m above floor level.

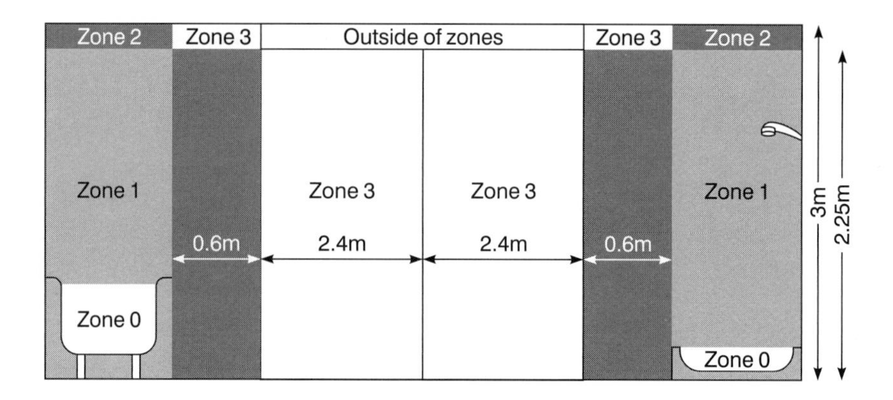

Fig. 15.46

Points to note

- Only SELV socket outlets or shaver units to BS EN 60742 may be installed.
- Equipment other than that specifically designed for use in this zone, must be protected by an r.c.d. of rated tripping current 30 mA or less.
- Equipment designed for use in this zone must be to at least IP X5 if water jets are likely to be used for cleaning purposes.

Supplementary equipotential bonding

Supplementary bonding must be carried out and will connect together the c.p.c. terminals of Class I and Class II circuits in zones 1, 2 and 3 and any extraneous conductive parts within these zones.

Such extraneous conductive parts will include:

- metallic gas, water, waste and central heating pipes
- metallic structural parts that are accessible to touch
- metal baths and shower basins.

This bonding may be carried out in the bath or shower room or in close proximity to it.

Electric floor units may be installed below any zone provided that they are covered with an earthed metal grid or metallic sheath and connected to the local supplementary bonding.

Where a cabinet or cubicle containing a bath or shower is installed in a room other than a bathroom, any socket outlet must either be a SELV or shaver socket or if it is outside zone 3 it can be a BS 1363 socket and must be protected by an r.c.d. of rated tripping current 30 mA or less. Also, local supplementary bonding is not required, outside zone 3.

Swimming pool installations

These are clearly hazardous areas with the presence of so much water. Such installations are fed from an SELV source at 18 V or 12 V.

Any surface wiring and accessories should be non-metallic.

Temporary and construction site installations

Installations such as these always present a hazardous situation and special attention must be paid to guarding against mechanical damage, damp and corrosion, and earth leakage.

Any such installation should conform with BSCP 1017 and be in the charge of a competent person who will ensure that it is safe, and his or her name and designation must be displayed clearly adjacent to the main switch of the installation. The associated distribution board should conform with BS 4363.

The installation should be tested at intervals of no more than 3 months, this period being shown clearly on any completion certificate issued.

As an added precaution against shock risk, portable tools should be supplied from a double-wound transformer with reduced secondary voltage and the secondary winding centre tapped to earth. This ensures that any fault to earth presents only 55 V to the user (Fig. 15.47).

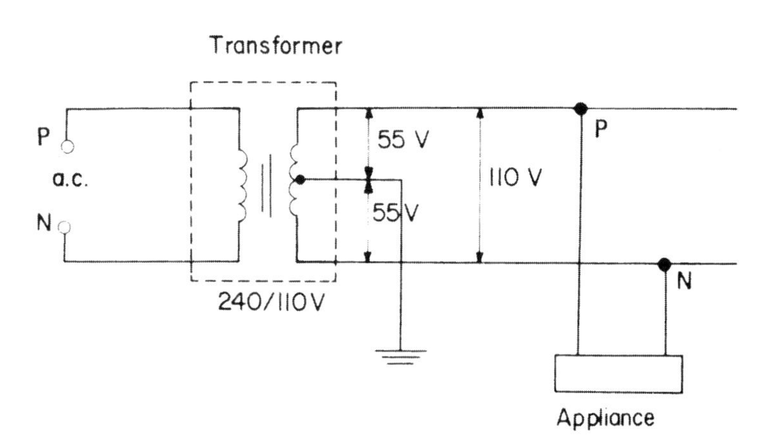

Fig. 15.47

Installations in flammable and/or explosive situations

In premises such as petrol stations, gas works, flour mills, etc., using electricity involves an obvious risk. Serious arcing at contacts or in faulty conductors or equipment could cause an explosion and/or fire. It is therefore important to ensure that such a dangerous situation does not arise, by installing suitable cable and fittings. An outline of the hazards and suitable wiring systems is given in CP 1003. The following paragraphs are extracts from that Code of Practice, reproduced by permission of the BSI, 2 Park Street, London W1A 2BS, from whom complete copies of the Code of Practice may be obtained.

Extracts from BS Code of Practice 1003 Part 1, 1964:

'Electrical apparatus and associated equipment for use in explosive atmospheres of gas or vapour (other than mining)' Part 1: 'Choice, installation and maintenance of flameproof and intrinsically safe equipment'.

Foreword

In dealing with the risk of fire or explosion from the presence of flammable liquids, gases or vapours, three sets of conditions are recognized:

Division 0

An area or enclosed space within which any flammable or explosive substance, whether gas, vapour or volatile liquid, is continuously present in concentrations within the lower and upper limits of flammability.

Division 1

An area within which any flammable or explosive substance, whether gas, vapour or volatile liquid, is processed, handled or stored, and where during normal operations an explosive or ignitable concentration is likely to occur in sufficient quantity to produce a hazard.

Division 2

An area within which any flammable or explosive substance, whether gas, vapour or volatile liquid, although processed or stored, is so well under conditions of control that the production (or release) of an explosive or ignitable concentration in sufficient quantity to constitute a hazard is only likely under abnormal conditions.

The conditions described as appertaining to Division 0 are such as normally to require the total exclusion of any electrical equipment, except in such special circumstances as to render this impracticable, in which case recourse may be possible to special measures such as pressurization or the use of intrinsically safe equipment.

A risk of the nature described under Division 1 can be met by the use of flameproof or intrinsically safe equipment, with which Part 1 is concerned, or by the use of the means described in Part 2 of the Code which is concerned with methods, other than the use of flameproof or intrinsically safe equipment, of securing safety in flammable and explosive atmospheres.

The certifying authority for flameproof apparatus is the Ministry of Power. The certifying authority for intrinsically safe electrical apparatus for use in factories coming within the scope of the Factories Act, is the Ministry of Labour. The recognized testing authority for flameproof enclosures and intrinsically safe circuits and apparatus is the Ministry of Power*.

Types of hazard

Two main types to be considered: gases and vapours or flammable liquids.

(a) *Explosive gases* or vapours are grouped according to the grade of risk and four groups are recognized.

Group I   Gas encountered in coal mining.
Group II   Various gases commonly met within industry.

* The Ministry of Power is now part of the Department of Trade and Industry, and the Factories Act is now overseen by the Health and Safety Executive.

Group III Ethylene, diethyl ether, ethylene oxide, town gas and coke oven gas.
Group IV Acetylene, carbon disulphide, ethyl nitrate, hydrogen and water-gas.

(i) Flameproof apparatus. Apparatus with flameproof enclosures, certified appropriately for the gas group which constitutes the risk, should be used. It should be noted that no apparatus is certified for Group IV; other techniques, e.g. pressurization, must therefore be applied if electrical apparatus has to be installed where gases in this group may be present in dangerous concentrations.

In general, apparatus certified for the higher groups will cover situations where gases from the lower groups are present.

(ii) Selection of apparatus for diverse risks. If, in an installation capable of subdivision, Group III apparatus is required for some parts while Group II would suffice for other parts, it is recommended that the former, which could cover the risks in a lower Group, should be used throughout lest apparatus of the latter group should inadvertently be transferred to a place where Group III is required.

(b) *Flammable liquids*. Flammable liquids give rise, in a greater or lesser degree according to their flashpoints and the temperature to which they are subjected, to explosive vapours which should be treated as under (a) above.

The liquid, however, constitutes a further risk in that fires may occur as a result of unvaporized liquid in the form of spray or otherwise, coming into contact with electrical equipment and then being ignited by a spark or other agency.

Types of wiring

Danger areas:

(a) Cables drawn into screwed solid-drawn steel conduit.
(b) Lead-sheathed, steel-armoured cable.
(c) Mineral-insulated, metal-sheathed cable.
(d) PVC insulated and armoured cable with an outer sheath of PVC.
(e) Polyethylene-insulated, PVC-covered overall and armoured.
(f) Cables enclosed in a seamless aluminium sheath with or without armour.

Automatic electrical protection

(a) All circuits and apparatus within a danger area should be adequately protected against overcurrent, short circuit and earth leakage current.
(b) Circuit breakers should be of the free-handle trip-free type to preclude misuse, such as tying-in or holding-in under fault conditions against the persistence of which they are designed to afford protection, and an indicator should be provided in all cases to show clearly whether the circuit breaker is open or closed.

Portable and transportable apparatus and its connections

Portable electrical apparatus should only be permitted in any hazardous area in the most exceptional circumstances which make any other alternative extremely impracticable, and then only if it is of a certified type.

Agricultural and horticultural installations

Any sort of farm or smallholding requiring the use of electricity presents hazards due to dampness, corrosion, mechanical damage, shock risk and fire/explosion risk.

It is therefore important to install wiring systems and apparatus that will reduce these hazards.

Switchgear

Main switchgear should be accessible at all times even in the event of livestock panic and should be located out of reach of livestock.

Most farms and smallholdings have several buildings and/or glasshouses fed from the installation, and each building should therefore have its own control switchgear located in or adjacent to it.

Any items of equipment remote from the main installation should also have control switchgear located close by.

It is very likely with large installations that both three-phase and single-phase circuits will exist, and switchgear and distribution board covers should be marked to show the voltage present.

All points including socket outlets must be switched and switches controlling machinery should have the ON and OFF positions clearly marked.

Cables

Cables are the parts of an installation most susceptible to damage and great care should be taken to choose and position them correctly.

All cables should be kept out of the reach of livestock and clear of vehicles. If a long run is to be placed along the side of a building it should, if possible, be run on the outside and as high as is practicable.

Ideally, only non-metallic conduit should be used, but when the use of metal tubing is unavoidable, only heavy-gauge galvanized conduit should be installed.

When it is necessary to wire between buildings, the cable may be underground at a depth sufficient to avoid damage by farm implements, or overhead, supported by buildings or poles, but not housed in steel conduit or pipes.

PVC or h.o.f.r.-sheathed cables (h.o.f.r. stands for heat-resisting, oil-resisting and flame-retardant) should not be installed where there is a chance of contact with liquid creosote, as this substance will damage the sheathing. Rubber-sheathed cables should be used indoors only in clean, dry situations.

Non-sheathed flexible cords of the twisted or parallel twin types must not be used, nor must cable couplers.

Earthing

In environments such as farms an 'all-insulated' type of installation is preferable. However, if this is not practicable the main regulations on earthing must be strictly observed. In addition, no metallic conduit must be used as the sole CPC, and earthing conductors must be protected against damage.

Electric fences

Mains-operated fence controllers can be a source of danger as there is always a chance that the output could, under fault conditions, be connected to the low-voltage supply. Hence they should not be placed where there is a likelihood of mechanical damage or interference by unauthorized persons. They should not be fixed to poles carrying power or telecommunication lines. They can, however, be fixed to a pole used solely for carrying an insulated supply to themselves.

Any earth electrode associated with a controller must be outside the resistance area of any electrode used for normal protective earthing, and only one controller should be used for a fence system.

When a battery-operated fence controller is being charged, the battery must be disconnected from the controller.

General

In some agricultural situations such as grain-drying areas, there is a risk of explosion from flammable dusts and in such situations all apparatus should be selected very carefully to prevent dangerous conditions from arising.

Edison-type screw lampholders must be of the drip-proof type and persons using wash troughs and sterilizing equipment, etc., must not be able to come into contact with switchgear or non-earthed bonded metalwork.

Finally, testing should be carried out at least once every 3 years.

16 Earthing and bonding

Earth: what it is, and why and how we connect to it

The thin layer of material which covers our planet, be it rock, clay, chalk or whatever, is what we in the world of electricity refer to as earth. So, why do we need to connect anything to it? After all, it is not as if earth is a good conductor.

Perhaps it would be wise at this stage to revise potential difference (p.d.). A potential difference is exactly what it says it is: a difference in potential (volts). Hence two conductors having p.d. of, say, 20 V and 26 V have a p.d. between them of 26 – 20 = 6 V. The original p.d.s, i.e. 20 V and 26 V, are the p.d.s between 20 V and 0 V and 26 V and 0 V.

So where does this 0 V or zero potential come from? The simple answer is, in our case, the earth. The definition of earth is therefore the conductive mass of earth, whose electric potential at any point is conventionally taken as zero.

Hence if we connect a voltmeter between a live part (e.g. the phase conductor of, say, a socket outlet) and earth, we would probably read 240 V; the conductor is at 240 V, the earth at zero. Of course it must be remembered that we are discussing the supply industry in the UK, where earth potential is very important. We would measure nothing at all if we connected our voltmeter between, say, the positive 12 V terminal of a car battery and earth, as in this case the earth plays no part in any circuit. Fig. 16.1 illustrates this difference.

Note the connection of the supply neutral in Fig. 16.1a, to earth, which makes it possible to have complete circuit via the earth. Supply authority neutrals should be at around zero volts, and in order to maintain this condition they are connected to the zero potential of earth.

This also means that a person in an installation touching a live part whilst standing on the earth would take the place of the voltmeter in Fig. 16.1a, and could suffer a severe electric shock. Remember that the accepted *lethal* level of shock current passing through a person is only 50 mA or 1/20 A. The same situation would arise if the person were touching, say, a faulty appliance and a gas or water pipe (Fig. 16.2).

One method of providing some measure of protection against these effects is to join together (bond) all metallic parts and connect them to earth. This ensures that all metalwork in a healthy situation is at or near zero volts, and under fault conditions all metalwork will rise to the same potential. So, simultaneous contact with two such metal parts would not result in a shock,

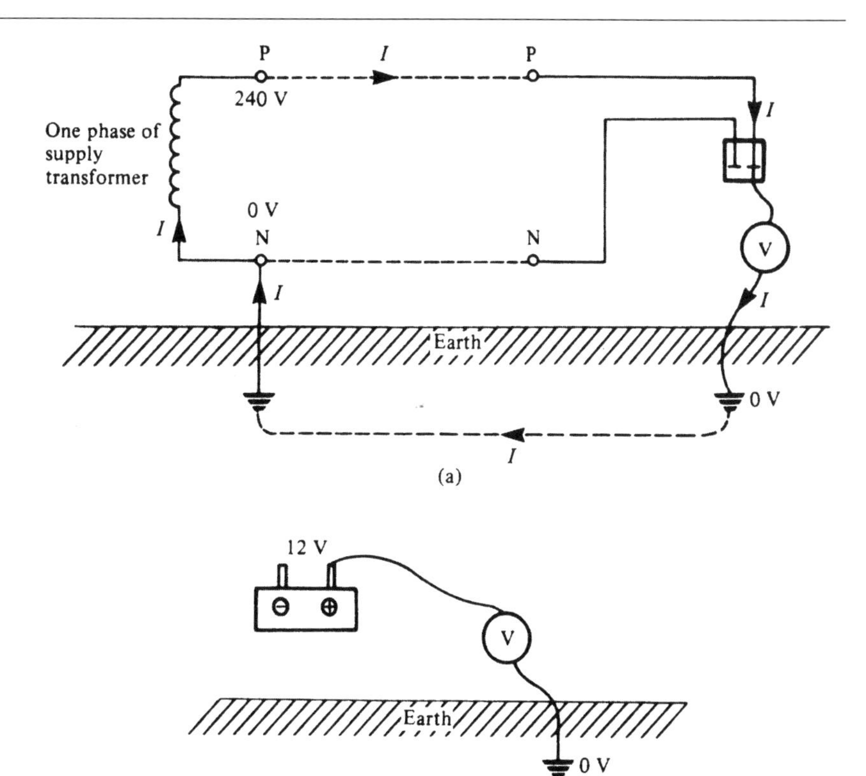

Fig. 16.1

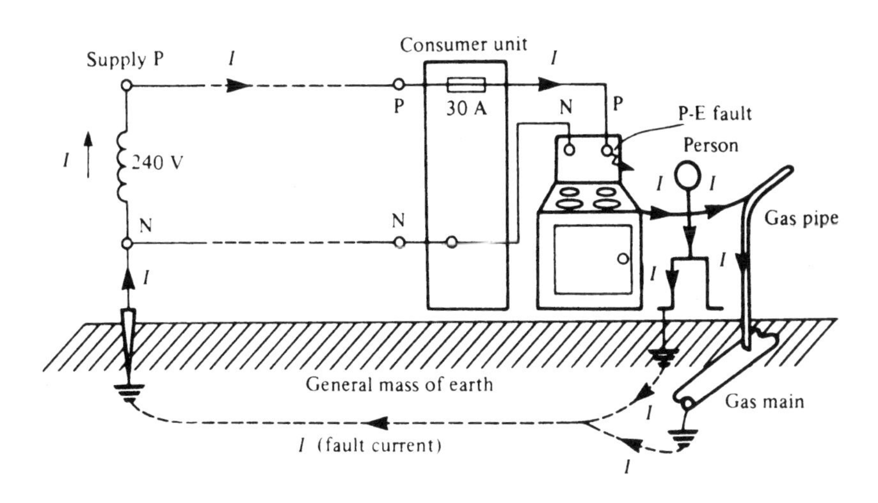

Fig. 16.2

as there will be no p.d. between them. This method is known as earthed equipotential bonding.

Unfortunately, as previously mentioned, earth itself is not a good conductor unless it is very wet, and therefore it presents a high resistance to the flow of fault current. This resistance is usually enough to restrict fault current to a level well below that of the rating of the protective device, leaving a faulty circuit uninterrupted. Clearly this is an unhealthy situation. The methods of overcoming this problem will be dealt with later.

In all but the most rural areas, consumers can be connected to a metallic earth return conductor which is ultimately connected to the earthed neutral of the supply. This, of course, presents a low-resistance path for fault currents to operate the protection.

Summarizing, then, connecting metalwork to earth places that metal at or near zero potential, and bonding between metallic parts puts such parts at the same potential even under fault conditions.

The Faraday cage

In one of his many experiments, Michael Faraday (1791–1867) placed himself in an open-sided cube which was then covered in a conducting material and insulated from the floor. When this cage arrangement was charged to a high voltage, he found that he could move freely within it touching any of the sides, with no adverse effects. He had in fact created an equipotential zone, and of course in a correctly bonded installation we live and/or work in Faraday cages!

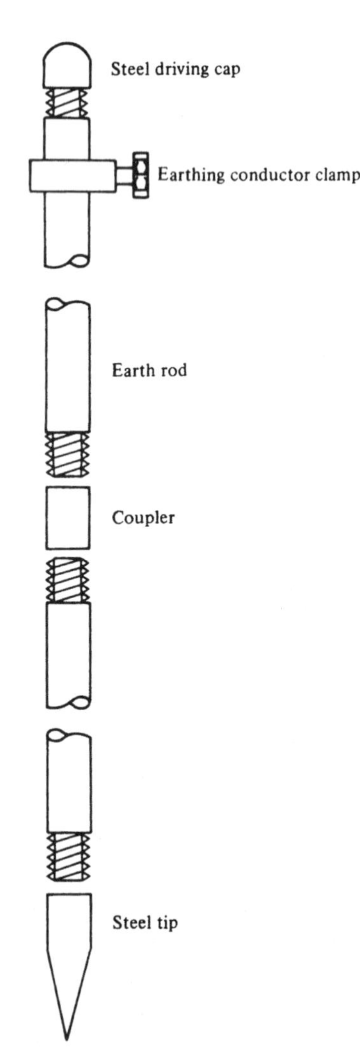

Fig. 16.3

Connecting to earth

In the light of previous comments, it is obviously necessary to have as low an earth path resistance as possible, and the point of connection to earth is one place where such resistance may be reduced. When two conducting surfaces are placed in contact with each other, there will be a resistance to the flow of current dependent on the surface areas in contact. It is clear, then, that the greater surface contact area with earth that can be achieved, the better.

There are several methods of making a connection to earth, including the use of rods, plates and tapes. By far the most popular method in everyday use is the rod earth electrode. The plate type needs to be buried at a sufficient depth to be effective and, as such plates may be 1 or 2 metres square, considerable excavation may be necessary. The tape type is predominantly used in the earthing of large electricity substations, where the tape is laid in trenches in a mesh formation over the whole site. Items of plant are then earthed to this mesh.

Rod electrodes

These are usually of solid copper or copper-clad carbon steel, the latter being used for the larger-diameter rods with extension facilities. These facilities comprise: a thread at each end of the rod to enable a coupler to be used for connection of the next rod; a steel cap to protect the thread from damage when the rod is being driven in; a steel driving tip; and a clamp for the connection of an earth tape or conductor (Fig. 16.3).

The choice of length and diameter of such a rod will, as previously mentioned, depend on the soil conditions. For example, a long thick electrode is used for earth with little moisture retention. Generally, a 1–2 m rod, 16 mm in diameter, will give a relatively low resistance.

Earth electrode resistance

If we were to place an electrode in the earth and then measure the resistance between the electrode and points at increasingly larger distances from it, we would notice that the resistance increased with distance until a point was reached (usually around 2.5 m) beyond which no increase in resistance was seen (Fig. 16.4).

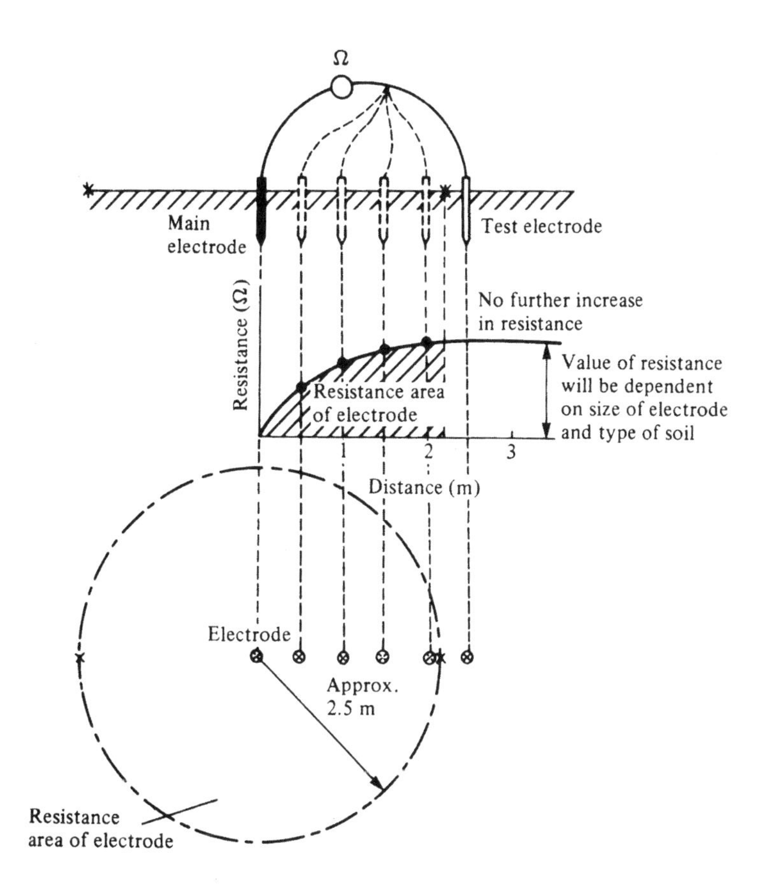

Fig. 16.4 *Resistance area of electrode*

The value of this *electrode resistance* will depend on the length and cross-sectional area of the electrode and the type of soil.

This resistance area is particularly important with regard to voltage at the surface of the ground (Fig. 16.5).

For a 2 m earth rod, with its top at ground level, 80%, to 90% of the voltage appearing at the electrode under fault conditions is dropped across the earth in the first 2.5 to 3 m. This is particularly dangerous where livestock are present as the hind and fore legs of an animal can be respectively inside and outside the resistance area: 25 V can be lethal. This problem can be overcome by ensuring that the whole of the electrode is well below ground level and by providing protection that will operate in a fraction of a second (earth leakage circuit breaker) (Fig. 16.6).

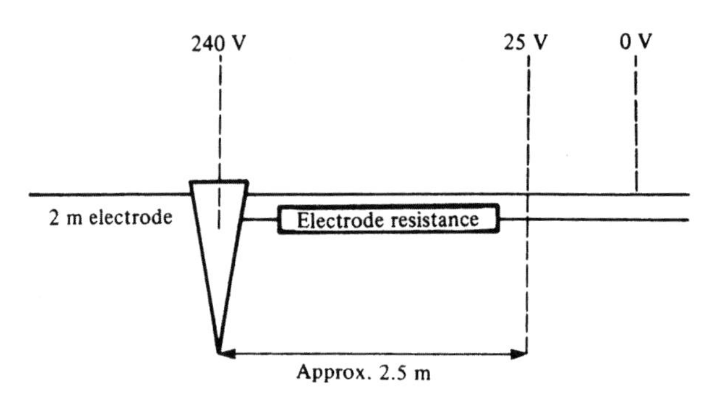

Fig. 16.5

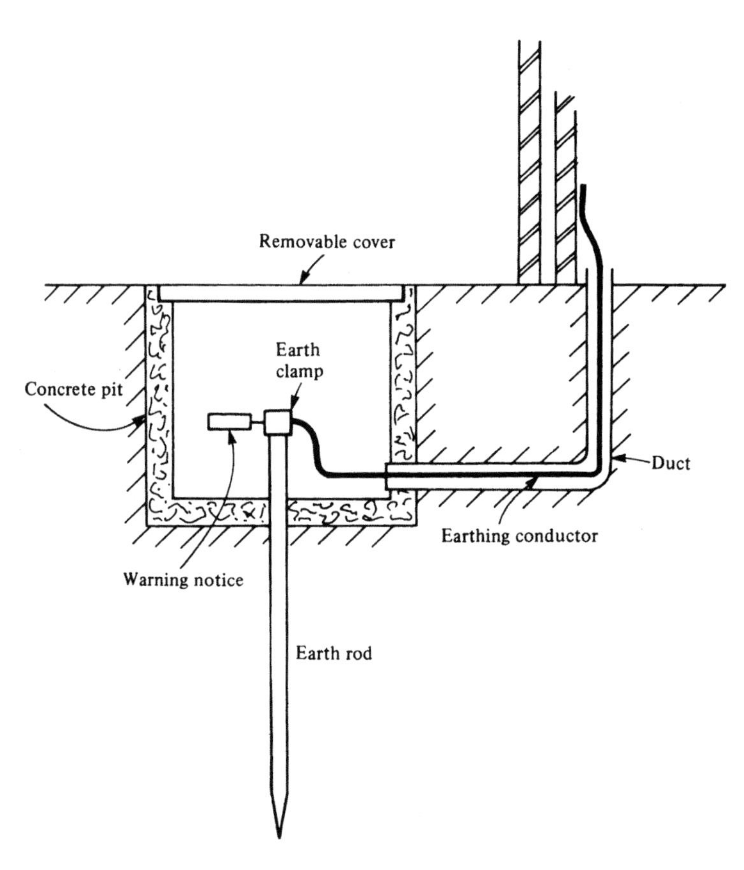

Fig. 16.6

Of course, values of resistance will depend on the type of soil and its moisture content. Any soil that holds moisture such as clay or marshy ground has a relatively low resistivity, whereas gravel or rock has a high resistivity. Typical values for soil resistivity in ohms are as follows:

Garden soil	5 to 50
Clay	10 to 100
Sand	250 to 500
Rock	1000 to 10 000

Earthing systems

Contact with metalwork made live by a fault is called *indirect contact*. One popular method of providing some measure of protection against such contact is by earthed equipotential bonding and automatic disconnection of supply. This entails the bonding together and connection to earth of:

1 All metalwork associated with electrical apparatus and systems, termed exposed conductive parts. Examples include conduit, trunking and the metal cases of apparatus.
2 All metalwork liable to introduce a potential including earth potential, termed extraneous conductive parts. Examples are gas, oil and water pipes, structural steelwork, radiators, sinks and baths.

The conductors used in such connections are called *protective conductors*, and they can be further subdivided into:

1 Circuit protective conductors, for connecting exposed conductive parts to the main earthing terminal.
2 Main equipotential bonding conductors, for bonding together main incoming services, structural steelwork, etc.
3 Supplementary bonding conductors, for bonding together sinks, baths, taps, radiators, etc., and exposed conductive parts in bathrooms and swimming pools.

The effect of all this bonding is to create a zone in which all metalwork of different services and systems will, even under fault conditions, be at a substantially equal potential. If, added to this, there is a low-resistance earth return path, the protection should operate fast enough to prevent danger.

The resistance of such an earth return path will depend upon the system. These systems have been designated in the IEE Regulations using the letters T, N, C and S. These letters stand for:

T terre (French for earth) and meaning a direct connection to earth.
N neutral
C combined
S separate

When these letters are grouped they form the classification of a type of system. The first letter in such a classification denotes how the supply source is earthed. The second denotes how the metalwork of an installation is earthed. The third and fourth indicate the functions of neutral and protective conductors. Hence:

1 A TT system has a direct connection of the supply source to earth and a direct connection of the installation metalwork to earth. An example is an overhead line supply with earth electrodes, and the mass of earth as a return path (Fig. 16.7).
2 A TN–S system has the supply source directly connected to earth, the installation metalwork connected to the earthed neutral of the supply source via the metal sheath of the supply cable, and the neutral and protective conductors throughout the whole system performing separate functions (Fig. 16.8).

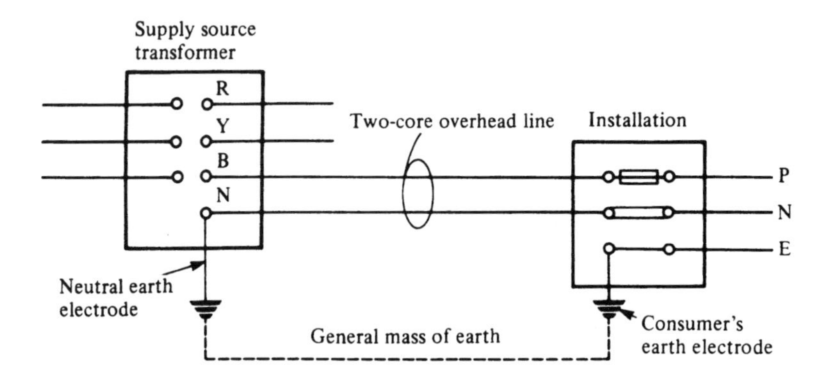

Fig. 16.7 *TT system*

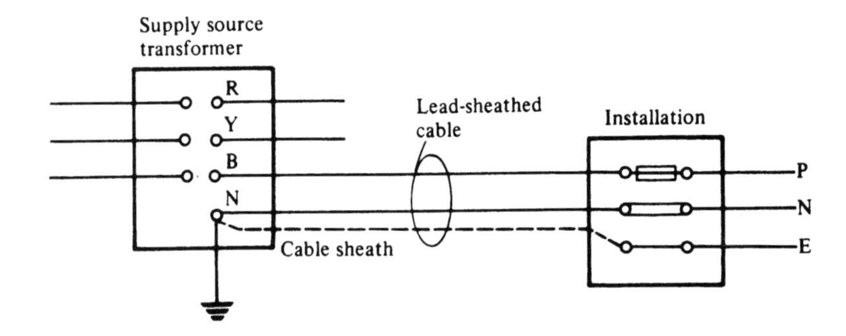

Fig. 16.8 *TN–S system*

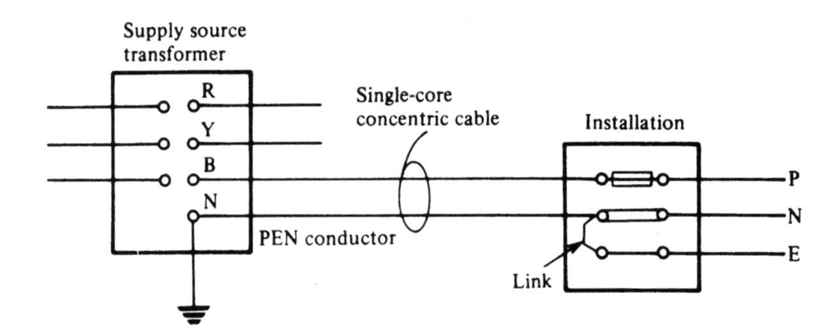

Fig. 16.9 *TN–C–S system*

3 A TN–C–S system is as the TN–S system but the supply cable sheath is also the neutral, i.e. it forms a combined earth/neutral conductor known as a PEN (Protective Earthed Neutral) conductor (Fig. 16.9). The installation earth and neutral are separate conductors. This system is also known as PME (Protective Multiple Earthing).

Note that only single-phase systems have been shown, for simplicity.

With this system (PME system), it is important to ensure that the neutral is kept at earth potential by earthing it at many points along its length (hence 'multiple' earthing). If this is not done, a fault to neutral in one installation could cause a shock risk in all the other installations connected to that system.

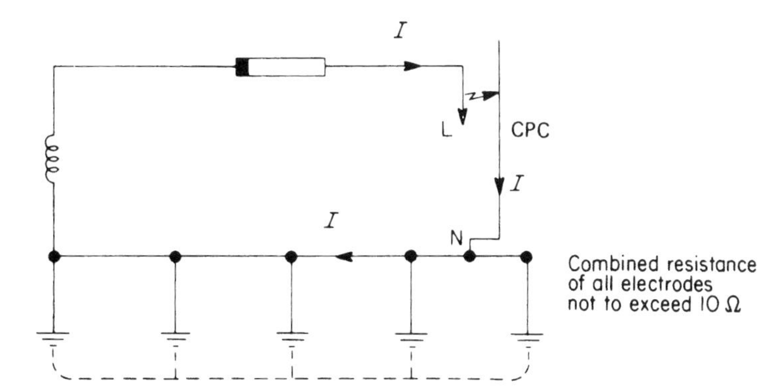

Fig. 16.10 *Protective multiple earthing or TN–C–S system*

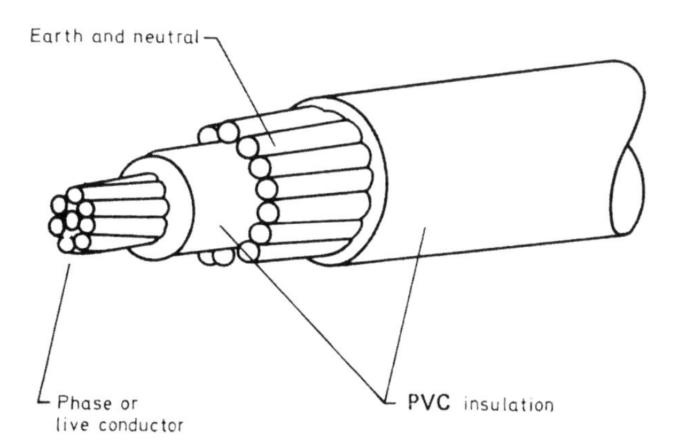

Fig. 16.11 *Single-core concentric cable*

The cable usually used for such a system is concentric cable which consists of a single-core cable (for single phase) surrounded by armouring which is the earth and neutral conductor. Three-core with concentric neutral would be used for three-phase, four-wire cable.

There are, however several hazards associated with the use of a PME system. These include:

1 Shock risk if neutral is broken. In this case, a fault on a PME system with a broken neutral would result in the neutral becoming live to earth either side of the break. This situation is more likely to occur with overhead supplies.
2 Fire risk. As heavy currents are encouraged to flow, there is a risk of fire starting during the time it takes for the protective devices to operate.

The chance of a broken neutral is lessened in underground cable to some extent by the use of concentric cable, as it is unlikely that the neutral conductor in such a cable could be broken without breaking the live conductor.

In view of the hazards of such a system there are strict regulations for its use, and approval from the Department of Trade and Industry must be obtained before it can be installed.

Summary

In order to avoid the risk of serious electric shock, it is important to provide a path for earth leakage currents to operate the circuit protection, and to endeavour to maintain all metalwork at the same potential. This is achieved by bonding together metalwork of electrical and non-electrical systems to earth. The path for leakage currents would then be via the earth itself in TT systems or by a metallic return path in TN–S or TN–C–S systems.

Earth fault loop impedance

As we have seen, circuit protection should operate in the event of a direct fault from phase to earth. The speed of operation of the protection is of extreme importance and will depend on the magnitude of the fault current, which in turn will depend on the impedance of the earth fault loop path.

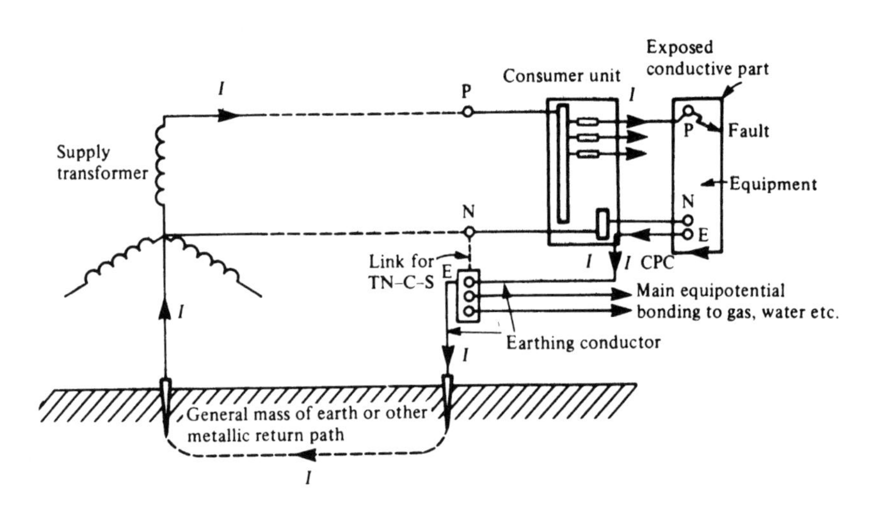

Fig. 16.12

Fig. 16.12 shows this path. Starting at the fault, the path comprises:

1 The circuit protective conductor (CPC).
2 The consumer's earthing terminal and earth conductor.
3 The return path, either metallic or earth.
4 The earthed neutral of the supply transformer.
5 The transformer winding.
6 The phase conductor from the transformer to the fault.

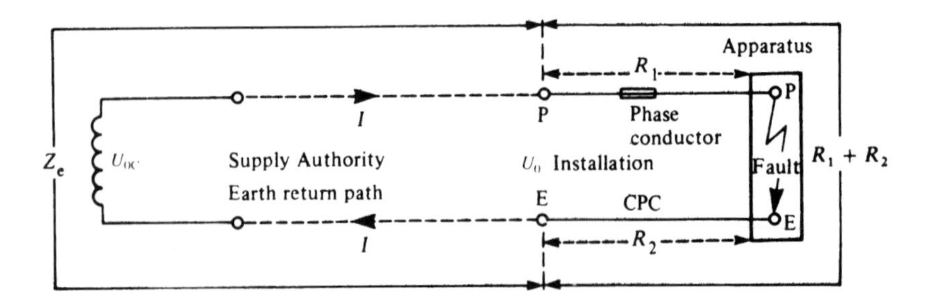

Fig. 16.13

Fig. 16.13 is a simplified version of this path. We have

$$Z_s = Z_e + R_1 + R_2$$

where Z_s is the actual total loop impendance, Z_e is the impedance external to the installation, R_1 is the resistance of the phase conductor, and R_2 is the resistance of the CPC. We also have

$$I = U_{0C}/Z_s$$

where I is the fault current and U_{0C} is the supply transformer open circuit voltage (usually 240 V). U_O is the voltage to earth at consumer terminals.

Determining the value of total loop impedance Z_s

The IEE Regulations require that when the general characteristics of an installation are assessed, the loop impedance Z_e external to the installation shall be ascertained.

This may be measured in existing installations using a phase-to-earth loop impedance tester. However, when a building is only at the drawing board stage it is clearly impossible to make such a measurement. In this case, we have three methods available to assess the value of Z_e:

1 determine it from details (if available) of the supply transformer, the main distribution cable and the proposed service cable; or
2 measure it from the supply intake position of an adjacent building having service cable of similar size and length to that proposed; or
3 use maximum likely values issued by the supply authority as follows:

TT system:	21 ohms maximum
TN–S system:	0.8 ohms maximum
TN–C–S system	0.35 ohms maximum.

Method 1 will be difficult for anyone except engineers. Method 3 can, in some cases, result in pessimistically large cable sizes. Method 2, if it is possible to be used, will give a closer and more realistic estimation of Z_e. However, if in any doubt, use method 3.

Having established a value for Z_e, it is now necessary to determine the impedance of that part of the loop path internal to the installation. This is, as we have seen, the resistance of the phase conductor plus the resistance of the CPC, i.e. $R_1 + R_2$. Resistances of copper conductors may be found from tables in the guidance notes to the Regulations, which give values of resistance/metre for copper and aluminium conductors at 20°C in milliohms/metre.

It should be noted that a copper conductor has a set resistance, no matter by what name it is called. Hence the phase conductor figures given without a CPC size, e.g. 16.00, will also be the value for a 16 mm^2 CPC. This enables us to find $R_1 + R_2$ for non standard arrangements. For example, a 25 mm^2 phase conductor with a 4 mm^2 CPC has $R_1 = 0.727$ and $R_2 = 4.61$, giving $R_1 + R_2 = 0.727 + 4.61 = 5.337$ m Ω/m.

So, having established a value for $R_1 + R_2$ we must now multiply it by the length of the run and divide by 1000 (the values given are in milliohms per metre). However, this final value is based on a temperature of 20°C, but conductor operating temperature is much higher, e.g. for PVC-sheathed cables it is usually 70°C. In order to get an approximate value of resistance at the higher temperature, a multiplier is used.

Hence, for a 20 m length of PVC insulated 16 mm^2 phase conductor with a 4 mm^2 CPC, the value of $R_1 + R_2$ would be

$$R_1 + R_2 = (1.15 + 4.61) \times 20 \times 1.2/1000 = 0.138 \text{ ohms.}$$

The factor 1.2 is the multiplier for PVC insulation.

We are now in a position to determine the total earth fault loop impedance Z_s from

$$Z_s = Z_e + R_1 + R_2$$

As previously mentioned, this value of Z_s should be as low as possible to allow enough fault current to flow to operate the protection as quickly as possible. Tables 41B1, B2 and D of the IEE Regulations give maximum values of loop impedance for different sizes and types of protection for both socket outlet circuits and bathrooms, and circuits feeding fixed equipment.

Provided that the actual values calculated do not exceed those tabulated, socket outlet circuits will disconnect under earth fault conditions in 0.4 s or less, and circuits feeding fixed equipment in 5 s or less. The reasoning behind these different times is based on the time that a faulty circuit can reasonably be left uninterrupted. Hence socket outlet circuits from which hand-held appliances may be used, and bathrooms with their high water content, clearly present a greater shock risk than circuits feeding fixed equipment.

It should be noted that these times, i.e. 0.4 s and 5 s, do not indicate the duration that a person can be in contact with a fault. They are based on the probable chances of someone being in contact with exposed or extraneous conductive parts at the precise moment that a fault develops.

See also Tables 41A, 604A and 605A of the IEE Regulations.

Example

Let us now have a look at a typical example of, say, a shower circuit run in an 18 mm length of 6.0 mm^2 (6242Y) twin cable with CPC, and protected by a 30 A BS 3036 semi-enclosed rewirable fuse. A 6.0 mm^2 twin cable has a 2.5 mm^2 CPC. We will also assume that the external loop impedance Z_e is measured as 0.27 ohms. Will there be a shock risk if a phase-to-earth fault occurs?

The total loop impedance $Z_s = Z_e + R_1 + R_2$. We are given $Z_e = 0.27$ ohms. For a 6.0 mm^2 phase conductor with a 2.5 mm^2 CPC, $R_1 + R_2$ is 10.49 mΩ/m. Hence, with a multiplier of 1.2 for PVC,

$$\text{total } R_1 + R_2 = 18 \times 10.49 \times 1.2/1000 = 0.226 \text{ ohms}$$

Therefore, $Z_s = 0.27 + 0.226 = 0.496$ ohms. This is less than the 1.14 ohms maximum given in Table 41B1 for a 30 A BS 3036 fuse. Hence the protection will disconnect the circuit in less than 0.4 s. In fact it will disconnect in less than 0.1 s, but the determination of this time will be dealt with in Chapter 18.

Residual current devices

We have seen how very important the total earth loop impedance Z_s is in the reduction of shock risk. However, in TT systems where the mass of earth is part of the fault path, the maximum values of Z_s given in Tables 41B1, B2 and D may be hard to satisfy. In addition, climatic conditions will alter the resistance of the earth in such a way that Z_s may be satisfactory in wet weather but not in very dry.

The IEE Regulations recommend therefore that protection for socket outlet circuits in a TT system be achieved by a residual current device (RCD), such that the product of its residual operating current and the loop impedance will not exceed a figure of 50 V. Residual current breakers (RCBs), residual current circuit breakers (RCCBs) and RCDs are one and the same thing.

For construction sites and agricultural environments this value is reduced to 25 V.

Principle of operation of an RCD

Fig. 16.14 illustrates the construction of an RCD. In a healthy circuit the same current passes through the phase coil, the load, and back through the neutral coil. Hence the magnetic effects of phase and neutral currents cancel out.

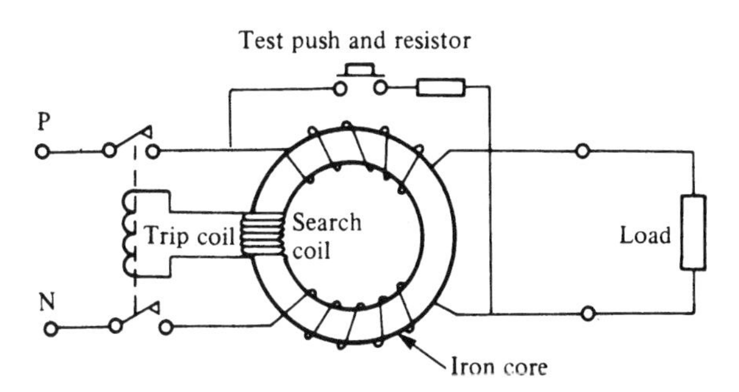

Fig. 16.14

In a faulty circuit, either phase to earth or neutral to earth, these currents are no longer equal; therefore the out-of-balance current produces some residual magnetism in the core. As this magnetism is alternating, it links with the turns of the search coil, inducing an e.m.f. in it. This e.m.f. in turn drives a current through the trip coil, causing operation of the tripping mechanism.

It should be noted that a phase-to-neutral fault will appear as a load, and hence the RCD will not operate for this fault.

The test switch creates an out-of-balance condition which tips the breaker. Its only purpose is to indicate that the breaker is in working order. **It does not check the condition of any part of the earth system.**

Out-of-balance currents as low as 5 mA to 30 mA will be detected, and therefore a person touching unearthed live metalwork would cause the breaker to operate before the lower lethal limit of 50 mA was reached (Fig. 16.15). It is still necessary, however, to ensure that the earth system of an installation is connected to a suitable earth electrode.

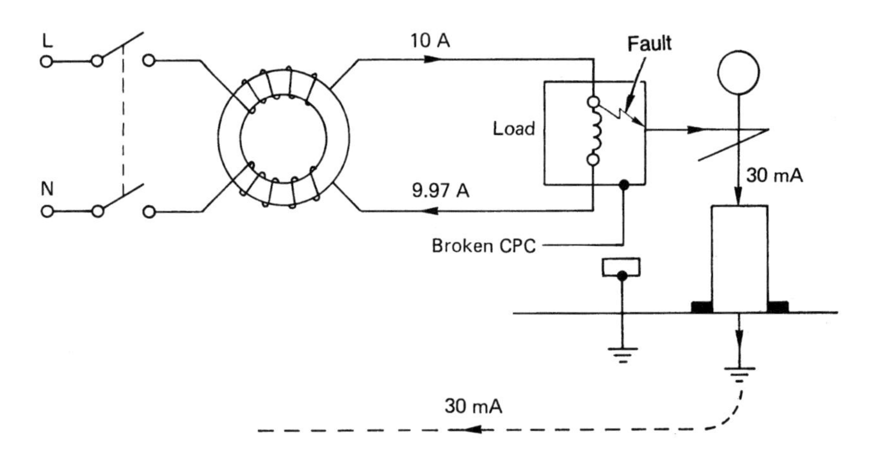

Fig. 16.15

Nuisance tripping

Certain appliances such as cookers, water heaters and freezers tend to have, by the nature of their construction and use, some leakage currents to earth. These are quite normal, but could cause the operation of an RCD protecting an entire installation. This can be overcome by using split-load consumer units, where socket outlet circuits are protected by a 30 mA RCD, leaving all other circuits controlled by a normal mains switch. Better still, especially in TT systems, is the use of a 100 mA RCD for protecting circuits other than socket outlets.

Modern developments in MCB, RCD and consumer unit design now make it easy to protect any individual circuit with a combined MCB/RCD, making the use of split-load boards unnecessary.

One area where the use of 30 mA RCDs is required is in the protection of socket outlets intended for the connection of portable appliances for use outside the main equipotential zone. Hence socket outlets in garages or even within the main premises which are likely to be used for supplying portable tools such as lawn mowers and hedge trimmers must be protected by a 30 mA RCD. All other equipment outside the main equipotential zone should, in the event of an earth fault, disconnect in 0.4 s.

An exception to the RCD requirement is where fixed equipment is connected to the supply via a socket outlet, provided that some means of preventing the socket outlet being used for hand-held appliances is ensured.

Supplementary bonding

This is perhaps the most debated topic in the IEE Regulations. The confusion may have arisen because of a lack of understanding of earthing and bonding. Hopefully, this chapter will rectify the situation.

By now we should know why bonding is necessary; the next question, however, is to what extent bonding should be carried out. This is perhaps answered best by means of question and answer examples:

1 Q: Do I need to bond the hot and cold taps and a metal sink together? Surely they are all joined anyway?

A: There is no requirement in BS 7671 to carry out such bonding.

2 Q: Do I have to bond all the radiators in a premises with a protective conductor run back to the main earth terminal?

A: Supplementary bonding is only necessary when extraneous conductive parts are simultaneously accessible with other extraneous conductive parts and where they are not already connected to the main equipotential bonding by a metallic path of a permanent and reliable nature, which can be an extraneous conductive part. So, even if two radiators are simultaneously accessible with each other or with, say, a metal-clad socket outlet, providing that the pipework feeding them is metal and has soldered connections and the socket outlet is correctly connected to a CPC, there is no need to bond.

3 Q: Do I need to bond metal window frames?

A: In general, no. Apart from the fact that most window frames will not introduce a potential from anywhere, the part of the window most likely to be touched is the opening portion, to which it would not be practicable to bond. There may be a case for the bonding of patio doors, which could be considered earthy with rain running from the lower portion to the earth. However, once again the part most likely to be touched is the sliding section, to which it is not possible to bond. In any case there would need to be another simultaneously accessible part to warrant considering any bonding.

4 Q: What about bonding in bathrooms?

A: Bathrooms are particularly hazardous areas with regard to shock risk, as body resistance is drastically reduced when wet. Hence bonding between the earthing terminals of circuits supplying Class I and Class II equipment and extraneous conductive parts must be carried out. Also of course, taps and metal baths need bonding together, and to other simultaneously accessible extraneous and exposed conductive parts. It may be of interest to note that in older premises a toilet basin may be connected into a cast iron collar which then tees outside into a cast iron soil pipe. This arrangement will clearly introduce earth potential into the bathroom, and hence the collar should be bonded to any simultaneously accessible conductive parts. This may require an unsightly copper earth strap. (For more details see page 298.)

5 Q: What size of bonding conductors should I use?

A: Main equipotential bonding conductors should be not less than half the size of the main earthing conductor, subject to a minimum of 6.0 mm^2 or, where PME (TNCS) conditions are present, 10.0 mm^2. For example, most new domestic installations now have a 16.00 mm^2 earthing conductor, so all main bonding will be in 10.00 mm^2. Supplementary bonding conductors are subject to a minimum of 2.5 mm^2 if mechanically protected or 4.0 mm^2 if not. However, if these bonding conductors are connected to exposed conductive parts, they must be the same size as the CPC connected to the exposed conductive part, once again subject to the minimum sizes mentioned. It is sometimes difficult to protect a bonding conductor mechanically throughout its length, and especially at terminations, so it is perhaps better to use 4.0 mm^2 as the minimum size.

6 Q: Do I have to bond free-standing metal cabinets, screens, workbenches, etc.?

A: No. These items will not introduce a potential into the equipotential zone from outside, and cannot therefore be regarded as extraneous conductive parts.

Points to note

1 An earthing terminal must be provided adjacent to the consumer's terminals. This is usually in the form of a rectangular metal block with cable entries and screws (Fig. 16.16).

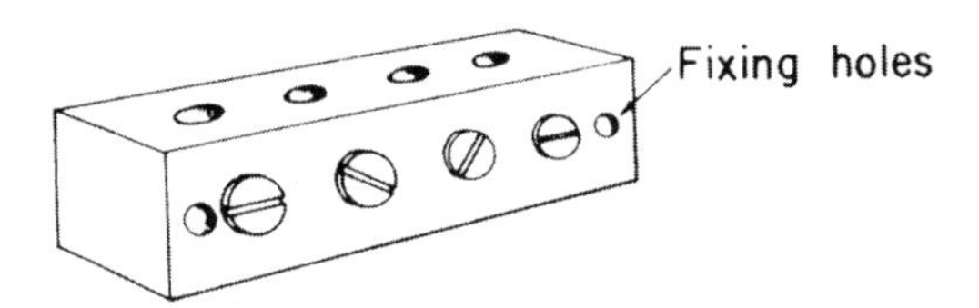

Fig. 16.16 *Consumer's earth terminal*

The earthing conductor from the earth terminal is connected to the cable sheath (if this system is used for earth return) by means of an earth clamp usually of the type shown in (Fig. 16.17).

2 All the exposed conductive parts of wiring systems and apparatus not intended to carry current shall be connected to the appropriate circuit protective conductors. This includes things such as metal boxes for socket outlets and metal casings of fires. There are some exceptions to these requirements, as follows:

(i) Short isolated lengths of metal used for the mechanical protection of cables (conduit used to carry cables overhead between buildings is *not* exempt).

(ii) Metal cable clips.

(iii) Metal lamp caps.

(iv) Metal screws, rivets or nameplates isolated by insulating material.

(v) Metal chains used to suspend luminaires.

(vi) Metal luminaires (such as lampholders) using filament lamps, provided that they are installed above a non-conducting floor and are screened, or positioned so that they cannot be touched by a person able to come into contact with earthed metal.

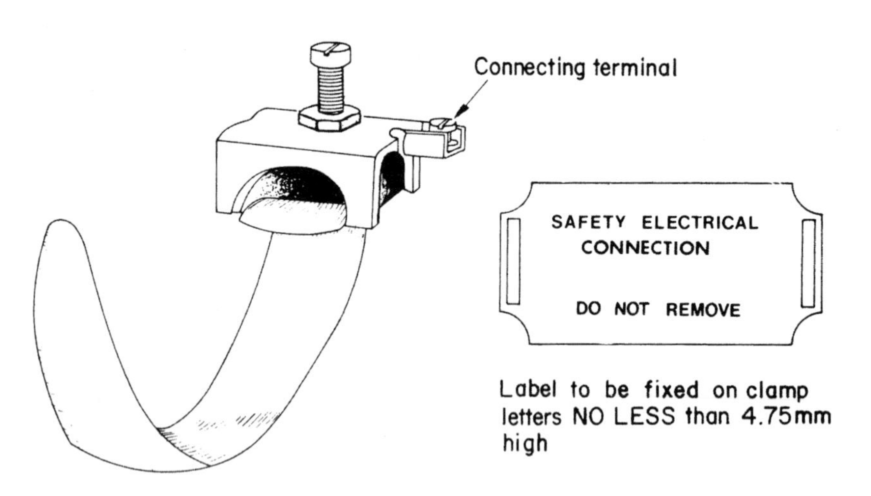

Fig. 16.17 *Earth clamp*

(vii) Catenary wires where insulated hangers are used.
If there is any metalwork in an installation that is likely to come into contact with earthed metal accidentally, then it must be either effectively bonded to, or segregated from, that earthed metal. This includes metal baths, sinks, exposed pipes, radiators, tanks, any structural steelwork that is accessible, and the framework of any mobile equipment such as cranes and lifts which have electrical apparatus fitted. The minimum size of copper bonding lead is $2.5\,mm^2$, with mechanical protection or $4.0\,mm^2$ without.

3 Gas and water services must be bonded to the consumer's earthing terminal as near as possible to the point of entry into the building, and on the consumer's side of the service. It is not permissible to use a gas or water pipe as an earth electrode, the minimum size of copper bonding lead is $6.00\,mm^2$. The bonding may be achieved by using the clamp illustrated in Fig. 16.17.

4 A circuit protective conductor other than copper strip must be insulated throughout its length, and where insulation is removed at terminations (i.e. stripping back twin with earth cable) the resulting bare CPC must be sleeved.

Self-assessment questions

1 What is the resistance of a 10 m length of $6.0\,mm^2$ copper phase conductor if the associated CPC is $1.5\,mm^2$?

2 What is the length of a $6.0\,mm^2$ copper phase conductor with a $2.5\,mm^2$ CPC if the overall resistance is 0.189 ohms?

3 If the total loop impedance of a circuit is 0.96 ohms and the cable is a 20 m length of $4.0\,mm^2$ copper with a $1.5\,mm^2$ CPC, what is the external loop impedance?

4 Will there be a shock risk if a double-socket outlet, fed by a 23 m length of $2.5\,mm^2$ copper conductor with a $1.5\,mm^2$ CPC, is protected by a 20 A BS 3036 rewirable fuse and the external loop impedance is measured as 0.5 ohms?

5 A cooker control unit incorporating a socket outlet is protected by a 30 A BS 3871 type 2 MCB, and wired in $6.0\,mm^2$ copper with a $2.5\,mm^2$ CPC. The run is some 30 m and the external loop impedance of the TN–S system cannot be measured. Is there a shock risk, and if so how could it be rectified?

17 Protection

Protection

What is protection?

The meaning of the word 'protection' as used in the electrical industry, is no different to that in everyday use. People protect themselves against personal or financial loss by means of insurance and from injury or discomfort by the use of the correct protective clothing. They further protect their property by the installation of security measures such as locks and/or alarm systems. In the same way, electrical systems need the following.

1 To be protected against mechanical damage, the effects of the environment and electrical overcurrents.
2 To be installed in such a fashion that persons and/or livestock are protected from the dangers that such an electrical installation may create.

Let us now look at these protective measures in more detail.

Protection against mechanical damage

The word 'mechanical' is somewhat misleading in that most of us associate it with machinery of some sort. In fact a serious electrical overcurrent left uninterrupted for too long can cause distortion of conductors and degradation of insulation; both of these effects are considered to be mechanical damage.

However, let us start by considering the ways of preventing mechanical damage caused by physical impact and the like.

Cable construction

A cable comprises one or more conductors each covered with an insulating material. This insulation provides protection from shock by direct contact and prevents the passage of leakage currents between conductors.

Clearly, insulation is very important and in itself should be protected from damage. This may be achieved by covering the insulated conductors with a protective sheathing during manufacture, or by enclosing them in conduit or trunking at the installation stage.

The type of sheathing chosen and/or the installation method will depend on the environment in which the cable is to be installed. For example, metal conduit with PVC singles or mineral-insulated (m.i.) cable would be used in

Fig. 17.1 *Mineral-insulated cable. On impact, all parts including the conductors are flattened, and a proportionate thickness of insulation remains between conductors, and conductors and sheath, without impairing the performance of the cable at normal working voltages*

preference to PVC sheathed cable clipped direct, in an industrial environment. Fig. 17.1 shows the effect of physical impact on m.i. cable.

Protection against corrosion

Mechanical damage to cable sheaths and metalwork of wiring systems can occur through corrosion, and hence care must be taken to choose corrosion-resistant materials and to avoid contact between dissimilar metals in damp situations.

Protection against thermal effects

This is the subject of Chapter 42 of the IEE Regulations. It basically requires common-sense decisions regarding the placing of fixed equipment, such that surrounding materials are not at risk from damage by heat.

Added to these requirements is the need to protect persons from burns by guarding parts of equipment liable to excessive temperatures.

Polyvinyl chloride

Polyvinyl chloride (PVC) is a thermoplastic polymer widely used in electrical installation work for cable insulation, conduit and trunking. General-purpose PVC is manufactured to the British Standard BS 6746.

PVC in its raw state is a white powder; it is only after the addition of plasticizers and stabilizers that it acquires the form that we are familiar with.

Degradation

All PVC polymers are degraded or reduced in quality by heat and light. Special stabilizers added during manufacture help to retard this degradation at high temperatures. However, it is recommended in the IEE Regulations that PVC-sheathed cables or thermo-plastic fittings for luminaires (light fittings) should not be installed where the temperature is likely to rise above 60°C. Cables insulated with high-temperature PVC (up to 80°C) should be used for drops to lampholders and entries into batten holders. PVC conduit and trunking should not be used in temperatures above 60°C.

Embrittlement and cracking

PVC exposed to low temperatures becomes brittle and will easily crack if stressed. Although both rigid and flexible PVC used in cables and conduit can reach as low as −5°C without becoming brittle, the Regulations recommend that general-purpose PVC-insulated cables should not be installed in areas where the temperature is likely to be consistently below 0°C. They further recommend that PVC-insulated cable should not be handled unless the ambient temperature is above 0°C and unless the cable temperature has been above 0°C for at least 24 hours.

Where rigid PVC conduit is to be installed in areas where the ambient temperature is below –5°C but not lower than –25°C, type B conduit manufactured to BS 4607 should be used.

When PVC-insulated cables are installed in loft spaces insulated with polystyrene granules, contact between the two polymers can cause the plasticizer in the PVC to migrate to the granules. This causes the PVC to harden and although there is no change in the electrical properties, the insulation may crack if disturbed.

Protection against ingress of solid objects and liquid

In order to protect equipment from damage by foreign bodies or liquid, and also to prevent persons from coming into contact with live or moving parts, such equipment is housed inside an enclosure.

The degree of protection offered by such an enclosure is indicated by an index of protection (IP) code, as shown in Table 17.1. It will be seen from this table that, for instance, an enclosure to IP56 is dustproof and waterproof.

Protection against electric shock

There are two ways of receiving an electric shock: by direct contact, and by indirect contact. It is obvious that we need to provide protection against both of these conditions.

Protection against direct contact

Direct contact: This is the contact of persons or livestock with live parts which may result in electric shock.

Clearly, it is not satisfactory to have live parts accessible to touch by persons or livestock. The IEE Regulations recommend five ways of minimizing this danger:

1 By covering the live part or parts with insulation which can only be removed by destruction, e.g. cable insulation.
2 By placing the live part or parts behind a barrier or inside an enclosure providing protection to at least IP2X or IPXXB. In most cases, during the life of an installation it becomes necessary to open an enclosure or remove a barrier. Under these circumstances, this action should only be possible by the use of a key or tool, e.g. by using a screwdriver to open a junction box. Alternatively access should only be gained after the supply to the live parts

Table 17.1 IP codes

First numeral: mechanical protection

0 No protection of persons against contact with live or moving parts inside the enclosure. No protection of equipment against ingress of solid foreign bodies
1 Protection against accidental or inadvertent contact with live or moving parts inside the enclosure by a large surface of the human body, for example a hand, but not protection against deliberate access to such parts. Protection against ingress of large solid foreign bodies
2 Protection against contact with live or moving parts inside the enclosure by fingers. Protection against ingress of medium-size solid foreign bodies
3 Protection against contact with live or moving parts inside the enclosure by tools, wires or such objects of thickness greater than 2.5 mm. Protection against ingress of small foreign bodies
4 Protection against contact with live or moving parts inside the enclosure by tools, wires or such objects of thickness greater than 1 mm. Protection against ingress of small solid foreign bodies
5 Complete protection against contact with live or moving parts inside the enclosure. Protection against harmful deposits of dust. The ingress of dust is not totally prevented, but dust cannot enter in an amount sufficient to interfere with satisfactory operation of the equipment enclosed
6 Complete protection against contact with live or moving parts inside the enclosures. Protection against ingress of dust

Second numeral: liquid protection

0 No protection
1 Protection against drops of condensed water. Drops of condensed water falling on the enclosure shall have no harmful effect
2 Protection against drops of liquid. Drops of falling liquid shall have no harmful effect when the enclosure is tilted at any angle up to 15° from the vertical
3 Protection against rain. Water falling in rain at an angle equal to or smaller than 60° with respect to the vertical shall have no harmful effect
4 Protection against splashing. Liquid splashed from any direction shall have no harmful effect
5 Protection against water jets. Water projected by a nozzle from any direction under stated conditions shall have no harmful effect
6 Protection against conditions on ships' decks (deck with watertight equipment). Water from heavy seas shall not enter the enclosures under prescribed conditions
7 Protection against immersion in water. It must not be possible for water to enter the enclosure under stated conditions of pressure and time
8 Protection against indefinite immersion in water under specified pressure. It must not be possible for water to enter the enclosure
X Indicates no *specified* protection

IPXXB denotes the standard finger

has been disconnected, e.g. by isolation on the front of a control panel where the cover cannot be removed until the isolator is in the 'off' position. An intermediate barrier of at least IP2X or IPXXB will give protection when an enclosure is opened: a good example of this is the barrier inside distribution fuse boards, preventing accidential contact with incoming live feeds.

3 By placing obstacles to prevent unintentional approach to or contact with live parts. This method must only be used where skilled persons are working.

4 By placing out of arm's reach: for example, the high level of the bare conductors of travelling cranes.
5 By using an RCD. Whilst not permitted as the sole means of protection, this is considered to reduce the risk associated with direct contact, provided that one of the other methods just mentioned is applied, and that the RCD has a rated operating current of not more than 30 mA and an operating time not exceeding 40 ms at 150 mA.

Protection against indirect contact

Indirect contact: This is the contact of persons or livestock with exposed conductive parts made live by a fault.

The IEE Regulations suggest five ways of protecting against indirect contact. One of these – earthed equipotential bonding and automatic disconnection of supply – has already been discussed. The other methods are as follows.

Use of class 2 equipment

Often referred to as double-insulated equipment, this is typical of modern DIY tools where there is no provision for the connection of a CPC. This does not mean that there should be no exposed conductive parts and that the casing of equipment should be of an insulating material; it simply indicates that live parts are so well insulated that faults from live to exposed conductive parts cannot occur.

Non-conducting location

This is basically an area in which the floor, walls and ceiling are all insulated. Within such an area there must be no protective conductors, and socket outlets will have no earthing connections.

It must not be possible simultaneously to touch two exposed conductive parts, or an exposed conductive part and an extraneous conductive part. This requirement clearly prevents shock current passing through a person in the event of an earth fault, and the insulated construction prevents shock current passing to earth.

Earth-free local equipotential bonding

This is in essence a Faraday cage, where all metal is bonded together but *not* to earth. Obviously great care must be taken when entering such a zone in order to avoid differences in potential between inside and outside.

The areas mentioned in this and the previous method are very uncommon. Where they do exist, they should be under constant supervision to ensure that no additions or alterations can lessen the protection intended.

Electrical separation

This method relies on a supply from a safety source such as an isolating transformer to BS 3535 which has no earth connection on the secondary side. In the event of a circuit that is supplied from such a source developing a live

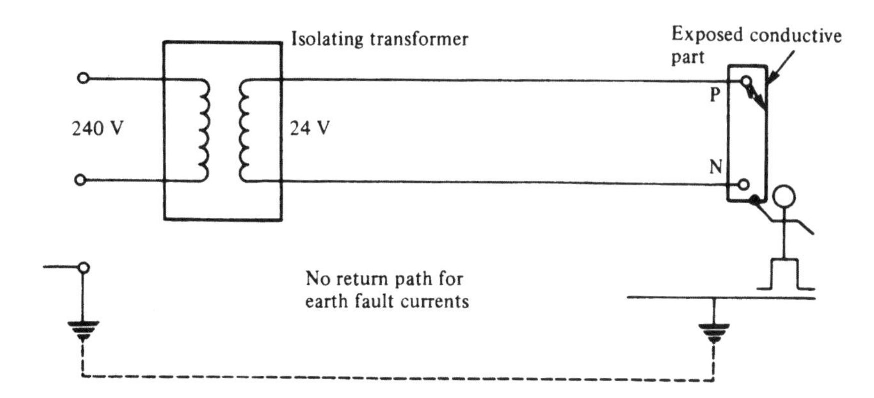

Fig. 17.2

fault to an exposed conductive part, there would be no path for shock current to flow: see Fig. 17.2.

Once again, great care must be taken to maintain the integrity of this type of system, as an inadvertent connection to earth, or interconnection with other circuits, would render the protection useless.

Exemptions

As with most sets of rules and regulations, there are certain areas which are exempt from the requirements. These are listed quite clearly in IEE Regulations and there is no point in repeating them all here. However, one example is the dispensing of the need to earth exposed conductive parts such as small fixings, screws and rivets, provided that they cannot be touched or gripped by a major part of the human body (not less than 50 mm), and that it is difficult to make and maintain an earth connection.

Protection against direct and indirect contact

So far we have dealt separately with direct and indirect contact. However, we can protect against both of these conditions with the following methods.

Separated extra-low voltage (SELV)

This is simply extra-low voltage (less than 50 V a.c.) derived from a safety source such as a class 2 safety isolating transformer to BS 3535; or a motor generator which has the same degree of isolation as the transformer, or a battery or diesel generator, or an electronic device such as a signal generator.

Live or exposed conductive parts of SELV circuits should not be connected to earth, or protective conductors of other circuits, and SELV circuit conductors should ideally be kept separate from those of other circuits. If this is not possible, then the SELV conductors should be insulated to the highest voltage present.

Obviously, plugs and sockets of SELV circuits should not be interchangeable with those of other circuits.

SELV circuits supplying outlets are mainly used for hand lamps or soldering irons, for example in schools and colleges. Perhaps a more common example of a SELV circuit is a domestic bell installation, where the transformer is to BS 3535. Note that bell wire is usually only suitable for 50–60 V, which means that it should *not* be run together with circuit cables of higher voltages.

Reduced voltage systems

The Heath and Safety Executive accepts that a voltage of 65 V to earth, three phase, or 55 V to earth, single phase, will give protection against severe electric shock. They therefore recommend that portable tools used on construction sites etc. be from a 110 V centre-tapped transformer to BS 4343. Fig. 17.3 shows how 55 V is derived.

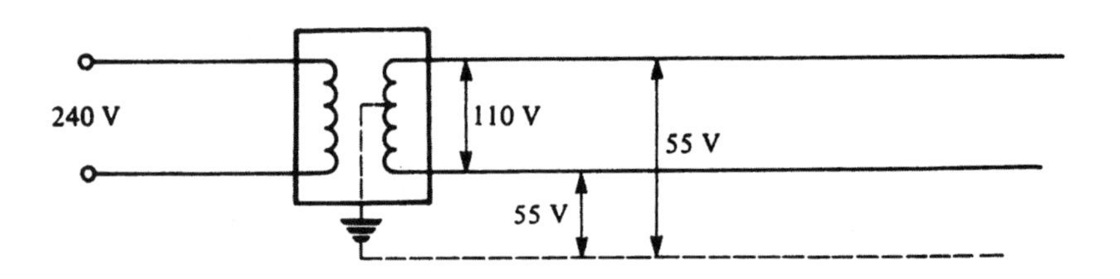

Fig. 17.3

Protection against overcurrent

An overcurrent is a current greater than the rated current of a circuit. It may occur in two ways:

1 As an overload current.
2 As a short circuit or fault current.

These conditions need to be protected against in order to avoid damage to circuit conductors and equipment. In practice, fuses and circuit breakers will fulfil both of these needs.

Overloads

Overheads are overcurrents, occurring in healthy circuits. They may be caused, for example, by faulty appliances or by surges due to motors starting.

Short circuits

A short-circuit is the current that will flow when a 'dead short' occurs between live conductors (phase to neutral for single-phase; phase to phase for three-phase). Prospective short-circuit current is the same, but the term is usually used to signify the value of short-circuit at fuse or circuit breaker positions.

Prospective short-circuit current is of great importance. However, before discussing it or any other overcurrent further, it is perhaps wise to refresh our memories with regard to fuses and circuit breakers and their characteristics.

Fuses and circuit breakers

As we all know, a fuse is the weak link in a circuit which will break when too much current flows, thus protecting the circuit conductors from damage.

It must be remembered that the priority of the fuse is to protect the *circuit conductors*, not the appliance or the user. Calculation of cable size therefore automatically involves the correct selection of protective devices.

There are many different types and sizes of fuse, all designed to perform a certain function. The IEE Regulations refer to only four of these: BS 3036, BS 88, BS 1361 and BS 1362 fuses. It is perhaps sensible to include, at this point, circuit breakers to BS 3871, BS EN 60898 and RCBOs to BS EN 61009, although the BS 3871 MCBs are no longer included in the IEE Regulations.

Fuses

A fuse is simply a device which carries a metal element, usually tinned copper, which will melt and break the circuit when excessive current flows.

There are three types of fuse:

1 the rewirable or semi-enclosed fuse;
2 the cartridge fuse and fuse link; and
3 the high-rupturing-capacity (h.r.c.) fuse.

The rewirable fuse (BS 3036)

A rewirable fuse consists of a fuse, holder, a fuse element and a fuse carrier, the holder and carrier being made of porcelain or Bakelite (Fig. 17.4). The

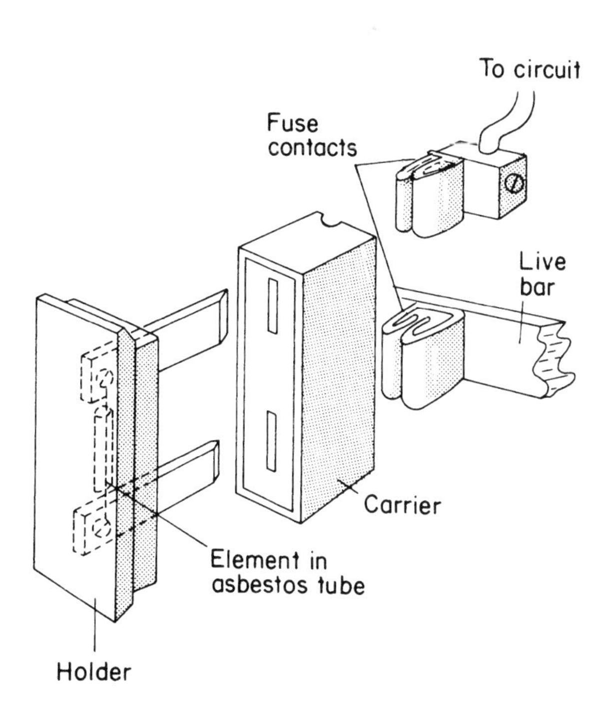

Fig. 17.4 *Typical rewirable fuse assembly*

circuits for which this type of fuse is designed have a colour code which is marked on the fuse holder and is as follows:

45 A – green
30 A – red
20 A – yellow
15 A – blue
5 A – white

Although this type of fuse is very popular in domestic installations, as it is cheap and easy to repair, it has serious disadvantages.

1 The fact that it is repairable enables the wrong size of fuse wire (element) to be used.
2 The elements become weak after long usage and may break under normal conditions.
3 Normal starting-current surges (e.g. when motors etc. are switched on) are 'seen' by the fuse as an overload and will therefore break the circuit.
4 The fuse holder and carrier can become damaged as a result of arcing in the event of a heavy overload.

Cartridge fuse (BS 1361 and BS 1362)

A cartridge fuse consists of a porcelain tube with metal and caps to which the element is attached. The tube is filled with silica (Fig. 17.5).

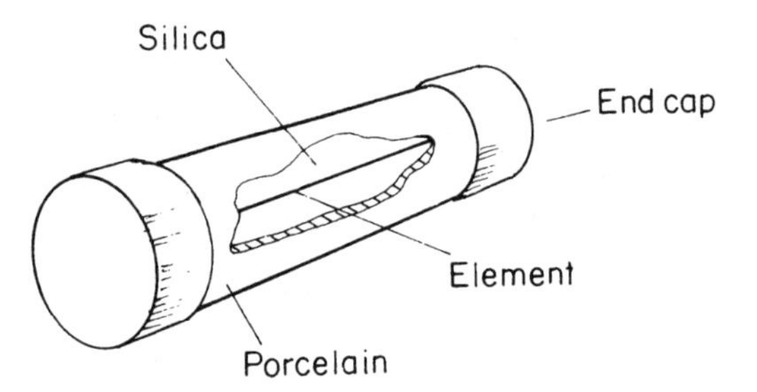

Fig. 17.5 *Cartridge fuse*

These fuses are found generally in modern plug tops used with 13 A socket outlets, and in some distribution boards and at mains intake positions (Electricity Board fuse). They have the advantage over the rewirable fuse of not deteriorating, accuracy in breaking at rated values and not arcing where interrupting faults. They are, however, expensive to replace.

High-rupturing-capacity fuses

The h.r.c. fuse is a sophisticated variation of the cartridge fuse and is normally found protecting motor circuits and industrial installations.

It consists of a porcelain body filled with silica with a silver element and lug type and caps. Another feature is the indicating element which shows when the fuse has blown.

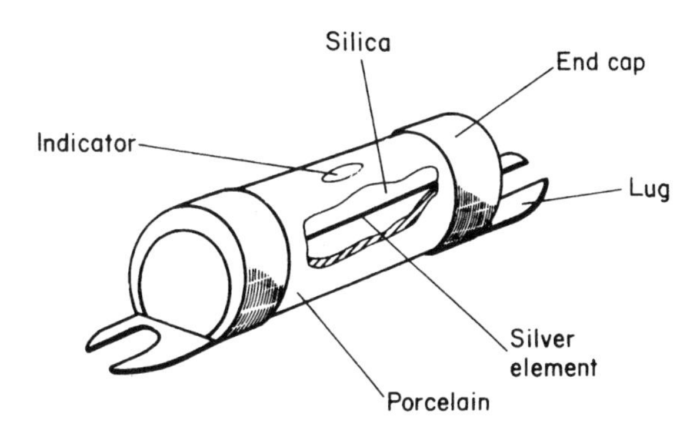

Fig. 17.6 *HRC fuse*

It is very fast-acting and can discriminate between a starting surge and an overload (Fig. 17.6).

Miniature circuit breakers (MCBs)

These protective devices have two elements, one thermal and one electromagnetic. The first looks after overloads and the second, short circuits.

Circuit breakers have one great advantage over the fuse in that, once having operated, they can be reset. They are also very accurate (tripping current) and fast and therefore provide a high degree of discrimination.

Class of protection

It will be evident that each of the protective devices just discussed provides a different level of protection, i.e. rewirable fuses are slower to operate and less accurate than m.c.b.s. In order to classify these devices it is important to have some means of knowing their circuit-breaking and 'fusing' performance. This is achieved for fuses by the use of a fusing factor:

$$\text{Fusing factor} = \frac{\text{fusing current}}{\text{current rating}}$$

where the *fusing current* is the minimum current causing the fuse to blow, and the *current rating* is the maximum current which the fuse can sustain without blowing. For example a 5 A fuse which blows only when 9 A flows will have a fusing factor of 9/5 = 1.8.

Rewirable fuses have a fusing factor of about 1.8.
Cartridge fuses have a fusing factor of between 1.25 and 1.75.
HRC fuses have a fusing factor of up to 1.25 (maximum).
Circuit breakers are designed to operate at no more than 1.5 times their rating.

Breaking capacity of fuses and circuit breakers

When a short circuit occurs, the current may, for a fraction of a second, reach hundreds or even thousands of amperes. The protective device must be able to

Table 17.2 British Standards for fuse links

Standard	Current Rating	Voltage Rating
1 BS 2950	Range 0.05 to 25 A	Range 1000 V (0.05A) to 32 V (25 A) a.c. and d.c.
2 BS 646	1, 2, 3 and 5 A	Up to 250 V a.c. and d.c.
3 BS 1362 cartridge	1, 2, 3, 5, 7, 10 and 13 A	Up to 250 V a.c.
4 BS 1361 HRC cut-out fuses	5, 15, 20, 30 and 45 A 60 A	Up to 250 V a.c.
5 BS 88 motors	Four ranges, 2 to 1200 A	Up to 600 V, but normally 250 or 415 V a.c. and 250 or 500 V d.c.
6 BS 2692	Main range from 5 to 200 A; 0.5 to 3 A for voltage transformer protective fuses	Range from 2.2 kV to 132 kV
7 BS 3036	5, 15, 20, 30, 45, 60, 100, 150 and 200 A	Up to 250 V to earth
8 BS 4265	500 mA to 6.3 A 32 mA to 2 A	Up to 250 V a.c.

British Standards for fuse links (*continued*)

Breaking capacity	Notes
1 Two or three times current rating	Cartridge fuse links for telecommunication and light electrical apparatus. Very low breaking capacity
2 1000 A	Cartridge fuse intended for fused plugs and adapters to BS 546: 'round-pin' plugs
3 6000 A	Cartridge fuse primarily intended for BS 1363: 'flat-pin' plugs
4 16 500 A	Cartridge fuse intended for use in domestic consumer units. The dimensions prevent interchangeability of fuse links which are not of the same current rating
5 Ranges from 10 000 to 80 000 A in four a.c. and three d.c. categories	Part 1 of Standard gives performance and dimensions of cartridge fuse links, whilst Part 2 gives performance and requirements of fuse carriers and fuse bases designed to accommodate fuse links complying with Part 1
6 Ranges from 25 to 750 MVA (main range) 50 to 2500 MVA (VT fuses)	Fuses for a.c. power circuits above 660 V
7 Ranges from 1000 to 12 000 A	Semi-enclosed fuses (the element is a replacement wire) for a.c. and d.c. circuits
8 1500 A (high breaking capacity) 35 A (low breaking capacity)	Miniature fuse links for protection of appliances of up to 250 V (metric standard)

break or make such a current without damage to its surroundings by arcing, overheating or the scattering of hot particles.

Table 17.2 indicates the performance of the more commonly used British Standard fuse links.

The breaking capacity of BS 3871 MCBs is indicated by an 'M' number, i.e. M3–3 KA, M6–6 KA and M9–9 KA.

Fuse and circuit breaker operation

Let us consider a protective device rated at, say, 10 A. This value of current can be carried indefinitely by the device, and is known as its nominal setting I_n. The value of the current which will cause operation of the device, I_2, will be larger than I_n, and will be dependent on the device's *fusing factor*. This is a figure which, when multiplied by the nominal setting I_n, will indicate the value of operating current I_2.

For fuses to BS 88 and BS 1361 and circuit breakers to BS 3871, this fusing factor is approximately 1.45; hence our 10 A device would not operate until the current reached $1.45 \times 10 = 14.5$ A. The IEE Regulations require co-ordination between conductors and protection when an overload occurs, such that:

1 The nominal setting of the device I_n is greater than or equal to the design current of the circuit I_b ($I_n \geqslant I_b$).
2 The nominal setting I_n is less than or equal to the lowest current-carrying capacity I_z, of any of the circuit conductors ($I_n \leq I_z$).
3 The operating current of the device, I_2, is less than or equal to 1.45 I_z ($I_2 \leq 1.45\ I_z$).

So, for our 10 A device, if the cable is rated at 10 A then condition 2 is satisfied. Since the fusing factor is 1.45, condition 3 is also satisfied: $I_2 = I_n \times 1.45 = 10 \times 1.45$, which is also 1.45 times the 10 A cable rating.

The problem arises when a BS 3036 semi-enclosed rewirable fuse is used, as it may have a fusing factor of as much as 2. In order to comply with condition 3, I_n should be less than or equal to $0.725 I_z$. This figure is derived from $1.45/2 = 0.725$. For example, if a cable is rated at 10 A, then I_n for a BS 3036 should be less than or equal to $0.725 \times 10 = 7.25$ A. As the fusing factor is 2, the operating current $I_2 = 2 \times 7.25 = 14.5$, which conforms with condition 3, i.e. $I_2 \leqslant 1.45 \times 120 = 14.5$.

All of these foregoing requirements ensure that conductor insulation is undamaged when an overload occurs.

Under short-circuit conditions it is the conductor itself that is susceptible to damage and must be protected. Fig. 17.7 shows one half-cycle of short-circuit current if there were no protection.

The r.m.s. value ($0.7071 \times$ maximum value) is called the prospective short-circuit current. The cut-off point is where the short-circuit current is interrupted and an arc is formed; the time t_1 taken to reach this point is called the pre-arcing time. After the current has been cut off, it falls to zero as the arc is being extinguished. The time t_2 is the total time taken to disconnect the fault.

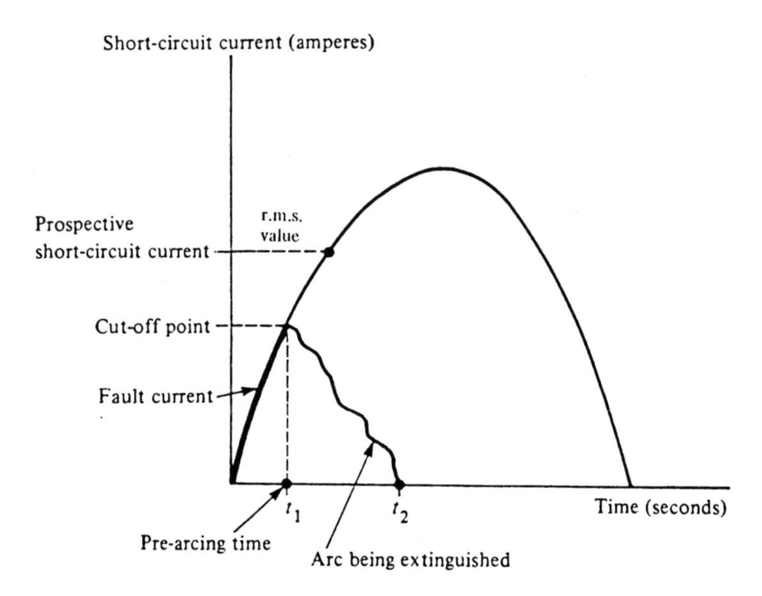

Fig. 17.7

During the time t_1, the protective device is allowing energy to pass through to the load side of the circuit. This energy is known as the pre-arcing let-through energy and is given by $I_f^2\, t_1$, where I_f is the short-circuit current. The total let-through energy from start to disconnection of the fault is given by $I_f^2\, t_2$ (see Fig. 17.8).

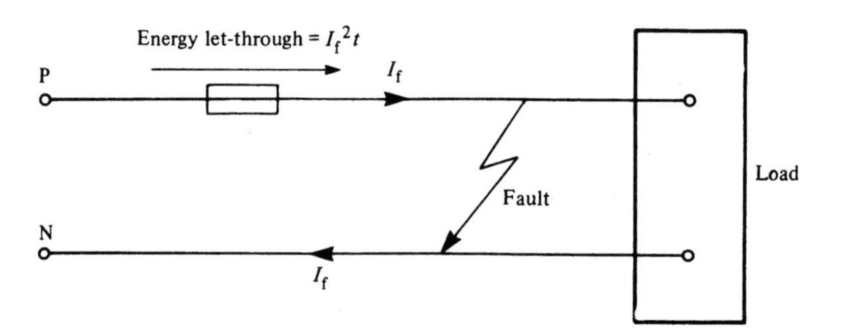

Fig. 17.8

For faults of up to 5 s duration, the amount of heat energy that a cable can withstand is given by k^2s^2, where s is the cross-sectional area of the conductor and k is a factor dependent on the conductor material. Hence the let-through energy should not exceed k^2s^2, i.e. $I_f^2t = k^2s^2$. If we transpose this formula for t, we get $t = k^2s^2/I_f^2$, which is the maximum disconnection time in seconds.

Remember that these requirements refer to short-circuit currents only. If in fact the protective device has been selected to protect against overloads and has a breaking capacity not less than the prospective short-circuit current I_p at the point of installation, it will also protect against short-circuit currents. However, if there is any doubt the formula should be used.

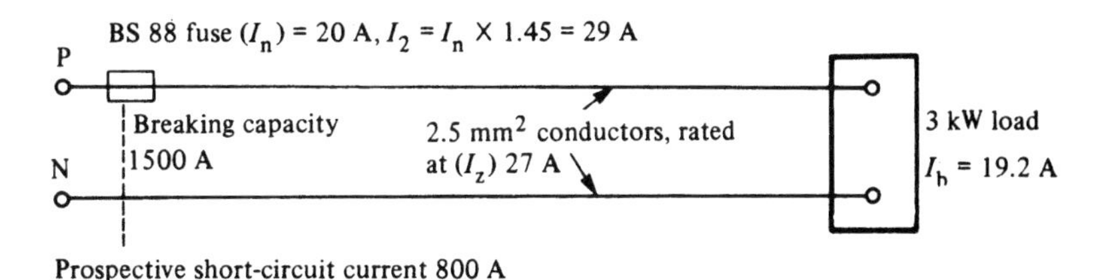

Fig. 17.9

For example in Fig. 17.9, if I_n has been selected for overload protection, the questions to be asked are as follows:

1 Is $I_n \geqslant I_b$? Yes.
2 Is $I_n \leqslant I_z$? Yes.
3 Is $I_2 \leqslant 1.45I_z$? Yes.

Then if the device has a rated breaking capacity not less than I_p, it can be considered to give protection against short-circuit current also.

When an installation is being designed, the prospective short-circuit current at every relevant point must be determined, by either calculation or measurement. The value will decrease as we move farther away from the intake position (resistance increases with length). Thus if the breaking capacity of the lowest rated fuse in the installation is greater than the prospective short-circuit current at the origin of the supply, there is no need to determine the value except at the origin.

Discrimination

Where more than one fuse protects a circuit (Fig. 17.10), it is clearly sensible that the correct fuse should blow under fault conditions. A fault on the appliance should cause fuse C to blow. If fuse B blew, although it would break the circuit to the faulty appliance, it would unnecessarily render the whole radial circuit dead. If fuse A blew, instead of B or C, all circuits from the distribution would be pointlessly disconnected.

The arrangement of fuses to protect the correct part of a circuit is called *discrimination*.

Simply because protective devices have different ratings, it cannot be assumed that discrimination is achieved. This is especially the case where a mixture of different types of device is used. However, as a general rule a 2:1 ratio with the lower-rated devices will be satisfactory. Table 17.3 shows how fuse links may be chosen to ensure discrimination.

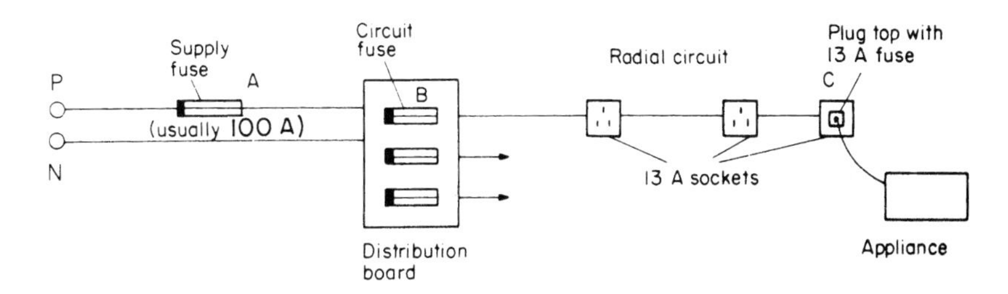

Fig. 17.10

Table 17.3 $I_f^2 t$ characteristics: 2–800 A fuse links. Discrimination is achieved if the total $I_f^2 t$ of the minor fuse does not exceed the pre-arcing $I_f^2 t$ of the major fuse

Rating (A)	$I_f^2 t$ pre-arcing	$I_f^2 t$ total at 415 V
2	0.9	1.7
4	4	12
6	16	59
10	56	170
16	190	580
20	310	810
25	630	1700
32	1200	2800
40	2 000	6000
50	3600	11 000
63	6500	14 000
80	13 000	36 000
100	24 000	66 000
125	34 000	120 000
160	80 000	260 000
200	140 000	400 000
250	230 000	560 000
315	360 000	920 000
350	550 000	1 300 000
400	800 000	2 300 000
450	700 000	1400 000
500	900 000	1 800 000
630	2 200 000	4 500 000
700	2 500 000	5 000 000
800	4 300 000	10 000 000

Fuses will give discrimination if the figure in column 3 does not exceed the figure in column 2. Hence:

a 2 A fuse will discriminate with a 4 A fuse.
a 4 A fuse will discriminate with a 6 A fuse.
a 6 A fuse will *not* discriminate with a 10 A fuse.
a 10 A fuse will discriminate with a 16 A fuse.

All other fuses will *not* discriminate with the next highest fuse, and in some cases, several sizes higher are needed, e.g. a 250 A fuse will only discriminate with a 400 A fuse.

Position of protective devices

When there is a reduction in the current-carrying capacity of a conductor, a protective device is required. There are, however, some exceptions to this requirement; these are listed quite clearly in the IEE Regulations. As an example, protection is not needed in a ceiling rose where the cable size changes from 1.0 mm^2 to say, 0.5 mm^2 for the lampholder flex. This is permitted as it is not expected that lamps will cause overloads.

Protection against undervoltage

From the point of view of danger in the event of a drop or loss of voltage, the protection should prevent automatic restarting of machinery etc. In fact such protection is an integral part of motor starters in the form of the control circuit.

The essential part of a motor control circuit that will ensure undervoltage protection is the 'hold-on' circuit (Fig. 17.11) (see also p. 286).

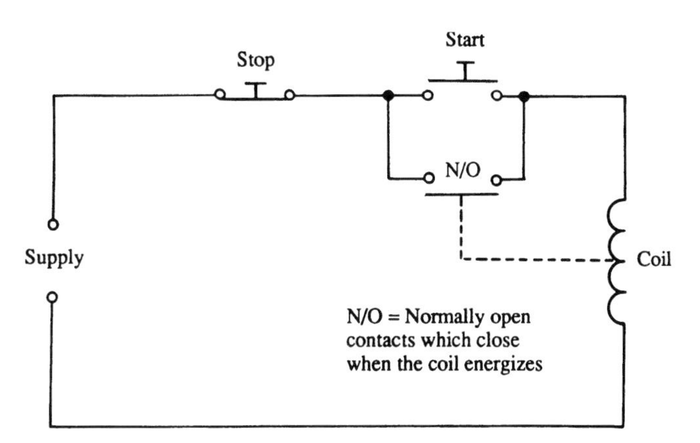

Fig. 17.11 *Hold-on circuit*

When the start button is pushed, the coil becomes energized and its normally open (N/O) contacts close. When the start button is released the coil remains energized via its own N/O contacts. These are known as the 'hold-on' contacts.

The coil can only be de-energized by opening the circuit by the use of the stop button or by a considerable reduction or loss of voltage. When this happens, the N/O contacts open, and, even if the voltage is restored or the circuit is made complete again, the coil will remain de-energized until the start button is pushed again. Fig. 17.12 shows how this 'hold-on' facility is built into a typical single-phase starter.

Control

Having decided how we are going to earth an installation, and settled on the method of protecting persons and livestock from electric shock, and conductors and insulation from damage, we must now investigate the means of controlling the installation. In simple terms, this means the switching of the installation or any part of it 'on' or 'off'. The IEE Regulations refer to this topic as 'isolation and switching'.

Isolation and switching

By definition, isolation is the cutting off of electrical energy from every source of supply, and this function is performed by a switch, a switch fuse or a fuse switch.

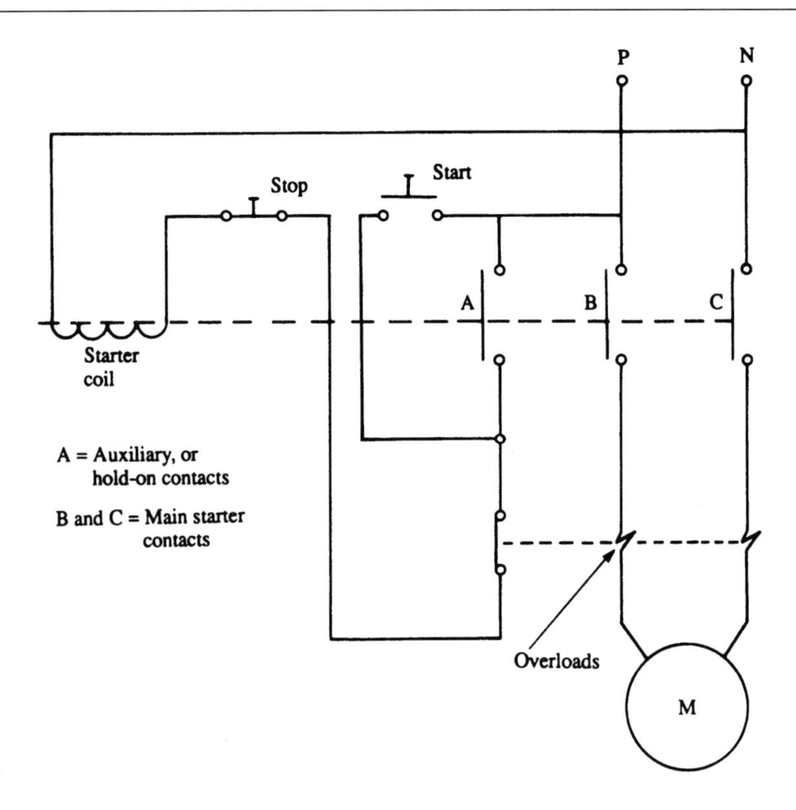

Fig. 17.12 *Single-phase motor starter*

Isolator

This is simply a double- or triple-pole switch in which the moving switch contacts are mechanically linked. In this way both live and neutral or all phases in a three-phase system are disconnected from the supply (Fig. 17.13).

Switch fuse

This is an extension of the isolator, in that the load side of the supply is interrupted by a fuse in the phase conductor.

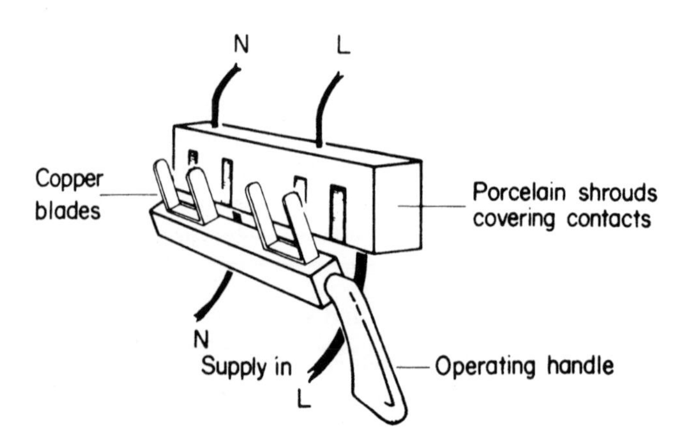

Fig. 17.13 *Internal arrangement of single-phase isolator*

Fuse switch

Fuse switches are used on three-phase systems. Hence the fuse forms part of the moving operating blade.

Consumer unit

The consumer unit is found in nearly all domestic installations. It consists of a double-pole isolator and a distribution board in one assembly.

Circuit breaker

The means of isolation must be double or triple pole. Some modern consumer units have a residual current device installed in place of the usual isolator.

Sequence of control

Fig. 17.14a–d illustrates some typical control sequences.

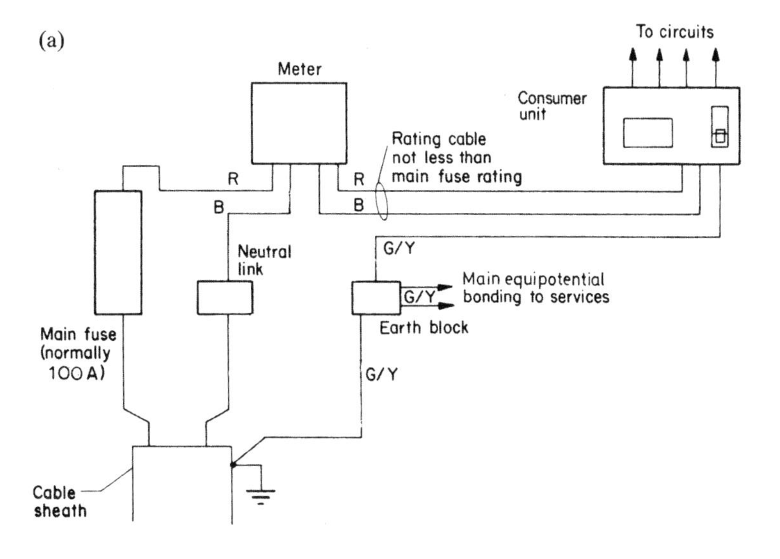

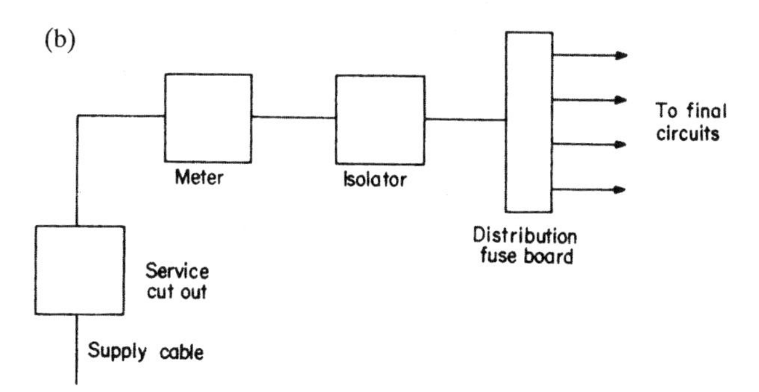

Fig. 17.14a and b

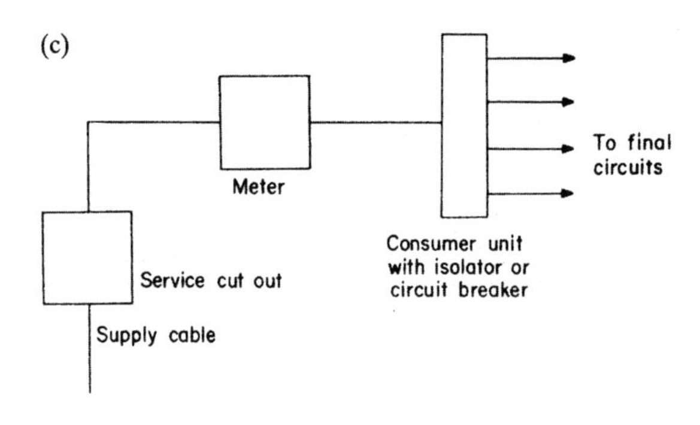

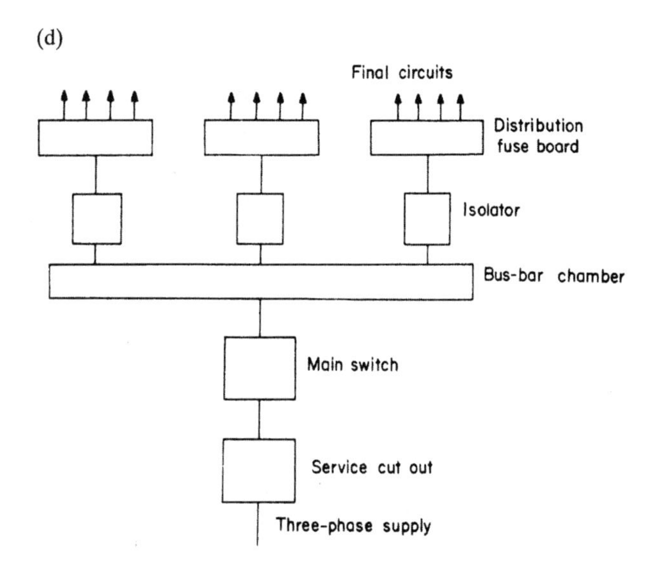

Fig. 17.14c and d

With a domestic installation, the main switch in a consumer unit is considered to be a means of isolation for the whole installation, and each fuse or circuit breaker to be isolators for the individual circuits. Ideally all of these devices should have some means of preventing unintentional re-energization, either by locks or by interlocks. In the case of fuses and circuit breakers, these can be removed and kept in a safe place.

In many cases, isolating and locking off come under the heading of switching off for mechanical maintenance. Hence a switch controlling a motor circuit should have, especially if it is remote from the motor, a means of locking in the 'off' position (Fig. 17.15).

A one-way switch controlling a lighting point is a functional switch, but could be considered as a means of isolation, or a means of switching off for mechanical maintenance (changing a lamp). A two-way switching system, however, does not provide a means of isolation, as neither switch cuts off electrical energy from all sources of supply.

In an industrial or workshop environment it is important to have a means of cutting off the supply to the whole or parts of the installation in the event of an emergency. The most common method is the provision of stop buttons

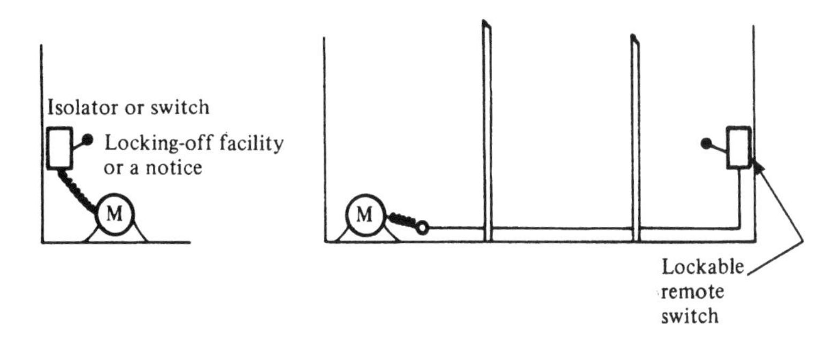

Fig. 17.15

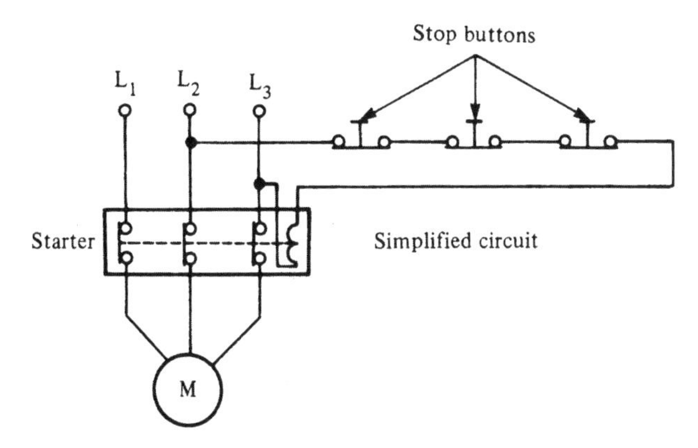

Fig. 17.16

suitably located and used in conjunction with a contactor or relay (Fig. 17.16).

Pulling a plug from a socket to remove a hazard is not permitted as a means of emergency switching. It is, however, allowed as a means of functional switching, e.g. switching off a hand lamp by unplugging.

Whilst we are on the subject of switching, it should be noted that a switch controlling discharge lighting (this includes fluorescent fittings) should, unless it is specially designed for the purpose, be capable of carrying at least twice the steady load of the circuit. The reason for this is that discharge lighting contains chokes which are highly inductive and cause arcing at switch contacts. The higher rating of the switch enables it to cope with such arcing.

Points to note

1 Every consumer's installation must incorporate a means of overcurrent protection which should comprise either a fuse or circuit breaker inserted in each phase conductor of the supply.
2 Every method of overcurrent protection should be based on the prospective short-circuit current (i.e. the current that would flow in the event of a dead short between phases on the load side of the protection).
3 Fuses and circuit breaker must be arranged in an installation such that discrimination is ensured.

4 Protective devices must be inserted only in the phase conductor of a two-wire system in which one wire is connected to earth (i.e. neutral is earthed at the supply transformer).
5 Every circuit conductor must be protected by a fuse or circuit breaker installed at the origin of the circuit. The protective device must have a setting not less than the design current of the circuit (e.g. a 5 A lighting circuit is protected by no less than 5 A protection). Also, the setting should not exceed the lowest current-carrying capacity of any conductor in the circuit, and the value at which the protection operates must not exceed 1.45 times the lowest current-carrying capacity of any circuit conductor.
6 Effective and accessible means of isolation must be provided at the origin of the installation to cut off all voltage as may be necessary to prevent danger.
7 When an installation serves two or more detached buildings (i.e. house and detached garage) a means of isolation must be provided in each building.
8 If the purpose of a switch or circuit breaker is not clear, it should be labelled to show which apparatus it controls.

18 Circuit and design

Design procedure

The requirements of IEE Regulations make it clear that circuits must be designed and the design data made readily available. In fact this has always been the case with previous editions of the Regulations, but it has not been so clearly indicated.

How then do we begin to design? Clearly, plunging into calculations of cable size is of little value unless the type of cable and its method of installation is known. This in turn will depend on the installation's environment. At the same time, we would need to know whether the supply was single or three phase, the type of earthing arrangements, and so on. Here then is our starting point, and it is referred to in the Regulations, Chapter 3, as 'Assessment of general characteristics'.

Having ascertained all the necessary details, we can decide on an installation method, the type of cable, and how we will protect against electric shock and overcurrents. We would now be ready to begin the calculation part of the design procedure.

Basically there are eight stages in such a procedure. These are the same whatever the type of installation, be it a cooker circuit or a submain cable feeding a distribution board in a factory. Here then are the eight basic steps in a simplified form:

1 Determine the design current I_b.
2 Select the rating of the protection I_n.
3 Select the relevant correction factors (CFs).
4 Divide I_n by the relevant CFs to give tabulated cable current-carrying capacity I_t.
5 Choose a cable size to suit I_t.
6 Check the voltage drop.
7 Check for shock risk constraints.
8 Check for thermal constraints.

Let us now examine each stage in detail.

Design current

In many instances the design current I_b is quoted by the manufacturer, but there are times when it has to be calculated. In that case there are two formulae involved, one for single-phase and one for three-phase:

Single-phase:

$$I_b = \frac{P}{V} \qquad (V \text{ usually } 240\,\text{V although declared at } 230\,\text{V})$$

Three-phase:

$$I_b = \frac{P}{\sqrt{3} \times V_L} \qquad (V_L \text{ usually } 415\,\text{V although declared at } 400\,\text{V})$$

Current is in amperes, and power P in watts.

If an item of equipment has a power factor (PF) and/or has moving parts, efficiency (*eff*) will have to be taken into account. Hence:

Single-phase:

$$I_b = \frac{P \times 100}{V \times \text{PF} \times eff}$$

Three-phase:

$$I_b = \frac{P \times 100}{\sqrt{3} \times V_L \times \text{PF} \times eff}$$

Nominal setting of protection

Having determined I_b we must now select the nominal setting of the protection I_n such that $I_n \geqslant I_b$. This value may be taken from IEE Regulations. Tables 41B1, B2 or D, or from manufacturers' charts. The choice of fuse or MCB type is also important and may have to be changed if cable sizes or loop impedances are too high.

Correction factors

When a cable carries its full-load current it can become warm. This is no problem unless its temperature rises further due to other influences, in which case the insulation could be damaged by overheating. These other influences are: high ambient temperature; cables grouped together closely; uncleared overcurrents; and contact with thermal insulation.

For each of these conditions there is a correction factor (CF) which will respectively be called C_a, C_g, C_f and C_i, and which derates cable current-carrying capacity or conversely increases cable size.

Ambient temperature C_a

The cable ratings in the IEE Regulations are based on an ambient temperature of 30°C, and hence it is only above this temperature that an adverse correction is needed. Table 4C1 of the Regulations gives factors for all types of protection other than BS 3036 semi-enclosed rewirable fuses, which are accounted for in Table 4C2.

Grouping C_g

When cables are grouped together they impart heat to each other. Therefore the more cables there are the more heat they will generate, thus increasing the temperature of each cable. Table 4B of the Regulations gives factors for such groups of cables or circuits. It should be noted that the figures given are for cables of the same size, and hence *no* correction need be made for cables grouped at the outlet of a domestic consumer unit, for example, where there is a mixture of different sizes.

A typical situation where correction factors need to be applied would be in the calculation of cable sizes for a lighting system in a large factory. Here many cables of the same size and loading may be grouped together in trunking and could be expected to be fully loaded all at the same time.

Protection by BS 3036 fuse C_f

As we have already discussed in Chapter 17, because of the high fusing factor of BS 3036 fuses, the rating of the fuse I_n, should be less than or equal to $0.725I_z$. Hence 0.725 is the correction factor to be used when BS 3036 fuses are used.

Thermal insulation C_i

With the modern trend towards energy saving and the installation of thermal insulation, there may be a need to derate cables to account for heat retention.

The values of cable current-carrying capacity given in Appendix 4 of the IEE Regulations have been adjusted for situations when thermal insulation touches one side of a cable. However, if a cable is totally surrounded by thermal insulation for more than 0.5 m, a factor of 0.5 must be applied to the tabulated clipped direct ratings. For less than 0.5 m, derating factors (Table 52A) should be applied.

Application of correction factors

Some or all of the onerous conditions just outlined may affect a cable along its whole length or parts of it, but not all may affect it at the same time. So, consider the following:

1 If the cable in Fig. 18.1 ran for the whole of its length, grouped with others of the same size in a high ambient temperature, and was totally surrounded with thermal insulation, it would seem logical to apply all the CFs, as they

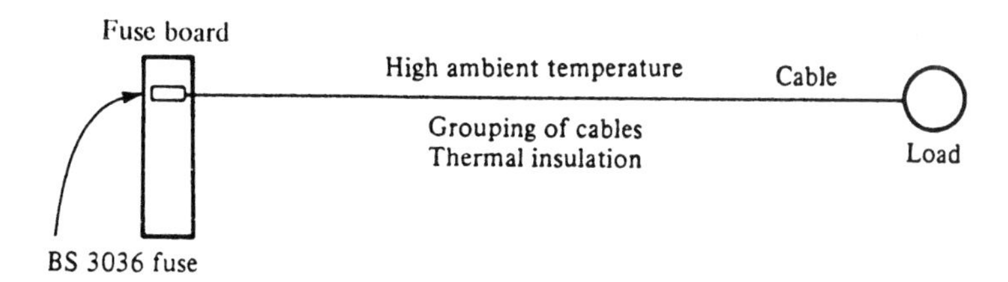

Fig. 18.1

all affect the whole cable run. Certainly the factors for the BS 3036 fuse, grouping and thermal insulation should be used. However, it is doubtful if the ambient temperature will have any effect on the cable, as the thermal insulation, if it is efficient, will prevent heat reaching the cable. Hence apply C_g, C_f and C_i.

2 In Fig. 18.2a the cable first runs grouped, then leaves the group and runs in high ambient temperature, and finally is enclosed in thermal insulation. We therefore have three different conditions, each affecting the cable in different areas. The BS 3036 fuse affects the whole cable run and therefore C_f must be used, but there is no need to apply all of the remaining factors as the worse one will automatically compensate for the others. The relevant factors are shown in Fig. 18.2b: apply only C_f = 0.725 and C_i = 0.5. If protection was *not* by BS 3036 fuse, then apply only C_i = 0.5.

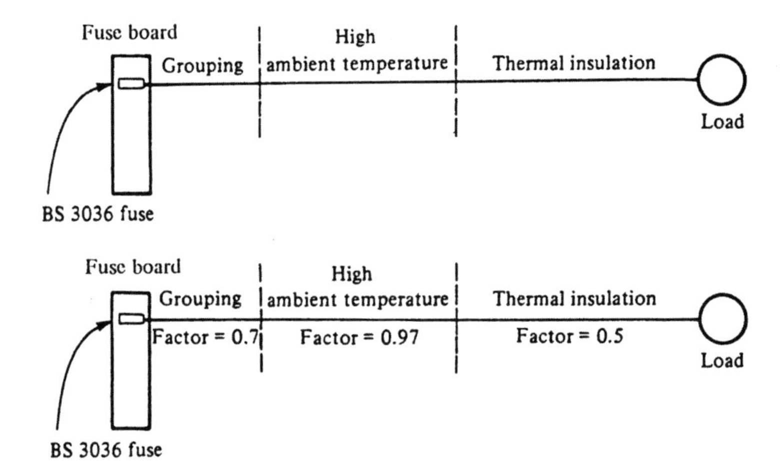

Fig. 18.2

3 In Fig. 18.3 a combination of cases 1 and 2 is considered. The effect of grouping and ambient temperature is 0.7 × 0.97 = 0.69. The factor for thermal insulation is still worse than this combination, and therefore C_i is the only one to be used.

Having chosen the *relevant* correction factors, we now apply them to the nominal rating of the protection I_n as divisors in order to calculate the tabulated current-carrying capacity I_t of the cable.

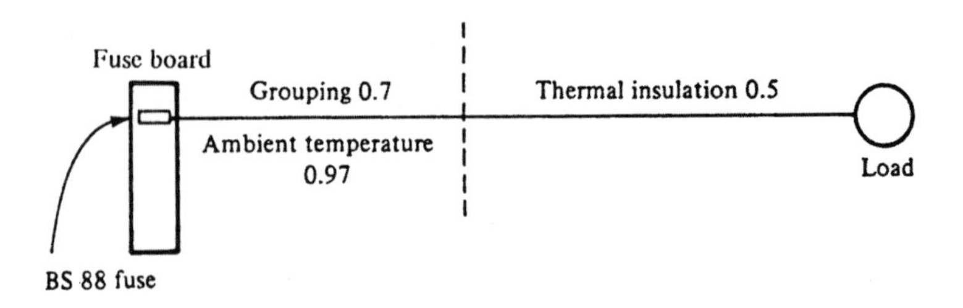

Fig. 18.3

Current-carrying capacity

The required formula for tabulated current-carrying capacity I_t is

$$I_t \geqslant \frac{I_n}{\text{relevant CFs}}$$

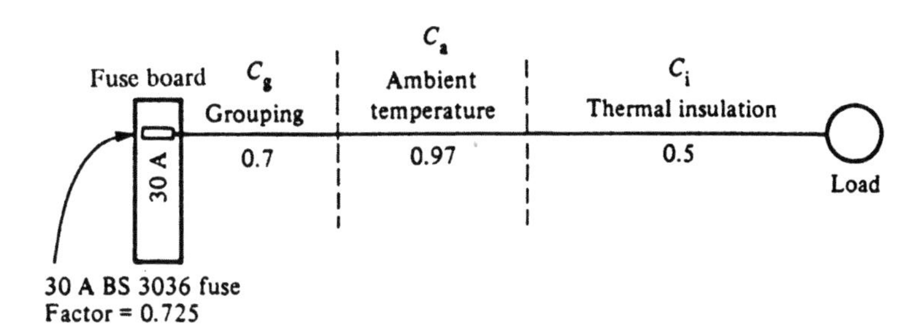

Fig. 18.4

In Fig. 18.4 the current-carrying capacity is given by

$$I_t \geqslant \frac{I_n}{C_f C_i} \geqslant \frac{30}{0.725 \times 0.5} \geqslant 82.75 \text{ A}$$

or, without the BS 3036 fuse.

$$I_t \geqslant \frac{30}{0.5} \geqslant 60 \text{ A}$$

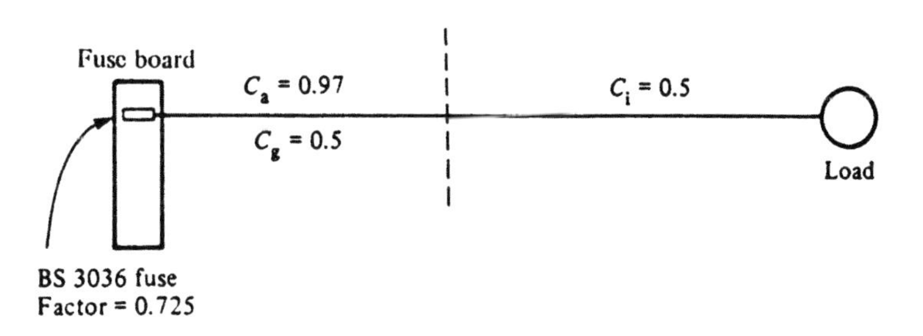

Fig. 18.5

In Fig. 18.5 $C_a\ C_g = 0.97 \times 0.5 = 0.485$, which is worse than C_i (0.5). Hence

$$I_t \geqslant \frac{I_n}{C_f\ C_a\ C_g} \geqslant \frac{30}{0.725 \times 0.485} \geqslant 85.3 \text{ A}$$

or, without the BS 3036 fuse,

$$I_z \geqslant \frac{30}{0.485} \geqslant 61.85 \text{ A}$$

Choice of cable size

Having established the tabulated current-carrying capacity I_t of the cable to be used, it now remains to choose a cable to suit that value. The tables in Appendix 4 of the IEE Regulations list all the cable sizes, current-carrying capacities and voltage drops of the various types of cable. For example, for PVC-insulated singles, single phase, in conduit, having a current-carrying capacity of 45 A, the installation is by reference method 3 (Table 4A), the cable table is 4DIA and the column is 4. Hence the cable size is 10.0 mm^2 (column 1).

Voltage drop

The resistance of a conductor increases as the length increases and/or the cross-sectional area decreases. Associated with an increased resistance is a drop in voltage, which means that a load at the end of a long thin cable will not have the full supply voltage available (Fig. 18.6).

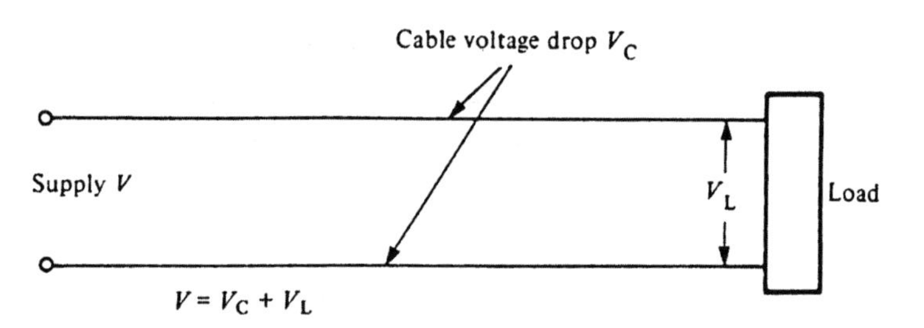

Fig. 18.6

The IEE Regulations require that the voltage drop V_c should not be so excessive that equipment does not function safely. They further indicate that a drop of no more than 4% of the nominal voltage at the *origin* of the circuit will satisfy. This means that:

1 For single-phase 230 V, the voltage drop should not exceed 4% of 230 = 9.2 V.
2 For three-phase 400 V, the voltage drop should not exceed 4% of 400 = 16 V.

For example, the voltage drop on a circuit supplied from a 230 V source by a 16.0 mm two-core copper cable 23 m long, clipped direct and carrying a design current of 33 A, will be

$$V_c = \frac{\text{mV} \times I_b \times L}{1000} \quad \text{(mV from Table 4D2B)}$$

$$= \frac{2.8 \times 33 \times 23}{1000} = 2.125\ \text{V}$$

As we know that the maximum voltage drop in this instance (230 V) is 9.2 V, we can determine the maximum length by transposing the formula:

$$\text{maximum length} = \frac{V_c \times 1000}{\text{mV} \times I_b}$$

$$= \frac{9.2 \times 1000}{2.8 \times 23} = 143\,\text{m}$$

There are other constraints, however, which may not permit such a length.

Shock risk

This topic has already been discussed in full in Chapter 17. To recap, the actual loop impedance Z_s should not exceed those values given in Tables 41B1, B2 and D of the IEE Regulations. This ensures that circuits feeding socket outlets, bathrooms and equipment outside the equipotential zone will be disconnected, in the event of an earth fault, in less than 0.4 s, and that fixed equipment will be disconnected in less than 5 s.

Remember: $Z_s = Z_e + R_1 + R_2$.

Thermal constraints

The IEE Regulations require that we either select or check the size of a CPC against Table 54G, or calculate its size using an adiabatic equation.

Selection of CPC using table 54G, IEE regulations

Table 54G simply tells us that

1 For phase conductors up to and including 16 mm^2, the CPC should be at least the same size.
2 For sizes between 16 mm^2 and 35 mm^2, the CPC should be at least 16 mm.
3 For sizes of phase conductor over 35 mm^2, the CPC should be at least half this size.

This is all very well, but for large sizes of phase conductor the CPC is also large and hence costly to supply and install. Also, composite cables such as the typical twin with CPC 6242Y type have CPCs smaller than the phase conductor and hence do not comply with Table 54G.

Calculation of CPC using the adiabatic equation

The adiabatic equation

$$s = \sqrt{(I^2\,t)}/k$$

enables us to check on a selected size of cable, or on an actual size in a multicore cable. In order to apply the equation we need first to calculate the earth fault current from

$$I = U_{0c}/Z_s$$

where U_{0c} is the open circuit voltage of the supply transformer (usually 240 V) and Z_s is the actual earth fault loop impedance. Next we select a k factor from Tables 54B to F, and then determine the disconnection time t from the relevant curve.

For those unfamiliar with such curves, using them may appear a daunting task. A brief explanation may help to dispel any fears. Referring to any of the curves for fuses in Appendix 3 of the IEE Regulations, we can see that the current scale goes from 2 A to 1200 A, and the time scale from 0.004 S to 10 000 s. One can imagine the difficulty in drawing a scale between 2 A and 1200 A in divisions of 1 A, and so a logarithmic scale is used. This cramps the large scale into a small area. All the numbered subdivisions between the major divisions increase in equal amounts depending on the major division boundaries; for example, all the numbered subdivisions between 100 and 1000 are in amounts of 22 (Fig. 18.7).

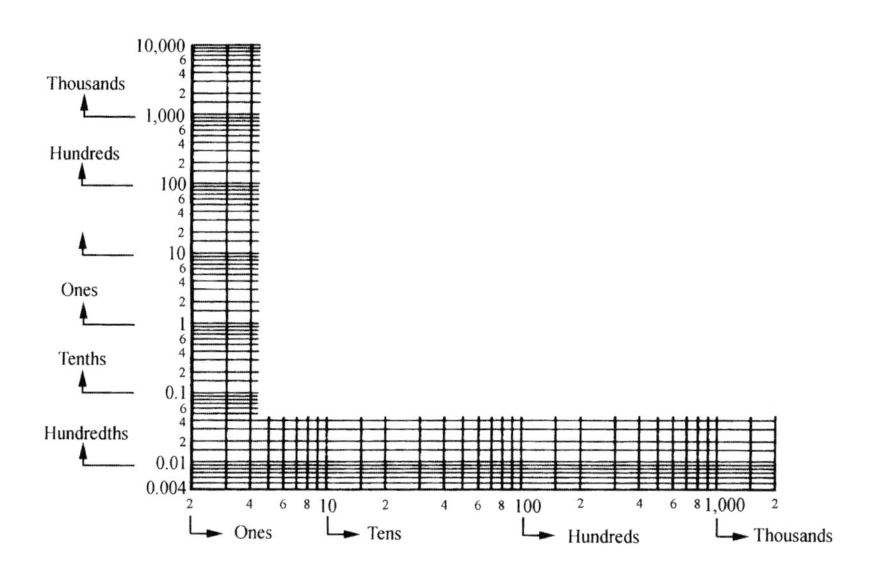

Fig. 18.7

Figs. 18.8 and 18.9 give the IEE Regulations time/current curves for BS 88 fuses. Referring to the appropriate curve for a 32 A fuse (Fig. 18.8), we find that a fault current of 200 A will cause disconnection of the supply in 0.6 s.

Where a value falls between two subdivisions, e.g. 150 A, an estimate of its position must be made. Remember that even if the scale is not visible, it would be cramped at one end; so 150 A would *not* fall half-way between 100 A and 200 A (Fig. 18.10).

It will be noted in Appendix 3 of the Regulations that each set of curves is accompanied by a table which indicates the current that causes operation of the protective device for disconnection times of 0.1 s, 0.4 s and 5 s.

The curves for MCBs to BS 3871 may give some cause for concern, but they are in fact easily explained (Figs. 18.11–18.13). It will be noted that there are two parts to each curve. This is because MCBs provide protection against overload and short-circuit currents, each of which is performed by a different

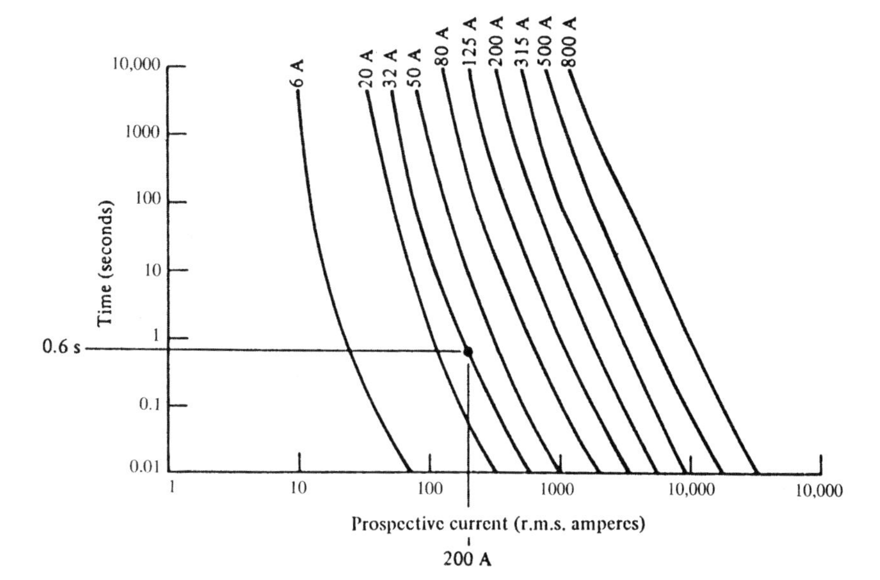

Fig. 18.8

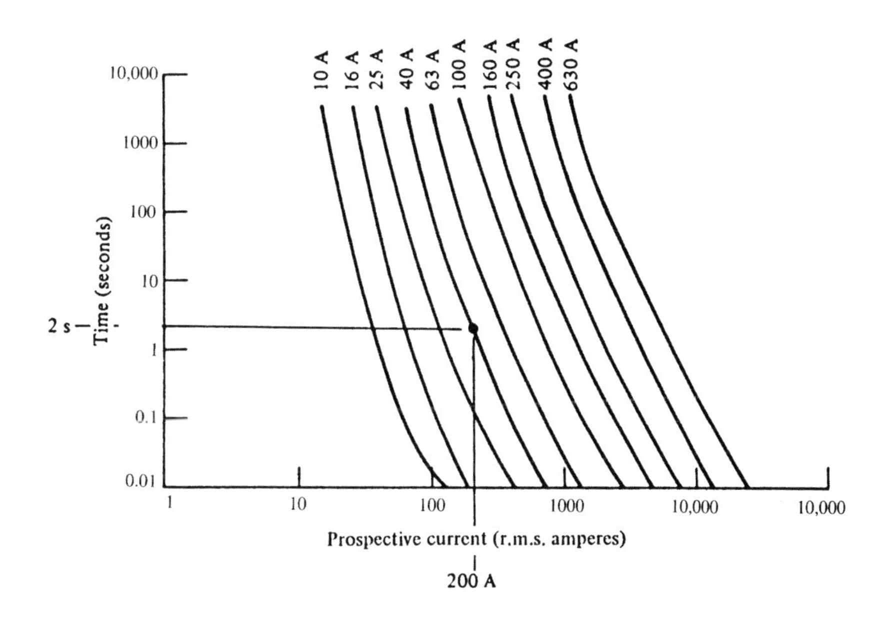

Fig. 18.9

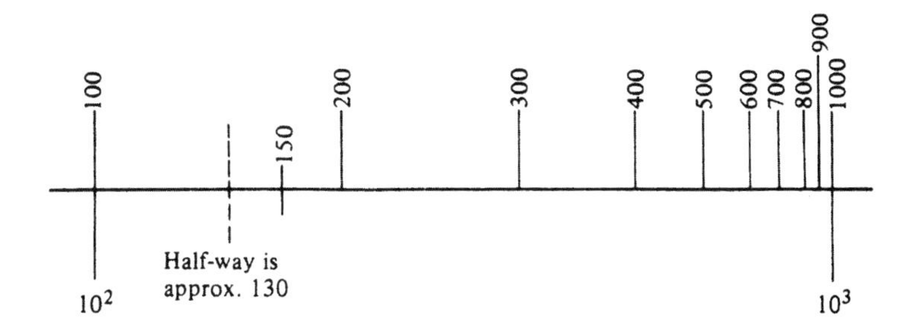

Fig. 18.10

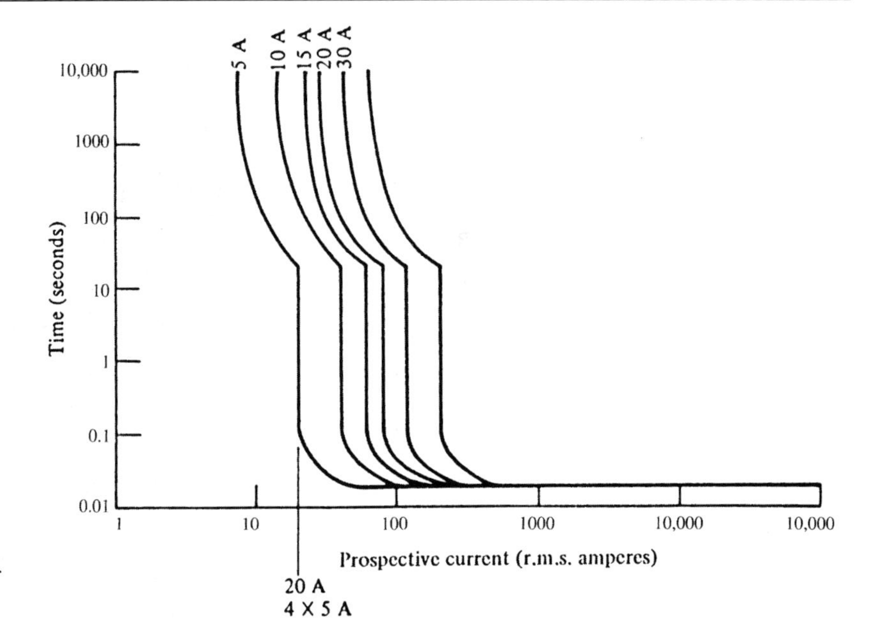

Fig. 18.11 *Time/current characteristics for type 1 MCBs to BS 3871. Example for 20 A superimposed. For times less than 20 ms, the manufacturer should be consulted*

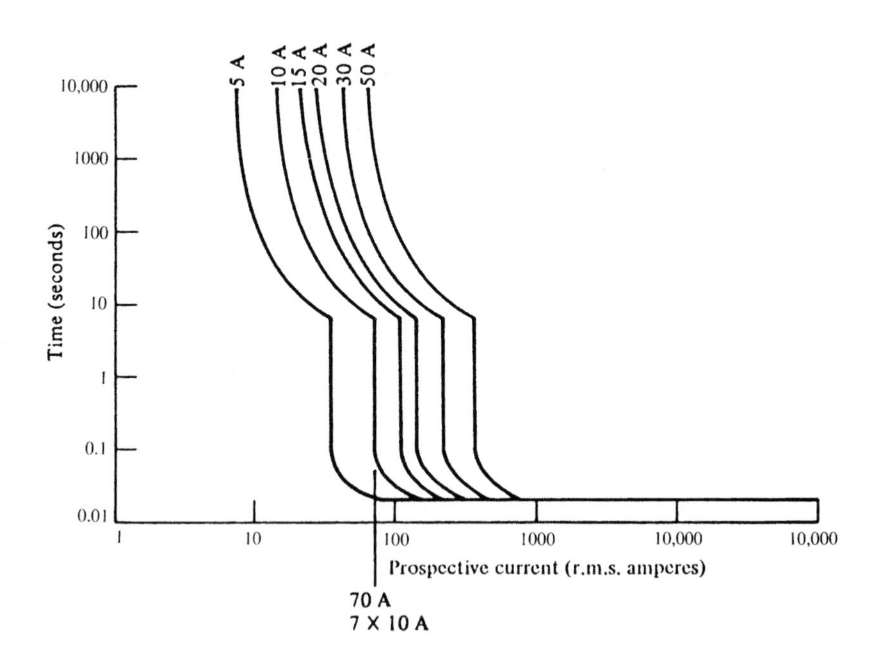

Fig. 18.12 *Time/current characteristics for type 2 MCBs to BS 3871. Example for 70 A superimposed. For times less than 20 ms, the manufacturer should be consulted*

part of the MCB. An overload is dealt with by a bimetallic mechanism, and a short circuit by an electromagnetic mechanism.

A type 1 MCB can take up to four times its rated current on overload before operating instantaneously. The magnetic part of the MCB operates at higher currents. A type 2 MCB can take up to seven times its rating, a type 3 up to ten times its rating, and types B and C up to five times their ratings.

Having found a disconnection time, we can now apply the formula.

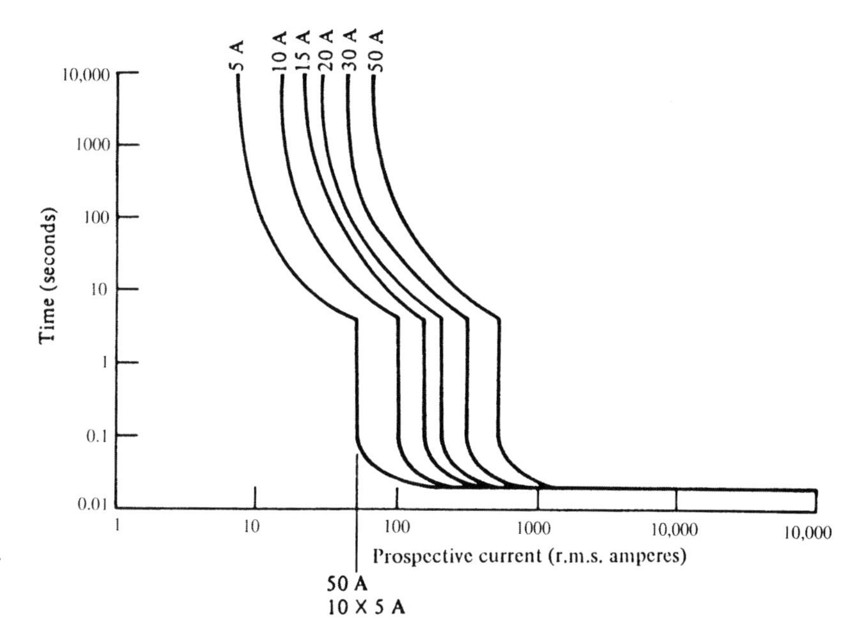

Fig. 18.13 *Time/current characteristics for type C cb to BS EN 60898. Example for 50 A superimposed. For times less than 20 ms, the manufacturer should be consulted*

Example of the use of the adiabatic equation

Suppose that in a design the protection was by 40 A BS88 fuse; we had chosen a 4 mm^2 copper CPC running with our phase conductor; and the loop impedance Z_s was 1.2 ohms. Would the chosen CPC size be large enough to withstand damage in the event of an earth fault?

We have

$$I = U_{0c}/Z_s = 240/1.2 = 200 \text{ A}$$

From the appropriate curve, we obtain a disconnection time t of 2 s. From Table 54C of the Regulations, $k = 115$. Therefore the minimum size of CPC is given by

$$s = \sqrt{(I^2\, t)}/k = \sqrt{200^2 \times 2)}/115 = 2.46 \text{ mm}^2$$

So our 4 mm^2 CPC is acceptable. Beware of thinking that the answer means that we could change the 4 mm^2 for a 2.5 mm^2. If we did, the loop impedance would be different and hence I and t would change; the answer for s would probably tell us to use a 4 mm^2. In the example shown, s is merely a check on the actual size chosen.

Having discussed each component of the design procedure, we can now put all eight together to form a complete design.

Example of circuit design

A consumer lives in a bungalow with a detached garage and workshop, as shown in Fig. 8.14. The building method is traditional brick and timber.

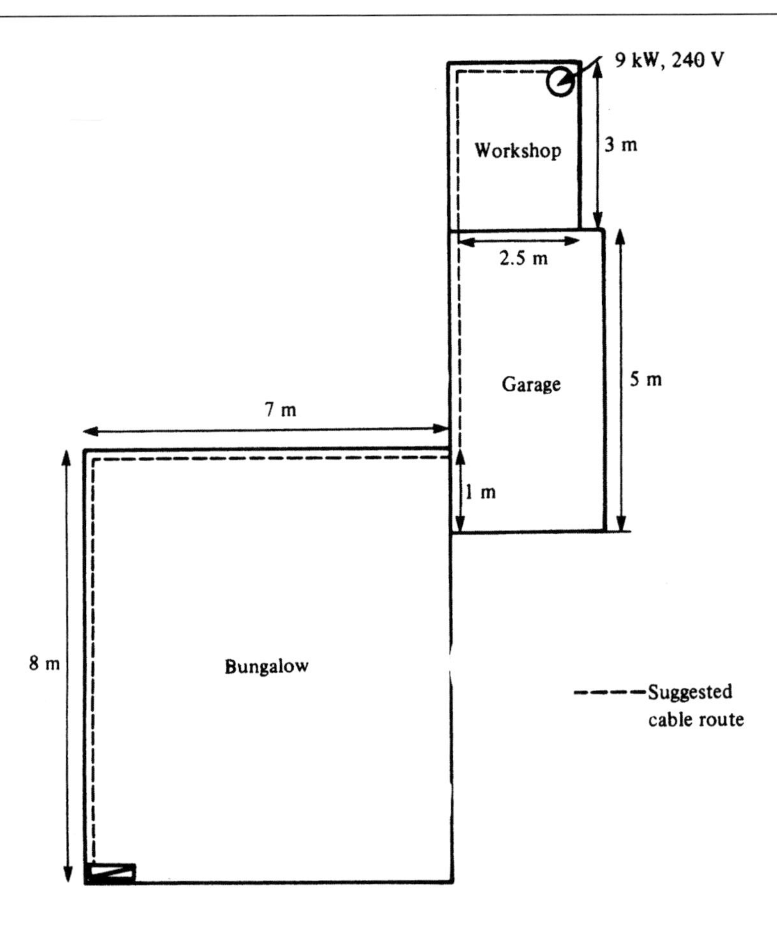

Fig. 18.14

The mains intake position is at high level, and comprises an 80 A BS 1361 240 V main fuse, an 80 A rated meter and a six-way 80 A consumer unit housing BS 3036 fuses as follows:

Ring circuit	30 A
Lighting circuit	5 A
Immersion heater circuit	15 A
Cooker circuit	30 A
Shower circuit	30 A
Spare way	

The cooker is 40 A, with no socket in the cooker unit.

The main tails are 16 mm^2 double-insulated PVC, with a 6 mm earthing conductor. There is no main equipotential bonding. The earthing system is TN–S, with an external loop impedance Z_e of 0.3 ohms. The prospective short-circuit current (PSC) at the origin of the installation has been measured at 800 A. The roof space is insulated to the full depth of the ceiling joists, and the temperature in the roof space has been noted to be no more than 40°C.

The consumer wishes to convert the workshop into a pottery room and install a 9 kW, 240 V electric kiln. The design procedure is as follows.

Assessment of general characteristics

Diversity

In any installation it is unlikely that every power and lighting point and other appliances will be used simultaneously and so the total current drawn from the supply is unlikely to be the total possible.

There are tables of recommended percentages of the total connected load for each circuit. It is from these that an estimated likely maximum load can be calculated, and the size of the main cable established.

Note: **The distribution board must be capable of taking the whole load without the application of diversity**.

Present maximum demand

Applying diversity, we have:

Ring	30 A
Lighting (66% of 5 A)	3.3 A
Immersion heater	15 A
A cooker (10 A + 30% of 30 A)	19 A
Shower	30 A
Total	97.3 A

Reference to Table 4D1A of the Regulations will show that the existing main tails are too small and should be uprated. Also, the consumer unit should be capable of carrying the full load of the installation *without* the application of diversity. So the addition of another 9 kW of load is not possible with the present arrangement.

New maximum demand

The current taken by the kiln is 9000/240 = 37.5 A. Therefore the new maximum demand is 97.3 + 37.5 = 134.8 A.

Supply details

Single phase
240 A, 50 Hz
Earthing: TN–S
PSC at origin (measured): 800 A

Decisions must now be made as to the type of cable, the installation method and the type of protective device. As the existing arrangement is not satisfactory, the supply authority must be informed of the new maximum demand, as a larger main fuse and service cable may be required. It would then seem sensible to disconnect, say, the shower circuit, and to supply it and the new kiln circuit via a new two-way consumer unit, as shown in Fig. 18.15.

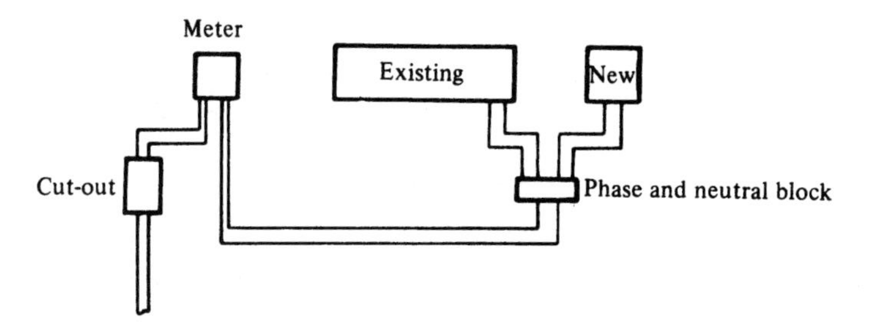

Fig. 18.15

Sizing the main tails

1 The new load on the existing consumer unit will be the old load less the shower load; 97.3 – 30 = 67.3 A. From Table 4D1A, the cable size is 16 mm^2.
2 The load on the new consumer unit will be the kiln load plus the shower load: 37.5 + 30 = 67.5 A. From Table 4D1A, the cable size is 16 mm^2.
3 The total load is 67.3 + 67.5 = 134.8 A. From Table 4D1 A, the cable size is 35 mm^2.
4 The earthing conductor size, from Table 54G, will be 16 mm^2. The main equipotential bonding conductor size, from Regulation 547-02, will be 10 mm^2.

For a domestic installation such as this, a PVC flat twin cable clipped direct through the loft space and the garage etc. would be most appropriate.

Sizing the kiln circuit cable

Design current

$$I_b = \frac{P}{V} = \frac{9000}{240} = 37.5 \text{ A}$$

Rating and type of protection

In order to show how important this choice is, it is probably best to compare the values of current-carrying capacity resulting from each type of protection.

As we have seen, the requirement for the rating I_n is that $I_n \geq I_b$. Therefore, using Tables 41B1 and 41B2 IEE Regulations, I_n will be as follows for the various fuse types:

BS 88	40 A	BS 3036	45 A
BS 1361	45 A	BS 3871 MCB	40 A
BS EN 60898 MCB	40 A		

Correction factors

C_a 0.87 or 0.94 if fuse is BS 3036
C_g not applicable
C_f 0.725 *only* if fuse is BS 3036
C_i 0.5 if cable is totally surrounded in thermal insulation.

Table 18.1

	BS 88 40 A	**BS 1361 45 A**	**BS 3036 45 A**	**MCB to BS 3871 40 A and BS EN 60898**
Surrounded by thermal insulation	$\frac{40}{0.5 \times 0.87} = 92$ A	$\frac{45}{0.5 \times 0.87} = 103.4$ A	$\frac{45}{0.5 \times 0.94 \times 0.725} = 132$ A	$\frac{40}{0.5 \times 0.87} = 92$ A
Not touching thermal insulation	$\frac{40}{0.87} = 46$ A	$\frac{45}{0.87} = 51.7$ A	$\frac{45}{0.94 \times 0.725} = 66$ A	$\frac{40}{0.87} = 46$ A
	BS 88	**BS 1361**	**BS 3036**	**MCB to BS 3871 and BS EN 60898**
Cable size with thermal insulation	25.0 mm^2	25.0 mm^2	35.0 mm^2	25.0 nmm^2
Cable size without	6.0 mm^2	10.0 mm^2	16.0 mm^2	6.0 mm^2
Cable size with half thermal insulation	16.6 mm^2	16.0 mm^2	25.0 mm^2	16.0 mm^2

* See item number 15. Table 4 A IEE Regulations.
In method 4, correction has already been made for cables touching thermal insulation on one size only.

Current-carrying capacity of cable

For each of the different types of protection, the tabulated current-carrying capacity I_t will be as shown below.

Cable size based on current-carrying capacity

Table 18.1 shows the sizes of cable for each type of protection (taken from Table 4D2 of the IEE Regulations).

Clearly BS 88 fuses and MCBs give the smallest cable size if the cable is kept clear of thermal insulation.

Check on voltage drop

The actual voltage drop is given by

$$\frac{\text{mV} \times I_b \times L}{1000} = \frac{7.3 \times 37.5 \times 24.5}{1000} = 6.7\,\text{V}$$

This voltage drop, whilst not causing the kiln to work unsafely, may mean inefficiency, and it is perhaps better to use a 10 mm^2 cable. This also gives us a wider choice of protection type, except BS 3036 rewirable. This decision we can leave until later.

For a $10\,mm^2$ cable, the voltage drop is checked as

$$\frac{4.4 \times 37.5 \times 24.5}{1000} = 4.04\,V$$

So at this point we have selected a $10\,mm^2$ twin cable. We have at our disposal a range of protection types, the choice of which will be influenced by the loop impedance.

Shock risk

The CPC inside a $10\,mm^2$ twin 6242 cable is $4\,mm^2$. Hence the total loop impedance will be

$$Z_s = Z_e + R_1 + R_2$$

For our selected cable, $R_1 + R_2$ for 24.5 m will be (from tables of conductor resistance):

$$\frac{6.44 \times 1.2 \times 24.5}{1000} = 0.189\,\text{ohms}$$

Note: The multiplier 1.2 takes account of the conductor resistance under operating conditions.

We are given that $Z_e = 0.3$ ohms: Hence

$$Z_s = 0.3 + 0.189 = 0.489\,\text{ohms}$$

This means that all but a 40 A type D MCB could be used (by comparison of values in Table 41B2 of the Regulations).

Thermal constraints

We still need to check that the $4\,mm^2$ CPC is large enough to withstand damage under earth fault conditions. We have

$$I = U_{0c}/Z_s = 24/0/0.489 = 490\,A$$

The disconnection time t for each type of protection from the relevant curves in the IEE Regulations is as follows:

40 A BS 88	0.05 s	40 A MCB type 3	0.025
45 A BS 1361	0.18 s	40 A MCB type B	0.015
40 A MCB type 1	0.02 s	40 A MCB type C	0.045
40 A MCB type 2	0.035 s		

From Table 54C of the Regulations, $k = 115$. Now

$$s = \sqrt{(I^2\,t)}/k$$

Therefore for each type of protection we have the following sizes s:

40 A BS 88	$0.9\,mm^2$	40 A MCB type 3	$0.6\,mm^2$
45 A BS 1361	$1.8\,mm^2$	40 A MCB type B	$0.42\,mm^2$
40 A MCB type 1	$0.6\,mm^2$	40 A MCB type C	$0.85\,mm^2$
40 A MCB type 2	$0.8\,mm^2$		

Hence our $4\,mm^2$ CPC is of adequate size.

Protection

It simply remains to decide on the type of protection. Probably a type 2 or type C MCB is the most economical. However, if this is chosen a check should be made on the shower circuit to ensure that this type of protection is also suitable.

Design problem

In a factory it is required to install, side by side, two three-phase 415 V direct on-line motors, each rated at 19 A full-load current. There is spare capacity in a three-phase distribution fuseboard housing BS 3036 fuses, and the increased load will not affect the existing installation. The cables are to be PVC-insulated singles installed in steel conduit, and a separate CPC is required (note Regulation 543–01–02). The earthing system is TN–S with a measured external loop impedance of 0.47 ohms, and the length of the cable run is 37 m. The worst conduit section is 7 m long with one bend. The ambient temperature is not expected to exceed 40°C.

Determine the minimum sizes of cable and conduit.

19 Testing

Measurement of electrical quantities

Instruments

Instruments play an important part in installation work enabling the measurement of the current, voltage, resistance, power and power factor.

The basic ammeter and voltmeter work on either the moving-iron or moving-coil principle, whereas the modern digital instrument uses complex electronics.

The moving-iron instrument (repulsion type)

Fig. 19.1 illustrates a moving-iron instrument of the repulsion type.

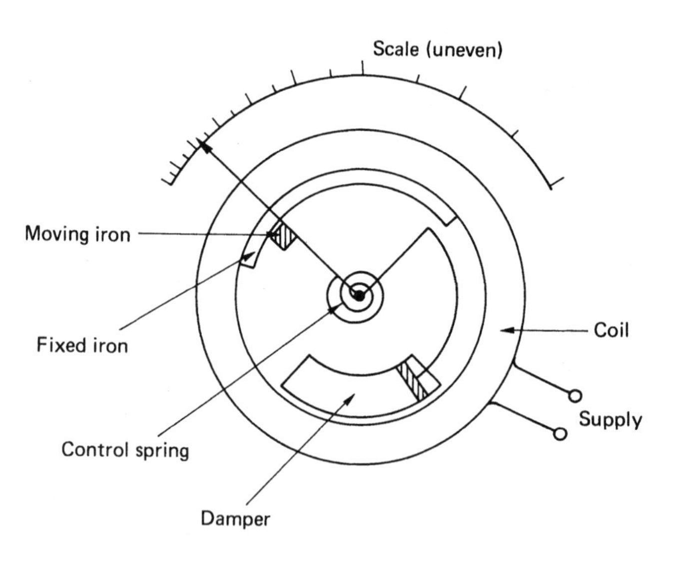

Fig. 19.1

It comprises a coil, with a fixed iron, a pointer with an iron vane attached (moving iron) and a sampling device inside it.

When a supply, either a.c. or d.c., is applied to the coil, both fixed and moving irons are magnetized to the same polarity and will therefore repel each other.

The design of the irons ensures that the repulsion is always in the same direction.

The damper ensures a slow and even movement of the pointer. It consists of a cylinder closed at one end and a light piston inside it. The pointer, which is attached to the piston, is slowed down by the air pressure which builds up in the cylinder, resisting the movement of the piston. A spring returns the pointer to zero when the supply is removed.

As the amount of movement depends on the square of the supply current, a small current produces a small movement and a large current a larger movement. Hence the scale tends to be cramped at the lower values of the current.

The moving-coil instrument

These work on the motor principle of a current-carrying coil in a magnetic field. Fig. 19.2a and b shows two variations of this type of instrument.

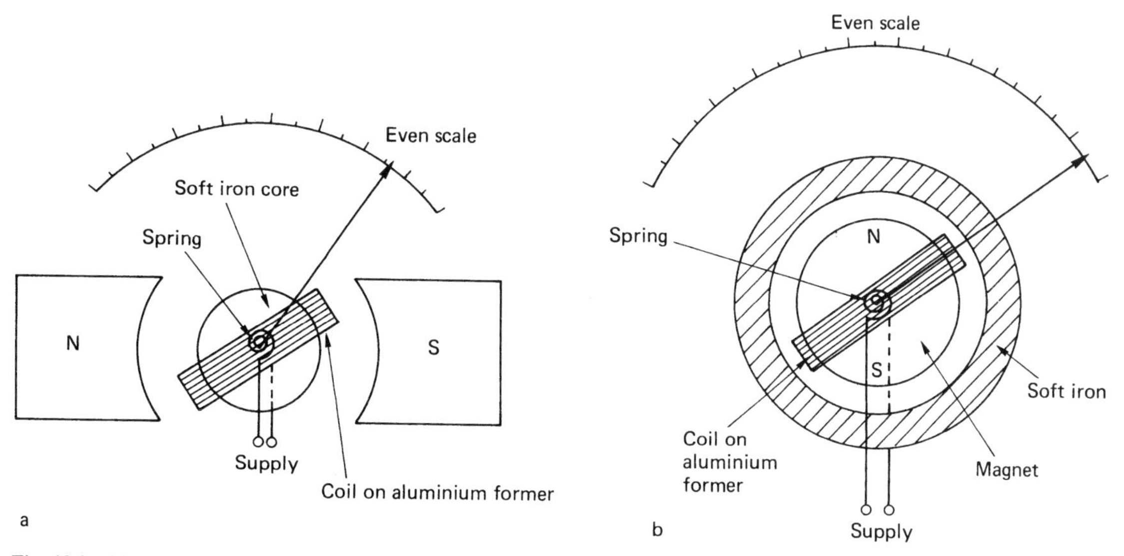

Fig. 19.2 *Moving-coil instruments*

The basic components of both the systems are a magnetic field, a core or shell of soft iron and a coil wound on an aluminium former; connection to the coil is made via the control springs.

Damping is achieved by eddy currents in the aluminium coil former. These currents cause small magnetic fields to flow which interact with the main field and cause the movement of the coil to slow down.

The digital instrument

The theory of operation is too complex to deal with here and hence only basic details will be considered.

A digital instrument is basically an electronic voltmeter with four sections:

1 The power supply and reference generators.
2 The signal-conditioning circuitry (current, resistance, voltage, etc.).

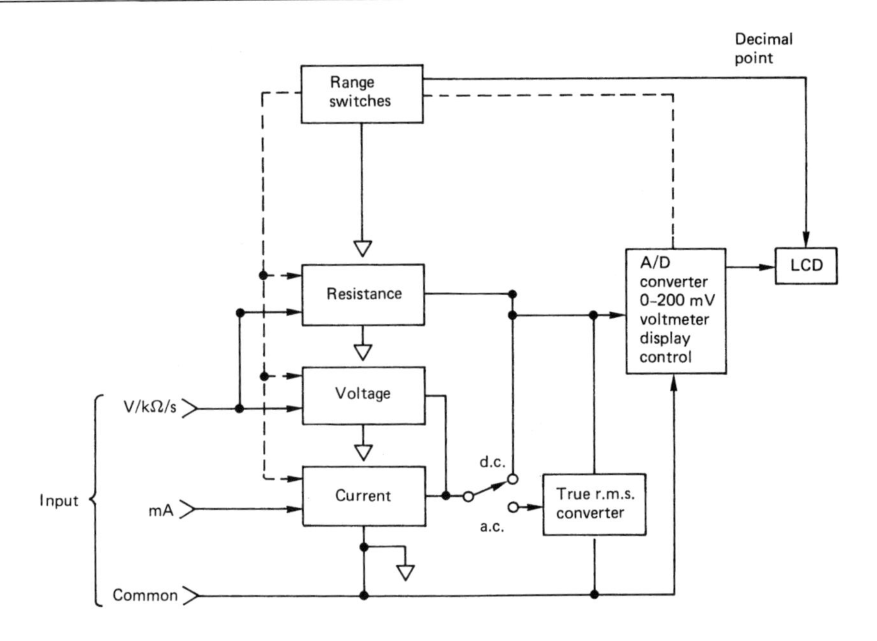

Fig. 19.3

3 The analogue-to-digital (A/D) converter.
4 The count/display module.

Fig. 19.3 shows a block diagram of a digital multi-meter.
A comparison of types of instruments is shown in Table 19.1.

Table 19.1

	Advantages	Disadvantages
Moving iron	Cheap, strong, can be used on a.c. and d.c.	Uneven scale, affected by heat and stray magnetic fields
Moving coil	Even scale, very accurate, unaffected by stray magnetic fields.	Fragile, expensive, can only be used on d.c.
Digital	Robust, no moving parts, accurate, easy reading.	Requires regular battery changes

Measurement of current

It is often necessary to extend the range of an ammeter to read values of current higher than the instrument's movement is designed for, and for this purpose, shunts or current transformers are used.

Ammeter shunts

As Fig. 19.4 shows, a shunt is simply a low-value resistor connected in parallel with the instrument.

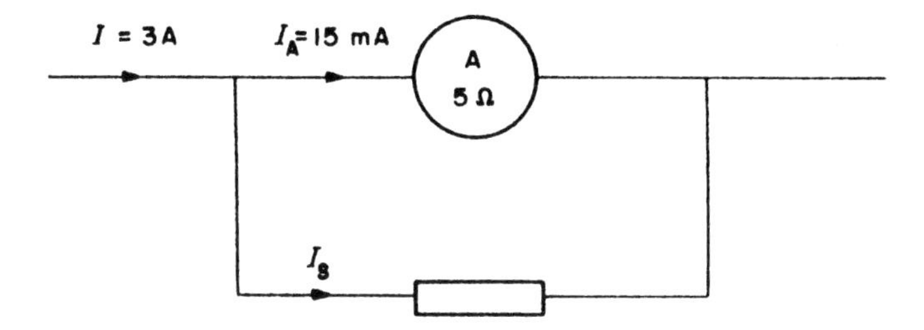

Fig. 19.4

Example

A moving-coil ammeter gives full-scale deflection (f.s.d.) at 15 mA. If the instrument resistance is 5 Ω, calculate the value of shunt required to enable the instrument to read currents up to 3 A.

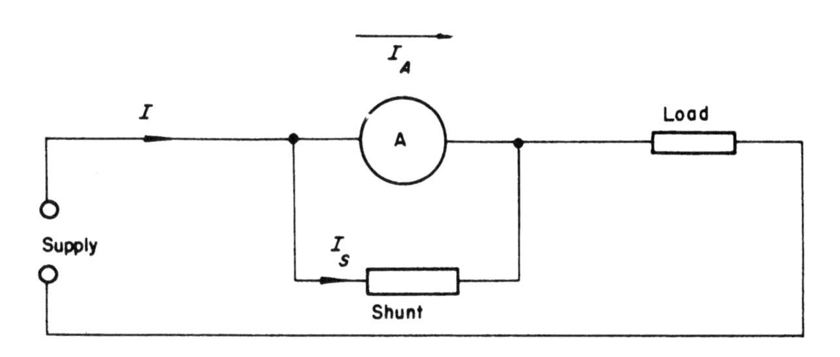

Fig. 19.5

For full-scale deflection Fig. 19.5 gives:

$$\text{Potential difference across meter} = I_A R_A = 15 \times 10^{-3} \times 5 = 75\,\text{mV}$$

$$\therefore \text{Potential difference across shunt} = 75\,\text{mV}$$

$$\text{Shunt current} = I - I_A = 3 - (15 \times 10^{-3}) = 3 - 0.015 = 2.985\,\text{A}$$

$$\therefore \text{Shunt resistance} = \frac{V_S}{I_S} = \frac{75 \times 10^{-3}}{2.985} = 0.025\ \Omega$$

The shunt may be used in conjunction with either a.c. or d.c. instruments. For measuring high a.c. currents, however, a current transformer is used.

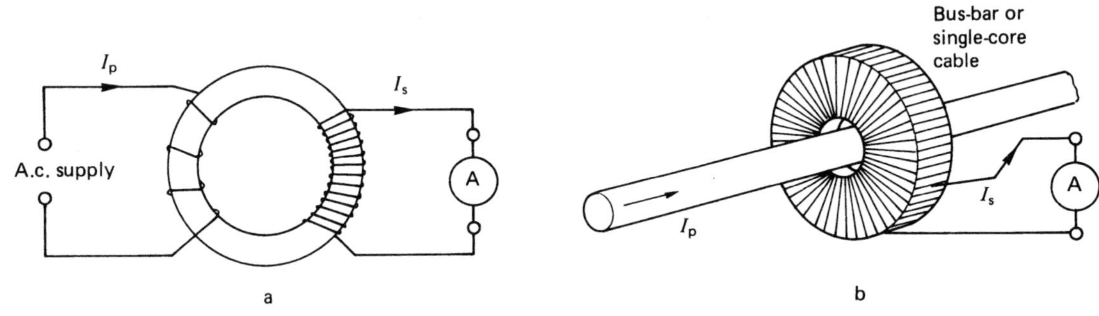

Fig. 19.6 *Current transformers: (a) wound type primary; (b) bar-type primary*

Current transformer

Current transformers (CTs) are usually of the wound or bar type shown in Fig. 19.6. As in any transformer, the secondary current will depend on the transformer ratio, i.e.

$$\frac{I_p}{I_s} = \frac{N_s}{N_p}$$

Example

An ammeter capable of taking 2.5 A is to be used in conjunction with a current transformer to measure a bus-bar current of up to 2000 A. Calculate the number of turns on the transformer.

$$\frac{I_p}{I_s} = \frac{N_s}{N_p}$$

$$\frac{2000}{2.5} = \frac{N_s}{1}$$

$$\therefore N_s = \frac{2000}{2.5}$$

$$= 800 \text{ turns}$$

Great care must be taken when using CTs, as high voltages normally associated with high currents will be stepped up on the secondary side, creating a potentially dangerous situation.

Before removing an ammeter or load (burden) from a CT, the secondary terminals must be shorted out.

Measurement of voltage

As with current measurement, moving-iron and moving-coil instruments are used.

The extension of the range of a voltmeter is achieved by using a multiplier or, for high a.c. voltages, a voltage transformer.

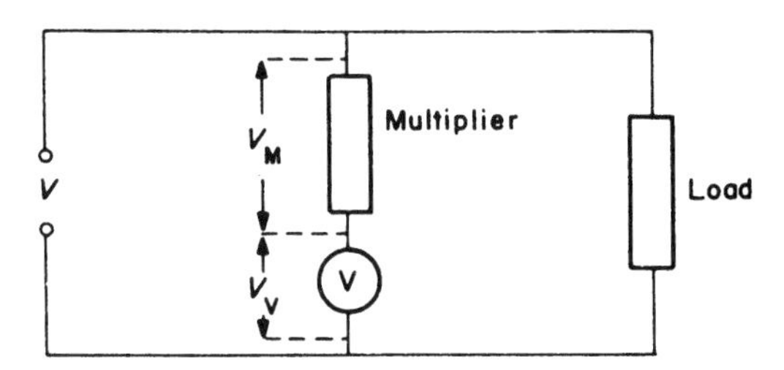

Fig. 19.7 *Voltmeter multiplier*

Voltmeter multiplier

A voltmeter multiplier is simply a resistance in series with the instrument as shown in Fig. 19.7.

Example

A moving-coil instrument of resistance 5Ω and f.s.d. at 20 mA is to be used to measure voltages up to 100 V. Calculate the value of the series multiplier required.

$$\text{Instrument voltage at f.s.d.} = I_V \times R_V$$
$$= 20 \times 10^{-3} \times 5$$
$$= 0.1\,\text{V}$$
$$\therefore \text{Voltage to be dropped across multiplier} = V - V_V$$
$$= 100 - 0.1$$
$$= 99.9\,\text{V}$$
$$\therefore \text{Value of resistance} = \frac{V_m}{I_m}$$
$$= \frac{99.9}{20 \times 10^{-3}}$$
$$= 4995\ \Omega$$

Voltage transformer

A voltage transformer (VT) is simply a typical double-wound step-down transformer with a great many turns on the primary and a few on the secondary.

Instruments in general

Multimeters

There are many types of multimeter now available, the more expensive usually giving greater accuracy. They all work on the moving-coil principle and use rectifiers for the d.c. ranges. Shunts, multipliers, VTs and CTs are switched in or out when ranges and scales are changed by the operator.

Wattmeters

A wattmeter is simply a combination of an ammeter and a voltmeter in one instrument, usually a dynamometer.

Fig. 19.9 shows how high-voltage connections can be made to a wattmeter.

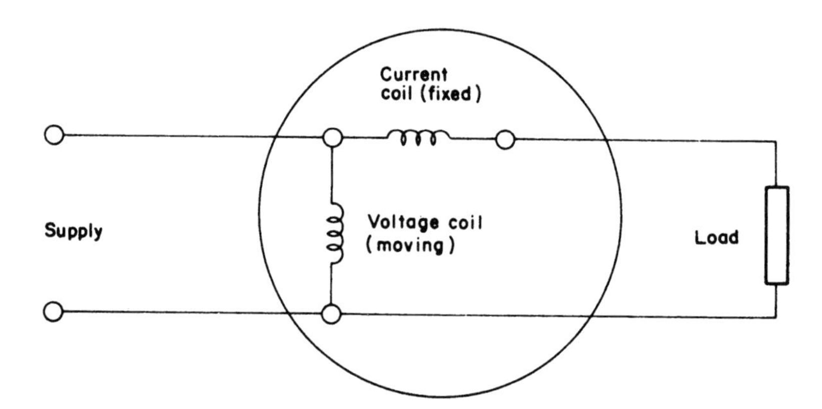

Fig. 19.8 *Wattmeter connection*

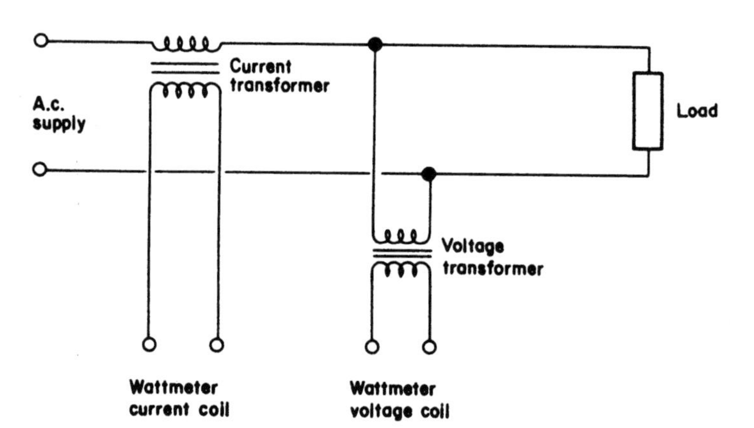

Fig. 19.9 *Use of current transformer and voltage transformer*

Tong tester

The tong tester or clip-on ammeter is a variation of the bar-primary current transformer. It consists of an insulated iron core in two parts that can be separated (like tongs), on one end of which is the secondary winding and an ammeter.

The core is clipped round a bus-bar or single-core cable and the current is registered on the ammeter (Fig. 19.10).

Phase-rotation indicator

The phase-rotation indicator is a simple three-phase induction motor. When connected to a three-phase supply, a disc, connected to the motor, rotates in the direction of the supply sequence. It is used when two three-phase systems are to be paralleled together (Fig. 19.11).

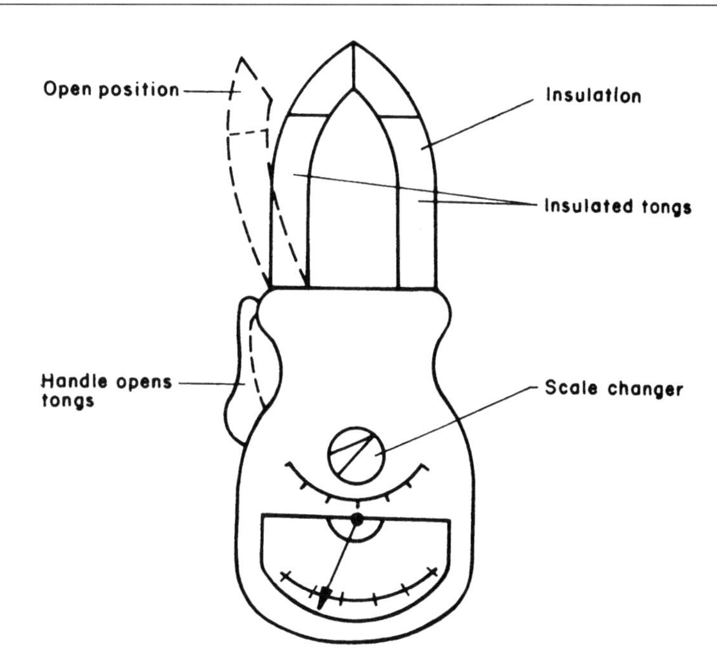

Fig. 19.10 *Clip-on ammeter*

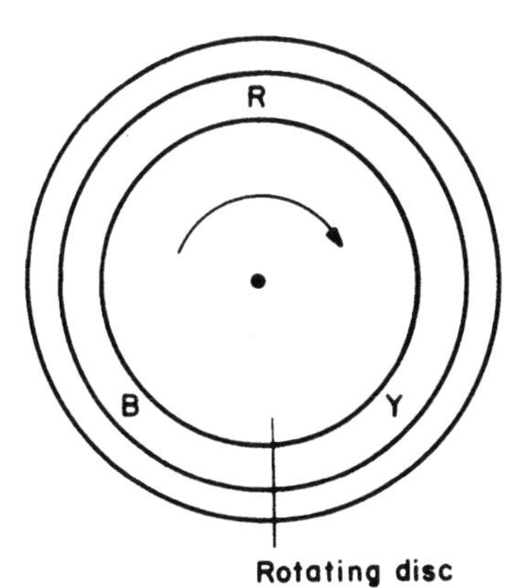

Fig. 19.11 *Phase rotation indicator*

Selection of test instruments

Let us now look at the range of quantities that an electrician is likely to encounter in the normal practice of his or her profession. If we take the sequence of the more commonly used tests prescribed by the IEE Wiring Regulations, and assign typical values to them, we can at least provide a basis for the choice of the most suitable instruments. It will be seen from Table 19.2

Table 19.2

Test	Range	Type of Instrument
1 Continuity of protective conductors	2 to 0.005 ohms or less	Low-reading ohmmeter
2 Continuity of ring final conductors	0.05 to 0.08 ohms	Low-reading ohmmeter
3 Insulation resistance	Infinity to less than 1 megohm	High-reading ohmmeter
4 Polarity	None	Ohmmeter, bell, etc.
5 Earth fault loop impedance	0 to 2000 ohms	Special ohmmeter
6 Earth electrode resistance	Any value over about 3 or 4 ohms	Special ohmmeter
7 Operation of RCD	5 to 500 mA	Special instrument
8 Prospective short-circuit current	2 A to 20 kA	Special instrument

that all that is required is an ohmmeter of one sort or another, a residual current device (RCD) tester and an instrument for measuring prospective short-circuit current.

It is clearly most sensible to purchase instruments from one of the established manufacturers rather than to attempt to save money by buying cheaper, lesser known brands. Also, as the instruments used in the world of installation are bound to be subjected to harsh treatment, a robust construction is all important.

Many of the well-known instrument companies provide a dual facility in one instrument e.g. prospective S/C current and loop impedance, or insulation resistance and continuity. Hence it is likely that only three or four instruments would be needed, together with an approved test lamp.

Approved test lamps and indicators

Search your tool boxes; find, with little difficulty one would suspect, your 'neon screwdriver' or 'testascope'; locate a very deep pond; and drop it in!

Imagine actually allowing electric current at low voltage (50 to 1000 V a.c.) to pass through one's body in order to activate a test lamp! It only takes around 10 to 15 mA to cause a severe electric shock, and 50 mA (1/20th of an ampere) to kill.

Apart from the fact that such a device will register any voltage from about 5 V upwards, the safety of the user depends entirely on the integrity

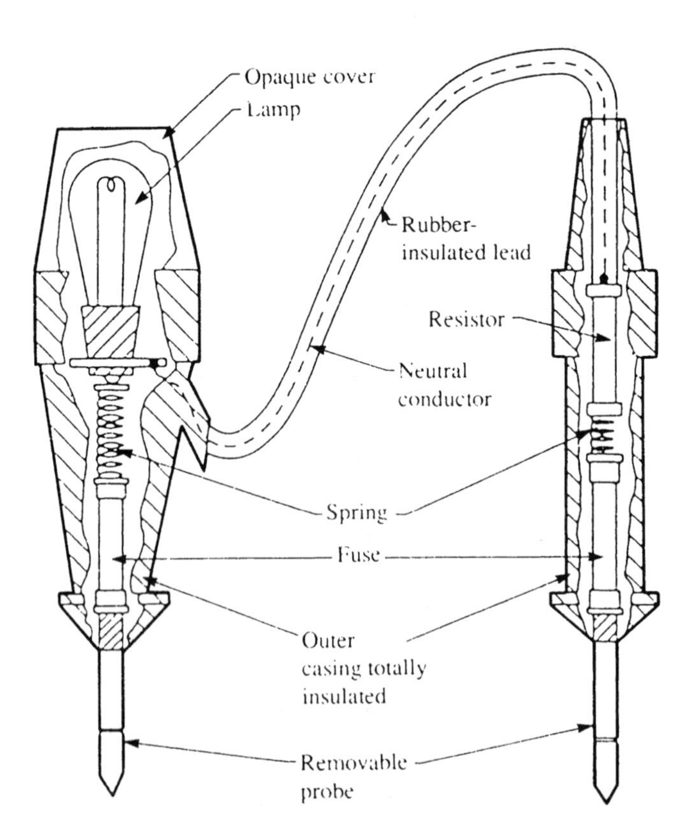

Fig. 19.12 *Approved test lamp*

of the current-limiting resistor in the unit. An electrician received a considerable shock when using such an instrument after his apprentice had dropped it in a sink of water, simply wiped it dry and replaced it in the tool box. The water had seeped into the device and shorted out the resistor.

An approved test lamp should be of similar construction to that shown in Fig. 19.12.

Accidental RCD operation

It has long been the practice when using a test lamp to probe between phase and earth for indication of a live supply on the phase terminal. However, this can now present a problem where RCDs exist in the circuit, as of course the test is applying a deliberate phase to earth fault.

Some test lamps have LED indicators, and the internal circuitry of such test lamps limits the current to earth to a level below that at which the RCD will operate. The same limiting effect applies to multimeters. However, it is always best to check that the testing device will have no effect on RCDs.

Calibration, zeroing and care of instruments

Precise calibration of instruments is usually well outside the province of the electrician, and would normally be carried out by the manufacturer or a local service representative. A check, however, can be made by the user to determine whether calibration is necessary by comparing readings with an instrument known to be accurate, or by measurement of known values of voltage, resistance, etc.

It may be the case that readings are incorrect simply because the instrument is not zeroed before use, or because the internal battery needs replacing. Most modern instruments have battery condition indication, and of course this should never be ignored.

Always adjust any selection switches to the off position after testing. Too many instrument fuses are blown when, for example, a multimeter is inadvertently left on the ohms range and then used to check for mains voltage.

The following set procedure may seem rather basic but should ensure trouble-free testing:

1 Check test leads for obvious defects.
2 Zero the instrument.
3 Select the correct range for the values anticipated. If in doubt, choose the highest range and gradually drop down.
4 Make a record of the test results, if necessary.
5 When a zero reading is expected and occurs (or, in the case of insulation resistance, an infinite reading), make a quick check on the test leads just to ensure that they are not open-circuited.
6 Return switches/selectors to the off position.
7 Replace instrument and leads in carrying case.

Initial inspection

Inspection and testing

Circumstances which require an initial verification

New installations or Additions or Alterations

General reasons for initial verification

1 To ensure equipment and accessories are to a relevant standard.
2 To prove compliance with BS 7671.
3 To ensure that the installation is not damaged so as to impair safety.

Information required

Assessment of general characteristics sections 311, 312 and 313 together with information such as drawings, charts etc. in accordance with Reg. 514–09–01.

Documentation required and to be completed

Electrical Installation Certificate signed or authenticated for the design and construction (could be the same person) and then for the inspection and test. A schedule of test results and an inspection schedule must accompany an Electrical Installation Certificate.

Sequence of tests

1 Continuity of all protective conductors
2 Continuity of ring final circuit conductors
3 Insulation resistance
4 Site applied insulation
5 Protection by separation of circuits
6 Protection against direct contact by barriers and enclosures provided during erection
7 Insulation of non-conducting floors and walls
8 Polarity
9 Earth electrode resistance
10 Earth fault loop impedance
11 Prospective fault current
12 Functional testing

Before any testing is carried out, a detailed physical inspection must be made to ensure that all equipment is to a relevant British or Harmonized European Standard, and that it is erected/installed in compliance with the IEE Regulations, and that it is not damaged such that it could cause danger. In order to comply with these requirements, the Regulations give a check list of some eighteen items that, where relevant, should be inspected.

However, before such an inspection, and test for that matter, is carried out, certain information *must* be available to the verifier. This information is the result of the Assessment of General Characteristics required by IEE Regulations Part 3, Sections 311, 312 and 313, and drawings, charts, and similar information relating to the installation. It is at this point that most readers who work in the real world of electrical installation will be lying on the floor laughing hysterically.

Let us assume that the designer and installer of the installation are competent professionals, and all of the required documentation is available.

Interestingly, one of the items on the check list *is* the presence of diagrams, instructions and similar information. If these are missing then there is a departure from the Regulations.

Another item on the list is the verification of conductors for current carrying capacity and voltage drop in accordance with the design. How on earth can this be verified without all the information? A 30 A Type B circuit breaker (CB) protecting a length of 4 mm^2 conductor may look reasonable, but is it correct, and are you prepared to sign to say that it is unless you are sure? Let us look then at the general content of the check list.

1. **Connection of conductors** Are terminations electrically and mechanically sound, is insulation and sheathing removed only to a minimum to allow satisfactory termination?
2. **Identification of conductors** Are conductors correctly identified in accordance with the Regulations?
3. **Routing of cables** Are cables installed such that account is taken of external influences such as mechanical damage, corrosion, heat etc.?
4. **Conductor selection** Are conductors selected for current carrying capacity and voltage drop in accordance with the design?
5. **Connection of single pole devices** Are single pole protective and switching devices connected in the phase conductor only?
6. **Accessories and equipment** Are all accessories and items of equipment correctly connected?
7. **Thermal effects** Are fire barriers present where required and protection against thermal effects provided?
8. **Protection against shock** What methods have been used to provide protection against direct and indirect contact?
9. **Mutual detrimental influence** Are wiring systems installed such that they can have no harmful effect on non-electrical systems, or that systems of different currents or voltages are segregated where necessary?
10. **Isolation and switching** Are there appropriate devices for isolation and switching correctly located and installed?
11. **Undervoltage** Where undervoltage may give rise for concern, are there protective devices present?
12. **Protective devices** Are protective and monitoring devices correctly chosen and set to ensure protection against indirect contact and/or overcurrent?
13. **Labelling** Are all protective devices, switches (where necessary) and terminals correctly labelled?
14. **External influences** Have all items of equipment and protective measures been selected in accordance with the appropriate external influences?
15. **Access** Are all means of access to switchgear and equipment adequate?

16 **Notices and signs** Are danger notices and warning signs present?
17 **Diagrams** Are diagrams, instructions and similar information relating to the installation available?
18 **Erection methods** Have all wiring systems, accessories and equipment been selected and installed in accordance with the requirements of the Regulations, and are fixings for equipment adequate for the environment?

So, we have now inspected all relevant items, and provided that there are no defects that may lead to a dangerous situation when testing, we can now start the actual testing procedure.

Testing continuity of protective conductors

All protective conductors, including main equipotential and supplementary bonding conductors must be tested for continuity using a low resistance ohmmeter.

For main equipotential bonding there is no single fixed value of resistance above which the conductor would be deemed unsuitable. Each measured value, if indeed it is measurable for very short lengths, should be compared with the relevant value for a particular conductor length and size. Such values are shown in Table 19.3.

Where a supplementary equipotential bonding conductor has been installed between *simultaneously accessible* exposed and extraneous conductive parts, because circuit disconnection times cannot be met, then the resistance (R) of the conductor, must be equal to or less than 50/Ia. In the case of construction sites and agricultural or horticultural installations the 50 is replaced by 25.

So, $R \leqslant$ 50/Ia, or 25/Ia, where 50, and 25, are the voltages above which exposed metalwork should not rise, and Ia is the minimum current causing operation of the circuit protective device within 5 secs.

For example, suppose a 45 A BS 3036 fuse protects a cooker circuit, the disconnection time for the circuit cannot be met, and so a supplementary bonding conductor has been installed between the cooker case and an adjacent

Table 19.3 Resistance (Ω) of copper conductors at 20°C

CSA					*Length (m)*					
(mm²)	**5**	**10**	**15**	**20**	**25**	**30**	**35**	**40**	**45**	**50**
1	0.09	0.18	0.27	0.36	0.45	0.54	0.63	0.72	0.82	0.9
1.5	0.06	0.12	0.18	0.24	0.3	0.36	0.43	0.48	0.55	0.6
2.5	0.04	0.07	0.11	0.15	0.19	0.22	0.26	0.03	0.33	0.37
4	0.023	0.05	0.07	0.09	0.12	0.14	0.16	0.18	0.21	0.23
6	0.02	0.03	0.05	0.06	0.08	0.09	0.11	0.13	0.14	0.16
10	0.01	0.02	0.03	0.04	0.05	0.06	0.063	0.07	0.08	0.09
16	0.006	0.01	0.02	0.023	0.03	0.034	0.04	0.05	0.05	0.06
25	0.004	0.007	0.01	0.015	0.02	0.022	0.026	0.03	0.033	0.04
35	0.003	0.005	0.008	0.01	0.013	0.016	0.019	0.02	0.024	0.03

central heating radiator. The resistance (R) of that conductor should not be greater than 50/Ia, and Ia in this case is 145 A (see Figure 3.2B of the IEE Regulations)

i.e. $50/145 = 0.34\,\Omega$

How then, do we conduct a test to establish continuity of main or supplementary bonding conductors? Quite simple really, just connect the leads from a low resistance ohmmeter to the ends of the bonding conductor (Figure 19.13). One end should be disconnected from its bonding clamp, otherwise any measurement may include the resistance of parallel paths of other earthed metalwork. Remember to zero the instrument first or, if this facility is not available, record the resistance of the test leads so that this value can be subtracted from the test reading.

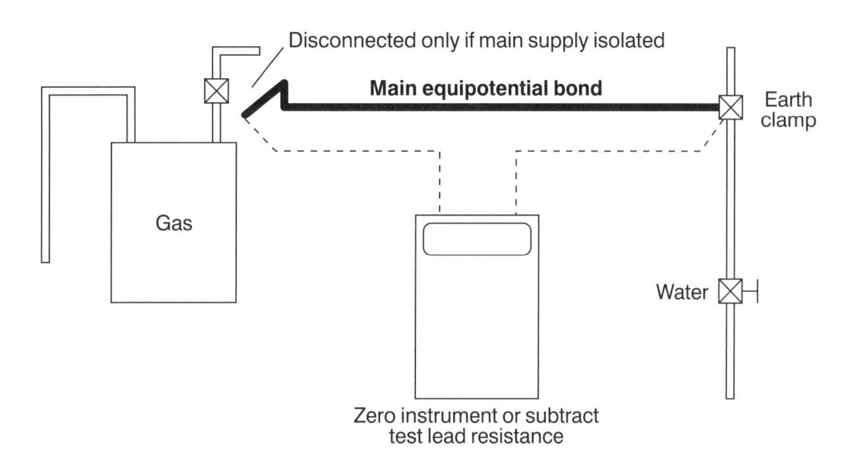

Fig. 19.13

IMPORTANT NOTE: If the installation is in operation, then *never* disconnect main bonding conductors unless the supply can be isolated. Without isolation, persons and livestock are at risk of electric shock.

The continuity of circuit protective conductors may be established in the same way, but a second method is preferred, as the results of this second test indicate the value of $(R_1 + R_2)$ for the circuit in question.

The test is conducted in the following manner:

1 Temporarily link together the phase conductor and CPC of the circuit concerned in the distribution board or consumer unit.
2 Test between phase and CPC at *each* outlet in the circuit. A reading indicates continuity.
3 Record the test result obtained at the furthest point in the circuit. This value is $(R_1 + R_2)$ for the circuit.

Figure 19.14 illustrates the above method.

There may be some difficulty in determining the $(R_1 + R_2)$ values of circuits in installations that comprise steel conduit and trunking, and/or SWA and m.i.m.s. cables because of the parallel earth paths that are likely to exist. In these cases, continuity tests may have to be carried out at the installation stage before

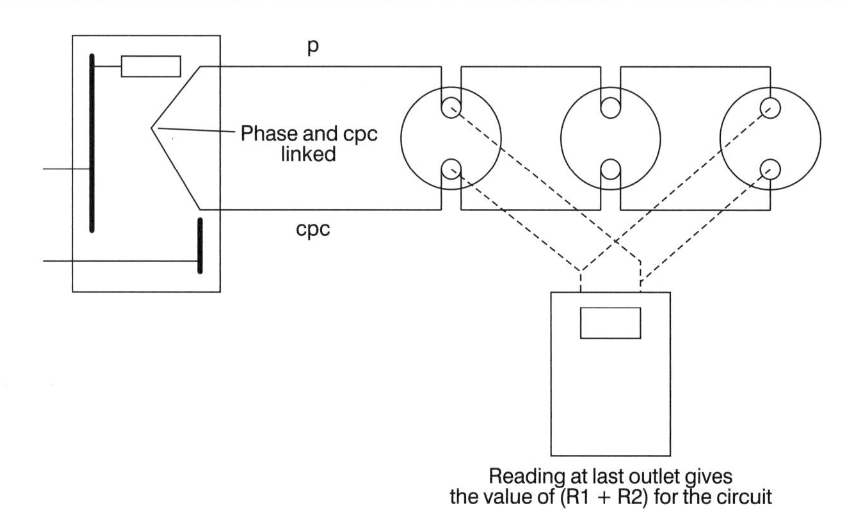

Fig. 19.14

accessories are connected or terminations made off as well as after completion.

Although it is no longer considered good working practice to use steel conduit or trunking as a protective conductor, it is permitted, and hence its continuity must be proved. The enclosure must be inspected along its length to ensure that it is sound and then the standard low resistance test is performed. If the verifier has any doubt as to the soundness of the conductor, a further test is made using a high current test instrument which has a test voltage not exceeding 50 V and can deliver up to 1.5 times the design current of the circuit up to a maximum of 25 A. This test can cause arcing at faulty joints and hence should not be carried out if there is any chance of danger.

Testing continuity of ring final circuit conductors

There are two main reasons for conducting this test:

1 To establish that interconnections in the ring do not exist.
2 To ensure that the CPC is continuous, and indicate the value of $(R_1 + R_2)$ for the ring.

What then are interconnections in a ring circuit, and why is it important to locate them? Figure 19.15 shows a ring final circuit with an interconnection.

The most likely cause of the situation shown in Figure 19.15 is where a DIY enthusiast has added sockets P, Q, R and S to an existing ring A, B, C, D, E and F.

In itself there is nothing wrong with this. The problem arises if a break occurs at, say, point Y, or the terminations fail in socket C or P. Then there would be four sockets all fed from the point X which would then become a spur.

So, how do we identify such a situation with or without breaks at point 'Y'? A simple resistance test between the ends of the phase, neutral or circuit protective conductors will only indicate that a circuit exists, whether there are interconnections or not. The following test method is based on the philosophy

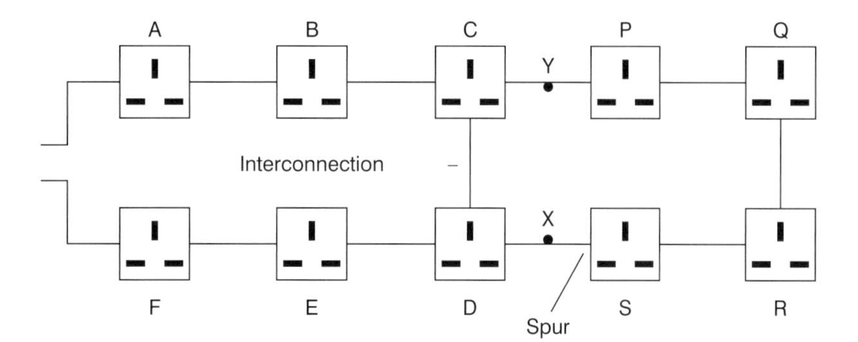

Fig. 19.15

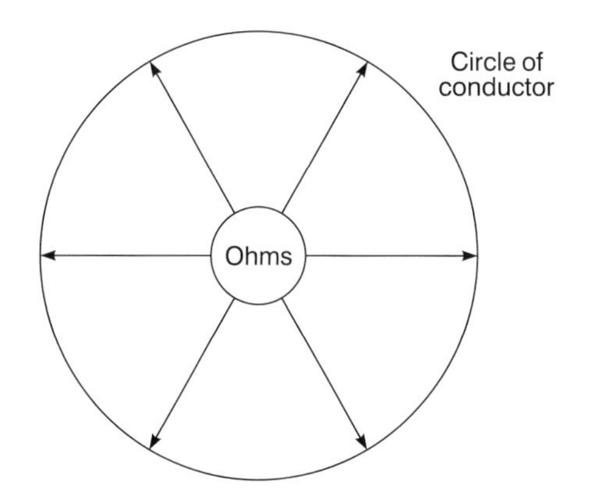

Fig. 19.16 Same value whatever diameter is measured

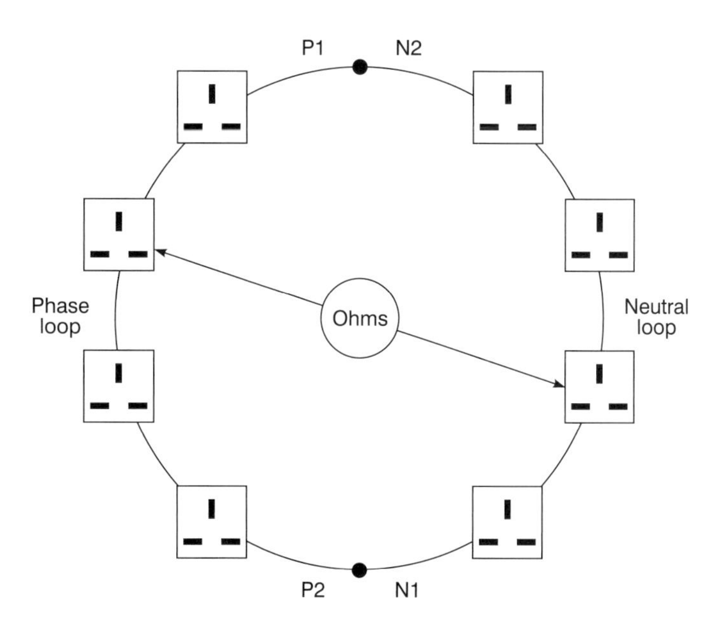

Fig. 19.17

that the resistance measured across any diameter of a perfect circle of conductor will always be the same value (Figure 19.16).

The perfect circle of conductor is achieved by cross connecting the phase and neutral legs of the ring (Figure 19.17).

The test procedure is as follows:

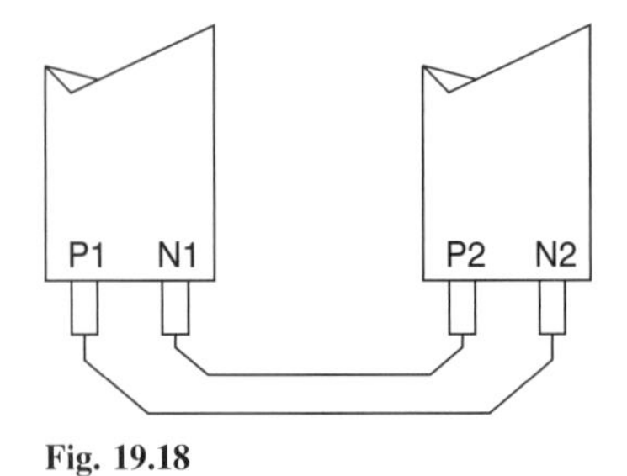

Fig. 19.18

1 Identify the opposite legs of the ring. This is quite easy with sheathed cables, but with singles, each conductor will have to be identified, probably by taking resistance measurements between each one and the closest socket outlet. This will give three high readings and three low readings thus establishing the opposite legs.
2 Take a resistance measurement between the ends of each conductor loop. Record this value.
3 Cross connect the opposite ends of the phase and neutral loops (Figure 19.18).
4 Measure between phase and neutral at each socket on the ring. The readings obtained should be, for a perfect ring, substantially the same. If an interconnection existed such as shown in Figure 19.15, then sockets A to F would all have similar readings, and those beyond the interconnection would have gradually increasing values to approximately the mid point of the ring, then decreasing values back towards the interconnection. If a break had occurred at point Y then the readings from socket S would increase to a maximum at socket P. One or two high readings are likely to indicate either loose connections or spurs. A null reading, i.e. an open circuit indication, is probably a reverse polarity, either phase-CPC or neutral-CPC reversal. These faults would clearly be rectified and the test at the suspect socket(s) repeated.
5 Repeat the above procedure, but in this case cross connect the phase and CPC loops (Figure 19.19).

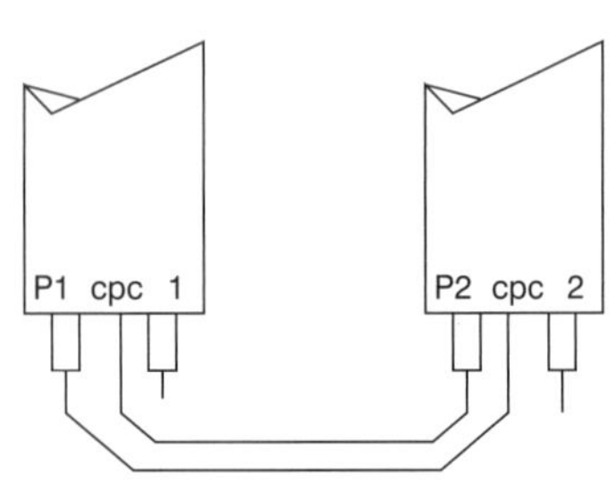

Fig. 19.19

In this instance, if the cable is of the flat twin type, the readings at each socket will increase very slightly and then decrease around the ring. This difference, due to the phase and CPC being different sizes, will not be significant enough to cause any concern. The measured value is very important, it is $R_1 + R_2$ for the ring.

As before, loose connections, spurs and, in this case, P – N cross polarity, will be picked up.

The details in Table 19.4 are typical approximate ohmic values for a healthy 70 m ring final circuit wired in 2.5/1.5 flat twin and CPC cable. (In this case the CPC will be approximately 1.67× the P or N resistance.)

Table 19.4

Initial measurements	P1–P2 0.52	N1–N2 0.52	cpc 1 – cpc2 0.86
Reading at each socket	0.26	0.26	0.32–0.34
For spurs, each metre in length will add the following resistance to the above values	0.015	0.015	0.02

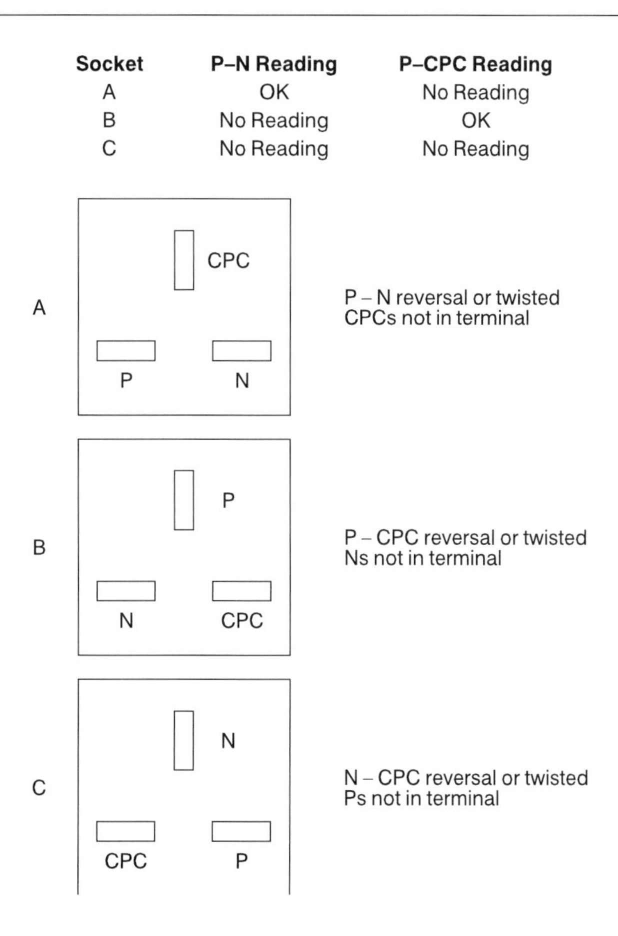

Fig. 19.20

As already mentioned null readings may indicate a reverse polarity. They could also indicate twisted conductors not in their terminal housing. The examples shown in Figure 19.20 may help to explain these situations.

Testing insulation resistance

This is probably the most used and yet abused test of them all. Affectionately known as ‘meggering’, an *insulation resistance test* is performed in order to ensure that the insulation of conductors, accessories and equipment is in a healthy condition, and will prevent dangerous leakage currents between conductors and between conductors and earth. It also indicates whether any short circuits exist.

Insulation resistance, as just discussed, is the resistance measured between conductors and is made up of countless millions of resistances in parallel (Figure 19.21).

The more resistances there are in parallel, the *lower* the overall resistance, and in consequence, the longer a cable the lower the insulation resistance. Add to this the fact that almost all installation circuits are also wired in parallel, it becomes apparent that tests on large installations may give, if measured as a whole, pessimistically low values, even if there are no faults.

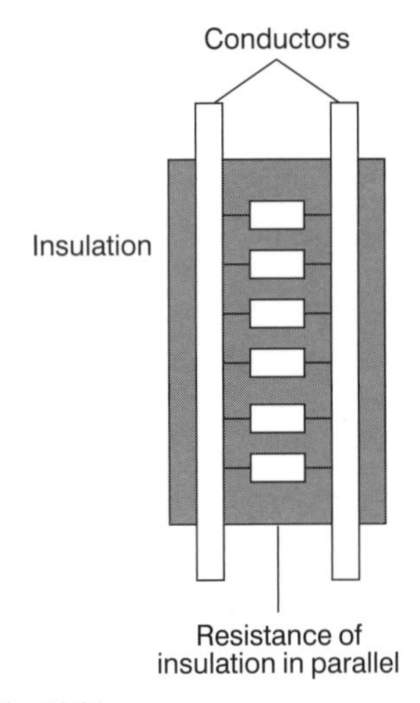

Fig. 19.21

Under these circumstances, it is usual to break down such large installations into smaller sections, floor by floor, sub-main by sub-main etc. This also helps, in the case of periodic testing, to minimize disruption. The test procedure is as follows:

1 Disconnect all items of equipment such as capacitors and indicator lamps as these are likely to give misleading results. Remove any items of equipment likely to be damaged by the test, such as dimmer switches, electronic timers etc. Remove all lamps and accessories and disconnect fluorescent and discharge fittings. Ensure that the installation is disconnected from the supply, all fuses are in place, and MCBs and switches are in the on position. In some instances it may be impracticable to remove lamps etc. and in this case the local switch controlling such equipment may be left in the off position.
2 Join together all live conductors of the supply and test between this join and earth. Alternatively, test between each live conductor and earth in turn.
3 Test between phase and neutral. For three phase systems, join together all phases and test between this join and neutral. Then test between each of the phases. Alternatively, test between each of the live conductors in turn. Installations incorporating two-way lighting systems should be tested twice with the two-way switches in alternative positions.

Table 19.5 gives the test voltages and minimum values of insulation resistance for ELV and LV systems.

Table 19.5

System	*Test voltage*	*Minimum insulation resistance*
SELV and PELV	250 V DC	0.25 MΩ
LV up to 500 V	500 V DC	0.5 MΩ
Over 500 V	1000 V DC	1.0 MΩ

If a value of less than 2 MΩ is recorded it may indicate a situation where a fault is developing, but as yet still complies with the minimum permissible value. In this case each circuit should be tested separately and each should be above 2 MΩ.

Example

An installation comprising six circuits have individual insulation resistances of 2.5 MΩ, 8 MΩ, 200 MΩ, 200 MΩ, 200 MΩ and 200 MΩ, and so the total insulation resistance will be:

$$\frac{1}{R_t} = \frac{1}{2.5} + \frac{1}{8} + \frac{1}{200} + \frac{1}{200} + \frac{1}{200} + \frac{1}{200}$$
$$= 0.4 + 0.125 + 0.005 + 0.005 + 0.005 + 0.005$$
$$= 0.545$$
$$R_t = \frac{1}{0.545}$$
$$= 1.83\,\text{M}\Omega$$

This is clearly greater than the 0.5 MΩ minimum but less than 2 MΩ, but as all circuits are greater than 2 MΩ the system could be considered satisfactory.

Special tests

The next four tests are special in that they are not often required in the general type of installation. They also require special test equipment. In consequence, the requirements for these tests will only be briefly outlined in this chapter.

Site applied insulation

When insulation is applied to live parts during the erection process on site in order to provide protection against direct contact, then a test has to be performed to show that the insulation can withstand a high voltage equivalent to that specified in the BS for similar factory built equipment.

If supplementary insulation is applied to equipment on site, to provide protection against indirect contact, then the voltage withstand test must be applied, and the insulating enclosure must afford a degree of protection of not less than IP2X or IPXXB.

Protection by separation of circuits

When SELV or PELV is used as a protective measure, then the separation from circuits of a higher voltage has to be verified by an insulation resistance test at a test voltage of 250 V and result in a minimum insulation resistance of 0.25 MΩ. If the circuit is at low voltage and supplied from, say, a BS 3535 transformer the test is at 500 V with a minimum value of 0.555 MΩ.

Protection by barriers or enclosures

If, on site, protection against direct contact is provided by fabricating an enclosure or erecting a barrier, it must be shown that the enclosure can provide a degree of protection of at least IP2X or IPXXB. Readily accessible horizontal top surfaces should be to at least IP4X.

An enclosure having a degree of protection IP2X can withstand the ingress of fingers and solid objects exceeding 12 mm diameter. IPXXB is protection against finger contact only. IP4X gives protection against wires and solid objects exceeding 1 mm in diameter.

The test for IP2X or IPXXB is conducted with a 'standard test finger' which is supplied at a test voltage not less than 40 V and no more than 50 V. One end of the finger is connected in series with a lamp and live parts in the enclosure. When the end of the finger is introduced into the enclosure, provided the lamp does not light then the protection is satisfactory.

The test for IP4X is conducted with a rigid 1 mm diameter wire with its end bent at right angles. Protection is afforded if the wire does not enter the enclosure.

Protection by non-conducting location

This is a rare location and demands specialist equipment to measure the insulation resistance between insulated floors and walls at various points.

Testing polarity

This simple test, often overlooked, is just as important as all the others, and many serious injuries and electrocutions could have been prevented if only polarity checks had been carried out.

The requirements are:

1 All fuses and single pole switches are in the phase conductor.
2 The centre contact of an Edison screw type lampholder is connected to the phase conductor.
3 All socket outlets and similar accessories are correctly wired.

Although polarity is towards the end of the recommended test sequence, it would seem sensible, on lighting circuits, for example, to conduct this test at the same time as that for continuity of CPCs (Figure 19.22).

As discussed earlier, polarity on ring final circuit conductors, is achieved simply by conducting the ring circuit test. For radial socket outlet circuits, however, this is a little more difficult. The continuity of the CPC will have already been proved by linking phase and CPC and measuring between the same terminals at each socket. Whilst a phase-CPC reversal would not have shown, a phase-neutral reversal would, as there would have been no reading at the socket in question. This would have been remedied, and so only phase-CPC reversals need to be checked. This can be done by linking together phase and neutral at the origin and testing between the same terminals at each

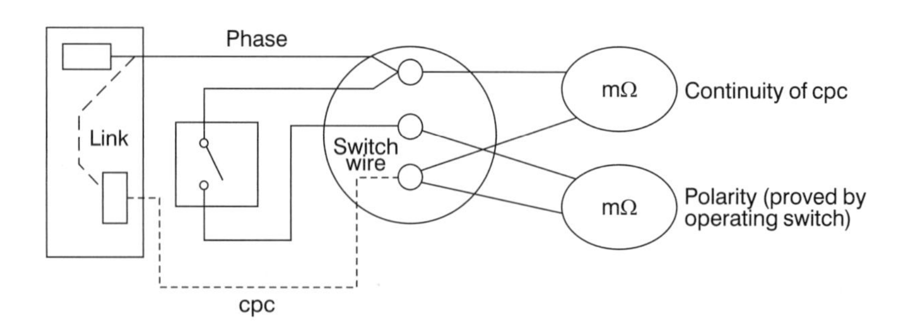

Fig. 19.22

socket. A phase-CPC reversal will result in no reading at the socket in question.

When the supply is connected, it is important to check that the incoming supply is correct. This is done using an approved voltage indicator at the intake position or close to it.

Testing earth fault loop impedance

This is very important but sadly, poorly understood. So let us remind ourselves of the component parts of the earth fault loop path (Figure 19.23). Starting at the point of fault:

1 The CPC.
2 The main earthing conductor and earthing terminal.
3 The return path via the earth for TT systems, and the metallic return path in the case of TN–S or TN–C–S systems. In the latter case the metallic return is the PEN conductor.
4 The earthed neutral of the supply transformer.
5 The transformer winding.
6 The phase conductor back to the point of fault.

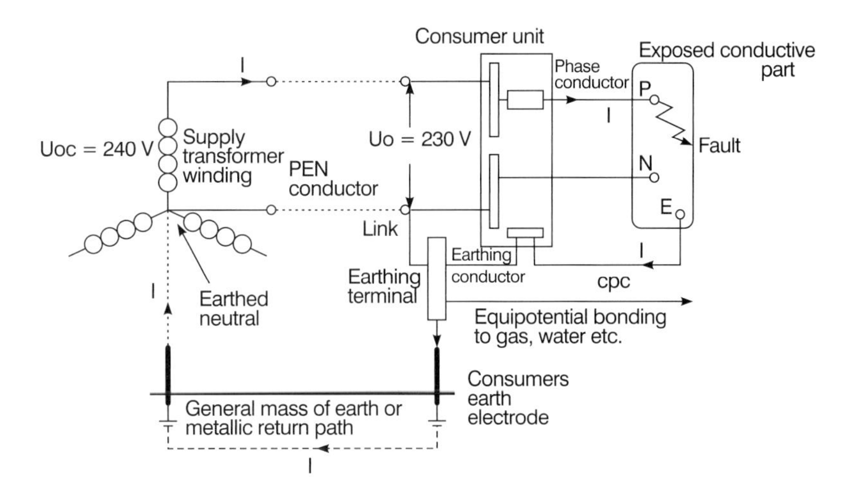

Fig. 19.23

Overcurrent protective devices must, under earth fault conditions, disconnect fast enough to reduce the risk of electric shock. This is achieved if the actual value of the earth fault loop impedance does not exceed the tabulated maximum values given in the IEE regulations.

The purpose of the test, therefore, is to determine the actual value of the loop impedance (Z_s), for comparison with those maximum values, and it is conducted as follows:

1 Ensure that all main equipotential bonding is in place.
2 Connect the test instrument either by its BS 4363 plug, or the ‘flying leads’, to the phase, neutral and earth terminals at the remote end of the circuit

Table 19.6 Values of loop impedance for comparison with test readings

Protection	*Disconnection time (s)*		*Rating of protection*																		
			5 A	6 A	10 A	15 A	16 A	20 A	25 A	30 A	32 A	40 A	45 A	50 A	60 A	63 A	80 A	100 A	125 A	160 A	200 A
BS 3036 fuse	0.4	Z_s max	7.5	–	–	2	–	1.38	–	0.85	–	–	0.46								
	5	Z_s max	13.9	–	–	4.1	–	3	–	2.97	–	–	1.25	–	0.87	–	–	0.42			
BS 88 fuse	0.4	Z_s max	–	6.66	4	–	2.11	1.38	1.12	–	0.82	0.64	–	0.47							
	5	Z_s max	–	10.5	5.8	–	3.27	2.28	1.8	–	1.44	1.05	–	0.82	–	0.64	0.45	0.33	0.26	0.2	0.14
BS 1361 fuse	0.4	Z_s max	8.17	–	–	2.57	–	1.33	–	0.9	–	–	0.45								
	5	Z_s max	12.8	–	–	3.9	–	2.19	–	1.44	–	–	0.75	0.54	–	0.39	0.28				
BS 3871 MCB Type 1	0.4 & 5	Z_s max	9	7.5	4.5	3	2.81	2.25	1.8	1.5	1.41	1.12	1	0.9	–	0.71					
BS 3871 MCB Type 2	0.4 & 5	Z_s max	5.14	4.28	2.57	1.71	1.6	1.28	1.02	0.85	0.8	0.64	0.57	0.51	–	0.4					
BS 3871 MCB Type 3	0.4 & 5	Z_s max	3.6	3	1.8	1.2	1.12	0.9	0.72	0.6	0.56	0.45	0.4	0.36	–	0.28					
BS EN 60898 CB Type B	0.4 & 5	Z_s max	–	6	3.6	–	2.25	1.8	1.44	–	1.12	0.9	0.8	0.72	–	0.57					
BS EN 60898 CB Type C	0.4 & 5	Z_s max	3.6	3	1.8	1.2	1.12	0.9	0.72	0.6	0.56	0.45	0.4	0.36	–	0.28					
BS EN 60898 CB Type D	0.4 & 5	Z_s max	1.8	1.5	0.9	0.6	0.56	0.45	0.36	0.3	0.28	0.22	0.2	0.18	–	0.14					

under test. (If a neutral is not available, e.g. in the case of a three-phase motor, connect the neutral probe to earth.)

3 Press to test and record the value indicated.

It must be understood, that this instrument reading is *not valid for direct comparison with the tabulated maximum values*, as account must be taken of the ambient temperature at the time of test, and the maximum conductor operating temperature, both of which will have an effect on conductor resistance. Hence, the $(R_1 + R_2)$ could be greater at the time of fault than at the time of test.

So, our measured value of Z_s must be corrected to allow for these possible increases in temperature occurring at a later date. This requires actually measuring the ambient temperature and applying factors in a formula.

Clearly this method of correcting Z_s is time consuming and unlikely to be commonly used. Hence, a rule of thumb method may be applied which simply requires that the measured value of Z_s does not exceed 3/4 of the appropriate tabulated value. Table 19.6 gives the 3/4 values of tabulated loop impedance for direct comparison with measured values.

In effect, a loop impedance test places a phase/earth fault on the installation, and if an RCD is present it may not be possible to conduct the test as the device will trip out each time the loop impedance tester button is pressed. Unless the instrument is of a type that has a built-in guard against such tripping, the value of Z_s will have to be determined from measured values of Z_e and $(R_1 + R_2)$, and the 3/4 rule applied.

IMPORTANT NOTE: Never short out an RCD in order to conduct this test.

As a loop impedance test creates a high earth fault current, albeit for a short space of time, some lower rated MCBs may operate resulting in the same situation as with an RCD, and Z_s will have to be calculated. It is not really good practice temporarily to replace the MCB with one of a higher rating.

External loop impedance Z_e

The value of Z_e is measured at the intake position on the supply side and with all main equipotential bonding disconnected. Unless the installation can be isolated from the supply, this test should not be carried out, as a potential shock risk will exist with the supply on and the main bonding disconnected.

Prospective fault current

This would normally be carried out at the same time as the measurement for Ze using a PFC or PSCC tester. If this value cannot be measured it must be ascertained by either enquiry or calculation.

Testing earth electrode resistance

In many rural areas, the supply system is TT and hence reliance is placed on the general mass of earth for a return path under earth fault conditions. Connection to earth is made by an electrode, usually of the rod type, and preferably installed as shown in Figure 19.24.

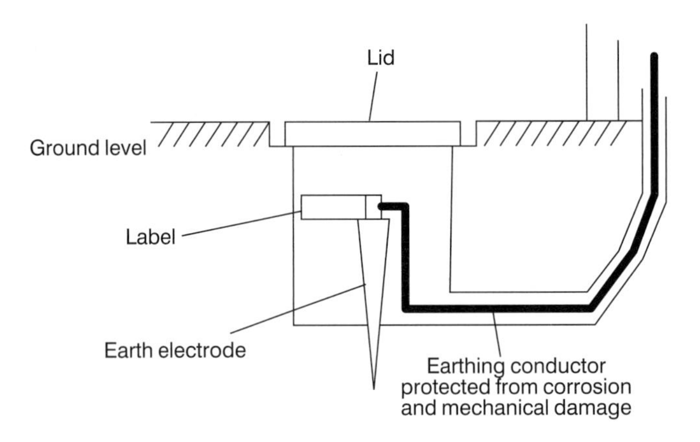

Fig. 19.24

In order to determine the resistance of the earth return path, it is necessary to measure the resistance that the electrode has with earth. If we were to make such measurements at increasingly longer distances from the electrode, we would notice an increase in resistance up to about 2.5–3 m from the rod, after which no further increase in resistance would be noticed (Figure 19.25).

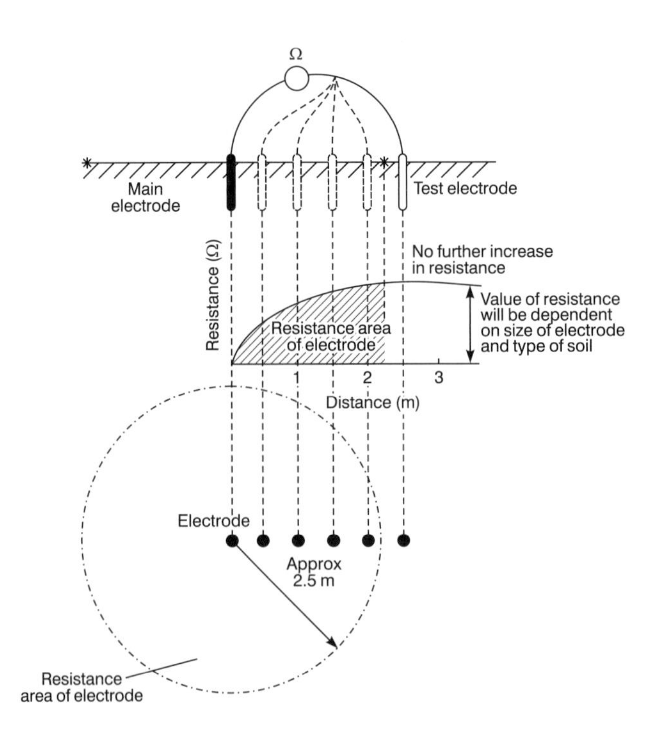

Fig. 19.25

The maximum resistance recorded is the electrode resistance and the area that extends the 2.5–3 m beyond the electrode is known as the earth electrode resistance area.

There are two methods of making the measurement, one using a proprietary instrument, and the other using a loop impedance tester.

Method 1 – protection by overcurrent device

This method is based on the principle of the potential divider (Figure 19.26).

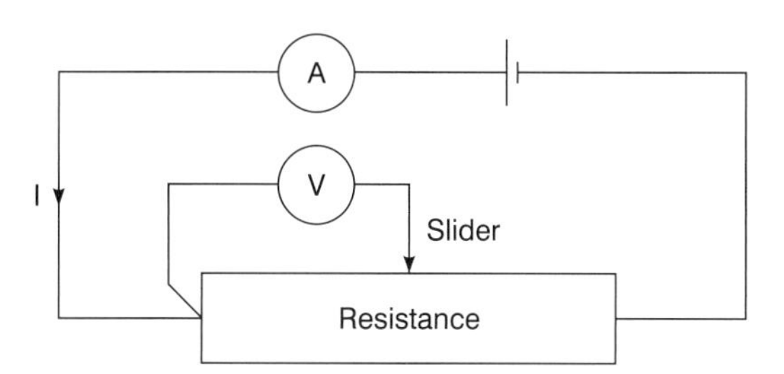

Fig. 19.26

By varying the position of the slider the resistance at any point may be calculated from $R = V/I$.

The earth electrode resistance test is conducted in a similar fashion, with the earth replacing the resistance and a potential electrode replacing the slider (Figure 19.27). In Figure 19.27 the earthing conductor to the electrode under test is temporarily disconnected.

The method of test is as follows:

1. Place the current electrode (C2) away from the electrode under test, approximately 10 times its length, i.e. 30 m for a 3 m rod.
2. Place the potential electrode mid way.
3. Connect test instrument as shown.
4. Record resistance value.

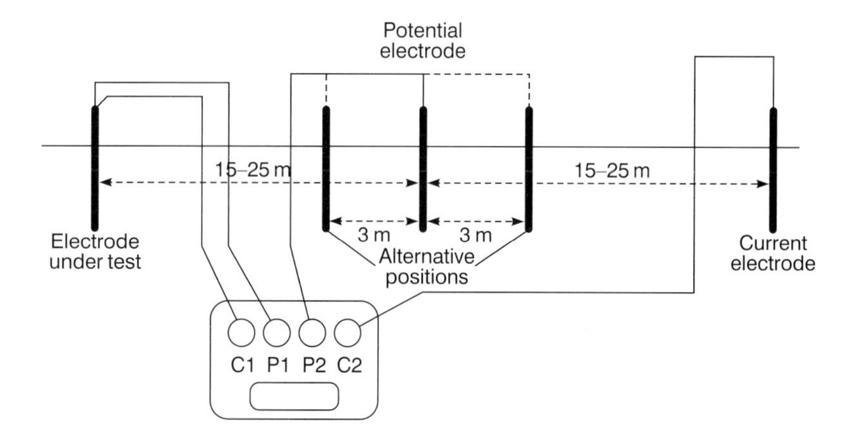

Fig. 19.27

5 Move the potential electrode approximately 3 m either side of the mid position, and record these two readings.
6 Take an average of these three readings (this is the earth electrode resistance).
7 Determine the maximum deviation or difference of this average from the three readings.
8 Express this deviation as a percentage of the average reading.
9 Multiply this percentage deviation by 1.2.
10 Provided this value does not exceed a figure of 5% then the accuracy of the measurement is considered acceptable.

If three readings obtained from an earth electrode resistance test were 181 Ω, 185 Ω and 179 Ω. What is the value of the electrode resistance and is the accuracy of the measurement acceptable?

$$\text{Average value} = \frac{181 + 185 + 179}{3}$$

$$= 181.67\,\Omega$$

$$\text{Maximum deviation} = 185 - 181.67$$

$$= 3.33$$

$$\text{Expressed as a percentage of the average} = \frac{3.33 \times 100}{181.67}$$

$$= 1.83\%$$

$$\text{Measurement accuracy} = 1.83\% \times 1.2 = 2.2\%$$

(which is acceptable)

For TT systems the result of this test will indicate compliance if the product of the electrode resistance and the operating current of the overcurrent device does not exceed 50 V.

Method 2 – protection by a residual current device

In this case, an earth fault loop impedance test is carried out between the incoming phase terminal and the electrode (a standard test for Z_e).

The value obtained is added to the CPC resistance of the protected circuits and this value is multiplied by the operating current of the RCD. The resulting value should not exceed 50 V. If it does, then Method 1 should be used to check the actual value of the electrode resistance.

Functional testing

RCD RCBO operation

Where RCDs/RCBOs are fitted, it is essential that they operate within set parameters. The RCD testers used are designed to do just this, and the basic tests required are as follows:

Table 19.7

RCD type	*½ rated*	*Full trip current*
BS 4239 and BS 7288 sockets	No trip	<200 ms
BS 4239 with time delay	No trip	½ time delay + 200 ms-time delay + 200 ms
BS EN 61009 or BS EN 61009 RCBO	No trip	300 ms
As above but Type S with time delay	No trip	130–500 ms

1 Set the test instrument to the rating of the RCD.
2 Set the test instrument to half rated trip.
3 Operate the instrument and the RCD should not trip.
4 Set the instrument to deliver the full rated tripping current of the RCD.
5 Operate the instrument and the RCD should trip out in the required time.

There seems to be a popular misconception regarding the ratings and uses of RCDs in that they are the panacea for all electrical ills and the only useful rating is 30 mA!!

Firstly, RCDs are not fail safe devices, they are electro-mechanical in operation and can malfunction. Secondly, general purpose RCDs are manufactured in ratings from 5 mA to 100 mA and have many uses. Let us first deal with RCDs rated at 30 mA or less. The accepted lethal level of shock current is 50 mA and hence RCDs rated at 30 mA or less would be appropriate for use where shock is an increased risk. BS 7671 indicates that RCDs of 30 mA or less should be used in the following situations:

1 To protect circuits supplying hand held equipment outside the equipotential zone.
2 To protect all socket outlet circuits in a TT system installation.
3 To protect all socket outlets in a caravan park.
4 To provide supplementary protection against Direct contact.
5 For fixed current using equipment in bathrooms.

In all these cases and apart from conducting the tests already mentioned, it is required that the RCD be injected with a current five times its operating current and the tripping time should not exceed 40 ms.

Where loop impedence values cannot be met, RCDs of an appropriate rating can be installed. Their rating can be determined from

$$I\,n = 50/Zs$$

Where I n is the rated operating current of the device
50 is the touch voltage
Zs is the measured loop impedence

RCDs can also be used for:

1 Discrimination e.g. a 100 mA device to protect the whole installation and a 30 mA for the sockets.
2 Protection against fire use, say, a 500 mA device.

All RCDs have a built-in test facility in the form of a test button. Operating this test facility creates an artificial out of balance condition that causes the device to trip. This only checks the mechanics of the tripping operation, it is not a substitute for the tests just discussed.

All other items of equipment such as switchgear, controlgear interlocks etc. must be checked to ensure that they are correctly mounted and adjusted and that they function correctly.

Periodic inspection

Periodic inspection and testing

Circumstance which require a periodic inspection and test

Test and inspection is due; insurance, mortgage, licensing reasons; change of use; change of ownership; after additions or alterations; after damage; change of loading; to assess compliance with current Regulations.

General reasons for a periodic inspection and test

1 To ensure the safety of persons and livestock.
2 To ensure protection of property from fire and heat.
3 To ensure that the installation is not damaged so as to impair safety.
4 To ensure that the installation is not defective and complies with the current Regulations.

General areas of investigation

Safety; wear and tear; corrosion; damage; overloading; age; external influences; suitability; effectiveness.

Documentation to be completed

Periodic test report, schedule of test results and an inspection schedule.

Sequence of tests

1 Continuity of all protective conductors
2 Continuity of ring final circuit conductors
3 Insulation resistance
4 Site applied insulation
5 Protection by separation of circuits
6 Protection against direct contact by barriers and enclosures provided during erection
7 Insulation of non-conducting floors and walls
8 Polarity
9 Earth electrode resistance
10 Earth fault loop impedance
11 Prospective fault current
12 Functional testing

or in any order to suit.

This could be so simple. As it is, periodic inspection and testing tends to be complicated and frustrating. On the domestic scene, I doubt if any house owner actually decides to have a regular inspection. They say, 'If it works it must be OK'. It is usually only when there is a change of ownership that the mortgage companies insist on an electrical survey. The worst cases are, however, industry and commerce. Periodic inspections are requested, reluctantly, to satisfy insurers or an impending visit by the HSE. Even then it is usually the case that 'you can't turn that off' or 'why can't you just test this bit and then issue a certificate for the whole lot'. Under the rare circumstances when an inspection and test is genuinely requested it is difficult to convince the client that, as there are no drawings, or information about the installation, and that no switchgear is labelled etc., you are going to be on site for a considerable time and at a considerable cost.

When there are no drawings or items of information, especially on a large installation, there may be a degree of exploratory work to be carried out in order to ensure safety whilst inspecting and testing. If it is felt that it may be unsafe to continue with the inspection and test, then drawings and information **must** be produced in order to avoid contravening the Health and Safety at Work Act Section 6.

However, let us assume, as with the initial inspection, that the original installation was erected in accordance with the 16th edition, and that any alterations and/or additions have been faithfully recorded on the original documentation which is, of course, readily available!

A periodic inspection and test under these circumstances should be relatively easy, as little dismantling of the installation will be necessary, and the bulk of the work will be inspection.

Inspection should be carried out with the supply disconnected as it may be necessary to gain access to wiring in enclosures etc. and hence, with large installations it will probably need considerable liaison with the client to arrange convenient times for interruption of supplies to various parts of the installation.

This is also the case when testing protective conductors, as these must *never* be disconnected unless the supply can be isolated. This is particularly important for main equipotential bonding conductors which need to be disconnected in order to measure Z_e.

In general an inspection should reveal:

1 Any aspects of the installation that may impair the safety of persons and livestock against the effects of electric shock and burns.
2 That there are no installation defects that could give rise to heat and fire and hence damage property.
3 That the installation is not damaged or deteriorated so as to impair safety.
4 That any defects or non-compliance with the Regulations, that may give rise to danger, are identified.

As was mentioned earlier, dismantling should be kept to a minimum and hence a certain amount of sampling will take place. This sampling would need to be increased in the event of defects being found.

From the testing point of view, not all of the tests carried out on the initial inspection may need to be applied. This decision depends on the condition of the installation.

The continuity of protective conductors is clearly important as is insulation resistance and loop impedance, but one wonders if polarity tests are necessary if the installation has remained undisturbed since the last inspection. The same applies to ring circuit continuity as the P–N test is applied to detect interconnections in the ring, which would not happen on their own.

It should be noted that if an installation is effectively supervised in normal use, then Periodic Inspection and Testing can be replaced by regular maintenance by skilled persons. This would only apply to, say, factory installations where there are permanent maintenance staff.

Certification

Having completed all the inspection checks and carried out all the relevant tests, it remains to document all this information. This is done on electrical installation certificates, inspection schedules, test schedules, test result schedules, periodic inspection and test reports, minor works certificates and any other documentation you wish to append to the foregoing. Examples of such documentation are shown in the IEE Guidance Notes 3 on inspection and testing.

This documentation is vitally important. It has to be correct and signed or authenticated by a competent person. Electrical installation certificates and periodic reports must be accompanied by a schedule of test results and an inspection schedule for them to be valid. It should be noted that three signatures are required on an electrical installation certificate, one in respect of the design, one in respect of the construction and one in respect of the inspection and test. (For larger installations there may be more than one designer, hence the certificate has space for two signatures, i.e. designer 1 and designer 2.) It could be, of course, that for a very small company, one person signs all three parts. Whatever the case, the original must be given to the person ordering the work, and a duplicate retained by the contractor.

One important aspect of the electrical installation certificate is the recommended interval between inspections. This should be evaluated by the designer and will depend on the type of installation and its usage. In some cases the time interval is mandatory, especially where environments are subject to use by the public. Guidance Notes 3 give recommended maximum frequencies between inspections.

A periodic report form is very similar in part to an electrical installation certificate in respect of details of the installation, i.e. maximum demand, type of earthing system, Z_e, etc. The rest of the form deals with the extent and limitations of the inspection and test, recommendations, and a summary of the installation. The record of the extent and limitations of the inspection is very important. It must be agreed with the client or other third party exactly what parts of the installation will be covered by the report and those that will not. The interval until the next test is determined by the inspector.

With regard to the schedule of test results, test values should be recorded unadjusted, any compensation for temperature etc. being made after the testing is completed.

Any alterations or additions to an installation will be subject to the issue of an electrical installation certificate, except where the addition is, say, a single point added to an existing circuit, then the work is subject to the issue of a minor works certificate.

Table 19.8

Symptom	Possible common cause	Diagnosis	Action
Complete loss of supply	1 Fault on suppliers (REC) main cable/equipment 2 Fault on service cable 3 Main fuse or CB operated 4 Main DB switch OFF 5 Main RCD operated	1 Check adjacent properties are also OFF 2 Check adjacent properties are ON 3 Check adjacent properties are ON 4 Visual check 5 Visual check	1 Contact REC 2 Contact REC 3 Contact REC 4 Switch back on 5 Re-set, if it trips, then switch off all CB's, re-set and turn on each CB until one causes the main RCD to operate. This is the likely faulty circuit
Loss of supply to a circuit	1 Circuit fuse or CB operated 2 Conductorbroken or out of terminal	1 Visual check 2 Check fuse/CB are OK	1 Replace or re-set as operation may be due to an overload. If Protection still operates, do NOT reset until fault has been found, usually by carrying out Insulation tests 2 Locate fault by carrying out visual check/continuity tests
Fire/burning	1 Overloaded cable 2 Damaged insulation 3 Water in fittings/accessories	1, 2 & 3 visual check and smell	1, 2 & 3 Turn off supply, to circuit(s) investigate fuse and cable sizes, check for water ingress, damaged insulation, visually and using an insulation resistance tester
Electric shock	1 Exposed live parts 2 Insulation breakdown 3 Earthing and bonding inadequate 4 Appliances incorrectly wired or damaged and with inappropriate fusing 5 Incorrect polarity in accessories	Use of an approved voltage indicator between exposed and/ or extraneous conductive parts	Turn off supply to circuit(s) check visually for covers missing etc. Carry out Insulation resistance and polarity tests on circuits and cables, and establish that all earthing and bonding is in place and that all protective devices are suitable for disconnection times

Table 19.9

General cause	Details
Insulation breakdown	1 Damage by installer 2 Damage by other trades Damage by user (misuse, nails in walls etc.) 3 Overloading
Fuse. CB or RCD operating instantly circuit is switched on	1 Short circuit caused by (a) damaged insulation (b) crossed polarity at terminations (c) water penetration in JB's, seals, glands etc. 2 Faulty appliances
Fuse or CB operating regularly after a period of time	1 Overload caused by too many loads on a circuit, or machinery stalling or with too much mechanical load 2 Slight water penetration or general dampness
Fuse or CB operates with no apparent fault	Transient over-voltage caused by switching surges, motor starting etc.

Summarising:

(i) The addition of points to existing circuits require a Minor Works Certificate.
(ii) A new installation or an addition or alteration that comprises new circuits requires an Electrical Installation Certificate.
(iii) An existing installation requires a Periodic Test Report.

Note: (ii) and (iii) must be accompanied by a schedule of test results and an inspection schedule.

Inspection and testing

As the client/customer is to receive the originals of any certification, it is important that *all* relevant details are completed correctly. This ensures that future inspectors are aware of the installation details and test results which may indicate a slow progressive deterioration in some or all of the installation.

These certificates, etc. will also form part of a 'sellers pack' when a client wishes to sell a property.

The following is a general guide to completing the necessary documentation and should be read in conjunction with the examples given in BS 7671 and the On-site-guide.

Electrical installation certificate

1 **Details of client**

Name: *Full name*
Address: *Full address and post code*
Description: *Domestic, industrial, commercial*
Extent: *What work has been carried out, e.g. full re-wire, new shower circuit, etc. Tick a relevant box.*

2 **Designer/constructor/tester**

Details of each or could be one person.

Note: Departures are not faults, they are systems/equipment, etc. that are not detailed in BS 7671 but may be perfectly satisfactory.

3 **Next test**

When the next test should be carried out is decided by the designer.

4 **Supply characteristics and earthing arrangements**

Earthing system: *Tick relevant box (TT, TN-S etc.)*
Live conductors: *Tick relevant boxes*
Nominal voltage: *Obtain from supplier, but usually 230 V single phase U and U_0 but 400 V U and 230 U_0 for three phase*
Frequency: *From supplier but usually 50*
PFC: *From supplier or measured. Supplier usually gives 16 kA*
Z_e: *From supplier or measurement. Supplier usually gives 0.8 Ω for TN-S; 0.35 Ω for TN-C-S and 21 Ω for TT systems*
Main fuse: *Usually BS 1361, rating depends on maximum demand.*

5 **Particulars of installation**

Means of earthing: *Tick 'suppliers facility' for TN systems, 'earth electrode' for TT systems*
Maximum Demand: *Value without diversity*
Earth electrode: *Measured value or N/A*
Earthing and bonding:
Conductors: *Actual sizes and material, usually copper*
Main Switch or Circuit breaker (could be separate units or part of a consumer control unit): *BS number; Rating, current and voltage; Location; 'not address', i.e. where is it located in the building; Fuse rating if in a switch-fuse, else N/A; RCD details only if used as a main switch.*

6 **Comments on existing installation**

Write down any defects found in other parts of the installation which may have been revealed during an addition or an alteration.

7 **Schedules**
Indicate the number of test and inspection 'schedules' that will accompany this certificate.

Periodic inspection report

1 **Details of client**

Name:	*Full name (could be a landlord, etc.)*
Address:	*Full address and post code (may be different to the installation address)*
Purpose:	*E.g. Due date; change of owner/tenant; change of use, etc.*

2 **Details of installation**

Occupier:	*Could be the client or a tenant*
Installation:	*Could be the whole or part (give details)*
Address:	*Full and post code*
Description:	*Tick relevant box*
Age:	*If not known, say so, or make an educated guess*
Alterations:	*Tick relevant box and insert age where known*
Last inspection:	*Insert date or 'not known'*
Records:	*Tick relevant box*

3 **Extent and limitations**
Full details of what is being tested (extent) and what is not (limitations). If there is not enough space on the form add extra sheets.

4 **Next inspection**
Filled in by inspector and signed, etc. under declaration.

5 **Supply details**
As per an Electrical Installation Certificate.

6 **Observations**
Tick relevant box, if work is required, record details and enter Relevant code (1, 2, 3 or 4) in space on right-hand side.

7 **Summary**
Comment on overall condition. Only common sense and experience can determine whether satisfactory or unsatisfactory.

8 **Schedules**
Attach completed schedules of inspections and test results.

Minor electrical installation works certificate

Only to be used when simple additions or alterations are made, *not when a new circuit is added.*

1 **Description:**	*Full description of work*
Address:	*Full address*
Date:	*Date when work was carried out*
Departures:	*These are not faults, they are systems/equipment, etc. that are not detailed in BS 7671 but may be perfectly satisfactory (this is usually N/A)*

2 **Installation details**

Earthing: *Tick a relevant box*

Protection against indirect contact: *99% of the time this will be EEBADS. Other methods should be recorded*

Protective device: *Enter type and rating. E.g. BS EN 60898 cb type B, 20 A*

Comments: *Note any defects/faults/omissions in other parts of the installation seen while conducting the minor works*

3 **Tests**

Earth continuity: *Measure and then tick in box if OK*

Insulation resistance: *Standard tests and results*

EFLI (Z_s): *Standard tests and results*

Polarity: *Standard tests, then tick in box if OK*

RCD: *Standard tests, record operating and current time*

4 **Declaration**

Name, address, signature, etc.

Schedule of test results (as per BS 7671)

1. Contractor: *Full name of tester*
2. Date: *Date of test*
3. Signature: *Signature of tester*
4. Method of protection against indirect contact: *99% EEBADS but could be SELV, etc.*
5. Vulnerable equipment: *Dimmers, electronic timers, CH controllers, etc., i.e. anything electronic*
6. Address: *Full, or if in a large installation, the location of a particular DB*
7. Earthing: *Tick the relevant box*
8. Z_e at origin: *Measured value*
9. PFC: *Record the highest value, i.e. PEFC or PSCC (should be the same for TN-C-S)*
10. Instruments: *Record serial numbers of each instrument, or one number for a composite instrument*
11. Description: *Suggest Initial or Periodic or whatever part of the installation is involved. E.g. Initial verification on a new shower circuit.*
12. kVA rating: *Taken from the device (difficult when there are different devices in an installation). Nothing to stop adding sheets to this form!*
13. Type and rating: *E.g. BS EN 60898 cb type B, 32 A, or BS 88 40 A, etc.*
14. Wiring conductors: *Size of Live and CPC, e.g. 2.5 mm²/1.5 mm²*
15. Test results: *Fill in all measured values (R1 + R2), etc. Tick box if ring P-N is OK. If any test does not appear on the sheet, e.g. 5 × $I_{\Delta n}$,* write the results in the remarks column

Schedule of inspections (as per BS 7671)

Do not leave boxes uncompleted.

N/A in a box if it is not relevant
✔ in a box if it has been inspected and is OK
✘ in a box if it has been inspected and is incorrect.

Fault finding

This is not an exact science as faults in electrical systems can be many, varied and difficult to locate. What we can state however, are the main symptoms of electrical faults, these are:

- loss of supply
- fire
- shock.

Table 19.8 indicates such symptoms, their possible common causes and the action to be taken. Column 2 illustrates, in general terms, the possible causes of faults. Table 19.9 summarises these in more detail.

Many faults are easily located, many are not, in all cases observe the following general procedure whenever possible:

1 Determine the nature/symptom of the problem.
2 Ask client/personnel for their recollections: how, when and where the problem occurred (this can save so much time).
3 Carry out relevant visual and instrument checks to locate the fault.
4 Rectify if possible.
5 Re-test.
6 Re-instate system.

20 Basic electronics technology

Most of us in the world of electrical installation work are familiar with values such as 240 V, 3 kW or 60 A. We tend to view the terms and quantities used in electronics with a certain unease, but the relationship between ohms, volts and amperes in the world of electronics is no different from that in installation work.

In order to begin to feel more at home with electronics, we first look at some of the many components used.

Electronics components

Resistors

The ohmic value of 240 V appliances rated at 60 W, 1000 W, 3 kW, etc., should be familiar to us by now, but the values of resistors used in electronics are many and varied. In order to identify readily all the different values, a colour code is used; the same code is used for capacitor values.

Resistor (or Capacitor) Colour Code	
Colour	Value
Black	0
Brown	1
Red	2
Orange	3
Yellow	4
Green	5
Blue	6
Violet	7
Grey	8
White	9

Tolerance Colour Code	
Colour	Percentage
Brown	1%
Red	2%
Gold	5%
Silver	10%
None	20%

Each resistor carries a series of coloured bands to indicate its value and tolerance:

Three bands for a resistor with tolerance of 20%.
Four bands for a resistor with tolerance of between 10% and 2%.
Five bands for a resistor with tolerance of 1%.

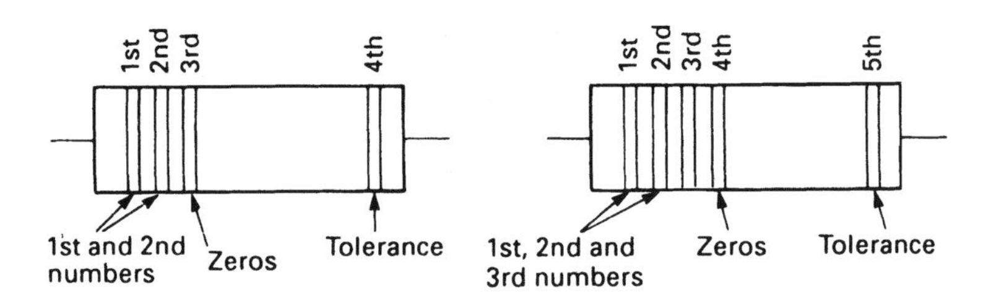

Fig. 20.1 *Coding bands*

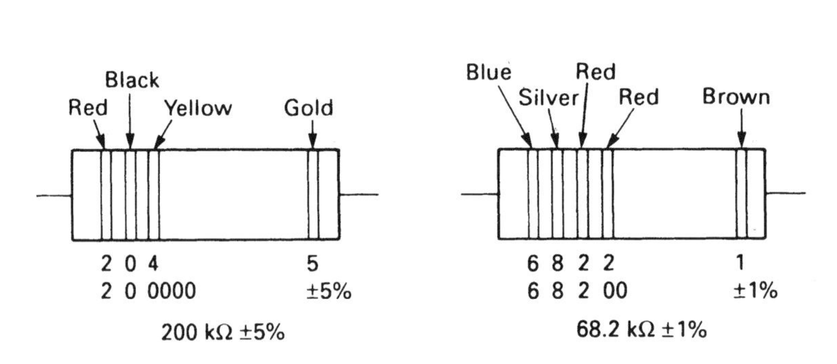

Fig. 20.2 *Colour values decoded*

These bands are interpreted as shown in Fig. 20.1. Fig. 20.2 shows two examples of how to 'read' the colour coding to ascertain the value and tolerance.

Sometimes a resistor code will use numbers and letters rather than colours. The letters used are as follows:

R, K, M indicate multiples of 1 Ω, 1000 Ω, 1 000 000 Ω respectively.
F, G, J, K, M indicate tolerances of 1%, 2%, 5%, 10%, 20% respectively.

The R, K, M code can also be used for decimal points, e.g.

3.3 kΩ may be shown as 3K3 Ω.
0.2 Ω may be shown as R2 Ω.

Other examples are:

2K2G indicates a value of 2.2 kΩ (2200 Ω) ± 2%.
6M8J indicates a value of 6.8 MΩ (6 800 000 Ω) ± 5%.
33KF indicates a value of 33.0 kΩ (33 000 Ω) ± 1%.
470RM indicates a value of 470.0 Ω ± 20%.

Resistor types

There are three types of resistor in common use: carbon, wirewound and carbon pre-set or variable.

The wire-wound type is usually chosen where high voltage is present; it is also more accurate than the carbon variety. The pre-set type, shown in Fig. 20.3 is used in the simple metal detectors used by electricians and to adjust EXIT and ENTRY times in alarm panels.

Fig. 20.3 *Pre-set resistor*

Capacitors

There are a number of ways of marking a capacitor with its value. The most common, apart from actually writing the value on the capacitor, is to use the same colour code as for resistors (Fig. 20.4). Remember, small capacitors are usually in the picofarad (pF) range (1 nanofarad (nF) = 1000 pF).

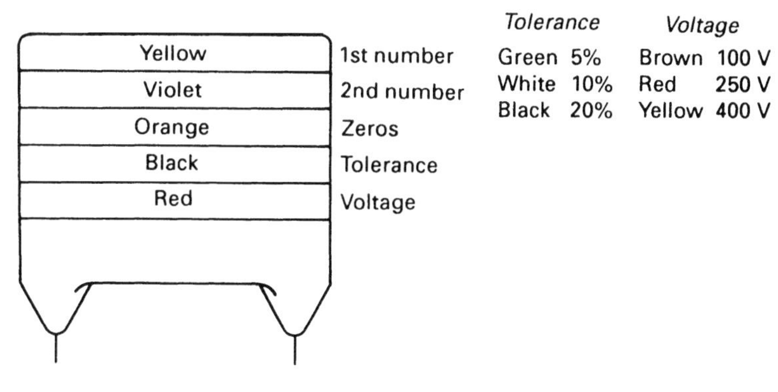

Fig. 20.4 *Capacitor colour coding*

One other method commonly used is the three-digit method, in which the third digit gives the number of zeros that follow the first two digits, to give the value in pF: e.g. 104 is the code for a capacitor of 100 000 pF or 100 nF.

Capacitor types

The many types of capacitor in use range from waxed paper and foil, electrolytic, polyester, mica and ceramic, to air – all of which have different applications depending on frequency, voltage, supply (a.c. or d.c.), losses, etc.

Inductors and transformers

Inductors and transformers have already been discussed in Chapter 4. The types used in electronics work on the same principles. Probably the main difference is the use of ferrite as a core for inductors that are used to tune for radio frequencies. This is a much more efficient material than iron as the hysteresis losses are much less.

Semi-conductors

These devices are neither strictly conductors nor insulators but, under certain circumstances, can become either.

Silicon is the most common semi-conductor material, and the addition of impurities such as aluminium or arsenic creates the circumstances under which it will conduct or insulate. For example, if we add aluminium to a sample of silicon it becomes what is known as a 'p'-type material; adding arsenic to the sample makes it an 'n'-type material.

The junction diode

If we now take a sample of 'p'-type and a sample of 'n'-type silicon and join them together we will have a 'junction diode'. By connecting a positive charge to the 'p'-type sample and a negative charge to the 'n'-type, the whole assembly will act as a conductor. Reversing the connections will result in the arrangement acting as an insulator (Fig. 20.5). Hence it is used commonly as a means of rectification.

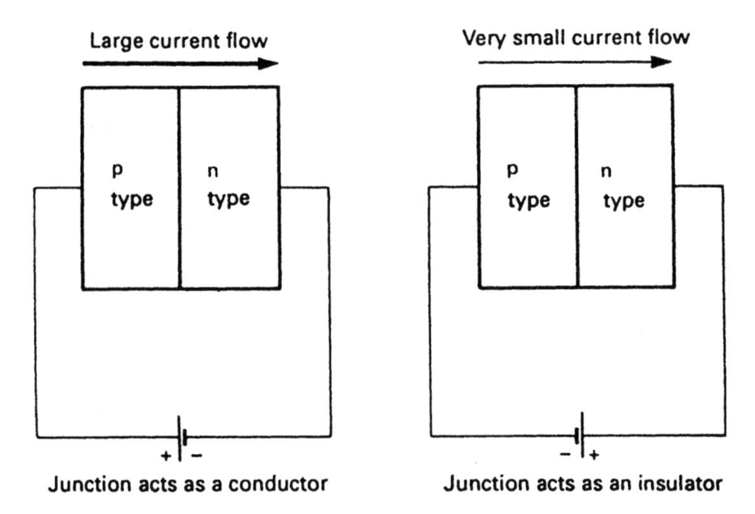

Fig. 20.5 *Semi-conductor*

Fig. 20.6

The symbol for the diode is shown in Fig. 20.6. The diode will conduct when current flow is in the direction of the arrow.

Fig. 20.7 shows how the forward and reverse current for a silicon diode vary with the applied voltage. At point X in the graph the reverse voltage is so great that the diode breaks down and conducts. This value is called the reverse breakdown voltage.

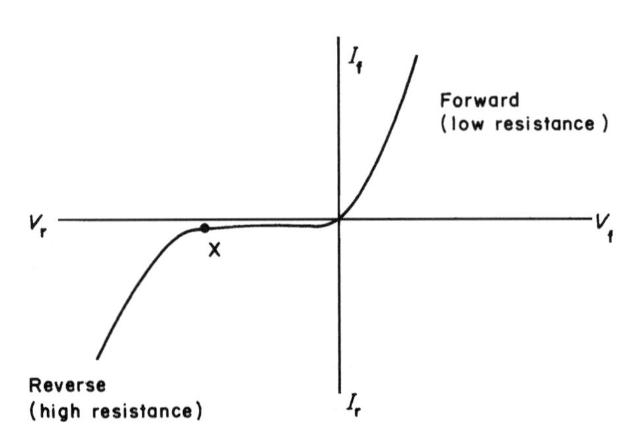

Fig. 20.7

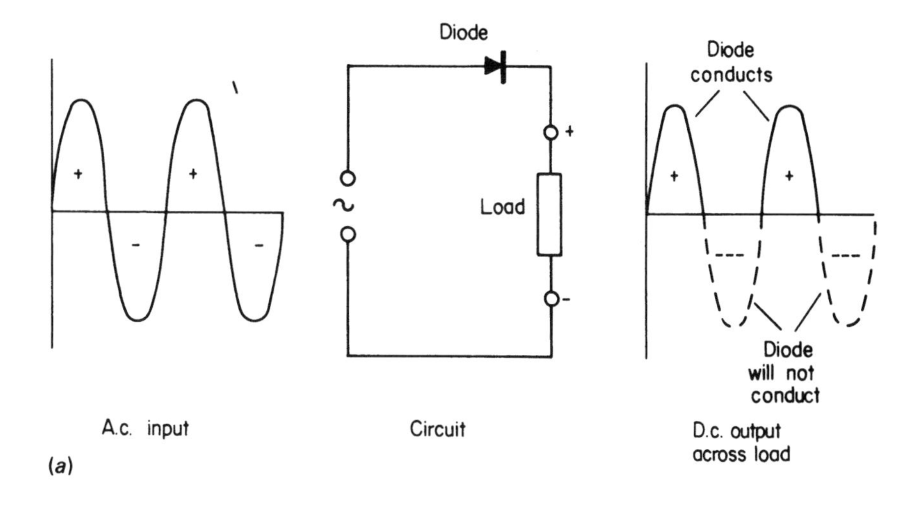

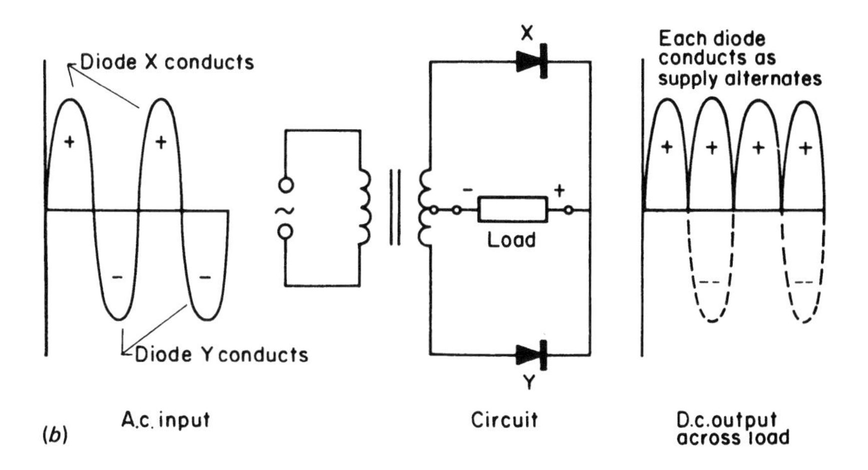

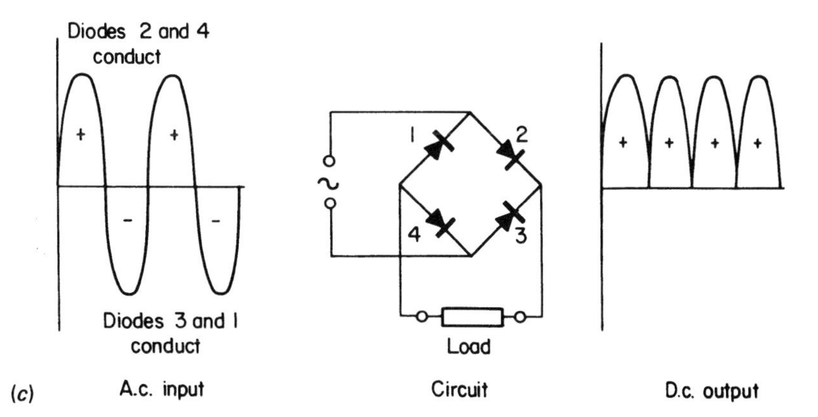

Fig. 20.8 *(a) Half-wave rectification; (b) full-wave rectification from transformer supply; (c) bridge-type full-wave rectification*

Rectification

Most electricity supply systems are a.c. and since many appliances require a d.c. supply it is necessary to change a.c. to d.c. This change is called *rectification*.

Fig. 20.8a illustrates how a diode or group of diodes can be used to rectify an a.c. supply. It can be seen that the rectified d.c. output is not true d.c., for which the waveform would be a straight line, but has something of a pulsating nature. This type of output is usually quite acceptable for most purposes in installation work. Should a more refined or smoothed output be required, the addition of capacitance and inductance (Fig. 20.9) can provide this.

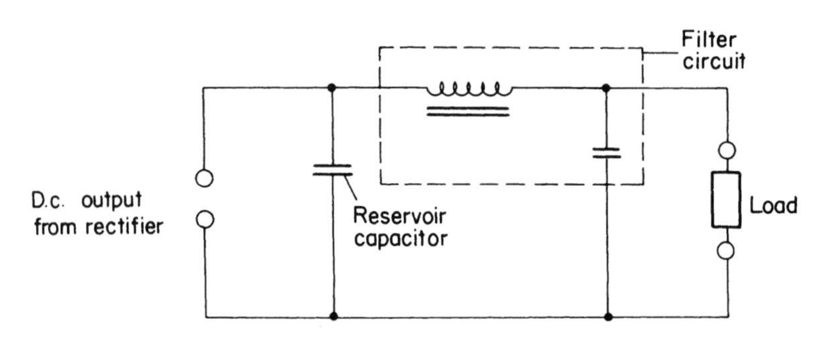

Fig. 20.9

Rectifier output

Since the output from a full-wave rectifier is a series of sinusoidal pulses, the average value of this output is given by

Average value = maximum or peak value × 0.637

and for half-wave

$$\text{Average value} = \frac{\text{maximum or peak value} \times 0.637}{2}$$

Note: When selecting a diode for a particular duty, ensure that it is capable of operating at the peak voltage. For example, consider a circuit that is required to operate at 12 V d.c. and a 240/12 V transformer is to be used in conjunction with diodes.

The 12 V output from the transformer is 12 V r.m.s. and therefore has a peak value of

$$\frac{12}{0.7071} = 16.97\text{ V}$$

The diode must be able to cope with this peak voltage.

Example

Calculate the average value of a full-wave rectified d.c. output if the a.c. input is 16 V.

$$\text{Peak value of a.c. input} = \frac{16}{0.7071} = 22.61\ \text{V}$$

Average value of d.c. output = 22.61 × 0.637 = 14.4 V

Applications

1. D.c. machines (supply).
2. Bell and call systems.
3. Battery charging.
4. Emergency lighting circuits.

Thyristors or silicon-controlled rectifiers

A thyristor is a four-layer *p–n–p–n* device with three connections (Fig. 20.10).

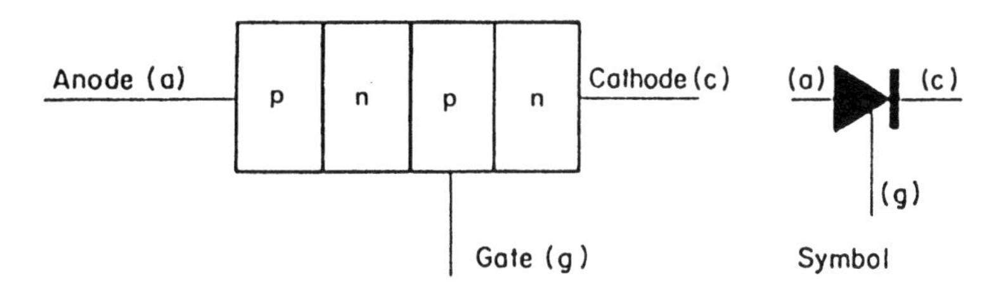

Fig. 20.10 *Thyristor*

Under normal circumstances, with positive on the anode and negative on the cathode, the thyristor will not conduct. If, however, a large enough positive firing potential is applied to the 'gate' connection, the thyristor will conduct and will continue to do so even if the signal on the gate is removed. It will cease to conduct when the anode potential falls below that of the cathode. The device, like the diode, will not conduct at all in the reverse direction.

Let us now consider what happens when a thyristor is wired in an a.c. circuit as shown in Fig. 20.11. Resistor R_1 ensures that the minimum gate potential required for firing is maintained and R_3 ensures that the gate potential does not rise to a level that could cause damage to the gate circuit. Variable resistor R_2 enables various gate potentials to be selected between maximum and minimum, and it will also be seen that, as the circuit is resistive, the applied voltage and the gate voltage are in phase (Fig. 20.12).

It will be seen from Fig. 20.13 that if the gate voltage is adjusted to a maximum, the gate will fire at point A. When it is at a minimum it fires at point B.

The current flowing in a resistive circuit is also in phase with the voltage and in this case it will flow only when the thyristor is conducting. Since this conduction takes place only when the thyristor is triggered by the gate, the

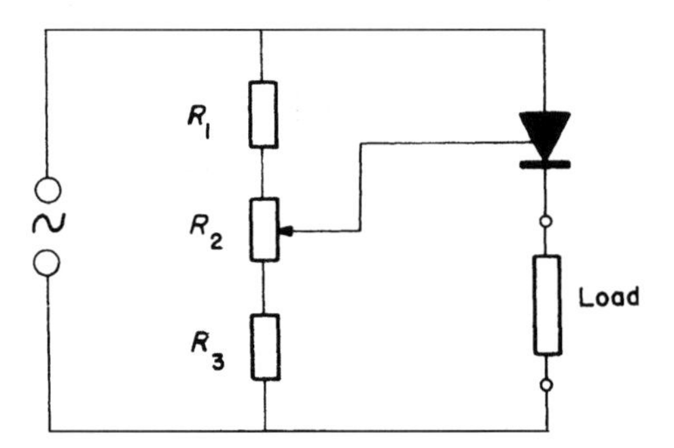

Fig. 20.11

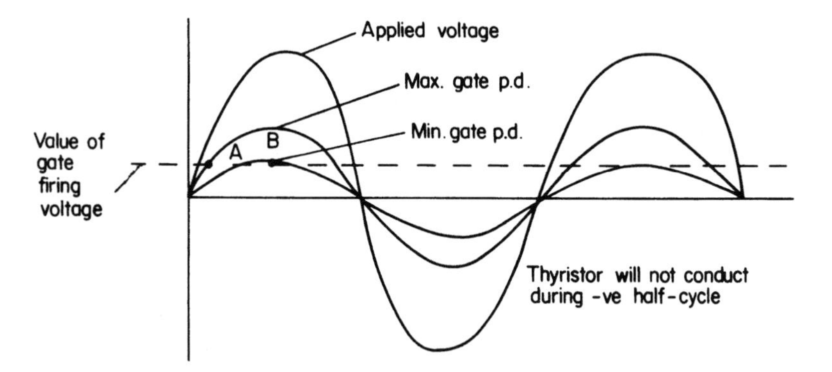

Fig. 20.12

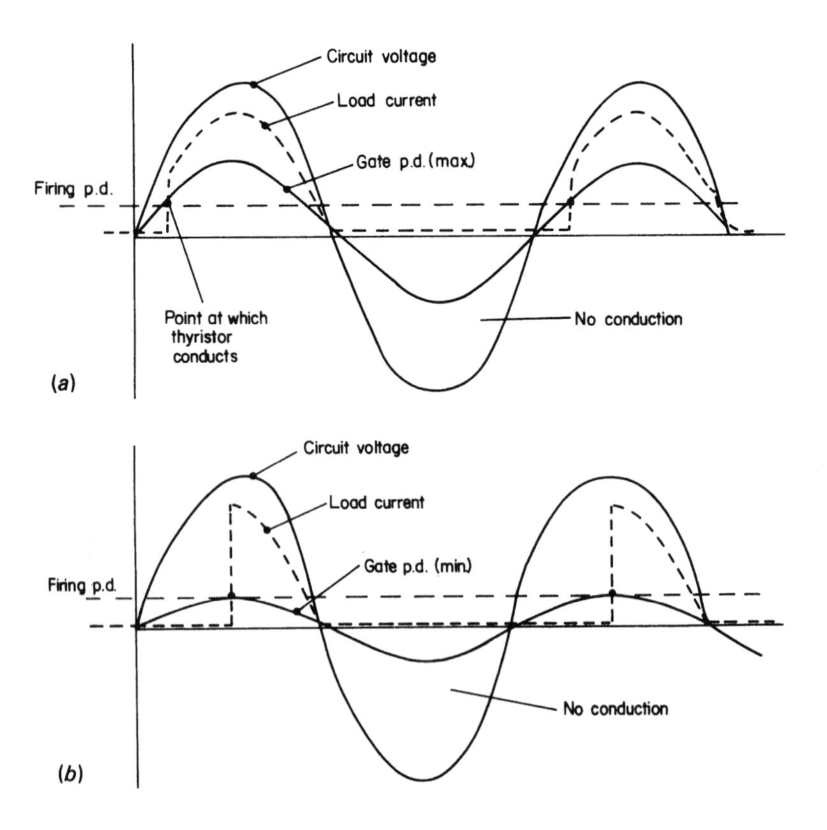

Fig. 20.13 *(a) Maximum gate potential; (b) minimum gate potential*

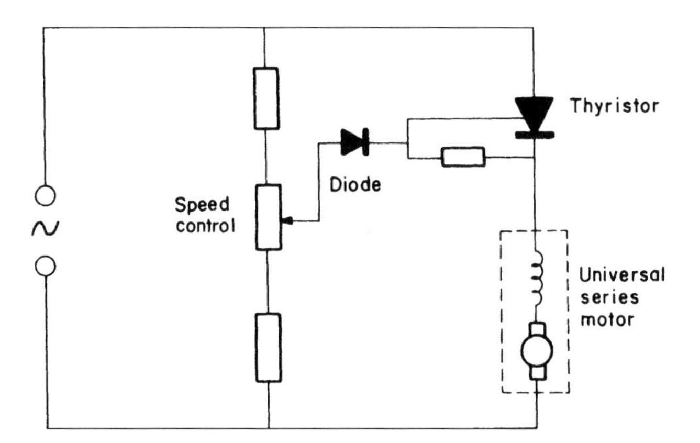

Fig. 20.14 *Basic speed control circuit for a motor*

amount of current flowing in any +ve half-cycle can be controlled by the gate potential (Fig. 20.13a and b).

This control of the amount of current flowing in each half-cycle can be used to control the speed of small motors such as those used in food mixers and hand drills. A simple circuit is shown in Fig. 20.14.

More complicated circuitry is now in use, utilizing thyristors, in order to control the speed of induction motors, something that in the past proved very difficult.

Transistor

The transistor is basically a semi-conductor which can act as a switch and an amplifier. It is manufactured from silicon just as diodes are, and could be thought of as two diodes 'back-to-back' (Fig. 20.15a and b).

The connections to a transistor are known as the emitter, collector and base and the symbols for transistors are as shown in Fig. 20.16a and b.

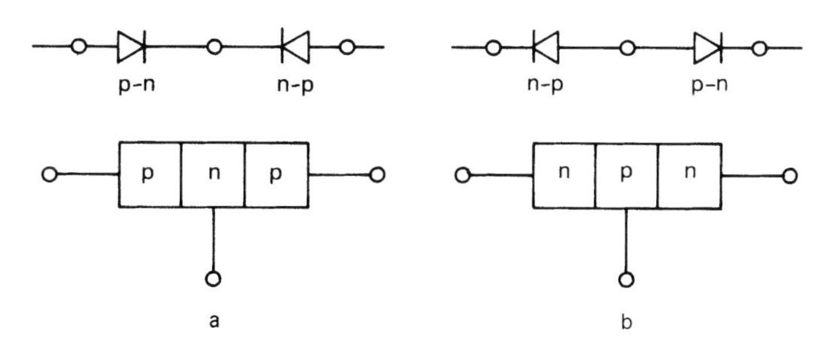

Fig. 20.15 *(a) P–n–p type; (b) n–p–n type*

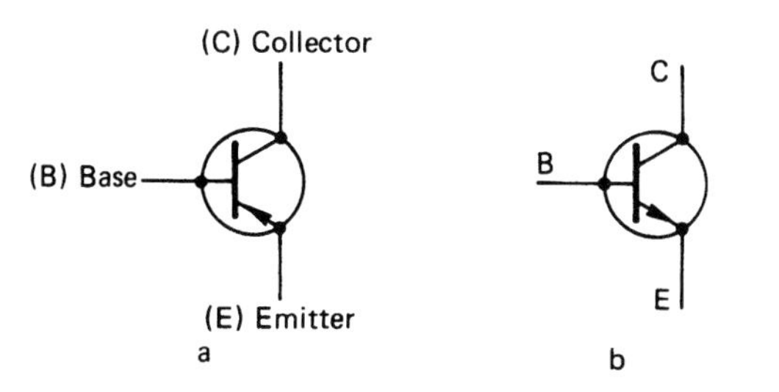

Fig. 20.16 *(a) P–n–p; (b) n–p–n*

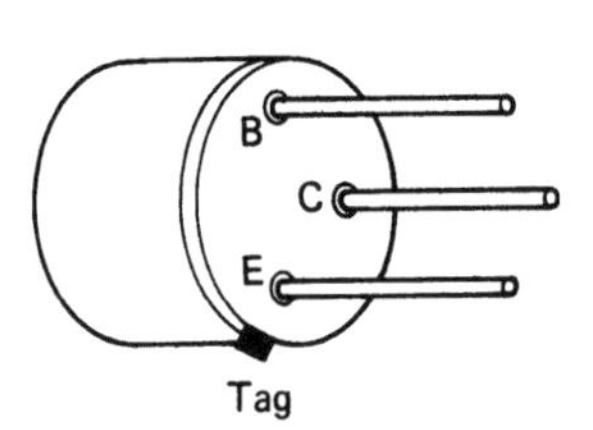

Fig. 20.17

To identify transistor connections, turn the device so that the connection leads are facing you and with the tag at the bottom left, the connections are as shown in Fig. 20.17.

There are two types of transistor: the field effect transistor (FET) and the bi-polar type. FETs are extremely small and are used mainly in ICs, whereas the bi-polar type, being much larger, is used extensively as a stand-alone component in circuitry.

Basis transistor action

When a small voltage is applied to the base, it switches the transistor on, and allows current to flow from the collector to the emitter and vice versa. If this current is passed through a resistor a voltage will be developed across that resistor which can be many times greater than the input voltage to the base; hence we have amplification.

Other commonly used components

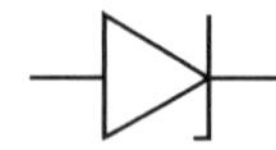

Zener diode

This is used in circuits to give voltage control/stabilization.

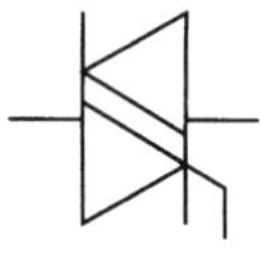

Triac

Triacs are often used 'back-to-back' to provide a smoother and more efficient thyristor effect. They will be found in dimmer switches.

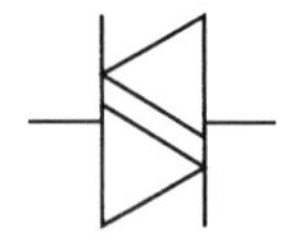

Diac

Used in conjunction with triacs as a triggering device.

Thermistor

Used as a means of sensing temperature change. Commonly found embedded in motor windings to detect overheating.

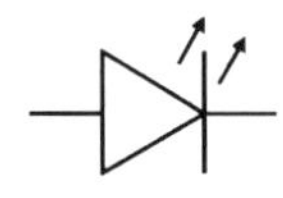

LED

Light-emitting diode; this is simply a semi-conductor signal lamp.

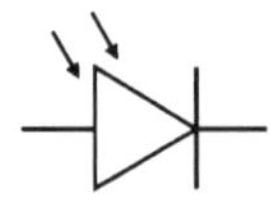

LSD or photodiode

Light-sensitive diode, used to activate a circuit in response to light.

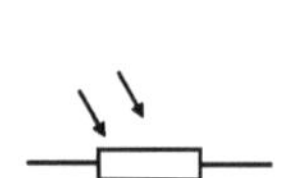

LDR

The light dependent resistor, is similar to the LSD but is able to handle much larger currents. It is used as a switching device in such areas as street lighting etc.

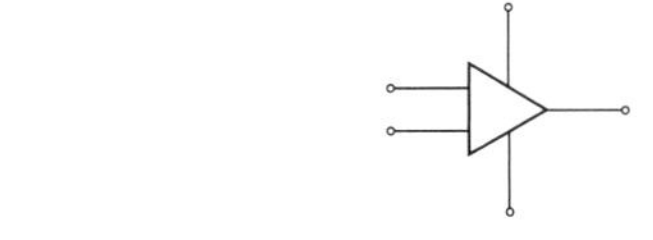

IC

The integrated circuit or *chip* as it is often called is a minute electronic circuit comprising many hundreds of thousands of microscopic components.

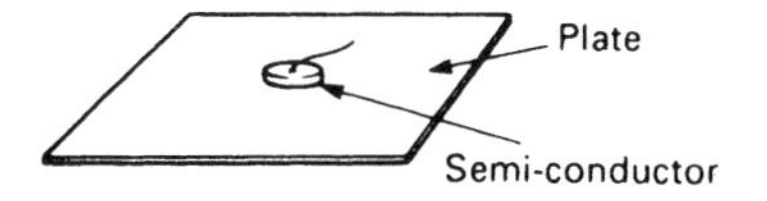

Fig. 20.18 *Heat sink*

Heat sinks

In many instances, semi-conductors will become hot when in operation and would be damaged if this heat were not dissipated. Embedding the device in the centre of a large plate or series of plates helps to dissipate the heat to the surrounding air (Fig. 20.18).

Electronics diagrams

Three main types of diagram are used:

1. **Block diagrams** (Fig. 20.19) are used to give a general indication of a complete system.

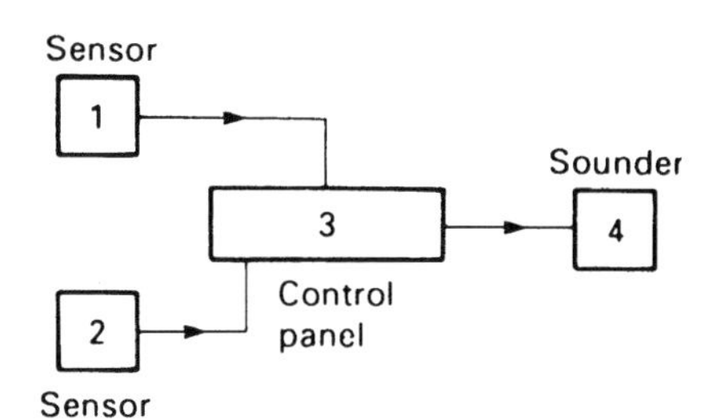

Fig. 20.19 *Block diagram of an intruder alarm system*

2. **Circuit diagrams** (Fig. 20.20), as in installation work, indicate how a circuit *works*.

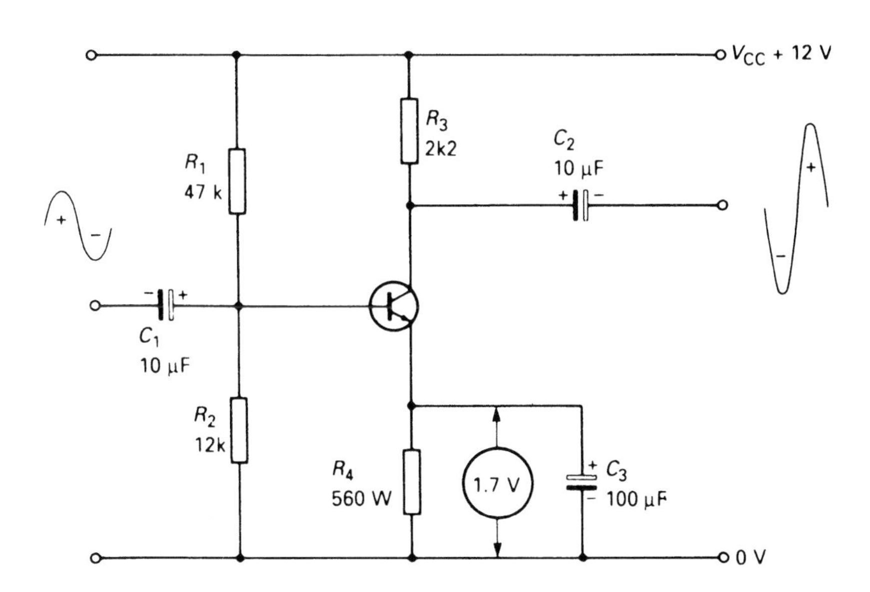

Fig. 20.20 *Simple transistor amplifier circuit*

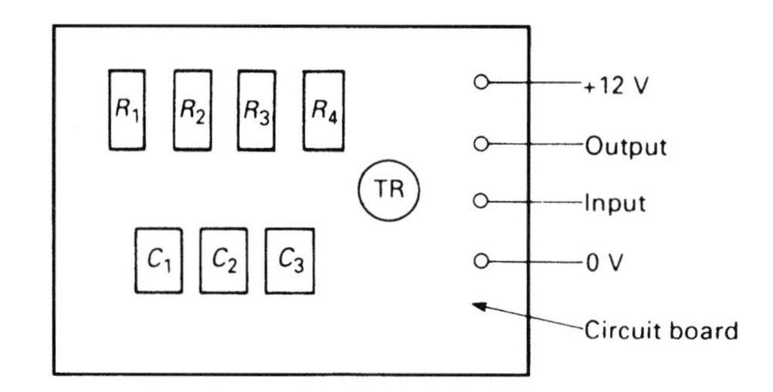

Fig. 20.21 *Layout diagram of an amplifier circuit*

3. **Layout diagrams** (Fig. 20.21) are similar to wiring diagrams in installation work: they show how a circuit should be *wired* and components are shown in their correct locations.

Note: Cross-reference should be possible between circuit and layout diagrams by means of pin numbers of a positional reference system.

Electronics assembly

Unlike installation circuits, electronic circuitry is almost entirely constructed using soldered joints. The formation of such joints is critical to ensure healthy circuit performance.

Soldering

A soldered joint comprises the surfaces to be joined and a material (solder) which is an alloy of 40% tin and 60% lead, melted on to the surfaces. To aid the soldering process a flux is used and usually this is incorporated in the solder.

Cleanliness is vital to ensure a good soldered joint – cleanliness not only of the surfaces to be joined but also of the 'bit' of the soldering iron.

Remember, when soldering, that *too much* solder will make a poor joint and that the iron should be at the correct temperature. Many modern irons have built-in temperature controls. Fig. 20.22 shows some effects of soldering faults.

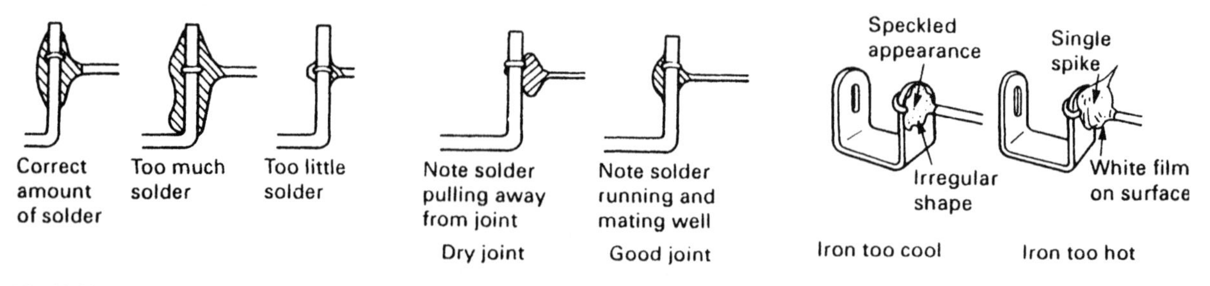

Fig. 20.22 *Soldered joints*

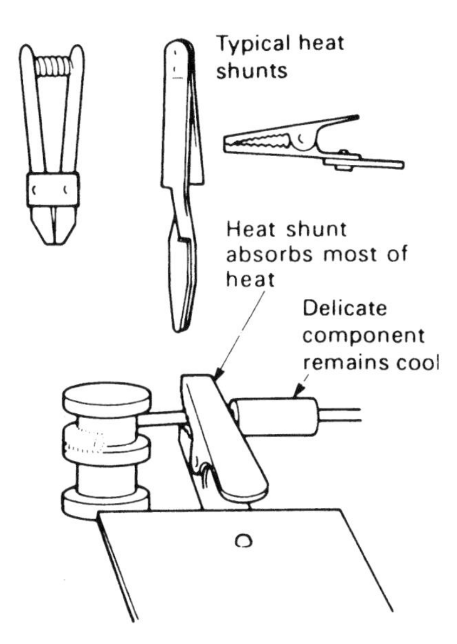

Fig. 20.23 *Heat shunts*

Heat shunts

As heat can damage electronics components, it is important to ensure that too much heat does not reach the component during the soldering process. Heat shunts are used for this purpose and can simply consist of a pair of pliers or a crocodile clip attached to the component lead. This helps to dissipate the heat before it can reach the component itself (Fig. 20.23).

De-soldering

It may be necessary on occasion to remove a component from a circuit. The process is similar to that of soldering; the joint is flooded with new hot solder and, as a heat balance is achieved, the old solder softens. When the whole lot is fluid, the solder is sucked away from the joint using either a copper braid or a specially designed solder-sucker.

Handling electronic circuits

Many electronic components are very sensitive to small electrostatic charges and often are damaged when subjected to these. In order to avoid such damage, work benches and the immediate floor area are supplied with anti-static mats and the operator has an earth-tag around his/her wrist to ensure there is a discharge of static.

Self-assessment questions

1 Write down the value of the resistors with the following colour codes:
(a) Orange–silver–red–silver
(b) Red–white–green–red
(c) Green–green–black–gold–brown
(d) Brown–red–gold–silver.

2 Give the colour code for the following values of resistor:
(a) 8.7 K 1%
(b) 15K8J
(c) 100RG
(d) 350PK 20%.

3 Indicate the values of the following colour-coded capacitors:
(a) Brown–silver–gold–green–brown
(b) Red–violet–gold–white–yellow
(c) Violet–green–gold–black–brown
(d) Brown–black–yellow–white–red.

4 Write down the colour code for the following capacitors:
(a) 25 nF 20% 400 V
(b) 53 nF 10% 100 V
(c) 15 nF 5% 250 V
(d) 200 nF 20% 250 V.

5 Explain with the aid of diagrams the difference between a heat sink and a heat shunt.

6 Draw the BS 3939 symbol for the following components:
(a) an LED
(b) a pnp transistor
(c) an iron-cored inductor
(d) a variable capacitor.

7 With the aid of drawings explain the difference between block, circuit and layout diagrams.

Answers to self-assessment questions

Chapter 1

Indices

1. 8^4
2. The whole numbers are *not* the same
3. 1
4. $10^1 \times 10^{-1}$; 10^0
5. 1 and 3^0
6. 1
7. (a) 1 (b) 10 (c) 25

Algebra

1. (a) $4X$
   (b) $5F$
   (c) $8Y + 4X$
   (d) $2M^2$
   (e) $6P^3$
   (f) 12
   (g) 2A
   (h) X

2. (a) $X = P + Q - Y$
   (b) $X = A + D - F$
   (c) $X = L - P - W - Q$
   (d) $X = 2$
   (e) $X = \frac{P.D}{M}$
   (f) $X = \frac{A}{W}$
   (g) $X = \frac{H.K}{2B}$
   (h) $X = \frac{M.Y}{A.B.C}$
   (i) $X = \frac{W}{(A + B)}$
   (j) $X = \frac{R(M + N)}{2P}$

Pythagoras and Trigonometry

1. A right-angled triangle
2. $H = \sqrt{B^2 + P^2}$
3. 19.416
4. 9.8
5. 18.33
6. Trigonometry
7. (a) $\sin \phi = \frac{P}{H}$; (b) $\cos \phi = \frac{B}{H}$; (c) $\tan \phi = \frac{P}{B}$
8. 9.46; 8.58φ
9. 51.31°; 12.49
10. 66.8°; 15.23

Chapter 2

1. +ve and −ve; protons are +ve, electrons are −ve
2. The number of electrons orbiting the nucleus of an atom
3. Those electrons which leave their orbits and wander through the molecular structure of the material
4. (a) A conductor has many random electrons, an insulator has very few
   (b) Conductor: copper, aluminium, silver, tungsten, steel, gold, etc.
   Insulator: PVC, rubber, mica, wood, glass, paxolin, etc.
5. (a) Amperes; (b) volts; (c) ohms; (d) coulombs; (e) μΩ mm
6. 160 m
7. 4.0 mm^2
8. 0.6 ohms
9. 64 ohms
10. (a) Current halves
    (b) Current doubles
    (c) Current remains the same
11. Nothing
12. The current in a circuit is directly proportional to the voltage and inversely proportional to the resistance, at constant temperature
13. (a) 24 ohms; (b) 6.25 A; (c) 24 V
14. 2200 ohms
15. Alternating current and direct current
16. Extra low: 50 V and below, between conductors, and between conductors and earth
    Low: between 50 V and 1000 V, conductor to conductor, and 50 V and 600 V conductor to earth

17.

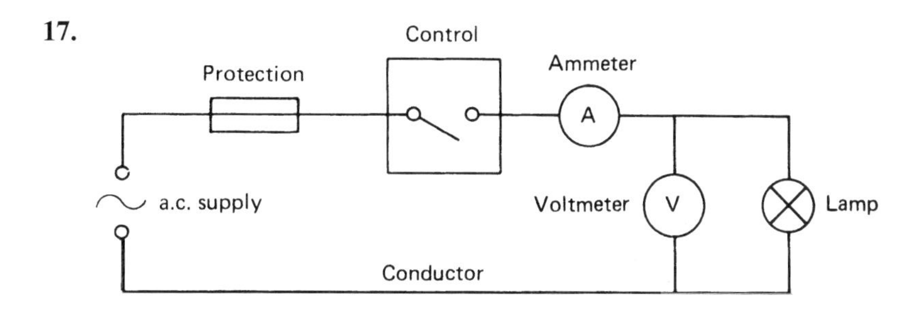

18.

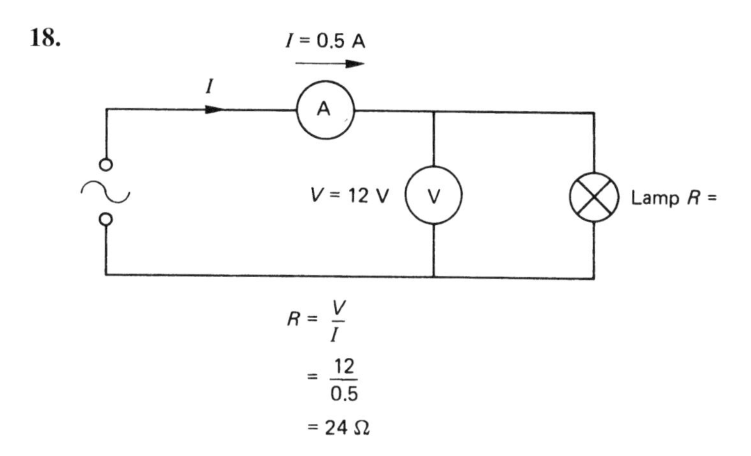

$$R = \frac{V}{I} = \frac{12}{0.5} = 24\ \Omega$$

Chapter 3

1. 97 Ω
2. 1440 Ω
3. 21.33 Ω
4. (a) 0.272 Ω
   (b) 38.672 Ω
5. (a) 4% of nominal
   (b) 9.2 V; yes
6. 8 A; 12 A; 4 A; 2 A; 26 A
7. (a) 19.2 Ω; 12.5 A
   (b) 96 Ω; 2.5 A
   (c) 57.6 Ω; 4.166 A
   (d) 1440 Ω; 0.166 A
8. 21.3 W; 8.88 W; 5.55 W
9. 6 Ω; 2 A; 3 A; 24 V; 12 V; 12 V
10. £274.35
11. 3 kW
12. 3 hours
13. 80%

Chapter 4

1. 40 cm^2
2. 60 T
3. 12 N
4. 5 Ω
9. 46.98 A; 35.36 A
13. 30 turns
14. 1 henry
15. 0.08 S; 20 A; 2 A
16. 5 A
17. 50 HZ
18. 240 V
19. 240 V
20. 20 A; 30 A; 36.05 A
21. 240 V; 144 V
22. 0.6
23. 330

Chapter 5

1. 7.3 mC
2. 1.92 mC
3. 10 μF
4. 50 μF; 12 mC; 4.8 mC
5. 80 cm^2
6. 50 Hz
7. 144 V; 192 V; 240 V
9. 9.6 A; 4.8 A; 10.7 A

Chapter 6

1. 6.25 A; 82.8 μF
2. 110 V; 0.95 leading
3. 66.4 μF
4. 5.57 μF
5. 7.8 A at 0.4 lagging
6. 10.8 kVA; 0.82 lagging

Chapter 7

3. 10 A; 240 V
4. 0.938 lagging
5. 5.77 A; 110 V
6. 0.8
7. 6 kW; 7.2 kVA
8. 22 A

Chapter 8

2. (b) 434 V; 390.6 V; 46.2 A
3. 28 A
4. (b) 8.33 revs/second
6. (b) 2.76%
9. 72 A; 443 μF
10. 118.7 Nm

Chapter 9

4. (b) 0.516 Ω; 1.55 V

Chapter 10

8. (b) 136 lx; 85.65 lx
9. (a) 64; (b) 45
10. (a) 44; (b) 41.25 A
14. 1.27 A; 0.94 lagging

Chapter 16

1. 151.8 Ω
2. 18 m
3. 0.5 Ω
4. No
5. Yes. Change the cooker control unit to one without a socket, thus converting the circuit to fixed equipment

Chapter 20

1. (a) 3.8 k 10%; (b) 29.5 k 5%;
(c) 550 k 1%; (d) 12 k 10%

2. (a) Silver–violet–red–brown
(b) Brown–green–silver–red–green
(c) Brown–black–black–silver
(d) Gold–green–black–gold

3. (a) 18 nF 5% 100 V
(b) 27 nF 10% 400 V
(c) 75 nF 20% 100 V
(d) 100 nF 10% 250 V

4. (a) Brown–green–gold–black–yellow
(b) Green–red–gold–white–brown
(c) Brown–green–gold–green–red
(d) Red–black–yellow–black–red

Index

A.c. circuits
in conduits, 255
in trunking, 259
A.c. generator, 66
three-phase, 71
A.c. motors, 137
single-phase, 141
capacitor-start, 142
capacitor-start capacitor-run, 143
reactance-start, 142
repulsion-start, 144
resistance-start, 143
shaded-pole, 141
universal or series, 144
three-phase, 137
squirrel-cage induction, 139
synchronous, 138
synchronous-induction, 138
wound-rotor type, 139
A.c. supply, 29
Access equipment, 215
Acts of Parliament,
Health and Safety at Work, 202
penalties of contravention, 204
statutory nature of, 202
Addition of waveforms, 69
Adiabatic equation, 347
Agricultural installations, 302, 315
Air thermostat, 277
Alarm systems, 286
Algebra, 6
All-insulated construction, 303
Ammeter, 30
Annealing, 264
Areas, 15
Armature, 64–65
Argon gas, 174
Armoured cable, 242, 246, 311
Atoms, 16
Average value of an alternating current, 71

Back e.m.f., 73, 127
Balanced three-phase systems, 72, 119
Balancing of loads, 188
Bar charts, 238
Bathrooms, 295
Batten holder, 269
Bearing extractor, 211
Bell and call system, 259
Bell (trembler), 64
Bill of quantities, 228
Bimetal strip, 276
British Standards and Codes of Practice, 225, 299
Burglar alarm, 286
open-circuit, 287
closed-circuit, 288
Bus-bars, 241
connection to, 247
trunking, 258, 259

Cable, 239, 320
clamping of, 247
PVC, 245, 265
selection, 346
supports for, 248
Calibration, 367
Call point, 287
Call system, 259, 290
Capacitance, 95
in a.c. circuits, 99
Capacitive reactance, 99
Capacitor, 94
charging, 99
dimensions, 95
discharging, 99
electrolytic, 95
energy stored in, 98
in electronics, 395
in parallel, 97
in series, 96
time constant, 99
working voltage, 101
Capacitance and resistance in parallel, 101
Capacitance and resistance in series, 100
Capacitor as suppressor, 101, 178
Capacitor for PF correction, 110, 156, 178, 186
Carbon, variation of resistance with temperature, 24
Ceiling rose, 269
Cells and batteries, 162
alkaline, 166
capacity of, 167
characteristics, 167
charging methods, 170
efficiency, 167
in series, 170
in parallel, 170
internal e.m.f., 168
internal resistance, 168
lead–acid, 163
maintenance, 164
primary, 162
secondary, 163
specific gravity, 164
terminal voltage, 165
Central heating systems, 292
Certification, 88
Charge, 20
Chemical effects, 162
Chime, 63
Choke, 87, 177
Circuit breaker, 326, 337
Circuit components, 31
Circuit diagram, as distinct from wiring diagram, 286
Circuits, installation of, 268
Circuit protective conductor, 309
Coercive force, 91
Colour coding, 241, 242
Cold-working, 265
Commutator, 65
Compounds, 16
Concentric cable, 311
Conduction, 274
Conductors, 19, 240
choice of, 346
construction of, 240
Conduit, 265
bending of, 250, 254
drainage, 255
jointing of, 251
metal, 249
supports for, 251
termination of, 251, 255
threading of, 253
Constantan, 22
Consumer unit, 337
Continuity, 370
Contracts, 227
Control, 31, 335
Convection, heating by, 275
Conventional current, 19
Cooker circuits, 273
Copper loss, 90
Correction factor, 342
Corrosion, 248, 302, 321

Cosine rule, 184
Coulomb, 20, 95
Cracking of PVC, 322
Crimping, 219, 244
Current, 20, 360
Current transformers, 89, 362
Cycle, 68

Damp and corrosive situations, 321
Daywork, 228
d.c. generators, 64, 135
  separately excited, 136
d.c. motors, 126
  face-plate starter, 132
  compound, 134
  reversing, 135
  series, 129
  shunt, 132, 133
d.c. supply, 29
Degradation of PVC, 321
Delta connections, 118, 121, 122
Depolarizer, 163
Design, 341
Diac, 404
Dielectric, 94
Digital instruments, 359
Diode, 398
Direct heating, 274
Direct contact, 322
Discharge lighting, 174
Discrimination, 329, 333
Diversity, 273, 353
Double insulation, 324
Ducted warm air, 276

Earth as a conductor, 304
Earth fault loop path, 312
Earth leakage, 316
Earth leakage circuit breakers, residual current-operated, 314
Earth electrodes, 306
Earth-electrode resistance, 307, 382
  measurement of, 382
Earth loop impedance, 312, 379
Earth terminal, 318
Earthing, of exposed metalwork, 309
Earthing clamp, 318
Earthing conductor, 312
Earthing systems, 309
Eddy currents, 65, 90
Efficiency, 54, 158
Effort, 213
Electric shock, 28, 223, 304, 322
Electrical safety, 221
Electrical Contractors' Association (ECA), 226
Electricity at Work Regulations 1989, 205, 225
Electricity Supply Regulations, 225
Electrolytes, 18, 24, 162
Electromagnet, 63
Electromagnetism, 58
Electromagnetic trip, 64, 329
Electromotive force, e.m.f., 20, 168
Electronic components, 407
Electronics, 395
Electrons, 19
  flow of, 18, 19
Elements, 17
Element of fluorescent lamp, 177
Emergency lighting, 291
Embrittlement of PVC, 322
Energy, 45, 51
Energy meter, 53
Energy stored in capacitor, 98
Energy stored in magnetic field, 78
Estimates, 233
Explosive situations, 299
Extra-low voltage, 325
  lighting, 294

Farad, 95
Faraday cage, 306
Fault finding, 394
Field around conductor, 59
Fire alarm, and Burglar alarm, 259, 286
Fire barriers, 259
Fire risk, 311
Fire safety, 220
First aid, 223
Fixed wiring, 241
Fixing, 260
Flammable atmospheres or materials, 299
Fleming's left-hand rule, 60
  right-hand rule, 62
Flexible conduit, 253
Flexible cord, 242
Floor warming, 276
Fluorescent lighting, 177
  rating of, 187
  stroboscopic effect, 188
Flux, 58
Flux density, 58
Force on conductor, 59, 60
Formulae, 7
Frequency, 68
Fuels, 196
Functional testing, 384
Fuses, 327
Fusing factor, 329

Generated e.m.f., 64, 65, 135
Generators,
  a.c., 65, 71
  d.c., 64
Generation, transmission and distribution systems, 199
Grid system, 198

Hardening, 264
Health and Safety at Work Act, 202
Heat shunt, 407
Heat sink, 405
Heat treatment, 264
Heating effects, 274
Henry, 72
'Hold-on' circuit, 286
Hole cutters, 257
Horse power, 158
Hydrogen atom, 16
Hydrometer, 165
Hysteresis, 90

IEE Regulations, 202, 225
Impedance, 81
Indices, 4
Indirect contact, 324
Inductance, 72
  mutual, 74
  self, 74
Induced e.m.f., 62, 73, 255
Inductive reactance, 78
Inductors in electronics, 397
Inclined plane, 211
Initial inspection, 368
Inspection and non-inspection fittings, 252
Installation systems, 268
Installation of motors, 149, 160
Installation work, R, L & C in, 103
Instantaneous value, 70
Instruments, 358
  digital, 359
  phase-earth loop tester, 365
  moving-coil, 359
  moving-iron, 358
  ohmmeter, 365
  selection, 365
  shunts and multipliers, 360, 363
Insulation resistance, 39, 375
Insulators, 19
Invar rod, 277
Inverse-square law, 182
Ionization, 178
IP codes, 240, 322, 323
Isolation, 335
Isolator, 336

Joint Industrial Board (JIB), 226
Joints, 233, 243
Joists, 248, 266
Joule, 46
Junction box, 243, 271
Junction diode, 398

Ladders, 215
Laminations, 65, 89
Lead–lag circuit, 188
Legislation, 203
LED, 404
Levers, 210
Lift shafts, 248
Lifting and handling, 210, 212
Light sources, 172

Lighting circuits,
 layouts, 270
 one-way, 268
 two-way, 269
 two-way and intermediate, 270
Lighting and illumination, 172
Lines of force, 58
Load, 31
Load force, 214
LSD, 404

Magnesium oxide, 242, 245
Magnetic effects, applications, 63
Magnetism, 58
Magnetizing force, 90
Making good, 248
Manganin, 22
Materials list, 234, 236
Measurement of power, 52, 123
Mercury vapour lamp, 176
Metalwork (cables through), 248
Mineral-insulated metal-sheathed cable, 241, 245, 265
Molecules, 16
Motor applications and fault diagnosis, 151
Motor enclosures, 150
Motor replacement, 150
Multimeters, 363
Mutual inductance, 74

NICEIC, 227
'Neon' tube, 174
'No-volt' protection, 132
Neutral conductor, 118
Neutrons, 16
Nuclear stations, 196
Nucleus, 16
Nuisance tripping, 316

Off-peak supplies, 284
Ohm's Law, 27
Oil dashpot, 146
Ovenstat, 278
Overcurrent protection, 326
Overhead lines, 200

Peak value, 67
Phase rotation indicator, 364
Phasors, 68, 80
Phenol-formaldehyde, 249
Plastics, 248
Polarity, 378
Pollution, 197
Polyvinyl chloride PVC, 249
Potential difference (p.d.), 19, 20, 169, 304
Power, 45, 120
Power factor, 85
Power factor correction, 108, 138, 186
 of a.c. motors, 156
Power in a.c. circuits, 84, 114, 120
Power-stations, 195
Prospective fault current, 381
Protection, 33, 320, 322
Protective conductors, 309
Protective multiple earthing (PME), 310
Proton, 16
Pulleys, 212
PVC, 249, 321
 conduit made from, 249
Pythagoras, 12

Radial circuits, 271
Radiation (heat), 274
Radioactive waste, 198
Rectification, 398
Regulations, statutory and non-statutory, 202–3
Relay, 286
Requisitions, 233
Reservoirs, 194
Residual current device, 315, 384, 367
Residual magnetism, 91
Resistance, 20
 in parallel, 37
 in series, 33
 in series and parallel, 40
 and inductance in series, 81
 and inductance in parallel, 83
Resistors, 395
Resistivity, 20, 240
Resuscitation, 224
Reverse breakdown voltage, 398
Ring circuits, 272
Rising mains, 240, 282
Riveting, 219
RKM coding, 396
r.m.s. value, 70
Rotor, 66, 138

Safety, electrical, 221
Safety and welfare, 202
Saturation, 91
SELV, 325
Scaffolding,
 independent, 218
 putlog, 219
 tower, 217
Scaffolding boards, 216
Schematic diagram, as distinct from wiring diagram, 287
Screw-on-seal (m.i.m.s.), 246
Screw jack, 211
Screw rule, 59
Security, 237
Segregation (of circuits), 256
Semi-conductors, 397
Sequence of control, 279, 337
Shaver unit, 297
Shock risk, 312, 314, 347, 366
Short-circuit, 326
Simmerstat, 278
Single loop generator, 65
Single-loop motor, 126
Siting,
 of power-stations, 195, 201
 of transmission lines, 201
Sleeving, 268
Slip, 139
Slip rings, 66, 149
Smoothing, 400
Sodium vapour lamp, 174, 175
Soil resistivity, 308
Solar energy, 198
Soldering, 219, 246, 406
Solenoid, 63
Space heating, 274
Spans, 218
Special locations, 295, 298, 299, 302
Specific heat, 53
Specifications, 228
Spurs, 272
Star connection, 118
Starters,
 auto-transformer, 148
 direct-on-line, 145
 for fluoresecent lamps, 178
 rotor-resistance, 149
 star–delta, 147
Starter (fluorescent), 178
Stator, 69, 138
Steam power, 195
Storage heating, 275
Strappers, 269
Stripping cable, 245
Stroboscopic effect, 188
Structure of firms, 232
Supply sources, 29
Supplementary bonding, 298, 309, 316
Swimming pools, 298
Synchronous speed, 137
Systems of earthing, 309

Tamper loop, 289
Tap-off trunking, 258
Tariffs, 52
Temperature coefficient, 23
Temperature limits for cables, 321
Tempering, 264
Temporary installations, 299, 315
Tenders, 227
Terminations, 243, 245
Terminal voltage, 165, 169
Testing, 358
Test lamp, 222, 366
Thermal effects, 321
Thermal trip, 329
Thermistor, 146, 404
Thermosetting polymers, 248
Thermos-softening, 248

Thermostats, 276
Three-heat switch, 50
Three-phase circuits, 118
Thyristors, 401
Time constant, 76, 99
Tong tester, 365
Tools, 260
Torque, 128, 159
Trade Unions, 226
Transformer, 87
  auto, 89
  current, 89, 362
  double-wound, 89
  in electronics, 395
  losses from, 90
  portable tools, use with, 222, 299, 326
  voltage, 363
Transformer ratios, 88
Transport of fuels, 196
Transistor, 403
Traywork, 260
Trestles, 217
Triac, 404
Trigonometry, 13
Trunking, 265
  cutting of, 256
  metal, 256
  PVC, 255
  supports for, 257
Tungsten filament lamp, 173

Ultra-violet light, 178
Unbalanced three-phase systems, 120
Underfloor heating, 276
Underground cables, 302
Undervoltage, 335
Unit (kWh), 51
Units, 1
  conversion of, 1

Variation order, 228
Volt drop on radial circuits, 283
Voltage, 20, 362
Voltage-bands, 29
Volt drop, 35, 199, 346
  maximum permissible, 35, 160
Voltmeter, 30
Volumes, 15

Waste disposal, 197
Water analogy, 25
Water heater,
  calculations, 53
  circuits, 273
Water power, 194
Watt, 45
Wattmeter, 52, 364
Wave power, 198
Waveform, 66, 68, 69
Wedge, 211
White meter, 285
Wind power, 198
Wiring diagram, as distinct from circuit diagram, 287
Work, 213
Work hardening, 264
Working voltage, 101

Zener diode, 404